AF698807

GEOTECHNICAL SPECIAL PUBLICATION NO. 94

PERFORMANCE CONFIRMATION OF CONSTRUCTED GEOTECHNICAL FACILITIES

PROCEEDINGS OF SESSIONS OF ASCE SPECIALTY CONFERENCE ON PERFORMANCE CONFIRMATION OF CONSTRUCTED GEOTECHNICAL FACILITIES

SPONSORED BY
The Geo-Institute of the American Society of Civil Engineers

April 9-12, 2000
Amherst, Massachusetts

EDITED BY
Alan J. Lutenegger
Don J. DeGroot

ARTHUR LAKES LIBRARY
COLORADO SCHOOL OF MINES
GOLDEN, CO 80401

GEO INSTITUTE

ASCE American Society of Civil Engineers
1801 ALEXANDER BELL DRIVE
RESTON, VIRGINIA 20191–4400

TA
703.5
.A85
2000

Abstract: This proceedings includes papers presented at the sessions of the ASCE Geo-Institute Specialty Conference on Performance Confirmation of Constructed Geotechnical Facilities held at the University of Massachusetts—Amherst, April 9-12, 2000. Geotech-nical engineering involves the evaluation of site conditions, the determination of soil and rock properties, and the application of design principles to produce solutions to ground engineering problems. Performance confirmation of the geotechnical aspects of a project provide a means for geotechnical engineers to validate design approaches, reinterpret property assessment, and improve the design process. The papers include performance and evaluation of shallow and deep foundations; roadway and railways; tunnels, excavations, and retaining structures; and fills, embankments, and slopes. Grouting in geotechnical engineering is also included, as are aspects of the statistical analysis involved in the evaluation.

Library of Congress Cataloging-in-Publication Data

ASCE Specialty Conference on Performance Confirmation of Constructed Geotechnical Facilities (2000 : Amherst, Mass.)

Performance confirmation of constructed geotechnical facilities : proceedings of sessions of ASCE Specialty Conference on Performance Confirmation of Constructed Geotechnical Facilities : April 9-12, 2000, Amherst, Massachusetts / sponsored by the Geo-Institute of the American Society of Civil Engineers ; edited by Alan J. Lutenegger, Don J. DeGroot.

p. cm. – (Geotechnical special publication ; no. 94)

Includes bibliographical references and index.

ISBN 0-7844-0486-0

1. Engineering geology–Evaluation–Congresses. I. Lutenegger, A.J. II. DeGroot, Don J. III. American Society of Civil Engineers. Geo-Institute. IV. Title. V. Series.

TA703.5. A85 2000
624.1'51--dc21 00-024611

Any statements expressed in these materials are those of the individual authors and do not necessarily represent the views of ASCE, which takes no responsibility for any statement made herein. No reference made in this publication to any specific method, product, process or service constitutes or implies an endorsement, recommendation, or warranty thereof by ASCE. The materials are for general information only and do not represent a standard of ASCE, nor are they intended as a reference in purchase specifications, contracts, regulations, statutes, or any other legal document. ASCE makes no representation or warranty of any kind, whether express or implied, concerning the accuracy, completeness, suitability, or utility of any information, apparatus, product, or process discussed in this publication, and assumes no liability therefore. This information should not be used without first securing competent advice with respect to its suitability for any general or specific application. Anyone utilizing this information assumes all liability arising from such use, including but not limited to infringement of any patent or patents.

Photocopies: Authorization to photocopy material for internal or personal use under circumstances not falling within the fair use provisions of the Copyright Act is granted by ASCE to libraries and other users registered with the Copyright Clearance Center (CCC) Transactional Reporting Service, provided that the base fee of $8.00 per article plus $.50 per page is paid directly to CCC, 222 Rosewood Drive, Danvers, MA 01923. The identification for ASCE Books is 0-7844-0486-0/00/ $8.00 + $.50 per page. Requests for special permission or bulk copying should be addressed to Permissions & Copyright Dept., ASCE.

Copyright © 2000 by the American Society of Civil Engineers, All Rights Reserved.
Library of Congress Catalog Card No: 00-024611
ISBN 0-7844-0486-0
Manufactured in the United States of America.

Geotechnical Special Publications

1 *Terzaghi Lectures*
2 *Geotechnical Aspects of Stiff and Hard Clays*
3 *Landslide Dams: Processes, Risk, and Mitigation*
4 *Tiebacks for Bulkheads*
5 *Settlement of Shallow Foundation on Cohesionless Soils: Design and Performance*
6 *Use of In Situ Tests in Geotechnical Engineering*
7 *Timber Bulkheads*
8 *Foundations for Transmission Line Towers*
9 *Foundations & Excavations in Decomposed Rock of the Piedmont Province*
10 *Engineering Aspects of Soil Erosion, Dispersive Clays and Loess*
11 *Dynamic Response of Pile Foundations–Experiment, Analysis and Observation*
12 *Soil Improvement: A Ten Year Update*
13 *Geotechnical Practice for Solid Waste Disposal '87*
14 *Geotechnical Aspects of Karst Terrains*
15 *Measured Performance Shallow Foundations*
16 *Special Topics in Foundations*
17 *Soil Properties Evaluation from Centrifugal Models*
18 *Geosynthetics for Soil Improvement*
19 *Mine Induced Subsidence: Effects on Engineered Structures*
20 *Earthquake Engineering & Soil Dynamics II*
21 *Hydraulic Fill Structures*
22 *Foundation Engineering*
23 *Predicted and Observed Axial Behavior of Piles*
24 *Resilient Moduli of Soils: Laboratory Conditions*
25 *Design and Performance of Earth Retaining Structures*
26 *Waste Containment Systems: Construction, Regulation, and Performance*
27 *Geotechnical Engineering Congress*
28 *Detection of and Construction at the Soil/Rock Interface*
29 *Recent Advances in Instrumentation, Data Acquisition and Testing in Soil Dynamics*
30 *Grouting, Soil Improvement and Geosynthetics*
31 *Stability and Performance of Slopes and Embankments II*
32 *Embankment of Dams–James L. Sherard Contributions*
33 *Excavation and Support for the Urban Infrastructure*
34 *Piles Under Dynamic Loads*
35 *Geotechnical Practice in Dam Rehabilitation*
36 *Fly Ash for Soil Improvement*
37 *Advances in Site Characterization: Data Acquisition, Data Management and Data Interpretation*
38 *Design and Performance of Deep Foundations: Piles and Piers in Soil and Soft Rock*
39 *Unsaturated Soils*
40 *Vertical and Horizontal Deformations of Foundations and Embankments*
41 *Predicted and Measured Behavior of Five Spread Footings on Sand*
42 *Serviceability of Earth Retaining Structures*
43 *Fracture Mechanics Applied to Geotechnical Engineering*
44 *Ground Failures Under Seismic Conditions*
45 *In Situ Deep Soil Improvement*
46 *Geoenvironment 2000*
47 *Geo-Environmental Issues Facing the Americas*
48 *Soil Suction Applications in Geotechnical Engineering*
49 *Soil Improvement for Earthquake Hazard Mitigation*
50 *Foundation Upgrading and Repair for Infrastructure Improvement*
51 *Performance of Deep Foundations Under Seismic Loading*
52 *Landslides Under Static and Dynamic Conditions–Analysis, Monitoring, and Mitigation*

53 *Landfill Closures–Environmental Protection and Land Recovery*
54 *Earthquake Design and Performance of Solid Waste Landfills*
55 *Earthquake-Induced Movements and Seismic Remediation of Existing Foundations and Abutments*
56 *Static and Dynamic Properties of Gravelly Soils*
57 *Verification of Geotechnical Grouting*
58 *Uncertainty in the Geologic Environment*
59 *Engineered Contaminated Soils and Interaction of Soil Geomembranes*
60 *Analysis and Design of Retaining Structures Against Earthquakes*
61 *Measuring and Modeling Time Dependent Soil Behavior*
62 *Case Histories of Geophysics Applied to Civil Engineering and Public Policy*
63 *Design with Residual Materials: Geotechnical and Construction Considerations*
64 *Observation and Modeling in Numerical Analysis and Model Tests in Dynamic Soil-Structure Interaction Problems*
65 *Dredging and Management of Dredged Material*
66 *Grouting: Compaction, Remediation and Testing*
67 *Spatial Analysis in Soil Dynamics and Earthquake Engineering*
68 *Unsaturated Soil Engineering Practice*
69 *Ground Improvement, Ground Reinforcement, Ground Treatment: Developments 1987-1997*
70 *Seismic Analysis and Design for Soil-Pile-Structure Interactions*
71 *In Situ Remediation of the Geoenvironment*
72 *Degradation of Natural Building Stone*
73 *Innovative Design and Construction for Foundations and Substructures Subject to Freezing and Frost*
74 *Guidelines of Engineering Practice for Braced and Tied-Back Excavations*
75 *Geotechnical Earthquake Engineering and Soil Dynamics III*
76 *Geosynthetics in Foundation Reinforcement and Erosion Control Systems*
77 *Stability of Natural Slopes in the Coastal Plain*
78 *Filtration and Drainage in Geotechnical/Geoenvironmental Engineering*
79 *Recycled Materials in Geotechnical Applications*
80 *Grouts and Grouting: A Potpourri of Projects*
81 *Soil Improvement for Big Digs*
82 *Risk-Based Corrective Action and Brownfields Restorations*
83 *Design and Construction of Earth Retaining Systems*
84 *Effects of Construction on Structures*
85 *Application of Geotechnical Principles in Pavement Engineering*
86 *Big Digs Around the World*
87 *Jacked Tunnel Design and Construction*
88 *Analysis, Design, Construction, and Testing of Deep Foundations*
89 *Recent Advances in the Characterization of Transportation Geo-Materials*
90 *Geo-Engineering for Underground Facilities*
91 *Special Geotechnical Testing: Central Artery/Tunnel Project in Boston, Massachusetts*
92 *Behavioral Characteristics of Residual Soils*
93 *National Geotechnical Experimentation Sites*
94 *Performance Confirmation of Constructed Geotechnical Facilities*

Preface

Geotechnical engineering involves the evaluation of site conditions, the determination of soil and rock properties, and the application of design principles to produce solutions to ground engineering problems. In nearly every design project, the results lead to a constructed facility. The level of geotechnical activity associated with that facility may range from limited, as in the case of a shallow foundation for a lightly loaded single story structure; to extensive, as in the case of a compacted embankment on soft clay. In many cases, the geotechnical engineer does not have the opportunity to evaluate the success of the design, and often can only rely on the fact that a project which does not experience a failure has generally been successful. There is seldom time or budget to analyze the performance of a particular project and to evaluate the consequences of that performance on subsequent design approaches. Both short term and long-term performance feedback of constructed geotechnical facilities provide valuable information to the geotechnical engineer. Performance confirmation of the geotechnical aspects of a project provide a means for geotechnical engineers to validate design approaches, reinterpret property assessment, and improve the design process.

This Geotechnical Specialty Publication (GSP) presents the results of papers presented at the ASCE GeoInstitute Specialty Conference on Performance Confirmation of Constructed Geotechnical Facilities held at the University of Massachusetts Amherst on April 9-12, 2000. The members of the conference steering committee were:

Carlton L. Ho, conference chair, University of Massachusetts Amherst
Marco Boscardin, GEI Consultants
John Christian, consultant
Al DiMillio, Federal Highway Administration
Scot Litke, Association of Drilled Shaft Contractors
Tom Papagiannakis, Washington State University
Dick Reynolds, consultant

Papers were presented in plenary sessions and individual breakout sessions at the conference. For the preparation of this GSP, the editors organized the papers in a topical format rather than by the sessions in which they were presented. All papers in these proceedings have been peer-reviewed by two anonymous reviewers in accordance with the standards of ASCE and the Geo-Institute. Required revisions were made by the authors prior to final acceptance and publication. All papers are elegible for discussion in the ASCE Journal of Geotechnical and Geo-Environmental Engineering and for ASCE awards. The editors wish to express sincere thanks to the session chairs for their diligence and to all authors for their patience in the preparation of this GSP.

Dan Brown
Michael Byle
Hubert Deets
Mohamad Gabr
Robert Gilbert
Youssef Hashash
Mohamad Hussein
Magued Iskander
K.K. Muraleetharan
Tom Papagiannakis
Steven Perkins
Ron Smith
Robert Stephens
C. Vipulanandan

The efforts of the anonymous reviewers are greatly appreciated .

A companion GSP on National Geotechnical Experimentation Sites was also produced from papers presented at this conference.

Alan J. Lutenegger
Professor of Civil and Environmental Engineering
University of Massachusetts
Amherst, MA

Don J. DeGroot
Associate Professor of Civil and Environmental Engineering
University of Massachusetts
Amherst, MA

Contents

Statistical Analyses

Performance and Evaluation of Roadways and Railways

Grouting in Geotechnical Engineering

Invited Plenary Papers

REPRESENTATIVENESS AND STATISTICS IN FIELD PERFORMANCE ASSESSMENT

Gregory B. Baecher[1]

ABSTRACT: Measurement of engineering performance is fundamental to empirical understanding, model development, and to the observational method. It is also expensive. Yet, how representative are field observations of the geotechnical conditions at work, and how informative are they of critical design assumptions? Interesting lessons for geotechnical practice are suggested from considering such issues, and from qualitative statistical reasoning about observing geotechnical performance on limited budgets. The discussion considers three interrelated topics: intuitive misperceptions about samples and sampling variations, the nature of uncertainty and randomness in modeling soil deposits, and simple lessons derived from statistical sampling theory.

INTRODUCTION

Monitoring field performance involves a great deal of sampling, and builds upon lessons learned in a number of other scientific and technical fields. In this regard, field monitoring differs little from other sampling activities, whether in testing pharmaceuticals, inspecting airplane engines, or conducting political polls. We make a limited number of observations and try to draw scientifically defensible conclusions from them. Clearly, the interpretation of field data requires a good deal of insight into geology and engineering mechanics, and also experience with construction practices. Still, intuition fails even the most sophisticated engineer or scientist faced with the vagaries of scattered experimental data, and too little time or too few resources with which to make more observations.

For simplicity, the present discussion limits consideration about the purpose of field monitoring to two objectives: (i) assessing site conditions, and (ii) testing the validity of an analytical model. Clearly, field monitoring has other objectives, for example, in quality control, and in observational methods, but these are deferred to another occasion. The discussion also ignores legal issues, despite their importance to practice.

[1] Professor and Chairman, Department of Civil and Environmental Engineering, University of Maryland, College Park, MD 20742. gbaecher@eng.umd.edu

INTUITION AND THE INTERPRETATION OF DATA

Surprisingly, even trained statisticians are easily led astray when using intuition to interpret sampling observations. Mere scientists and engineers, as a result, have little hope of accurately interpreting sample variations, errors, and biases simply based on inspection. Yet, that is usually the approach taken in practice. Interestingly—but maybe not surprisingly—the errors people make in intuitively interpreting data are remarkably similar from one person to another.

Sample variation

When measurements are made in the laboratory or field they exhibit scatter. These measurements might more generically be called, *observations*, to include things other than instrument readings. A set of observations is typically called a *sample*. If the sample comprises a large number of observations, the data scatter among the latter tends to exhibit regularity. That is, the scatter within a sample and from one sample to another tends to display regular patterns, and over the years statisticians have learned to categorize those patterns, and to use them to draw inferences about the population from which the sample comes, the *parent population*. Patterns of scatter within an individual sample are interpreted against what is known of the probable scatter among samples to make estimates about the parent population.

For convenience, the scatter within a sample or from one sample to another is described by a frequency distribution or histogram, and this in turn can be summarized by its low order statistical moments. The most useful of these are the first two moments, the *mean* or arithmetic average of the observations, m, and the *variance* or mean-square variation about the mean, s^2. The *standard deviation*, s, is the root-mean-square variation (square-root of the variance), and the ratio of standard deviation to mean is the *coefficient of variation*, $\Omega=s/m$. Such mathematical functions of the sample observations are said to be *statistics* of the data, or alternately, *sample statistics*, and form the basis for making inferences about the parent population.

The *law of large numbers*, a fundamental principle of statistical theory, implies that, as the sample becomes larger, the statistical properties of the sample become ever more to resemble the population from which the sample is taken.[2] The operative phrase here is, "as the sample becomes larger." For example, one is often interested in the average value of some parameter or performance measure in the field, across all the elements of the population that may not have been observed in the sample. If the number of observations in a sample is large, it seems reasonable, and the law of large numbers confirms, that one might use the sample average of the set of observed values as indicative of the population average in the field. But, what

---

[2] The Law of Large Numbers is more specific and limited than this colloquial interpretation (Feller, 1967), but the practical implication is quite broad. See, also, Maistrov (1974) for an historical sketch.

about the case where the sample is not large? This is almost always true in geotechnical practice.

The law of large numbers says that variations of the moments of a sample about their counterparts in the parent population become ever smaller as sample size increases, but for small samples these variations can be large. Presume we take many samples of size n from the same population, and for each sample we calculate the sample mean, m. The values of m across the many samples themselves exhibit scatter, and could be plotted as a histogram. This distribution of the sample mean, or of any other sample parameter, is called the *sampling distribution*. The sampling distribution is the frequency distribution of some sample statistic over repeated sampling. The theoretical variance of the scatter among the means of each of many samples of size n is $Var(m)=s^2/n$. Var(m) is the second moment of the sampling distribution of m. Correspondingly, the standard deviation of the sample means is, $s_m=s/(n)^{1/2}$.

The coefficient of variation of soil properties measured in the field can be as large as 100%, although values of 30-50% are more common (Kulhawy and Trautmann, 1996; Phoon and Kulhawy, 1996). Thus, if ten (10) tests are made, the variation of their sample average about the (unknown) population (soil deposit) average would have a standard deviation of 10-16%. Since, under very general assumptions, the sampling distribution of the mean is well approximated by a Normal distribution, the range for which one would be comfortable bracketing the population mean is, say, 2 to 2.5 standard deviations, or in this case between ±20-40% of the best estimate.[3] In other words, there is considerable uncertainty in the inference of even average soil properties, when reasonable sample sizes are taken into account. There is, of course, even more uncertainty about inferences of soil properties at specific locations within a soil deposit.

Representativeness

Despite the fact that sampling variations can be large—and in geotechnical practice, they are large—there is an intuitive tendency to treat sample results as representative of—or similar to—the population from which they are taken. Most people believe intuitively that samples should reflect the essential characteristics of the population out of which they arise, and thus the converse, that essential characteristics of the population should mimic those of the sample. People's intuition tells them that a sample should be similar to the population from which it comes, but that is only true in the limit, as sample sizes become large. This leads to errors. Speaking strictly, *representativeness* is a property of sampling plans, not of samples. A sampling plan is representative of the population being sampled if every element of the population has an equal chance of affecting the (weighted) properties of the sample (Cochran,

---

[3] The Normal limit to the sampling distribution follows from the Central Limit Theorem, which is closely related to the Law of Large Numbers, see, e.g., Feller (1967)

1977), and from this one speaks of "representative sampling."[4] A sample, in contrast can never be representative: it is a unique collection of particular elements within the population, and each such collection has different properties.

Take, for example, the string of sample outcomes deriving from six tosses of a fair coin, {H,H,H,H,H,H}. Most people intuitively think of this string as less likely to occur than the string, {H,T,T,H,T,H}, even though each has the same probably, $(½)^6$. This is akin to the Gambler's Fallacy that if *heads* has not appeared in some time, it is overdue and should occur with increased probability. Intuition tells us that the sample should represent the population, that is, be similar to the population in salient aspects, and in short runs as well as long. In this case, the sample should have about the same number of heads as tails, and the sequence of heads and tails should be "random," that is, erratic. That this is a misperception is obvious to anyone who thinks about it; yet, our intuition tells us otherwise.

The same thing is true of samples of geotechnical observations. We presume them to be representative of the geotechnical population out of which they arise. The averages within the sample ought to be about the same as the averages in situ. The variability of observations ought to be about the same as the variability in situ. Spatial patterns of variation among the observations ought to mirror spatial patterns in situ. All of these things are true in the limit, but for small samples they are compromised by sampling variability, and may be profoundly untrue. Small samples of the size typical in geotechnical practice seldom display the salient properties of the population; the variability among sample outcomes is simply too great.

Overconfidence

This intuitive belief in representativeness leads people to believe that important characteristics of a population should manifest in every sample, no matter the sample size. Yet, we know from statistical theory that this is not true: small samples exhibit large variation from one to another. This leads people to put too much faith in the results of small numbers of observations and to overestimate the replicability of such results. If tests are repeated, people have unreasonably high expectations that significant results will be replicated. Thus, the ten (10) observations of field performance above are made, and one is surprised that the next set of ten yields a 30% difference in average results. A person's typically response is not to ascribe this difference to expectable statistical variation, but to seek a cause. The engineering literature is filled with well-intentioned attempts to explain away differing sample results, when in fact, such explanations would be more in order had such differences not been observed.

---

[4] In some places the term, representative sampling, is used to mean that the probability of sampling sub-populations is set equal the relative frequency of those sub-populations within the overall population. This meaning is subsumed within the definition here.

A corollary to this belief in the representativeness of small samples is the overconfidence even scientifically trained people place in their inferences or estimates of unknown quantities. In a famous early study, Alpert and Raiffa (1982) demonstrated that when asked to place 25%:75% or 5%-95% confidence bounds on estimates of unknown quantities, the true values of the quantities being estimated fall outside the assessed bounds considerably more often that the nominal 50% or 10%, respectively. Often more than half the real values fall outside 5%-95% confidence bounds people estimate. This result has been replicated in another early study by Folayan et al (1970) involving engineers' estimates of the properties of San Francisco Bay Mud, and by Hynes and Vanmarcke (1976) involving predictions of embankment height at failure for the MIT I-95 Test Embankment. Data from Folayan et al. are recalculated in Table 1 to show 95% confidence intervals on the subjective assessments of the mean compression ratio, and to show 95% confidence intervals derived from 42 tests at the site. The lack of overlap between the subjects' intervals and that calculated from sample observations suggests strong overconfidence on the part of the subjects.

Subject	2.5% limit	97.5% limit
1	0.29	0.31
2	0.27	0.28
3	0.26	0.29
4	0.26	0.34
5	0.20	0.43
Sample	0.32	0.36

Table 1. 95% confidence intervals on average compression ratio for San Francisco Bay Mud at a particular construction site, subjectively estimated by five engineers. Confidence interval also shown based on n=42 observations at the site (after, Folayan, et al., 1970).

As reliability analysis becomes increasingly important to geotechnical practice, it is sometimes suggested that a field expedient way of assessing the standard deviation of an uncertain quantity is by eliciting the maximum and minimum bounds one could conceive the quantity having, and then assuming that this range spans a certain number of standard deviations of variation, typically, ±3s. The reasoning is that for a Normal variate, ±3 standard deviations spans 99.75% of the variation. But, if people are over confident of their estimates of uncertain quantities—which we know them to be—then people will frequently be surprised in practice to find their maximum and minimum bounds exceeded. Thus, the "six-sigma" rule is unconservative, and possibly quite significantly. This can also be seen in Figure 1, in which the expected range of sample values, $r_n=|x_{max}-x_{min}|$, for a Normal variate is plotted as a function of sample size. Even for samples as large as n=20, the range expected in a sample is less than 4 standard deviations. The reciprocal of this expected range, in fact, makes a useful estimator of standard deviation, and one with known sampling properties (Snedecor and Cochran, 1980).

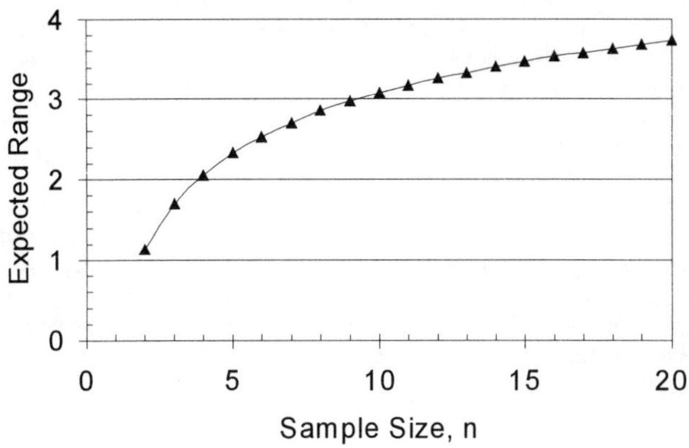

Figure 1. Expected range of Normal sample in standard deviation units

<u>"Law of small numbers"</u>

In a series of celebrated papers in the, 1970's, the late Amos Tversky and Daniel Kahneman, now of Princeton University, introduced the scientific world to the systematic differences between the way people perceive probability and the way statistical theory operates, and to the term *representativeness* as used above (1971, 1974, 1979). That body of work, and the explosion of studies that followed, are sometimes referred to as the "heuristics and biases" school of thought on subjective probability (see, e.g., Morgan and Henrion, 1990).

This body of work emphasizes that the use of representativeness (similarity) to judge probabilities is fraught with difficulty, because it is not affected by factors that should influence judgments of probability. Important among these are the overconfidence described above, a disregard for base rates (*a priori* probabilities), and ignorance of common regression effects. This concept that observers presume samples to be representative of the population seems benign, but leads to serious errors of judgment in practice. Tversky and Kahneman (1971), dubbed this, "The Law of Small Numbers," which states simply, that the Law of Large Numbers applies to small numbers as well.

This overlooking of sample size manifests even when a problem is stated so as to emphasize sample size, and in many different contexts. Consider, for example, a question that arose in a flood hazard damage reduction study. A river basin was analyzed in two different ways to assess levee safety. In the first case, the river was divided into 10miles (6 km) long reaches; in the second, the river was divided into 1 mile (0.6 km) long reaches. Would the average settlements within the levee reaches have greater variability in the first case, the second case, or about the same in each? Of an admittedly unscientific sample of 25 graduate students and engineers, 7 said the first (more variation among long reaches), 6 said the second (more variation among short reaches), and 12 said the last (about equal). But clearly, the long reaches have the least variation among their average settlements, because they are larger samples. Smaller samples are more erratic.

Prior probabilities

A second manifestation of representativeness is that people tend to overlook background rates and focus instead on the likelihood of the observed data when drawing conclusions. To review for a moment, Bayes' Theorem says that the probability one ascribes to an event or parameter estimate should be the product of two probabilities: the probability *a priori* to observing new data, and the likelihood (conditional probability) of the new data given the event or parameter value. This is summarized by the familiar expression,

$$\Pr\{\Theta \mid data) = \Pr\{\Theta\} L\{data \mid \Theta\} \tag{1}$$

in which Θ is an event or parameter (the state of nature), $\Pr\{\Theta\}$ is the probability of Θ prior to observing the data, $\Pr\{\Theta \mid data)$ is the probability after observing the data, and $L\{data \mid \Theta\}$ is the conditional probability of the data given Θ (i.e., the Likelihood). This relationship led DeFinetti (1937) to say, "data never speak for themselves," they tell us only how to modify what we thought before we saw them to what we should logically think afterward. What the data tell us is summarized in the likelihood function. What we thought before is summarized in the prior probabilities.

Sometimes representativeness leads people to place undue importance on sample data (because they "should be similar to the population"), and in so doing ignore, or at least downplay, prior probabilities (the latter sometimes referred to as *base-rates* in the heuristics and biases literature). As a simple example, in risk analyses for dam safety a geologist might be asked to assess the probability that faults exist undetected in the bottom of a valley. Noting different rock formations on the adjoining valley walls, he or she might assign a high probability to faulting, because of the association of this condition with faulting, in spite of the fact, say, that the base-rate of faulting in the region is low. The two sources of evidence, prior probability and likelihood, should each influence the *a posteriori* probability (Eqn. 1), but intuition leads us to focus on the sample likelihood and, to some extent, ignore the prior probability.

Regression to the mean

Today, we think of regression analysis as fitting lines to data, but when Francis Galton did his pioneering work in the 1870's, and coined the term, his interest was not in best-fit lines but in reversion to the mean (Stigler, 1999). Galton experimented with the sizes of peas, and noted that, on average, size is inherited. Large peas tend to have larger than average offspring, and small peas the reverse. He noted also that, while on average the offspring of large peas are larger than their counterparts, they are also on average smaller than their parents. The offspring revert part of the way back to the population average. The fitting of lines came into the picture because the average distance between the size of the offspring and the population average was a linear function of the distance between the size of the parent and the population average (Figure 2).

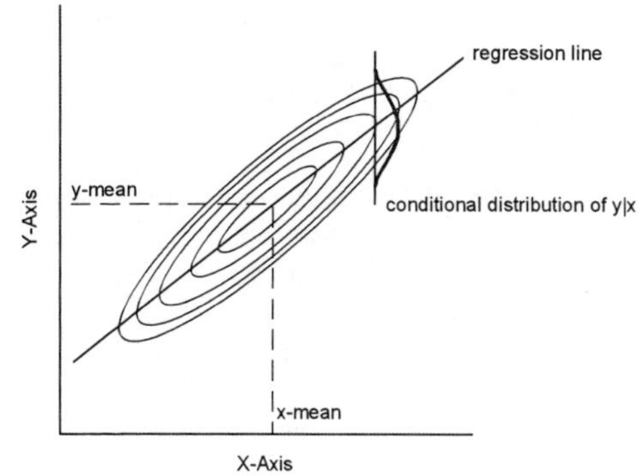

Figure 2. Regression line representing the expected values of y for given value of x. Note, because the regression line is less steep than the axis of the data ellipse, the conditional average of y for a given x is proportionately closer to the y-mean than the value of x is to the x-mean.

This regression effect occurs all the time in everyday life, and is related to the error people make in presuming representativeness. We look at the present or most recent sample or observation, and presume it is representative of the next; but, even eliminating sample size effects for the moment, this may not be the case. Consider that a numerical model with sophisticated constitutive equations is used to predict the performance of some earth structure. This model is applied to a randomly selected test section, and performs well. The predictions it makes of, say, deformations are closely matched by field measurements. Now, the model is applied to another test section. Will it perform as well? No, on average it will not, and one should not be surprised: it's basic regression.

Model predictions are based on theory, simplifications, and statistical parameter estimates. There is necessarily variation in how well a model predicts from one test section to another. Yet, if the model has predictive validity, it will on average be correlated to actual performance, and the accuracies of prediction from one test section to another should be correlated as well. If two predictions are correlated, there exists a regression relationship between them. Invoking Galton's observation, one should expect the second prediction to be less good than the first more than half the time. Of course, the converse is also true. If the first model prediction was not so good, the second will on average be better.

THE "RANDOM SOIL PROFILE"

In order to circumvent intuitive errors it has become more common to use formal statistical methods in analyzing field monitoring data, and indeed soil testing generally. This is part of a larger trend toward the use of risk and reliability methods in geotechnical engineering, a trend heralded by the emergence of load-resistance factor design (LRFD) in geotechnical codes (Kulhawy and Phoon, 1996), the increasing use of risk analysis in dam safety (Von Tunn, 1996) and flood hazard damage reduction studies (USACE, 1996a, 1996b), and the appearance of prominent lectures on practical applications of reliability.

These new approaches have introduced concepts into geotechnical engineering that are relatively new to practice, and perhaps not fully appreciated by those trying to use them. First, what does it mean for soil properties at a particular site and within a particular soil profile to be "random?" Clearly, unlike the weather, soil properties do not fluctuate erratically with time. In principle, the properties of the soil ought to be knowable everywhere. The only reason they are not known everywhere, and with precision up to our ability to measure, is that limited resources or limited testing technology has precluded them being observed.

Second, what does it mean for predictions of engineering performance to be "uncertain?" Uncertainty comes in many forms. Field measurements are scattered, so the ability to calibrate models to engineering performance is imprecise. Soil test data are biased, so estimates of soil engineering parameters that go into the models are inaccurate. The models used to predict performance are simplifications of reality, so forecasts are only approximations. Do all these different types of uncertainty affect predictions of engineering performance in the same way?

Third, what does it mean for uncertainties to be related to one another, that is, correlated? Some parameters, c and ϕ for example, are not actually separate physical properties but rather curve fitting numbers, and thus dependent on one another. Along a long reach of levee or long excavation, the variation of performance in space may have a systematic although uncertain pattern. Errors in estimating commonly shared parameters may mean that uncertainties in different types of engineering performance are implicitly related, even if mechanistically independent. Do these interdependencies significantly affect predictive uncertainty?

The nature of randomness

> Random *(adjective). Date: 1565. 1. a: lacking a definite plan, purpose, or pattern b: made, done, or chosen at random; 2. a: relating to, having, or being elements or events with definite probability of occurrence. b: being or relating to a set or to an element of a set each of whose elements has equal probability of occurrence. [Merriam-Webster, 1999].*

We use terms like *randomness, uncertainty, probability,* and *chance* all the time in the course of professional practice, yet without devoting much thought to what they mean, or to the larger philosophical questions their use implies. Most engineers, at least those who deal with the macroscopic world, think of nature as deterministic. For any effect there is a cause, and a cause and its effect are mechanistically linked. What then does it mean for something to be random? If the world is deterministic rather than random—at least at the scale of earth structures—what does it mean to speak of probabilities in relation to the world? Do probabilities describe some fundamental physical process, or do they have to do with limited information?

When we describe something as random, we normally mean that it is inherently unpredictable except probabilistically. Flood frequencies, for example, have been treated as an inherently random aspect of nature for many decades. In flood hazard studies we describe flood discharges only in exceedance probabilities (return periods). Thus, we treat flood discharges as if their magnitudes were generated by a celestial pair of dice. The peak discharge in a specific period of time, such as this year, cannot be predicted. All that can be said is that in a long series of years like this one, some fraction of the years will experience peak discharges larger than some fixed value.

Does this mean that rainfall and runoff are unpredictable processes of nature? No, not necessarily. Given advances in atmospheric science and hydrology, it is becoming ever more common for weather models to be used in predicting rainfall, and thus runoff and flood heights. Such models have also been used to predict probable maximum floods for dam safety studies (Salmon, 1999). When flood discharges are predicted by mechanistic modeling, they cease to be treated as random processes. The uncertainties surrounding predictions of flood flows change from those associated with random events to those associated with model and parameter errors. So, the assumption of randomness is only a convenience of modeling.

Randomness at the macro scale is an assumption, not an inherent part of the physical world. In principle, one ought to be able to predict whether a tossed coin lands heads-up or heads-down, but in practice, it is more convenient to assume that coin tossing is a random process resulting in a consistent frequency of each possible outcome as the experiment is repeated a large number of times. Randomness is not a property of the world; it is an artifact of modeling.

The nature of uncertainty

> Uncertain *(adjective). Date: 14th century. 1: Indefinite, indeterminate 2 : not certain to occur : Problematical 3: not reliable: Untrustworthy 4 a : not known beyond doubt : Dubious b: not having certain knowledge: Doubtful c: not clearly identified or defined 5: not constant: Variable, Fitful [Merriam-Webster, 1999].*

The word probability has several meanings in ordinary conversation. One is the interpretation of probabilities as relative frequencies of the random events discussed above, for which simple games involving coins, cards, dice, and roulette wheels provide examples. The distinctive feature of games of chance is that the outcome of a given trial cannot be predicted with certainty, although the collective results of a large number of trials display some regularity. There are many similar examples involving collections of people, molecules of a gas, genes, and so on.

A second interpretation is of probabilities in relation to propositions of uncertain veracity. Many uncertain propositions cannot be defined in the language of repeatable experiments. An individual can be uncertain about the truth of a scientific theory, a religious doctrine, or even about the occurrence of a specific historical event when inadequate or conflicting eyewitness accounts are involved. Using probability as a measure of uncertainty about propositions enlarges its domain of application to phenomena that do not meet the requirement of repeatability. On the other hand, this interpretation of probability as a measure of uncertainty implies that probability is subjective; it varies from one person to another.

Some aspects of geotechnical engineering can be treated as if they were random and thus describable by relative frequencies (e.g., flood frequency, the spatial variations of soil properties), but not all. Others may not have to do with real world processes that are repeatable; they may have to do with an unique event that we are unsure about. In this case, probability has a meaning of degree of belief or strength of opinion. Such strength of opinion may not be one-to-one identifiable with observed responses in the past, but may depend on qualitative experience, reasoning from first principles, and intuition.

Aleatory and epistemic probabilities

The above discussion leads to the conclusion that in modeling soil deposits and drawing statistical conclusions from sample data we deal with two types of uncertainty: that associated with the randomness of nature, and that associated with our incomplete knowledge of nature. The exact border between these two depends on modeling decisions, that is, the separation of uncertainty into randomness and ignorance is not a facet of nature, but a facet of analysis. Nonetheless, the distinction is of utmost importance to our understanding of natural processes, and to our ability to make sense of data and to make predictions with quantified reliability.

Increasingly, this distinction between randomness and ignorance is being described as the distinction between *aleatoric* and *epistemic* uncertainty. The notion of randomness, since the time of ancients, has concerned natural processes that are unpredictable. The role of dice, patterns of the weather, whether or not an earthquake occurs. Such unpredictable occurrences have been called aleatoric by Hacking (1975) and others, after the Latin *aleator*, meaning "gambler" or "die caster." The notion of uncertainty more generally, at least since the Enlightenment, has concerned what we know. The truth of a proposition, guilt of an accused, whether or not war will break out. Such unknown things have been called epistemic, after the Greek, meaning "knowledge." Both terms are now widely used in risk analysis, especially in applications dealing with seismic hazard, nuclear safety, and severe storms.

The term *probability*, when applied to aleatoric uncertainties, is usually taken to mean the frequency of occurrence in a long or infinite series of similar trials. In this sense, probability is a property of nature. We may or may not know what the value of the probability is, but the probability in question is a property of reality for us to learn. There is, presumably, a "true" value of this probability. We may know the true value only imprecisely, but there is a value to be known. Two observers, given the same evidence, and enough of it, should converge to the same numerical value.

The term *probability*, when applied to epistemic uncertainties, is usually taken to mean the degree of belief in the occurrence of an event or the truth of a proposition. In this sense, probability is a property of the individual. We may or may not know what the value of the probability is, but the probability in question can be learned by self-interrogation. There is, by definition, no "true" value of this probability. Probability is a mental state, and therefore unique to the individual. Two observers, given the same evidence, may arrive at different probabilities, and both be right!

In modern practice, reliability analysis incorporates probabilities of both the aleatoric and epistemic variety, and many that are both aleatoric and epistemic simultaneously. This has proved problematic, because it is confusing to separate out the two components of an individual probability assignment, and, unfortunately, the separation is important. Furthermore, the separation is not an immutable property of nature, but an artifact of analysis.

Spatial variation and its effects

How does aleatoric uncertainty—randomness—arise in soil testing and performance monitoring? Consider in more detail how we estimate soil parameters. We observe scatter in test data, and treat that scatter as if it derived from some random process. Presume for the moment that the test data are free from measurement errors. We then use statistical theory to summarize the data and to draw inferences about some hypothetical population of soil samples or test sections. Most people would agree that the things being observed, that is the soil properties or engineering performances, are not random. One may not know the properties or performances at every point in the soil deposit, but the properties and performances are knowable. They could, in principle, be observed. The variation in the data is spatial.

To say that the variability being observed is spatial rather than random is to liken the soil deposit to a deck of playing cards. Once the deck of cards has been shuffled and placed on the table, the order of cards in the deck is fixed. As with a soil deposit, the variation in the deck is spatial, not random. A player simply does not know the order of the cards before the game begins. Indeed, in many card games, such as Bridge or Blackjack, the winning strategy is for players to try to infer the order of cards remaining in the deck as play proceeds. In this way, card games are profoundly different from dice games, in which the sequential outcomes are assumed random. Thus, geotechnical engineering can be thought akin to card games, but not to dice.

To simplify modeling and inferences from data, the assumption is sometimes made to treat spatial variations within a soil deposit as if they were random—even though they are not. In this case, the variations are a function of space, and said to be *stochastic*, the term merely meaning a random process defined over some independent variable such as time or place. This is a modeling, and presumably simplifying assumption that transfers some of the uncertainty about soil properties or performance from the epistemic column to the aleatoric column, where it is easier to handle. Moving the boundary between the two types of uncertainty does not lessen the total. It does, however, create an irreducible level of uncertainty in the analysis and corresponding predictions, by presuming a fraction of the total uncertainty to be random, and thus unknowable. The trade off against this irreducible level of uncertainty is that one hopes the modeling assumption will allow more powerful methods of mathematics to be applied to the problem of inference and estimation, and thus in the end, a more precise outcome achieved.

An implication of making the trade off between aleatoric and epistemic uncertainty is that, what is meant by a predictive probability may change. Consider the "probability of excessive settlement" of a long levee, or by the same token, the "probability of excessive lateral deformation" in a long excavation. What does it mean to say, that this probability equals, say, 10%? Does it mean that 10% of the levee or excavation should be expected to behave adversely? Does it mean that there is a 10% chance that the entire levee or excavation will behave adversely? Does it mean something between these two? Confusion over this issue is frequent in the literature, where the spatial fraction of adverse performance of a large structure is many times used to verify a probabilistic prediction. The answer to the question depends on how the modeling assumptions are made; specifically, on how the total uncertainty is divided between aleatoric (spatial) and epistemic (parametric). To the extent that all the uncertainty is assumed aleatoric, the probability refers to a spatial fraction. To the extent that all the uncertainty is assumed epistemic, the probability refers to a chance of complete failure. Almost always, the uncertainty is apportioned between aleatoric and epistemic, so the probability itself is a mixture.

A second implication of the trade off between aleatoric and epistemic uncertainty is the variability of performance as a function of scale. To the extent that uncertainty is presumed to be aleatoric, the uncertainty averages over space and perhaps time. The variability of measured performance among long test sections will be less than the variability among short test sections. This was alluded to above. The variability of soil properties among large specimens will be less than among small specimens. The variability among in situ tests that mobilize large soil volumes will be less than the variability among in situ tests that mobilize small soil volumes. The converse is true of behaviors that rest on extreme soil or formation properties. Seepage conditions that depend on the most transmissive element of a formation become both more variable with scale and also on average more extreme. Rock slope failures that depend on the least favorably inclined joint become more variable and also more probable as the volume of rock mass considered becomes larger.

A simple-but-useful model of geotechnical uncertainty

There are many uncertainties in geotechnical predictions, and each is assessed somewhat differently from the others, and each affects the conclusions of an analysis in different ways. Important among these are,

- External loads (e.g., seismic accelerations, water elevations);
- Model and parameter uncertainty, including soil engineering properties;
- Undetected ("changed") site conditions;
- Poorly understood behavior (lack of adequate models); and
- Operational practices and human performance.

For the present, we ignore those uncertainties pertaining to external loads and human performance, and concentrate on those pertaining to geotechnical performance.

As above, one normally divides these geotechnical uncertainties into one set treated as aleatoric (naturally random) and one set treated as epistemic (poorly known). Uncertainties about soil engineering parameters, for example, are usually considered aleatoric. There is natural variability of soil properties within a formation, and this is characterized by a mean (or trend), variance, other moments, and a distribution function. Uncertainties about model representation, on the other hand, are usually considered epistemic. But the distinction becomes hazy when one is faced with actually assigning numbers to probabilities.

Consider in more detail the estimates of soil parameters. We observe scatter in test data, and treat that scatter as if deriving from some random process. We then use statistical theory to summarize the data and to draw inferences about some hypothetical population of soil samples. But, most people would agree that soil properties are not random. One may not know the properties at every point in a formation, but the properties are knowable. The variation is spatial, not random.

Having made the decision to treat some part of the variation in soil properties as aleatoric, the question becomes, how much? We may, on the one hand, model the randomness by a constant spatial mean, constant variance, and some probability distribution function (pdf) of variation about the mean. On the other hand, we may model the randomness by a polynomial trend, some residual variance, and pdf. In the second case, we have moved the boundary between what is modeled as aleatoric and what as epistemic. The polynomial trend explains more of the data scatter, and the variance of residuals around it is smaller than with the constant trend; but more parameters are needed to fit the trend, and the statistical error attending their estimation is larger because there are fewer degrees of freedom.

The scatter we observe in soil property data also comes in part from measurement errors. Measurement errors are of two types, (i) individually small, liable to be positive and negative, and cumulative; or (ii) large, consistently either positive or negative, and systematic. The former are sometimes called, "noise," and treated as aleatoric; the latter sometimes called, "bias," and treated as epistemic. The former are due to the sum effect of a large number of real disturbances, too many and individually too small to be treated separately. The latter are due (usually) to simplifications in the models used to interpret observations.

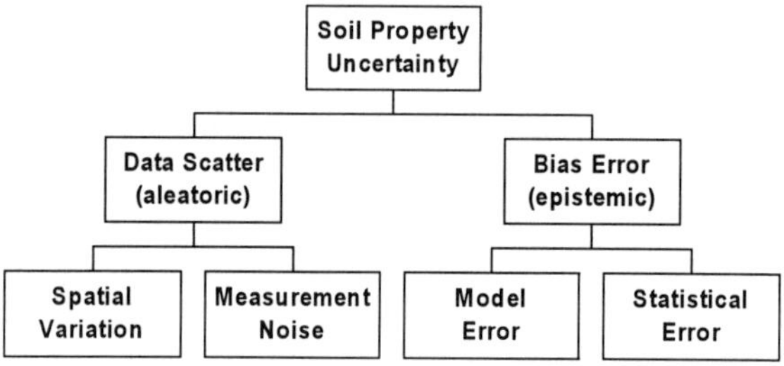

Figure 3. Simple structure of uncertainty in soil properties

Statistical error derives from limited numbers of observations. Having made a set of field measurements $\{x_1, \ldots, x_n\}$, an estimate of the mean in the field can be made by using the sample mean, $m_x = (1/n)\Sigma x_i$, as an estimator. Of course, were one to have made another set of n measurements at slightly different places, the numerical values of $\{x_1, \ldots, x_n\}$ would have differed somewhat from the original set, and m_x would be correspondingly different. So there is error due to statistical fluctuations among data sets, and this leads to error in the estimate of the pdf of the presumed aleatoric variation of the soil properties. Furthermore, this error is systematic. If the mean is in error at one location, it is in error by the same amount at every location. Even if one does assume that spatial variation and measurement noise of soil properties can be modeled as aleatoric, the statistical error is epistemic.

LESSONS FROM SAMPLING THEORY

Thus far, we have used sampling terms somewhat loosely. For the following, it is useful to be more precise. The objective of sampling is to infer characteristics of a large population without measuring all of its members. A group of elements—observations—selected from the large population according to some statistical plan is called, a *sample*. A sample has summarizing characteristics, for example, a sample mean and variance, called, *sample statistics*. Sample statistics can be any mathematical function of the individual observations within a sample. As long as the sample is taken according to a *statistical sampling plan*, estimates of the characteristics of the population can be made from the sample statistics, and the reliability of these estimates can be calculated.

The large population of interest might be called, the *target population*. The population available for sampling, and from which the sample observations are taken, is called, the *sampled population*. Ideally, the target population and the sampled population are the same, but many times this is not the case. For example, geological conditions at depth may not be available to a reconnaissance survey, and so conditions on the surface are sampled instead. The characteristics of the sampled population can be estimated from the characteristics of the sample using sampling theory, and the reliability of these estimates quantitatively determined. Lesson #1: The characteristics of the target population can only be estimated from the characteristics of the sampled population using professional judgment, and the reliability of these estimates quantitatively determined only subjectively.

In order for the reliability of these estimates of the characteristics of the sampled population to be quantitatively determined, the sampling plan must possess four qualities: (i) each element of the sampled population must have a non-zero chance of being observed, (ii) the relative probability of each element being observed must be known, (iii) each element appearing in the sample must be weighted in inverse proportion to its probability of being observed, and (iv) for any two possible samples the sums of the reciprocals of the relative weights of all elements in the samples are the same. Representativeness of the sampling plan is achieved not by giving each element an equal chance of being observed, but by compensating for the differences in probability by weighting. Thus, each element has an equal chance of affecting the weighted sample characteristics. An individual sample can never be representative, it is unique; only the sampling plan can be made representative.

In order for a sampling plan to adhere to the requirements above, and thus to be a statistical sampling plan, the elements within the sampled population observed in the sample must be chosen randomly. That is what the first requirement says. If this is not so, the sample may still have utility, but the reliability of estimates of the characteristics of the sampled population cannot be quantitatively determined. The typical sample of geotechnical observations is taken *purposively*, that is, by conscious selection by the engineer. Some places in the geotechnical literature one even reads that purposive sampling is to be preferred to random sampling, because only in that way can one be assured that the most critical sections are measured. Lesson #2: There is nothing improper about this view, but it precludes any quantitative statement about the accuracy or precision of potential estimation errors.

All samples exhibit variability. The variation within the sample, along with sample statistics such as the sample mean, are used to make estimates of population characteristics, but also to assess the reliability of those estimates. For many important cases, such as estimating the population mean or the standard deviation, and to a first-order approximation, the precision of an estimate made from a sample is proportional to the sample standard deviation and inversely proportional to the square root of the sample size. For example, the standard error (imprecision) in estimating the population mean is $s/\sqrt{n}$. Lesson #3: Estimate precision increases only as root-n.

Increasing the precision by a factor of two requires increasing the sample size by a factor of four.

Since any quantitative statement about the precision of estimates requires an estimate of the population variance, the degrees of freedom in making estimates from samples is reduced by at least one from the number of observations, because the variability of the population also needs to be estimated from the sample. What about the common case in geotechnical engineering where there is only one observation? This happens all the time in practice. Some even favor this case because there is no messy data scatter with which to grapple. In principle, of course, no reasonable statistical inference can be made from one observation. True, the single observation provides an estimate of the population mean, but it provides no information on variation within the population, and thus, we have no appraisal of the accuracy of the inference. Yet, we make inferences from single observations all the time, because we have an intuitive estimate of how much variability there might be in the population based on experience with other situations, and we apply those subjective notions of the variability within generic soil formations to the particular problem at hand. All the limitations discussed above about such intuitive information processing applies to this case as well.

CONCLUSIONS

As engineers and scientists, we make measurements of field performance or of in situ properties both to increase our understanding of the behavior of geological materials and to improve our ability to predict performance. The act of making these measurements is inherently statistical, and is subject to the same principles and the same constraints of statistical behavior as is any other sampling enterprise. Yet, in geotechnical practice, most such sampling activities are approached intuitively. There are qualitative as well as quantitative lessons to be learned from sampling theory, that allow us to draw stronger and less error-prone inferences from the results of our field measurements than we do now by intuition.

ACKNOWLEDGEMENTS

The author wishes to thank David Bowles, John Christian, Karl Dise, Desmond Hartford, Michael Klosterman, and Andy Zielinski for interesting and helpful discussions on the topics of this paper.

APPENDIX I. REFERENCES

Alpert, M. and Raiffa, H. (1982). “A progress report on the training of probability assessors,” in, *Judgment under uncertainty, heuristics and biases*, Kahneman, D., Slovic, P., and Tversky, A. (Eds.), Cambridge University Press, Cambridge, 294-306.

Baecher, G.B. (1987). “Uncertainty analysis for dam projects (summary),” Report GL-87-4; “Statistical analysis of geotechnical data,” Part 1 GL-87-1; “Statistical quality control for engineered embankments," Part 2 Report GL-87-2; "Error analysis for geotechnical engineering," Part 3 Report GL-87-3; US Army Corps of Engineers, Vicksburg, MS.

Cochran, W. G. (1977). *Sampling techniques*, Wiley, New York.

DeFinetti, B. (1937). "Foresight: Its logical laws, its subjective sources," in, *Studies in subjective probability*, Kyburg, H., and Smokler, H. (Eds.), John Wiley & Sons, NY, 93-158.

Feller, W. (1967). *An introduction to probability theory and its applications*, Wiley, New York.

Folayan, J., Höeg, K., and Benjamin, J. (1970). "Decision Theory Applied to Settlement Predictions." *Journal Soil Mechanics and Foundation Engineering, 96*(4), 1127-1141.

Hacking, I. (1975). *The emergence of probability*, Cambridge University Press, Cambridge.

Hynes, M., and Vanmarke, E. (1976). "Reliability of Embankment Performance Predictions." *Proceedings* Engineering Mechanics Division Specialty Conference, University of Waterloo Press.

Kahneman, D., and Tversky, A. (1979). "Prospect theory." *American Economic Review*, 47, 263-291.

Kulhawy, F. H., and Phoon, K. K. (1996). "Engineering Judgment in the Evolution from Deterministic to Reliability-Based Foundation Design." *Uncertainty in the Geological Environment*, Madison, WI, 29-49.

Kulhawy, F. H., and Trautmann, C. H. (1996). "Estimation of In-Situ Test Uncertainty." *Uncertainty in the Geologic Environment*, ASCE Specialty Conference, Madison, WI, 269-286.

Maistrov, L. E. (1974). *Probability theory; a historical sketch*, Academic Press, New York.

Matheron, G. (1989). *Estimating and Choosing: An essay on probability in practice*, Springer-Verlag, Berlin.

Morgan, M. G. and Henrion, M. (1990). *Uncertainty : a guide to dealing with uncertainty in quantitative risk and policy analysis*, Cambridge University Press, Cambridge.

Phoon, K. K., and Kulhawy, F. H. (1996). "On Quantifying Inherent Soil Variability." *Uncertainty in the Geologic Environment*, ASCE Specialty Conference, Madison, WI, 326-340.

Salmon, G. (1999). Remarks before the US Army Corps of Engineers Workshop on Risk Analysis for Dam Safety, Vicksburg.

Snedecor, G.W., and Cochran, W.G. (1980). *Statistical methods, 7th Ed.*, The Iowa State University Press.

Stigler, S. M. (1999). *Statistics on the table : the history of statistical concepts and methods, Harvard University Press*, Cambridge, MA.

Tversky, A., and Kahneman, D. (1971). "Belief in the "law of small numbers." *Psychological Bulletin, 76*, 105-110.

Tversky, A., and Kahneman, D. (1974). "Judgment under uncertainty: heuristics and biases." *Science*, 185, 1124-31.

USACE (1996a), "Risk-based analysis for evaluation of hydrology/hydraulics, geotechnical stability, and economics in flood damage reduction studies", ER1105-2-101, US Army Corps of Engineers, Washington, DC.

USACE (1996b), "Risk-Based analysis for flood damage reduction studies", EM 1110-2-1619, US Army Corps of Engineers, Washington, DC.

Von Thun, J.L. (1996). "Risk assessment of Nambe Falls Dam," *Uncertainty in the Geologic Environment*, ASCE Specialty Conference, Madison, WI.

APPENDIX II. SYMBOLS USED

m = mean

s = standard deviation

Cov[•,•] = covariance

Pr{Θ} = probability of Θ

Pr{Θ | *data*) = conditional probability of Θ given a set of data

$r = |x_{max}-x_{min}|$ = range

Var[•] = s^2 = variance

Ω = s/m = coefficient of variation

Θ = parameter or state of nature

L{*data* | Θ} = likelihood of Θ for a set of data = conditional probability of a set of data, given Θ

GEOENVIRONMENTAL PERFORMANCE MONITORING

Edward Kavazanjian, Jr.[1]

ABSTRACT: Performance observations and measurements serve the same essential functions on geoenvironmental projects as they do on traditional geotechnical projects. The classical observational approach to geotechnical projects expounded upon by Terzaghi, Casagrande, Peck, and Whitman is as viable and necessary on geoenvironmental projects as it is on traditional projects. However, on geoenvironmental projects, there is often an added emphasis on observation and measurement for regulatory compliance and regulatory approval compared to traditional projects. Furthermore, the limitations and challenges of the classical observational approach associated with measurement of extremely small quantities, uncertainty with respect to mechanisms and phenomena of interest, and emotional issues related to project impacts may be exacerbated on geoenvironmental projects. Case histories are used to illustrate the role of performance observation and measurement on geoenvironmental projects. Monitored natural attenuation is discussed as the penultimate example of the application of the observational approach on geoenvironmental projects.

INTRODUCTION

In order to protect both public safety and capital investment as well as to facilitate cost-effective design and continuous improvement in our practice, performance prediction, performance monitoring, and systematic evaluation and interpretation of monitoring data for confirmation of performance are of indispensable importance on geoenvironmental projects. The essential role played by observations and measurements for confirmation of performance (i.e., performance monitoring) in geotechnical practice has long been recognized by the geotechnical profession. Uncertainties inherent to the determination of subsurface stratigraphy and geotechnical properties and to the prediction of phenomena and quantities of interests combine to make geotechnical engineering a relatively imperfect discipline. Due to these uncertainties, performance monitoring of constructed facilities is an essential element of geotechnical practice, more so than in many other engineering disciplines. Compared to traditional geotechnical projects, on geoenvironmental projects the importance and challenges of performance monitoring may be accentuated and complicated by factors inherent to

[1]Principal, GeoSyntec Consultants, 2100 Main Street, Suite 150, Huntington Beach, CA 92648, EdKavy@GeoSyntec.com.

geoenvironmental practice. These inherent factors include the small quantities that must be measured, the long time frame over which monitoring must be conducted, uncertainty with respect to the phenomena of interest, and the emotional issues associated with human health and environmental protection considerations.

Performance monitoring is perhaps more common on geoenvironmental projects than on traditional geotechnical projects because on geoenvironmental projects performance observation and measurement is frequently mandated by regulatory requirements. Regulatory requirements can oftentimes relieve the engineer of the burden of convincing the owner of the need for performance monitoring. However, minimal regulatory compliance can be counterproductive. Blind or simplistic compliance with regulatory requirements can lead to wasted resources and misleading results, as can performance observations and measurements made without development of a sound rationale for the monitoring system coupled with *a priori* prediction of the monitoring parameters. Furthermore, regulatory requirements are minimum standards and do not relieve the engineer of his obligation to develop a comprehensive and rational performance monitoring system.

PERFORMANCE MONITORING OBJECTIVES

Table 1 lists the primary objectives of observation and measurement of the performance of geoenvironmental systems. These are essentially the same objectives of performance monitoring for traditional geotechnical projects, but with added emphasis on regulatory issues. Objectives of performance observation and measurement in geotechnical practice include: evaluation of properties (e.g., by back-analysis); regulatory compliance; validation or verification of models, including both input assumptions (e.g., stratigraphy for slope stability analyses) and model algorithms; protection of human health and the environment; regulatory acceptance; and optimization of design.

TABLE 1: Geoenvironmental Performance Monitoring Objectives

	Objectives
1.	Protection of Human Health and the Environment
2.	Validation/Verification of Model Algorithms and Modeling Assumptions
3.	Evaluation of Properties
4.	Regulatory Compliance
5.	Regulatory Acceptance
6.	Optimization of Design

The performance objectives identified in Table 1 are not mutually exclusive. One or more of these objectives may support attainment of one or more of the other objectives (e.g., evaluation of properties and validation/verification of models may support regulatory acceptance and/or optimization of design). The systematic application of performance monitoring to achieve the objectives listed in Table 1 (with the possible exception of regulatory compliance) is broadly referred to in geotechnical engineering as the Observational Approach (or Method).

THE OBSERVATIONAL APPROACH

Historical Perspective

The role of performance observation and measurement in geotechnical practice has been discussed at length in papers by many authors, including classical papers by Casagrande (1965), Peck (1969), and Whitman (1984). These authors, in their expositions of the observational approach and the role of calculated risks in geotechnical practice, eloquently describe the function of performance observation and measurement in optimizing design, protecting human health and the environment, and improving the state of practice. These authors also address, both directly and implicitly, the limitations and challenges associated with applying observational methods to geotechnical problems.

Like many other fundamental aspects of geotechnical practice, the role of observations and measurements for confirmation of the performance of geotechnical systems, generally referred to in our profession as the observational method or approach, was first elucidated by Karl Terzaghi. Casagrande (1965) and Peck (1969) both discuss Terzaghi's observational approach to geotechnical practice. Casagrande (1965) describes the types of risks inherent to earthwork and foundation engineering and the extent to which we are justified or forced to assume such risks in our practice. He also, importantly, discussed how to cope with such risks. In describing the types of risks, Casagrande distinguished between engineering risks and human risks. Engineering risks included calculated risks and unknown risks (risks that cannot be identified until they reveal themselves by failure or other events). Calculated engineering risks are taken when quantitative analyses are either non-extant or of doubtful validity. Casagrande illustrated the role of the observational approach with respect to managing calculated risks (minimizing the probability of adverse impact) with several case histories. Voluntary assumption of calculated risks is justified only when the consequences of adverse impact are acceptable or can be minimized by performance monitoring. However, on many projects, there are involuntary (unavoidable) risks that may have catastrophic consequences (e.g., loss of life). In these cases, performance monitoring to minimize the potential for catastrophic consequences becomes an essential element of the project.

Peck (1969) formalized the observational approach developed by Terzaghi. Peck defined the observational approach as follows: assess the most probable condition and the most adverse conceivable development; establish key parameters for observation; calculate the parameters of observation under both conditions; compare results to predictions; and change the design as necessary. Essential aspects of the observational approach as formalized by Peck include action plans for every conceivable unfavorable situation and the choice of key parameters for observation. Peck notes that failure to anticipate unforeseen conditions can result in disaster because either there may be no adequate solution should unfavorable conditions develop or the engineer may have to make decisions under pressure and may therefore make the wrong decision. Peck states that if you cannot solve the hypothetical problems, you cannot use the observational method. Furthermore, the selection of key parameters for observation requires a feel for the significant physical phenomena governing the project and reliable means for making the appropriate observations. Peck cautions that mistaken preconceptions about the phenomenon of interest can lead to omission of relevant observations and that unreliable measurements can lead to wrong actions. He also warns that the results of employing the observational approach may be disappointing and/or may lead to disaster because they are not part of a well-conceived monitoring program. Peck suggests that the observational approach should not be used if the cost or probability of an adverse outcome is too high. Presumably, this statement only applies when the risk of an adverse outcome is voluntary. The observational approach would seem to be the most prudent approach when there is an involuntary risk of an adverse outcome.

Whitman (1984) updated the concept of taking a calculated risk in implementing a geotechnical project to recognize recent advances in risk analysis and risk management. Whitman stated that the most important developments since Casagrande were the recognition that all risks cannot be eliminated and a growing willingness in the profession and in society to face questions as to what the allowable risk is in an open manner. However, he noted that in areas where emotional issues (e.g., issues involving life safety) are involved (e.g., dam safety), rational assessment of risk is difficult. When their health or welfare is threatened, some people are unwilling to assume any risk as a consequence of civil works or industrial development. When this type of unrealistic expectation enters the discourse, it is unlikely that either the observational approach or quantitative risk analysis can be formally applied.

Application to Geoenvironmental Practice

Observational methods and the observational approach are applied to geoenvironmental engineering projects for the same reasons as they are applied to traditional geotechnical engineering projects. However, development of performance monitoring systems and interpretation of the performance monitoring data on geoenvironmental projects is often complicated by the factors listed in Table 2. These factors include the extremely small quantities that must be measured, the long periods of time over which the quantities of interest must be monitored, imperfect knowledge of the phenomena to be monitored, difficulties in measuring the parameters of interest, and emotional issues associated with environmental risks. These factors combine to make performance observation, monitoring, and interpretation perhaps even more challenging in geoenvironmental practice than for many conventional geotechnical projects.

TABLE 2: Geoenvironmental Performance Monitoring Challenges

Challenge	Solution
1. Long Time Periods of Interest	"Early Warning" Systems, Formal Monitoring Plans with Financial Assurance
2. Imperfect Knowledge of Phenomenon and Impacts	Multi-Parameter Modeling, Periodic Review and Updating, Sensitivity Studies
3. Extremely Small Quantities	Improved Measurement Techniques, Monitoring of Surrogates, Statistical Analyses
4. Difficulties in Measuring Parameters of Interest	Indirect Measurements, Monitoring of Surrogates
5. Emotional Issues	Public Education, Open Communication

Many of the factors listed in Table 2 are associated with the limitations and challenges addressed by Casagrande, Peck, and Whitman in their discussions of the observational approach and calculated risks. Table 2 identifies some of the approaches used in geoenvironmental practice to accommodate these factors.

An important difference between performance observation and measurement for geoenvironmental projects and for traditional geotechnical projects is the long duration for which monitoring is often required. Terzaghi, as quoted by Casagrande (1965), said a project is not over until it is successfully constructed. On many geoenvironmental projects, for practical purposes, the project is never over; performance monitoring will continue for the indefinite future. To accommodate this long time frame, many geoenvironmental projects have formal monitoring and reporting plans. These plans detail monitoring and reporting procedures and criteria. These plans also frequently contain financial assurance to provide for monitoring and reporting costs for the foreseeable future (typically, 30 years, as the net present value of monitoring costs more than 30 years out is negligible) and for remediation of any foreseeable release. A prudent monitoring program also provides for "early warning" of an impending problem. Early warning may be provided by monitoring the important quantities for levels below "action" levels (i.e., below levels that require corrective action), by monitoring at points between the source and potential receptors or the point of compliance, and/or by monitoring at points which allow for implementation of cost-effective corrective actions.

Other strategies employed by geoenvironmental monitoring plans to accommodate the challenges cited in Table 2 include:

- monitoring of surrogates (e.g., innocuous chemicals which generally accompany chemicals of concern) for quantities which are of concern at very low values (e.g., parts per billion) or which cannot be directly monitored;

- statistical analysis, including trend analyses and non-parametric analysis of variance, to detect parameters or parameter variations with small thresholds;
- indirect monitoring of phenomena or parameters (e.g., monitoring of soil moisture content instead of liquid flux) when direct measurement is difficult;
- periodic review and revision of monitoring plans (e.g., comprehensive 5-year reviews) to incorporate new findings and advances in practice and to accommodate imperfect knowledge of phenomena and their impacts and improvements in monitoring techniques; and
- periodic multi-parameter monitoring (e.g., monitoring of broad, all-inclusive lists of chemicals of concern), usually in conjunction with periodic reviews of the monitoring plan, to accommodate imperfect knowledge of phenomena.

Dealing with the emotional issues surrounding geoenvironmental projects can be one of the most difficult challenges for the engineer. Training in rational engineering analysis and an in-depth understanding of technical issues may sometimes lead to an unsympathetic reaction to emotional issues raised by concerned citizens. Project opponents may also prey upon these emotional issues to prevent rational discourse and thereby disrupt the project. Public education and open disclosure on the risks and benefits associated with a project are the only recourse when these emotional issues are raised and are best employed from project inception, before reason is clouded by emotion.

The following case histories illustrate the application of observational methods on geoenvironmental projects. These case histories were chosen to illustrate how observations and measurements are used to achieve the objectives in Table 1 and the challenges in Table 2 and to illustrate the insights provided by Casagrande, Peck, and Whitman with respect to the benefits and pitfalls of the observational approach.

CASE HISTORIES

Landfill Groundwater Monitoring

The groundwater monitoring system for the City of Los Angeles Lopez Canyon Sanitary Landfill provides an excellent illustration of the use of geoenvironmental observations and measurements to protect human health and the environment. In achieving this objective, groundwater monitoring system development accommodated several of the other objectives in Table 1. This case history also illustrates approaches to dealing with the challenges listed in Table 2.

The Lopez Canyon Sanitary Landfill is a municipal solid waste landfill owned and operated by the City of Los Angeles (the City). In accordance with federal and state regulations, a multi-media (soil, groundwater, surface water, air) Systems Monitoring Plan (SMP) was prepared for the landfill to document compliance with environmental monitoring requirements (GeoSyntec, 1994). The SMP is incorporated in its entirety in the Final Post-Closure Maintenance Plan (FPCMP) for the landfill. The monitoring system described in the SMP includes groundwater monitoring wells, multi-point vadose zone monitoring probes, surface water monitoring stations, ambient air quality monitoring points, and a grid over the waste mass for instantaneous surface emissions measurement. The monitoring system is designed to give early warning of a release of substances from the landfill that could eventually endanger human health or degrade the environment so that appropriate corrective action may be taken. The FPCMP provides for financial assurance (in the form of a trust fund) for 30 years of post-closure monitoring and for foreseeable corrective actions that may be required over that 30-year period.

Figure 1 is a plan view of the landfill illustrating the site topography, the boundaries of the waste units, and the locations of the groundwater monitoring wells. The landfill is nestled among and astride a series of ridges that divide the topography into three canyons, labeled Canyons A, B, and C. Disposal Area C (in Canyon C) has a modern composite liner system. Disposal Areas A, B, and AB+ (in Canyons A and B and astride the ridge that separates these canyons from Canyon C) are unlined. Disposal Area AB+ is underlain by a leachate collection system (an underdrain collection system). In order to develop a rational and effective groundwater monitoring system, a conceptual hydrogeologic model of the site was developed. In developing the conceptual model, it was recognized that there might be both shallow (alluvial) and deep (bedrock) aquifers at the site. Following general precepts of hydrogeology, it was assumed the uppermost aquifer phreatic surface mimicked, in general, the natural ground surface topography. In accordance with this assumption, it was assumed that Canyon C drains west and then south out through the landfill gate and that Canyons A and B drain east into south-flowing Bartholomaus Canyon. These general directions of groundwater flow are shown on Figure 1.

The geologic strata under Lopez Canyon are considered "non-water-bearing" rocks by the California Department of Water Resources in that they yield low quantities of groundwater that is high in dissolved solids and minerals and is not suitable for either domestic or agricultural uses. However, there are domestic water supply wells in Kagel Canyon immediately east of Bartholomaus Canyon, there are agricultural wells in Lopez Canyon to the northwest of the landfill, and to the south of the landfill lies an alluvial basin that is a source of domestic water supply. Therefore, a release of contaminants from the landfill could impact domestic and/or agricultural water supply. Furthermore, state regulations consider any degradation in groundwater quality unacceptable, whether or not the groundwater has beneficial uses.

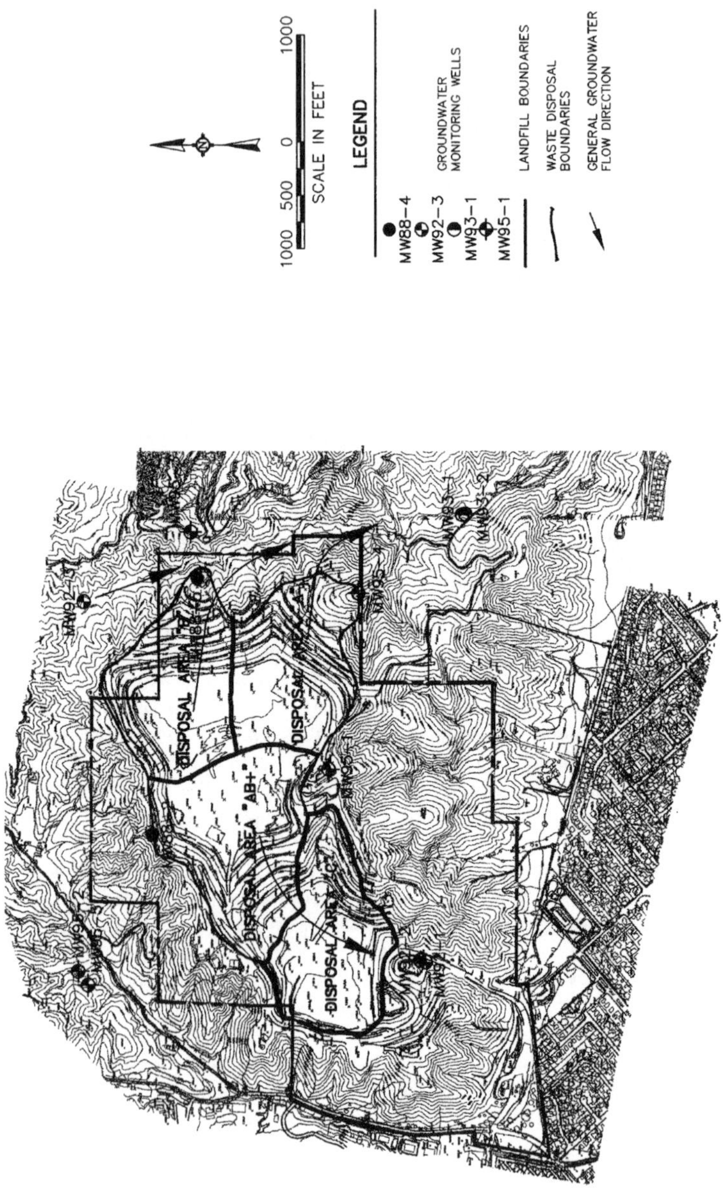

Figure 1. Lopez Canyon Landfill Groundwater Monitoring System

The configuration of the twelve wells of the groundwater monitoring network for the Lopez Canyon Landfill shown in Figure 1 is described as follows:

- MW 92-1 and MW 92-2 are paired deep and shallow wells that monitor groundwater in alluvium and bedrock flowing to the south and west out of Canyon C;
- MW 93-1 and MW 93-2 are paired deep and shallow wells that monitor groundwater in alluvium and bedrock flowing south out of Bartholomaus Canyon;
- MW 95-2 and MW 95-5 are paired deep and shallow wells that monitor groundwater flowing in alluvium and bedrock in Lopez Canyon north and west of the landfill;
- MW 95-1 and MW 88-2 are deep wells that monitor groundwater in bedrock at the north and south boundaries of the landfill, respectively;
- MW 88-5 and MW 95-4 are shallow wells that monitor alluvial groundwater flowing east out of Canyons A and B, respectively;
- MW 92-3 is a shallow alluvial well in Bartholomaus Canyon upgradient to the north and east of the landfill; and
- MW 95-3 is a well drilled into bedrock in the ridge between Bartholomaus and Kagel Canyon.

Based on the conceptual hydrogeologic model, only monitoring wells MW 92-1, MW 92-2, MW 93-1, and MW 93-2 are necessary to monitor for a release from the landfill for the most probable conditions. The other wells either provide early warning of a potential release so that the impact of a release can be minimized and the release can be cost-effectively controlled or monitor for unanticipated (less likely) migration pathways from the landfill. MW 88-5 and MW 95-4 are located at the mouths of Canyons A and B, where the waste is unlined, to provide early warning of a release of landfill-related compounds from these areas. A release detected by one of these wells could be cost-effectively addressed before impacting Bartholomaus Canyon or reaching a point of compliance by corrective action in the immediate vicinity of the impacted well. MW 95-3 was installed to provide additional protection for Kagel Canyon drinking water supplies. MW 95-2 and MW 95-5 monitor for unanticipated northward migration of landfill-related compounds into Lopez Canyon. MW 95-1 and MW 88-2 monitor deep bedrock groundwater for landfill-related compounds. MW 92-3 serves as a background well for alluvial flows along Bartholomaus Canyon and also monitors for upgradient migration of landfill-related compounds due to gas migration and/or diffusion. In addition to these monitoring wells, the three domestic water supply wells in Kagel Canyon that are closest to the landfill are also monitored.

The configuration of this monitoring system is entirely consistent with the observational approach and management of calculated risks. The Lopez Canyon Landfill groundwater monitoring system is configured to:

- account for the most likely release scenarios that could adversely impact human health and/or the environment;
- detect contaminant migration along unanticipated pathways before they adversely impact human health and/or the environment; and
- provide early warning of a release and facilitate corrective action.

The groundwater testing program at Lopez Canyon is designed to address several of the challenges and limitations cited in Table 2 that are posed by applying the observational approach to this project. For instance, one possible impact of the landfill on groundwater is the migration of heavy metals from the landfill. Heavy metal migration may be difficult to detect in the Lopez Canyon environment due to the small quantities that must be monitored and the high background concentrations of metals. Therefore, the monitoring wells at Lopez Canyon are currently sampled semi-annually for the following surrogate compounds for heavy metals: pH; total dissolved solids; chloride; sulfate; and nitrate. Statistically significant increases in one or more of these compounds should be a precursor of a heavy metal release. Every five years, the monitoring wells at Lopez Canyon are sampled for a suite of 16 heavy metals commonly associated with solid waste landfills. The Lopez Canyon monitoring wells are also tested semi-annually for 47 commonly occurring volatile organic compounds and for four additional less common volatile organic compounds. The four additional compounds were selected from a broad list of less-commonly occurring compounds because they have been found in the landfill's leachate. Every five years, the leachate is tested for the entire list of less-commonly occurring compounds to see if additional compounds should be added to the list of semi-annually monitored compounds. Furthermore, the entire monitoring plan is reviewed every five years for concurrence with current standards of practice.

In this manner, the groundwater sampling and testing program accounts for:

- the extremely small quantities of the constituents of concern;
- difficulties in measuring parameters of interest;
- the most likely compounds to be released from the landfill; and
- imperfect knowledge of phenomenon of interest.

To deal with the emotional issues surrounding the impact of a landfill on the surrounding community, the City has put in place a comprehensive community relations program that includes open communication and public education. Copies of all monitoring reports are placed in a document repository in the local public library for public access. A quarterly newsletter is distributed in the local community and quarterly workshops are held at the landfill to discuss issues of concern with the general public. Queries from local residents on monitoring results and other landfill-related issues are promptly responded to in detail by City staff. In this manner, the community is informed and educated and emotional issues associated with landfill impacts can be diffused without undue alarm.

Evapotranspirative Soil Covers

Soil moisture content monitoring in an evapotranspirative soil final cover (ET cover) for a municipal solid waste (MSW) landfill illustrates the role performance observations and measurements can play in gaining regulatory approval for new technology. In semi-arid climates, an ET cover (a cover which controls percolation into the waste mass by evaporation and transpiration) can provide an environmentally superior and cost-effective alternative final cover compared to a prescriptive clay barrier cover. However, because many regulators consider ET covers a new and unproven technology, acceptance of these alternative final cover systems to date has been conditioned upon post-construction monitoring of their effectiveness. Because there is no firm criteria on allowable percolation into the waste mass of a MSW landfill, acceptance criteria for ET covers has typically been based upon comparative analyses of percolation into the waste through a prescriptive clay barrier cover and through the ET cover.

Figure 2 shows cross sections for the prescriptive final cover designed in accordance with California regulations for unlined MSW landfills and a typical alternative ET final cover for southern California. The prescriptive cover is sometimes referred to as a "resistive" type of cover as it controls percolation with a low permeability barrier layer composed of at least 0.3 m of soil with a saturated hydraulic conductivity no greater than 1×10^{-6} cm/s. The ET cover consists of a vegetated homogeneous soil layer typically 1.2 m to 1.8 m in thickness with a saturated hydraulic conductivity between 1×10^{-4} to 1×10^{-5} cm/s. The ET cover may be described as a "capacitive" type of cover as it controls percolation by storing infiltration during wet periods in the soil pores by capillary action and releasing it to the atmosphere by evaporation and transpiration during dry periods. While the saturated hydraulic conductivity is not an essential property of the ET cover, it does serve as an index of the ability of the soil to store water in the pores by capillary action and to release water from the pores to the atmosphere by evaporation and transpiration.

ET covers are cheaper and easier to construct and maintain than prescriptive covers. ET covers are generally at least as thick as the prescriptive cover, thereby affording equivalent or superior isolation of the waste.

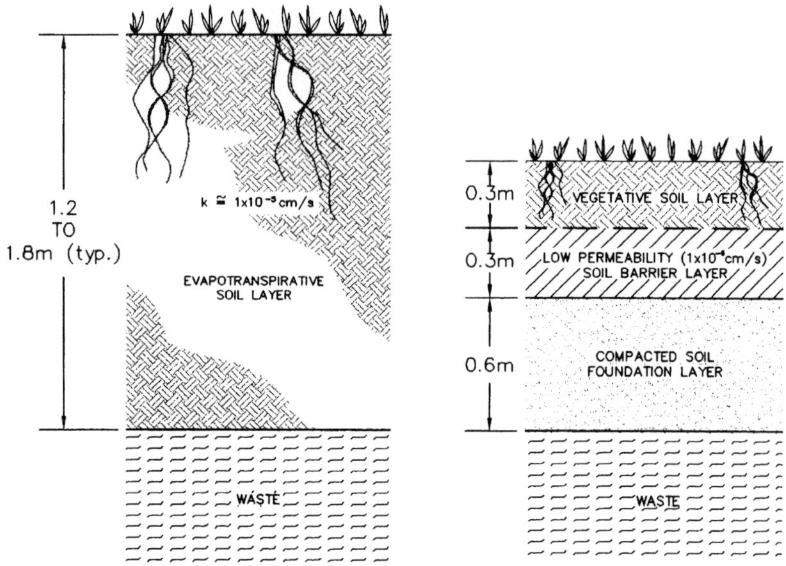

Evapotranspirative Cover Prescriptive Cover

Figure 2. Evapotranspirative and Prescriptive Final Covers

Therefore, if the ET cover can be shown to allow less percolation into the waste than the prescriptive cover, the ET cover may be considered to provide superior environmental protection compared to the prescriptive cover. Due to the difficulty in measuring percolation and the cost and time associated with building side by side test plots of the prescriptive clay barrier and ET covers, in several jurisdictions ET covers have been permitted based on soil moisture monitoring for a duration of two to five years.

Soil moisture measurements alone are not sufficient for quantitative evaluation of percolation. However, an unsaturated flow model that accurately predicts seasonal moisture content fluctuations with depth can be used with good confidence to evaluate the relative percolation performance of an ET cover compared to a prescriptive cover. Figure 3 shows theoretical predictions of the percolation through an ET cover compared to the prescriptive cover for a landfill in southern California. These predictions were made using an unsaturated flow model and site-specific climate data. The ten-year period with the greatest cumulative precipitation was selected from the historic record for development of Figure 3.

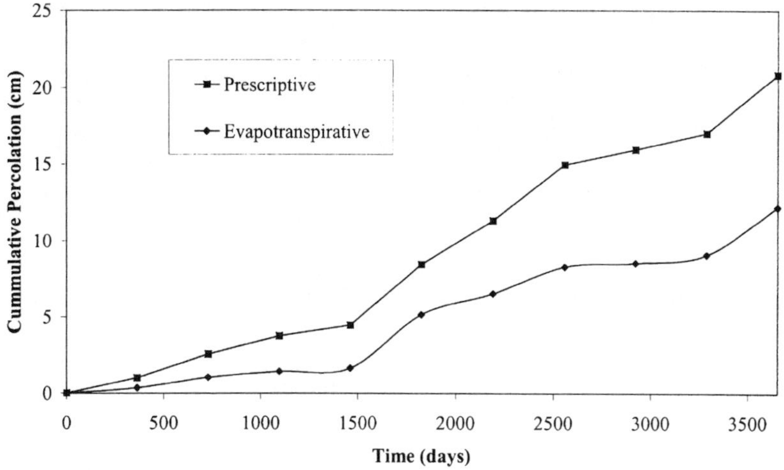

Figure 3. Predicted Cumulative Percolation for the Wettest 10-Year Period

On the basis of the analysis illustrated in Figure 3, conditional regulatory approval was obtained for use of the ET cover pending satisfactory field performance monitoring. To monitor the ET cover, a weather station and two soil moisture monitoring stations were installed at the site following the completion of ET cover construction. At each soil moisture monitoring station, time-domain reflectometry (TDR) probes were configured to give a continuous profile of soil moisture content in the final cover to a depth of 1.8 m. Figure 4 shows a comparison between the observed seasonal moisture content fluctuation at the same depth at each TDR monitoring station and the moisture content fluctuation predicted at that depth by the model used to generate Figure 3 and the site specific climate data collected by the on-site weather station. The data illustrated in Figure 4 was used to demonstrate to the regulators that the unsaturated flow model used to establish the superior percolation resistance of the ET cover compared to the prescriptive cover was reliable for predicting relative percolation through the ET and prescriptive covers. This data, combined with the "regular" monitoring data for air quality, water quality, and vadose zone gas concentrations that indicated that the final cover was performing satisfactorily, was sufficient to convince the regulators to remove the conditional approval granted when the ET cover was first proposed.

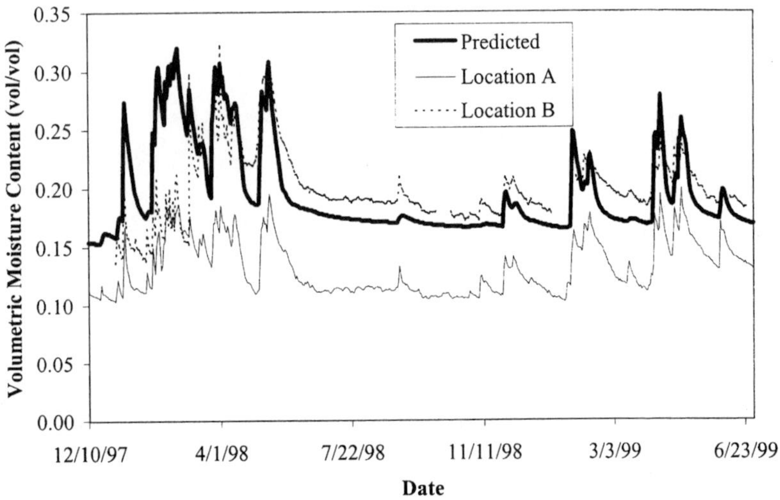

Figure 4. Volumetric Moisture Content at a Depth of 0.15 m

The moisture content measurements in the ET soil cover were used to:

- evaluate difficult to measure soil properties (e.g., unsaturated flow parameters);
- validate the numerical model for unsaturated flow;
- optimize the design of the cover; and
- gain regulatory approval of a new technology.

Landfill Seismic Stability

The seismic stability analysis of the Operating Industries, Inc. (OII) Landfill illustrates the role performance observations and measurements can play in evaluation of difficult to measure properties. Due to difficulties in sampling and testing solid waste materials, it is extremely difficult to reliably determine the cyclic stress-strain properties of solid waste using only conventional laboratory and in-situ testing. Small-strain stiffness can be determined in the field by non-invasive shear wave velocity measurements (Kavazanjian, et al., 1996). However, the best evidence for intermediate- and large-strain cyclic stress-strain properties for waste comes from field performance observations.

Figure 5 presents an aerial view illustrating the unique hazard posed by the OII landfill. The very steep, very tall non-engineered slopes of the landfill are immediately adjacent to a major freeway and residential development. Furthermore, the landfill sits on top of a major earthquake fault. Figure 6 (Kavazanjian, et al., 1995) presents a cross section through the steep slopes at the east end of the landfill. Due to concern over the static and seismic stability of the landfill, the U.S. Environmental Protection Agency (EPA) initiated a performance monitoring program at the landfill in 1988. Performance monitoring at the OII Landfill included slope inclinometers, surface settlement points, and strong motion instrumentation.

Figure 5. Aerial View of the OII Landfill

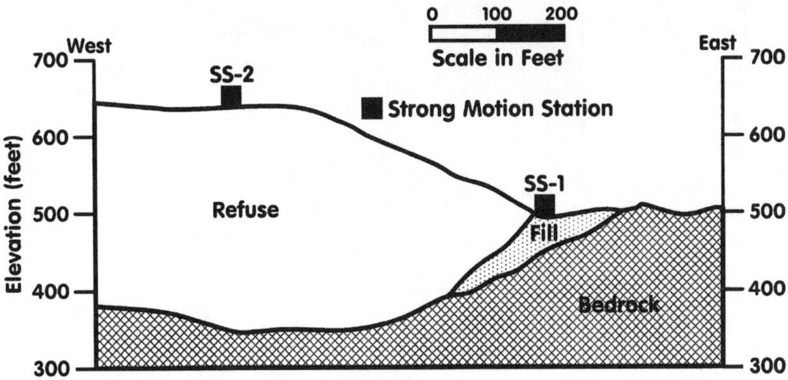

Figure 6. Cross Section Through the East End of the OII Landfill (after Hushmand Associates, 1994)

The data obtained by the OII Landfill strong motion instrumentation was essential to evaluating the seismic stability of the landfill. Data collected by the strong motion station between 1987 and 1995 was used to back-calculate the dynamic properties of the waste up to an intermediate cyclic shear strain of approximately 0.1 percent. Large-diameter cyclic laboratory testing was conducted on reconstituted samples of the waste to evaluate the cyclic stress-strain properties at larger strains (Kavazanjian, et al., 1999). Unfortunately, the laboratory data resulted in a relatively wide range of potential large-strain cyclic stress-strain behavior. However, by considering the compatibility between the observed behavior (for cyclic shear strains up to 0.1 percent) and the large-strain behavior determined in the laboratory testing, the best estimate (recommended) properties of the waste were established for use in the seismic response analysis of the landfill (Matasovic and Kavazanjian, 1998). Figure 7 illustrates the results of one of the back analyses of strong motion data used to evaluate the small- and intermediate-strain behavior of the waste. Figure 8 illustrates how the properties back calculated from the field performance measurements were combined with the laboratory data to evaluate the recommended dynamic properties of the waste. The waste properties shown in Figure 8 were used to evaluate landfill seismic response for input to the seismic slope stability analyses. Using field observations in this manner, the seismic stability of the OII Landfill was demonstrated to the satisfaction of all stakeholders, including state and federal regulatory agencies and two blue-ribbon technical review panels.

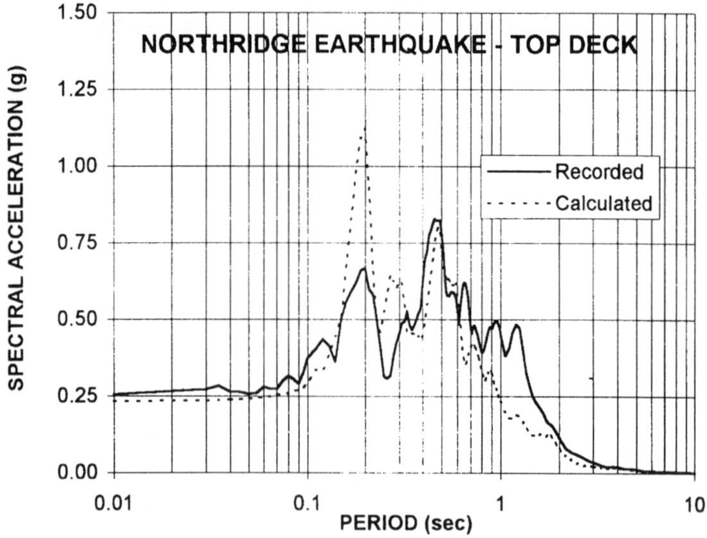

Figure 7. Back Analysis of OII Strong Motion Data (after Matasovic and Kavazanjian, 1998)

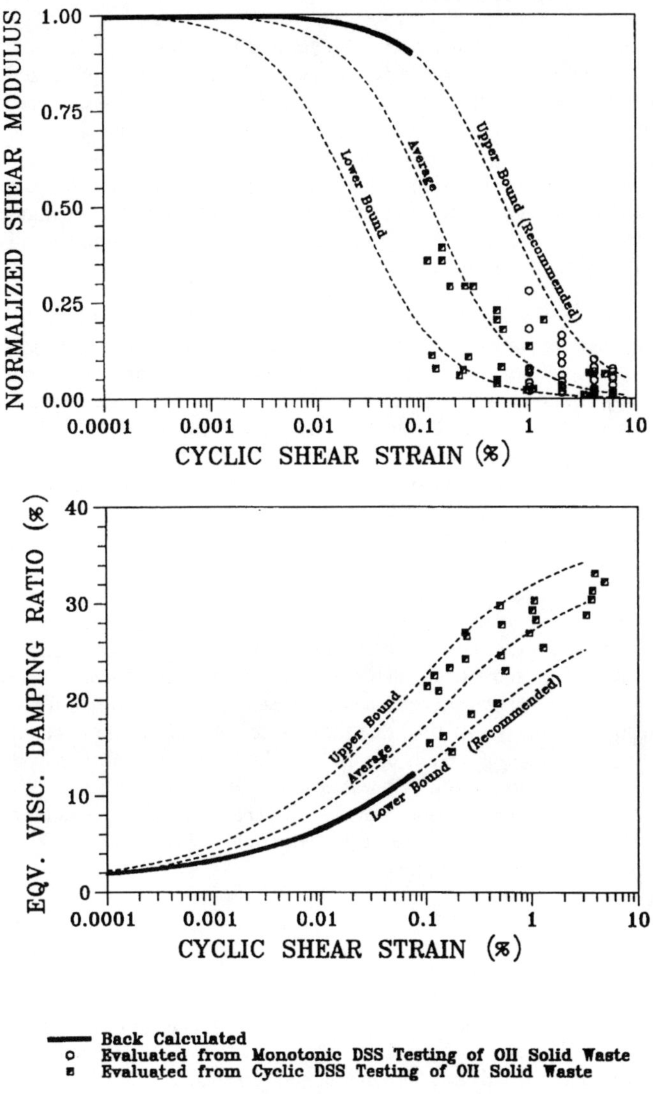

Figure 8. Dynamic Properties of OII Waste (Matasovic and Kavazanjian, 1998)

MONITORED NATURAL ATTENUATION

Monitored natural attenuation (MNA), perhaps the most significant advance in geoenvironmental practice in the past decade, is the penultimate application of the observational approach in geotechnical engineering. In MNA, performance monitoring is used to verify fate and transport analyses that justify no action or reduced action alternatives for corrective action projects. Successful implementation of MNA embodies all of the axioms of the role of the calculated risk in geotechnical practice as elucidated by Casagrande and Whitman and the observational method of Terzaghi as formalized by Peck. The application of MNA is subject to all of the challenges and limitations identified in Table 2.

Sometimes incorrectly referred to as a "do-nothing" solution, MNA requires a comprehensive program of observations and measurements designed to monitor for both expected conditions and less likely or unanticipated conditions that could impact human health or the environment. MNA requires a thorough understanding of contaminant transport and transformation phenomenon and local geologic conditions (i.e., potential contaminant pathways). If MNA is to be employed, contingency plans must be developed *a priori* in case monitoring indicates adverse conditions have developed. MNA requires a detailed evaluation of the risks involved and communication of those risks to all stakeholders. If there is no adequate solution should adverse conditions develop, MNA is not an acceptable corrective action approach. If the relevant quantities cannot be properly monitored due to measurement difficulties or the small quantities that must be measured, MNA is not a viable approach. Biological and chemical transformations of hazardous and potentially hazardous substances often represent a significant source of uncertainty on MNA projects. MNA observation and measurement programs often must be continued for long periods of time due these uncertainties, the slow rate at which many natural attenuation processes operate, and the long period of time over which some substances remain potentially hazardous. MNA holds great promise for cost-effective environmental protection provided the challenges and limitations of the observational approach and the role of the calculated risk are carefully considered when applying this technique in practice.

CONCLUSION

Observation and measurement for confirmation of performance (performance monitoring) serves the same functions on geoenvironmental projects as they do on traditional geotechnical projects. Functions of performance monitoring include protection of human health and the environment, model verification, evaluation of properties, regulatory compliance and acceptance, and optimization of design. Due to the regulatory considerations and inherent traits of geoenvironmental projects, observation and measurement for confirmation of performance may be even more important on geoenvironmental projects than on traditional geotechnical projects.

Inherent traits of geoenvironmental projects may also accentuate the challenges and limitations of performance monitoring on geoenvironmental projects. Geoenvironmental performance monitoring challenges and limitations include extremely small quantities of concern, long time period of interest, uncertainty with respect to phenomena of interest, difficulties in measurement of parameters of interest, and emotional issues associated with environmental impacts. However, these challenges and limitations are not insurmountable. By clearly defining the objectives of the monitoring program and by considering the associated limitations and challenges, performance observations and measurements can play a key role in providing cost-effective protection to human health and the environment. Monitored natural attenuation represents the penultimate application of performance monitoring on geoenvironmental projects. Provided the challenges and limitations of the observational approach and the role of the calculated risk in engineering practice are properly addressed, monitored natural attenuation can provide extremely cost-effective environmental protection.

APPENDIX-I. REFERENCES

Casagrande, A., "Role of the 'Calculated Risk' in Earthwork Engineering," *Journal of the Soil Mechanics and Foundation Division,* ASCE, Vol. 91, No. SM4, July, 1965, pp. 1-40.

GeoSyntec, "Systems Monitoring Report, Lopez Canyon Sanitary Landfill," *Technical Report* prepared for the City of Los Angeles, Department of Public Works, Bureau of Sanitation by GeoSyntec Consultants, Inc., Huntington Beach, California, August, 1994, 45 p.

Hushmand Associates, "Landfill Response to Seismic Events," *Report* prepared for the USEPA Region IX by Hushmand Associates, Laguna Niguel, CA, 1994.

Kavazanjian, E., Jr., Matasovic, N., Stokoe, K., and Bray, J.D., "In-Situ Shear Wave Velocity of Solid Waste from Surface Wave Measurements," Proceedings, *2nd International Congress on Environmental Geotechnics,* Osaka, Japan, Balkema, Vol. 1, pp. 97-104.

Kavazanjian, E., Jr., Matasovic, N., and Bachus, R.C., "Large-Diameter Static and Cyclic Laboratory Testing of Municipal Solid Waste," Proceedings, *Sardinia 99, Seventh International Waste Management and Landfill Symposium,* Environmental Sanitary Engineering Centre, Cagliari, Italy, 1999, Vol. III, pp. 437-444.

Kavazanjian, E., Jr., Matasovic, N., Bonaparte, R., Schmertmann, G. R., "Evaluation of MSW Properties for Seismic Analysis," Proceedings, Geo*Environment 2000 Conference, Geotechnical Special Publication No. 46,* American Society of Civil Engineers, New Orleans, Louisiana, 1995, Vol. II, pp. 1126-1141.

Matasovic, N. and Kavazanjian, E., Jr., "Cyclic Characterization of OII Landfill Solid Waste, *Journal of Geotechnical and Geoenvironmental Engineering,* ASCE, Vol. 124, No. 3, 1998, pp. 197-210.

Peck, R.B., "Advantages and Limitations of the Observational Method in Applied Soil Mechanics," *Geotechnique,* Vol. 19, No. 2, June, 1969, pp. 171-187.

Whitman, R.V., "Evaluating Calculated Risk in Geotechnical Engineering," *Journal of Geotechnical Engineering,* ASCE, Vol. 110, No. 2, February 1984.

MANAGING LARGE AMOUNTS OF GEOTECHNICAL PERFORMANCE DATA FOR THE CENTRAL ARTERY/TUNNEL PROJECT

W. Allen Marr, Ph.D., P.E.[1], F. ASCE

ABSTRACT

The Central Artery/Tunnel Project in Boston, Massachusetts is the largest, most complex highway project in American history. It involves the construction of 7.5 miles of highway through the center of the city with half the work occurring underground. Several conditions favored an extensive geotechnical monitoring program to provide engineers and contractors with up-to-date performance information. This paper describes the data acquisition and management processes used to collect, store, validate, reduce and report data from a large number of geotechnical instruments. It also describes some of the unique experiences presented by this challenging project.

PROJECT OVERVIEW

The Central Artery/Tunnel Project is the largest, most complex highway project in American history. The $10.8 billion project consists of 161 lane miles of urban highway -- about half underground in a 7.5-mile corridor -- through the heart of one of America's oldest and most historic cities. Figure 1 shows the layout of the project. The heart of the project is replacement of Boston's aging, elevated Central Artery (I-93) with a modern eight-to-ten lane underground expressway. The project will also extend the Massachusetts Turnpike (I-90) to Logan Airport, build a

---

[1]CEO, GEOCOMP Corporation/GeoTesting Express, Inc., 1145 Massachusetts Ave., Boxborough, MA 01719, 978.635.0012, wam@geocomp.com

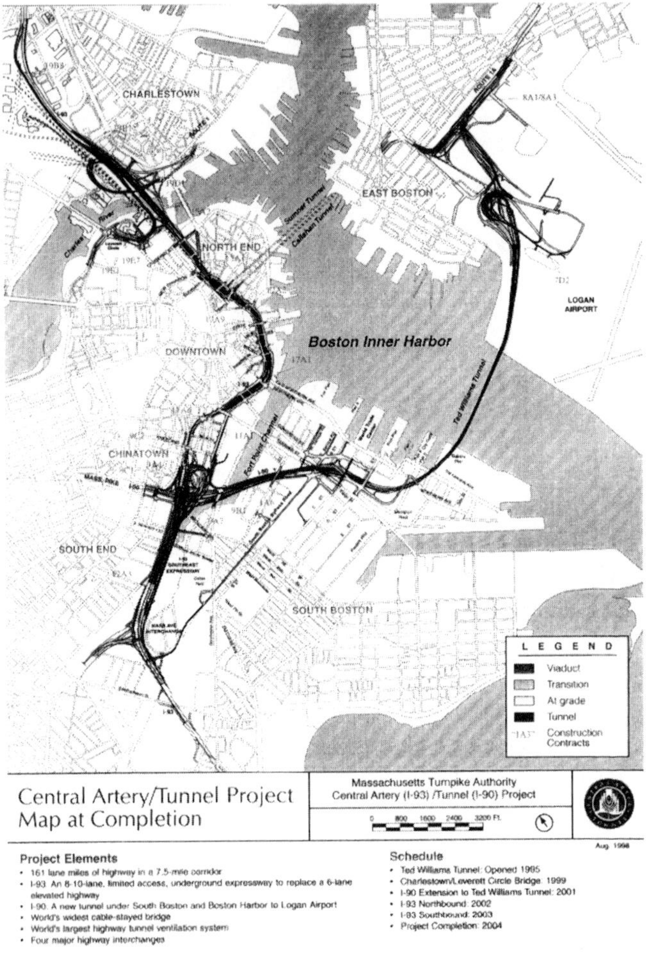

Figure 1 Layout of Central Artery/Tunnel Project (from MTA)

dramatic 10-lane cable-stayed bridge across the Charles River, create a state-of-the-art incident response and traffic management system, construct five major highway interchanges and have the largest vehicular tunnel ventilation system in the world. The CA/T is unprecedented since all construction is taking place in the middle of downtown Boston which must remain open and accessible for businesses, residents and tourists.

The project requires the excavation of more than 10 million cubic meters (13 million cubic yards) of soil - hence the name "The Big Dig." If the trucks hauling this soil away were lined up end to end, they would stretch from Boston to the Panama Canal. The project will place 3 million cubic meters (3.8 million cubic yards) of concrete - enough to build a continuous sidewalk four feet wide from Boston to Tokyo. Reinforcing steel placed in the concrete would form a one-inch steel bar that would wrap completely around the earth at the equator.

The existing, elevated Central Artery will remain in service until the underground highway is complete in 2004. The existing structure was designed in the 1950s to handle 75,000 vehicles per day; it now carries nearly 200,000 vehicles every day. If nothing is done, Artery traffic will be bumper to bumper for up to 16 hours a day by 2010. The regional economy loses more than $500 million annually in wasted fuel, late deliveries and other traffic-related delays. When finished, the downtown Central Artery (I-93) will be carrying 250,000 or more vehicles a day with normal urban peak periods of about two hours in the morning and evening. With eight to ten lanes on the underground expressway (compared to just six on today's elevated highway), fewer on and off ramps, and a new network of surface streets, through traffic will travel at high speed. Local traffic will distribute itself at street level rather than weaving back and forth, jockeying for position on the highway. The Ted Williams Tunnel will carry more than 90,000 vehicles a day (compared to a bit more than 20,000 today with limited access), reducing use of the Sumner and Callahan tunnels to about 70,000 vehicles a day from the 100,000 per day before construction.

Carbon monoxide levels in Boston's air will be reduced an estimated 12 percent because traffic will be moving rather than standing still on a grid-locked highway. Cross streets between downtown and the North End and waterfront that were severed or disrupted by the elevated highway since the 1950s will be reconnected. The green elevated highway will be gone, replaced by 110,000 m^2 (27 acres) of new open space along the Artery corridor. Three quarters of that new space will remain open, with the rest dedicated to controlled development.

The geotechnical setting for the project is complex. In many locations the upper 6-9 m (20-30 ft) of the subsurface consist of miscellaneous fill containing numerous utilities, abandoned foundations, infilled marine structures, and miscellaneous debris with contaminated materials in many locations. The underlying natural soils consist of various layers of organic silts and silty sands which overly a deposit of Boston Blue Clay. A dense layer of glacial till underlies the clay and overlies bedrock. The top of bedrock varies in quality, being highly weathered in some locations and competent hard rock in others. Part of the alignment passes through areas with no clay. Most of the project is located in soil above the rock. At its deepest point, the tunnel is approximately 36 m (120 ft) below the street level. Ground water is located about 3 m (10 ft) below the ground surface.

Figure 2 Illustration of Project Elements (courtesy of David Macauley)

The bulk of the downtown portion of the project uses the open-cut method of construction to build the tunnels. (Figure 2.) After relocating the numerous utilities, structural slurry walls are constructed on each side of the tunnel. In many locations, these walls support the excavation during construction, provide underpinning to the existing Central Artery Expressway, and become the permanent walls for the finished structure. The first level of bracing and a road deck is added to span between the slurry walls. The deck permits surface traffic to continue over the work area.

The critical geotechnical work begins after these walls are in place. As excavation proceeds downward, massive struts are placed across the excavation between the slurry walls to provide structural support and limit the lateral deformations of the slurry walls. Because of the closeness of many historic buildings and major structures, maximum lateral movements of the support walls are typically limited to less than 38 mm (1.5 inches). With excavations as wide as 36 m (120 ft) and as deep as 36 m (120 ft), the support system must be relatively stiff and properly installed to keep the deformations small. A primary role of the geotechnical instrumentation system is to determine that deformations are kept within allowable limits during the construction.

Construction in this complex geotechnical environment required the use of many new approaches brought from all over the world. The deep slurry walls in close working environments brought the latest soil milling machines from Europe. Extensive areas of soft soil required improvement by jet grouting and deep soil mixing with equipment from Europe and Japan. Ground freezing and jacked box structures were employed to place three large tunnels beneath a complex of active railroad lines. Much of this technology had not been previously used on a large scale in the United States.

The Central Artery/Tunnel Project is owned and managed by the Massachusetts Turnpike Authority. Design and construction management consulting is provided by Bechtel/Parsons Brinckerhoff, a joint venture of Bechtel Corporation of San Francisco and Parsons Brinckerhoff Quade & Douglas, Inc., of New York. The US Federal Highway Administration is providing 70 percent of the funding for the project.

ROLE OF GEOTECHNICAL INSTRUMENTATION

The combination of a complex geotechnical environment, numerous sensitive structures next to the work, novel methods of construction, and many involved stakeholders produce an environment where the potential costs of adverse or unexpected performance are high. Those costs may involve not only money but time and political considerations. Despite an extensive geotechnical investigation and comprehensive design effort, many uncertainties remained about the likely effects and performance of the constructed facilities. Consequently the project team made a comprehensive geotechnical instrumentation program a key part of the project plan.

The purposes of the geotechnical instrumentation program (Bechtel/Parsons Brinkerhoff, 1991) are to provide data to help the project team to:

- prevent or minimize damage to structures by providing data to determine the source of ground deformations
- develop protective and preventive measures to structures
- select appropriate remedial measures where required
- evaluate critical design assumptions where significant uncertainty exists
- determine the adequacy of the contractor's methods, procedures and equipment
- monitor the effectiveness of protective, remedial and mitigative measures
- provide feedback to the contractor on his performance
- assess the contractors' performance, assess contractor-initiated design

changes, change orders, changed conditions and disputes

- provide documentation for assessing damages sustained to adjacent structures allegedly resulting from ground deformations and other construction related activities
- advance the state-of-the-art by providing performance data to help improve future designs.

Key to the success of the project is limiting ground movements, changes in water level and water flows to within acceptable limits. Performance measurements play a key role in evaluating important design assumptions and performance predictions and controlling critical construction sequences.

There are hundreds of facility owners who can potentially be affected by the construction operations. Providing an accurate record of actual performance helps permit rapid responses to concerned owners and resolution of disputes that may arise during or after construction. The data must be accurate, reliable and have credibility to all participants.

One key goal of this project is to minimize claims. Data from the geotechnical instrumentation help document actual performance and quantify changed conditions, factors which help quantify the true elements of inflated claims.

UNIQUE INSTRUMENTATION ISSUES WITH THIS PROJECT

The project involves more than 140 separate design and construction contracts. To obtain consistency in instrumentation and data across the entire project, the Management Consultant (MC) defined a set of objectives for all instrumentation, developed a project wide instrumentation specification, and defined a method to collect and process data from the instrumentation. Of major concern to the project managers was to have in place a system that produces reliable data when it is needed in the form that it is needed. Designers for each section of the project used the information provided by the MC to develop specific instrument locations for their designs and establish permissible values for each type of measurement.

Bechtel Parsons Brinckerhoff established the idea of using Threshold and Limiting Response Values for each instrument. Readings below the Threshold Response Values indicate no potential for damage to adjacent structures, or readings indicating performance that is within the design parameters. Readings that exceed the Threshold Response Values indicate uncertainty exists regarding potential damage. The Contractor is alerted to the existence of this condition and all parties involved increase their level of vigilance for execution of the work. Means are sought to return the reading to less than the Threshold Response Value or stop the advancement of the reading past the Limiting Response Value. Readings above the Limiting Response Values may require the Contractor to stop all work until the cause of the high readings can be determined, evaluated, and remedial measures invoked. Table 1 illustrates a typical set of Threshold and Limiting Response Values for one contract.

Table 1: Typical Instrument Response Values

Instrument Type	Threshold Value	Limiting Value
Piezometers outside excavation in fill	1 ft decrease below HL during utility excavation	2 ft decrease below HL during utility excavation
Piezometers outside excavation in clay	20 ft within 100 ft of building 6 ft more than 100 ft from building	Not specified
Piezometers inside excavation in bedrock	Not specified	5 ft below that required to maintain FS for uplift > 1.1
Piezometers outside excavation in till and rock	20 ft decrease	40 ft decrease
Inclinometers in slurry wall and soil	0.001*D for north and south walls 0.0015*D for east and west walls	0.002*D for north and south walls 0.0025*D for east and west walls
Probe extensometers	1 inch settlement of any point	2.5 inches settlement of any point
Deformation monitoring points-Type 1	1 inch of settlement	2.5 inches of settlement
Deformation monitoring points-Type 2	Angular distortion of 1/1,200	Angular distortion of 1/1,000
Deformation points on manholes	Vertical: 1/4 inch	Vertical: 3/4 inch
Strain gages on struts	100 percent of allowable stress	125 percent of allowable stress
Seismographs	For historic structures: PPV≤0.2 ips For non-historic structures:PPV≤0.3 ips	For historic structures: PPV≤0.3 ips For non-historic structures:PPV≤0.5 ips

D=depth of excavation; HL= historic low water level; PPV=maximum peak particle velocity; ips=inches per sec; 1 ft=0.3 m; 1 inch=25.4 mm

Source: Becthel Parsons Brinckerhoff

The large size of the excavations, the closeness to numerous sensitive structures, the variable subsurface conditions, the first-time use of new construction equipment and methods, and the sensitive political environment were all factors pressing the need for rapid turnarounds of data from the instrumentation. The MC decided that they would take the responsibility for evaluating and interpreting the data with respect to performance of the design and execution of the construction. The MC established an internal geotechnical instrumentation group (GIG) to develop the reading schedule for each instrument and to examine the data coming from the instrumentation. The GIG also provides interpretation of the data for the Resident Engineers (REs). Each Contractor would be provided with the data associated with its contract and be responsible for interpreting that data to control his work. The data would also be provided on request to outside parties where appropriate. This approach would allow all groups to have immediate access to consistent, checked performance data and reduce the potential arguments caused by conflicting and incomplete data sets. The Contractors were given the right to collect their own readings from instruments on their contracts anytime. They could also install any additional instrumentation they required for their specific needs.

The MC determined that they would need all data collected on any given day processed and available to them by the start of the next day. This would permit all involved parties (engineers, contractors and abuttors) to have access to data within one day. Provisions were also made for situations returning a faster turnaround of data. In these instances, data would be required to within two hours of the reading time.

The Management Consultant and the Owner decided to award a single contract to monitor all geotechnical instrumentation, excluding those read by survey, for a six-year period. A separate contract was awarded to provide all surveying. Table 2 summarizes the various responsibilities of the various parties with regard to geotechnical instrumentation on this project.

During the planning and design phases of the project, the Management Consultant developed a database system to receive, store, retrieve and plot data from all geotechnical instruments using the project's Oracle based Geographic Information System (GIS) system. The Geotechnical Instrumentation Contractor (GIC) and the Surveying Contractor would place data into this system. Then project staff could retrieve the specific data they needed in graphical and tabular form on their computer screen or as a printout. The GIS system included provisions to plot groups of instruments over time and produce contours of data in plan and cross section. The goal was to provide the ability to view all instrumentation data at any location on the project immediately.

Table 2: Responsibilities for Geotechnical Instrumentation Activities

Management Consultant - (MC)
- -manage GIC contract
- -coordinate requests for readings
- -provide weekly and daily reading requests to GIC
- -maintain GIS database of instrumentation data and locations
- -receive and distribute GIC reports
- -evaluate and interpret data from instrumentation
- -respond to complaints about performance

Geotechnical Instrumentation Contractor - (GIC)
- -maintain and calibrate readout equipment
- -read instrumentation as requested
- -check and validate data
- -place data into GIS database
- -produce daily printed and electronic reports for previous 24 hrs of readings
- -produce weekly summary reports
- -notify specified party when instrument exceeds Response Value
- -maintain Ready Action Team to respond to emergency requests

Construction Contractors - (CC)
- -procure and install instrumentation
- -procure readout equipment
- -maintain instrumentation and replace if necessary
- -evaluate readings taken by GIC to control safety of work
- -take additional readings if required for safety of work

Section Designers -(SD)
- -specify locations for instrumentation
- -specify Threshold and Limiting Response Values
- -evaluate instrumentation data when Response Values are exceeded

Procurement and installation of the geotechnical instrumentation were made a part of the construction contracts. The contractor on each section must install and maintain the instrumentation in working order. Contractors also procure the readout equipment required for reading the instrumentation. He must collect three sets of initial readings that show stable, reproducible data. At that point the readout equipment and the responsibility for taking readings are transferred to the

Geotechnical Instrumentation Contractor. Any in place instrument that becomes damaged or inoperable must be repaired or replaced by the Contractor. The readout equipment is maintained and repaired by the GIC.

The Geotechnical Instrumentation Contract was developed as a unit price contract, i.e., the Contractor would be paid a set price for each successful reading on each type of instrument. The initial estimate of the work was a total of 1.3 million readings. The work was advertised for bid to a pre qualified group of firms.

In April 1997, the joint venture of GEOCOMP Corporation (Boxborough, MA) and TLB Associates (Millersville, Maryland), referred to as GCB, was awarded the geotechnical instrumentation contract based on the lowest submitted price. The total contract amount was approximately $10,000,000 for the six-year term of the contract. This price represented approximately 60% of the engineer's estimate of the contract value. Table 3 lists the typical types of instruments and numbers of readings contemplated for the six-year period.

Table 3: Anticipated Number of Instruments Over Life of Contract

Instrument	**Expected Total Number**
Observation Wells	787
Piezometers	1,181
Crack Gauges	145
Convergence Gages	65
Mechanical Strain Gauges	102
Vibrating Wire Strain Gauges by Data Logger	5,865
Inclinometers	501
In-place Slope Inclinometers	31
Multiple Point Heave Gauges	129
Single Point Extensometer	59
Multiple Point Extensometers	21
Probe Extensometers	405
In-Place Tiltmeters	343
Dial Gauges	2
Shear Displacement Gauges	10
Load Cells	170
TOTAL	9,816

Source: Becthel Parsons Brinckerhoff

The use of unit prices for a geotechnical monitoring contract is new as far as we can determine. This form of contract is attractive to the Owner because it provides a way to control costs compared with the traditional time and materials contract. It is also attractive to the Geotechnical Instrumentation Contractor who can control its costs.

The contract requires all data collected in a given day to be processed, checked, and entered into the GIS system by 8:00 a.m. of the following day. Three printed copies of a summary report are also required by the same time. These copies were intended for the MC, the Resident Engineer, and the Contractor. The printed reports contain a summary of all instruments that exceeded the established Threshold and Limiting Response Values, time history plots of all instruments whose current readings exceed the Threshold Response Values, time history plots for all inclinometers, and tabulations of all reduced data collected in the previous day. The contract contains penalty clauses for submitting incorrect data and for late reports.

The contract also provides for a Ready Action Team to be on call always. This team must be capable of reading and reporting data for any instrument on the project within two hours of receiving notice. This work item is paid on an hourly basis.

Soon after the start of work, the MC began receiving complaints about vibrations from building residents and owners. The seismograph records could provide a factual basis to decide whether the complainer had a valid issue; however, those records would not be available until the next day. If the record had been retrieved early in the day, it might not contain the data for the time corresponding to the cause of the complaint. GCB agreed to modify its work procedure to pick up and report data from all seismographs by 1:00 P.M. of the same day. This change permitted the MC to obtain the data to respond to any complaint about excessive vibrations within one day.

APPROACH TO THE WORK

GCB's approach to the job is to automate all aspects of the work that make financial sense. The largest component of our costs is labor; therefore every aspect of data collection, reduction and reporting was reviewed to determine what time saving measures could be adopted. Since we are paid only for correct readings, errors must be kept to a minimum. The requirements for rapid turnover of data caused us to look for methods to do our work quickly.

Since the data must be provided to the MC in electronic form, the data should obviously be put into electronic form as early in the work process as possible. After reviewing and testing the possible approaches, we decided to employ handheld industrial computers to collect the data from each instrument. We chose the PSION Workabout because of its ruggedness, its small size, its ability to run for weeks on a single battery, its ability to be programmed in a language similar to VisualBasic, its instant "ON" function, and its large data storage capacity. We call these devices

Personal Data Recorders or PDRs. Hawkes and Marr (1999) give more information on the development of this equipment for the CA/T project.

A second labor intensive function was identified in the effort required to check all readings for errors and validate the data before transmitting it to the project database. It had been envisioned that we would load data into a temporary location in the project database and use the project's GIS Oracle database system to review that data for correctness. Once a data point was determined to be valid, it would be marked as correct and transferred to the permanent project database. This activity would have to occur over a telephone line and be subject to the availability of the project's network system. We viewed this approach as time consuming, risky and involving important components over which we had no control. Instead we placed a networked PC system in our project office and built a database system to handle the checking of data and transfer of data to the project database. This system was expanded to process reading requests from the project, prepare work orders for the technicians, download information into the PDRs, upload data from the PDRs into a temporary database, and provide statistics on the performance of the system, of our equipment and of our technicians.

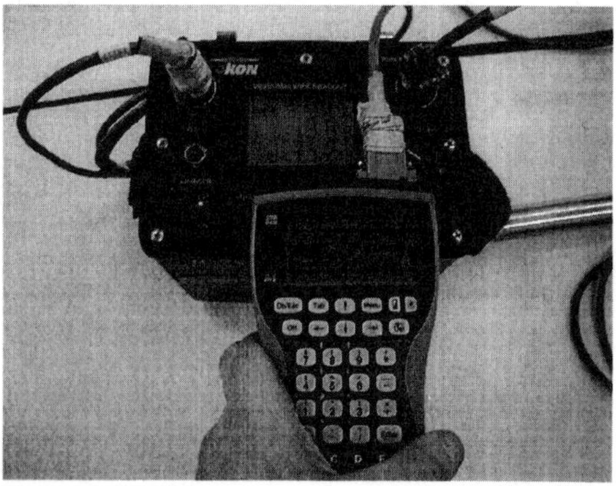

Figure 3 PDR connected to vibrating wire readout.

Each technician uses a PDR to enter the data from each instrument at the location of the instrument. If the readout device provides an electronic reading, the PDR is connected to the readout device and the reading placed into the PDR electronically, as shown in Figure 3. Data from vibrating wire strain gages and piezometers, inclinometers and tiltmeters are collected this way. If the readout device is a gage, dial or screen, the technician enters the reading into the PDR using its keypad. Data from tape measures, dial gages, observation wells, probe extensometers, convergence gages and crack gages are recorded this way. The technician may also enter remarks about the instrument or data into the PDR.

Custom remarks may be keyed in or the technician may select a preprogramed remark from a list of common ones.

Software for the PDR to collect and check data for all the instruments read on the project was developed by the GCB project team. The entire database of instrument readings is kept on each PDR and is used to validate new readings immediately after they are obtained.

At the beginning of each day each technician loads a list of instruments to be read that day into his or her PDR. As readings are collected, this work list is updated to reflect completed work. Instruments that cannot be reached or read are also noted in the PDR. At the end of the shift, the technicians upload the data from their PDRs to the temporary work file on a PC server. The PC server identifies any instruments that are unaccounted for on that day and requires the technician to enter an explanation. Only after this process is complete, can the technicians end their shift. At this point, a paper record of the reading status is printed out for the technician to check and initial to verify that the technician's duties have been accomplished for the shift. This is the first piece of paper generated by the data collection process.

Error-free data is absolutely crucial to the success of the CA/T instrumentation program. For the program to work, the entire project must believe in the reliability of the data they receive. Since most of the Threshold and Limiting Response Values involve relatively small changes, there is no room for error in our work. Given the rapid turnaround requirements for the project, there is little time for multiple layers of quality control checks by independent individuals. We took the approach of reducing the number of steps in the work process to the absolute minimum and using as many computer checks of the data as we could conceive. A key part of this process was employing the PDRs. Immediate entry of the reading into the PDR removes the traditional steps of writing a reading on paper and someone else manually entering that reading into a computer. Both steps present multiple opportunities for errors. We also use the computing abilities of the PDRs to compare the new reading with previous data from that instrument to decide whether the new reading is valid. If the new reading differs significantly from previous readings, the technician is immediately instructed to check her equipment and take a second reading. If the second reading agrees with the first, then the reading is considered valid. If the second reading differs from the first, then the technician must seek to identify and remove the cause of the difference. This procedure greatly reduces the time required to identify and correct most errors in the data. If the confirmed reading exceeds a Response Value, the technician must notify a project representative.

Figure 4 Typical daily reports.

Soon into the project we discovered that producing the paper reports required by the contract was consuming a large amount of time and making it difficult for the MC's geotechnical engineers to keep track of the data. The taller stack of paper in Figure 4 is one copy of one day's report produced in the early phases of the project when the number of instruments relatively small. We had to produce three copies of this stack each day. Working with the project staff, we devised some summary reports that focused on reports of instruments that have readings that exceed limits and instruments with problems. We replaced the detailed seismograph reports with summary reports of instances of exceedence and electronically submitted files. The result is illustrated in Figure 4 by the smaller stack of paper that reports on three times as many instruments.

Many people are surprised to learn that we do not use dataloggers to a significant degree on the project. Except for the seismographs and some vibrating wire strain gauges and piezometers provided with data loggers by the Contractor, all other instruments are read by having a technician physically go to the instrument. This approach is purely an economic one. Typically, instruments in the active work zone are read weekly. If the work is particularly sensitive, the reading schedule may go to one per day, or occasionally a few readings per day. The costs for us to procure and install data loggers far exceed any savings in labor we can gain by their use. We do maintain an inventory of portable data loggers to use on occasions where more frequent data are required. This is typically the case for pump tests on new dewatering systems.

SOME OF OUR EXPERIENCES

Scope of the Work - Some Interesting Statistics

Comprehending the scope of our activities on this project is difficult. Table 4 gives some interesting statistics. We project that over six years, we will collect, store, validate and report about 60,000,000 data points. Considering that each data point has an associated date, time, technician ID and remark field, the final database will consist of about 3 gigabytes of information. From a data storage perspective, the project is dominated by data from seismographs and inclinometers. From a labor perspective, inclinometers make up the largest part of our effort. By the end of our contract, we estimate that our technicians will have pulled enough inclinometer cable to stretch from Boston to San Francisco. The seismographs include data loggers that are set to record the maximum velocity experienced on three orthogonal directions during each five minute intervals. We currently monitor about 4,000 separate instruments on a monthly, weekly and daily basis.

Table 4: Geotechnical Instrument Readings for CA/T Project

Instrument Type	Number Presently Read	Readings to Date	Projected Total Readings for Contract
Seismographs	32	19,000,000	41,000,000
Inclinometers	842	6,000,000	13,000,000
Facade Monitoring Points	122	480,000	950,000
Vibrating Wire Strain Gauges	130	220,000	650,000
Vibrating Wire Piezometers	990	146,000	200,000
Probe Extensometers	226	75,000	166,000
Crack Gauges	1,043	71,000	157,000
Observation Wells	320	20,000	45,000
Convergence Gauges	215	20,000	44,000
Vibrating Wire Load Cells	55	11,000	24,000
Tiltmeters	48	8,500	19,000
Shear Displacement Gages	17	2,000	5,000
Multiple Point Probe Ext.	6	2,700	6,000
Single Point Probe Ext.	17	2,100	5,000
Dial Gauges	4	1,600	3,500
TOTALS	4,067	26,059,900	56,274,500

The entire effort is done by a total of 17 full-time-equivalent people. We have a Geotechnical Instrumentation Engineer, Superintendent, Mechanic, 12 full-time equivalent technicians, and two office support people. (We employ several co-op students from the local universities on a part-time basis.)

Data Validation

With an average of 27,000 readings taken every day, even a low error rate of 0.1% results in 27 errors per day. An error rate this high would destroy the project's confidence in the geotechnical instrumentation program. It is essential to the success of our work that we employ procedures to minimize errors.

As indicated earlier, a key part of our approach was to use PDRs to remove the number of steps required to get data into electronic form. A second part of this approach was to devise error checking routines that could be programmed into the PDRs to check a data point for possible errors immediately while the technician is still at the instrument. We looked for guidance on rules for checking the validity of a data point but did not find much useful information in the literature. We used our own experiences to devise a set of checking rules over time. We currently use the history of readings for each instrument to establish an anticipated reading range for the reading. A reading inside this range is considered valid. A reading outside this range is considered suspect. Anticipated reading ranges are determined using one of the following methods:

(1) plus/minus two standard deviations from the mean value, using the most recent eight readings to compute mean and standard deviation.
(2) maximum and minimum values of the most recent eight readings.
(3) predetermined anticipated value plus-minus an acceptable error. This is used for instruments with very small reading fluctuations, such as crack gages and tape extensometers.
(4) a range determined by the specific conditions for the instrument.

The anticipated reading range is computed instantly by the PDR using the database of previous readings. The technician can switch between different methods for setting the acceptable range. The chosen method is stored with the reading so that a later evaluation can be made of the data.

If the reading is determined to fall outside the accepted range, the PDR instructs the technician to check the readout equipment and take a second reading. If the second reading also falls outside the accepted range, the PDR instructs the technician what action to take. Possible actions include telephoning MC Geotechnical Instrumentation Group immediately, telephoning the instrumentation superintendent, or tagging the instrument in the database for further evaluation.

Readings that are tagged are so identified in the daily reports and checked the following day. Apart from inclinometers and seismographs, there have been few occasions where the field reading has exceed a Response Value.

This approach has drastically reduced the number of false alarms caused by errors in the data. However, it is still not foolproof. Instances where an instrument reading changes suddenly and drastically occur almost daily. Adjacent construction activities, such as jet grouting, pile driving, pumping, and excavation, account for many. A direct connection between these activities and an instrument reading may not be obvious to the technician. Dewatering pumps can cause head drops in piezometers of 6 to 9 m (20 to 30 feet) over a few hrs with no surface evidence for the technician to see. These changes are reported to the Resident Engineer who will notify the Contractor or request an additional reading be made immediately.

Problems with Readout Equipment

In the first two years of our work, we have experienced several problems with the readout boxes. These are the units that connect to the instrument at the surface to obtain a reading from the instrument. The readout boxes for vibrating wire piezometers and inclinometers tend to become erratic or fail during wet weather and at temperatures below -10°C (10°F). With technicians already hampered by poor working conditions, these failures are particularly troublesome. Productivity is greatly slowed as they try to dry out or warm up equipment. This is an area that I think deserves more attention from the instrument manufacturers.

We also find that differences between readout units can cause significant differences in the measured data. This has been a major problem with inclinometers. The expected error from reading a vertical inclinometer is 1/4" (6 mm) per 100 ft (30 m). This amounts to 10 mm (3/8") in a 45 m (150 ft) long hole, typical of many of the project inclinometers. Using two different readout units can potentially result in a difference in measured horizontal movement of as much as 10 mm (3/4") due to the measurement accuracy of the system alone. This equals the Threshold Response Value for most inclinometers on the project. We have sometimes experienced differences of inches in values obtained with two different readouts. The solution has been to assign a specific inclinometer probe, readout box and cable to each inclinometer where possible.

We are still vulnerable to problems that develop with a probe, readout box or cable and we have to use a replacement. We also have to use another probe when we return a unit to the manufacturer for an annual calibration, a process that takes several weeks. Our solution has been to assign a backup probe, readout box and cable to every inclinometer. A set of readings is taken on the same day with the primary system and with the backup system.

Even with these steps, we still experience unexpected results from the inclinometer systems. These appear to result from a variety of random causes, including corrosion of connectors, wear and tear on connectors, moisture in the connectors or readout equipment, and mechanical failures of the wheel bearings and in cables near the connectors. Manifestations of these problems can be inconsistent and non-reproducible behavior that complicates troubleshooting. We are overcoming this problem by tracking the mean and standard deviation of the bias of each instrument with time. Variations in these statistics from the historical trend can be indicators of equipment problems.

We also find some of the readout equipment sensitive to how it is used. One technician may obtain a different result than another due to slight differences in measurement technique. This has been a particular problem with tape extensometers and probe extensometers. We currently address this problem by assigning the same technician to the same instruments to the extent possible. However, illness, vacations and turnover require a replacement technician. This change can produce variations in the data that are sometimes enough to show that a Threshold or Limiting Response Value has been exceeded. We have also increased training of technicians to develop common measuring techniques among all technicians.

The contract requires that each readout unit be factory calibrated once per year. We have found this to be an expensive and time-consuming process. Up to six weeks are required to obtain calibrations on inclinometer probes and up to eight weeks for seismographs. To the extent possible we have developed on-site certification systems to provide instant calibration checks, help troubleshoot suspect equipment and determine when a unit needs to be factory calibrated. We have installed test stations for inclinometers, vibrating wire instruments, and seismographs at our field office so we can easily validate readout equipment and train our staff in the use of the equipment.

Much of the readout equipment originally provided to us is wearing out. Some inclinometer cables have lasted only a few months. Some inclinometer probes and readout boxes are reaching the end of their useful life. Most of the readouts are battery powered. Long term daily recharges of the units now exceed their design life. The contract had no provisions for us to replace worn out equipment. We have negotiated a contract change to correct this problem.

Performance of Instruments

Most of the instruments installed for the project have performed extremely well. There have been very few failures from malfunctioning of the instruments. Instruments have been damaged or lost from construction activities. These have to be replaced by the Contractors. A continual effort must be applied to keep the readout connectors in working order, the instruments labeled and covers in place. Freezing weather makes it difficult to locate instruments and remove covers to access the readout points.

At any particular time, up to 18% of the instruments are not read when they are scheduled to be read. Since we are paid only for successful readings, it is in our interest to reduce this number. Despite a concerted effort, we have not been very successful. Reasons for not reading instruments include: unable to find the instrument, unable to access the instrument, instrument damaged, equipment problems, or technician problems. We receive requests daily to read newly installed instruments with no information on where the instrument is located. Contractors are constantly moving readout cable, covering instruments with materials and otherwise obscuring our work. Some instruments are located in parking areas, on roadways, in active work areas, and on active rail lines where they cannot be reached by the technician. We are now devoting a full-time person to work on resolving these problems and improving our success rate.

Communications

An effective, credible instrumentation system is critical to the success of this project. If the project management staff begins to ignore the data from the system, its value to the project greatly decreases. Most of the serious problems we have experienced to date have roots in poor communications.

Initially, we failed to obtain many readings simply because the mechanism for assigning us work involved each Resident Engineer (RE). The MC geotechnical instrumentation group individually faxed their requests for their contract assignments based on their expectations of the REs' needs. Weekly requests were sent on Thursday supplemented with daily requests for immediate readings in critical areas. This required our superintendent to spend several hours each Friday trying to sort out the work effort for the coming week. Some requests were not clear. Some were not received. Some were misinterpreted. To deal with this we developed an electronic work order. Each member of the MC geotechnical instrumentation group enters their reading requests into a revised GIS instrumentation database. We download these requests from the project GIS system into our on-site data management system.

Early in the project we received complaints from some Resident Engineers because we were not responding as quickly to their immediate requests as they expected. It was not unusual for example to receive a request on Friday afternoon for readings to be taken over the weekend. Given the unit price nature of our contract, we have to balance our labor force to just that required for the anticipated work level carefully. Sudden requests for additional work are difficult to fulfill without something else going undone. We initiated an effort to educate the REs in our capabilities, limitations and dedication to the work. This resulted in a better understanding of the relative roles of each group and improved paths of communication.

Our project manager and superintendent hold monthly meetings with the MC Representative. These meetings provide the opportunity to identify problems quickly and jointly seek their resolution.

Another important component of our communication system is a periodic meeting between project staff and our technicians. These meetings are used to present recent developments on the project and to stress the importance of the geotechnical instrumentation to the overall success of the project. The objective of these meetings is to help keep the technician staff motivated to perform careful work.

A key component of our communication system is a partnering session held every three months with the MC and our management staff. These sessions are attended by senior management of both groups. They present an opportunity for both sides to identify problems and work out solutions well before those problems become the source of major disagreements.

Staffing Issues

A dedicated, conscientious staff is essential to our success on this project. All the computers and bright ideas in the world will not replace the consistent efforts of our technician staff to complete their daily rounds with a minimum of errors. Our principal staff concerns are training, safety, burnout and recruiting of new staff.

New technicians must meet the contract requirements for education and experience. They undergo approximately one week of training before being assigned to production work. They are partnered with a senior technician to develop an understanding of the job layout, the equipment and measurement principles. A new technician's initial assignment will be on a straightforward task with minimal chance for error, such as reading a group of observation wells. This gives time to learn the working environment, the equipment and our data collection processes before undertaking more demanding work. Within a short time, new technicians learn more skills and become adept at all phases of the work. The Instrumentation Superintendent works with each technician at least once every three months to observe that their techniques and methods are appropriate.

Safety of personnel receives top level attention. Each employee receives safety training from the project team, from the railroads and from the subway authority. We hold weekly "toolbox" meetings to review safety procedures. All staff is required to use a hard hat, safety glasses, a safety vest and hard toe shoes when on project property. This is a difficult requirement to enforce on hot days.

Burnout is a considerable problem on a project of this duration. Reading instruments can become monotonous quickly. This leads to errors and accidents. We try to hire people with a knack for this type of work. We try to take actions to reduce burnout. We rotate employees around assignments and around the project. We try to minimize overtime. We rotate people off the project when they have had enough. Our experience so far has been better than we expected at the start of the project.

WHAT WE HAVE LEARNED

The project is now reaching near the two-thirds mark of construction. Many of the excavations are at or near their bottom point. It is still too early to distill the detailed technical conclusions that may result from our work. So far, the geotechnical instrumentation confirms that project design is working as intended. Movements have generally been small and within allowable limits. This is remarkable when one considers that the Threshold Response Values are typically 20 mm (3/4 inch) of max. horizontal movement for 30 m (100 ft) deep excavations. Groundwater drawdown has generally been achieved where required and limited where required. Damage to adjacent facilities has been limited to minor architectural cracking overall.

Perhaps the major benefit of the geotechnical instrumentation program to date has come from its use to answer questions and complaints from abutters. Claims of excessive vibrations and construction damages are successfully and quickly countered with measured data that show compliance with project requirements. As abutters become less tolerant of the side effects of our underground activities, the primary role of geotechnical instrumentation may well become one to help control movements to small levels and provide proof that those levels are achieved.

The size and time demands of this project have clearly shown the need to check and verify questionable readings as close to the instrument as possible. We do not have the time, nor can we afford the effort, to return to an instrument to check a questionable reading. On the other hand, we cannot afford the potential damage to the image of the instrumentation system that may result from submitting wrong data. The electronic systems we have engaged on this project have been of tremendous help in reducing the number of questionable readings transmitted to the project staff. Based on our experience, we think the area of data validation is one that could benefit from additional research. How does one validate new data that departs from prior trends without ignoring a true change?

Sensitivity of the readout equipment to moisture, cold weather, wear and tear, and technician techniques suggest that there is value to be gained from manufacturers developing more robust readout systems. These systems should work in all weather conditions, sustain the abuse common to construction projects, and have built-in checks for indications of equipment malfunction.

This project shows that near-real-time geotechnical data can be provided to project staff without the need for expensive data loggers on all instruments. By near-real-time, I mean within 24 hours of the reading request. It is clear from our experience that data loggers become economically viable and almost necessary when the reading frequency and turnaround times decrease to less than one day.

Contrary to the beliefs expressed by some at the start of this project, a unit price contract for reading geotechnical instruments is working quite effectively. The Owner is receiving a quality service for 40% less than the Engineer's estimate. The project has reliable measurements whenever they require them. We, as the Geotechnical Instrumentation Contractor, have a financial incentive to invent better ways of performing our service. Examples of our innovation on this project performed at no added cost to the project include:

- Modifying reading schedule for seismographs to provide reports on same day data were collected.
- Developing software to receive weekly reading requests in a more organized way by electronic means.
- Developing a system to transfer seismograph records to project database electronically with a printed summary report only to cut down on the volume of paper required to report seismograph data.
- Developing an on-site database system to store data temporarily and check it for validity rather than use the Project GIS system as originally envisioned.
- Providing ability to validate readings in the field if an instrument exceeds a Limiting Value and initiate action immediately. This cuts down on the

reaction time in case of an emergency.

Frequent communications along multiple paths are essential for the success of our project. The partnering sessions, project meetings, and technician meetings with project staff provide ways to identify and resolve problems and differences at their early stages.

The Central Artery/Tunnel project geotechnical instrumentation program is meeting all of the objectives outlined at the start of the project. Our only shortcoming resides in our limited ability to evaluate the enormous quantity of performance data that we are collecting. An effective system is in place to use the data to manage the construction and respond to community concerns. This system meets the needs of the project. As engineers, we would like to compare the measured performance with that expected by the designers to see what can be learned for future projects. However, these desires reach beyond the needs of the project. We have not identified an external source of funding to do this evaluation.

ACKNOWLEDGMENTS

The author acknowledges and thanks the Massachusetts Turnpike Authority for permission to publish this paper and Bechtel/Parsons Brinckerhoff for their cooperation. In particular, I recognize the cooperation of Dr. Thom Neff, Mr. Charles Daugherty, Mr. Dave Druss and Mr. Dan Bobrow, all of the Bechtel/Parsons Brinckerhoff management team. I also acknowledge the dedicated efforts of our project instrumentation team, lead by Mr. Martin Hawkes and Dr. Rob Nyren of GEOCOMP Corporation and Mr. Terry Knox of TLB Associates, and carried out by a dedicated group of hard working technicians. This paper was prepared with the support of GEOCOMP Corporation without public funds. It contains the views and opinions of the author and does not represent the views or opinions of the CA/T project.

REFERENCES

Bechtel/Parsons Brinckerhoff (1991), *Geotechnical Instrumentation, Concept Report No. 2AB24, Central Artery (I-93)/Tunnel (I-90) Project*, report to Commonwealth of Massachusetts, Department of Public Works.

Hawkes, M. and Marr, W.A. (1999), "Data Acquisition and Management for Geotechnical Instrumentation on the Central Artery/Tunnel Project," *Field Instrumentation for Soil and Rock, ASTM STP 1358*, G. Durham and W.A. Marr, Eds., American Society for Testing and Materials, pp. 319-332.

Performance and Evaluation of Shallow and Deep Foundations

Defect Detection and Examination of Large Drilled Shafts Using a New Cross-Hole Sonic Logging System

Les R. Chernauskas[1] and Samuel G. Paikowsky [2]

ABSTRACT

Drilled shafts and other mixed or cast-in-place concrete deep foundation elements can be costly solutions. These foundations usually carry very high design loads, and often serve as a non-redundant, single load-carrying unit. These conditions have created a need for a high-level of quality assurance and control applied to each in-place constructed deep foundation element.

The non-destructive testing method, Cross-Hole Sonic Logging (CSL), currently offers the most reliable technique for assessing the integrity of in-place constructed deep foundation elements. Recent years have seen progress in CSL instrumentation, taking advantage of the available computer technology. The software applications, however, have greatly fallen behind, thereby limiting the effectiveness and potential of the CSL method and deep foundations integrity testing in general.

A new, original CSL testing system by the name of PISA (Pile Integrity Sonic Analyzer) makes use of an innovative software and data acquisition system, hence representing the state-of-the-art in deep foundation integrity testing. The PISA has the capability to show real-time graphical information during logging, including planar tomography, which can identify the boundaries of a compromised zone within the foundation element. The equipment operates completely in a Windows™ graphical environment allowing alphanumeric and graphical reports to be generated directly into word processing software. The real-time graphical representation during logging and the ease of reporting enables immediate, extensive on-site evaluation and decision making.

The PISA system was evaluated on several different construction sites. The obtained results from one particular site are presented, demonstrating the ease of use, accuracy of measurements and enhanced capabilities. The systems' abilities are shown to be superior to any other currently available commercial system.

INTRODUCTION

Deep foundations integrity testing mostly applies to foundations constructed from concrete or grout, such as drilled shafts, drilled mini piles, pressure-injected footings, and precast concrete piles. Drilled shaft foundations usually carry very high design loads, and often serve as non-redundant, single load-carrying units. The integrity testing is required for quality control during construction to detect flaws in the pile (e.g. necking, cracking, void, poor quality material, etc.). Such defects are not uncommon in these cast-in-place concrete piles. As a result of the ever-increasing and demanding design requirements on these foundations, a need for a high-level of quality assurance and control has been created.

Cross-Hole Sonic Logging (CSL), is one of the more common testing methods for determining the integrity of in-place constructed deep foundation elements, such as drilled shafts and caissons. A minor variation of this method, called Single-Hole Sonic Logging (SSL) can also be used on smaller diameter drilled mini-piles and augercast piles. These methods are both non-destructive testing (NDT) methods and involve generating a sonic pulse with one transducer (transmitter) and picking the signal up with another transducer (receiver). The transducers typically consist of a geophone or accelerometer. The methods differ only in the number of tests per pile and the location/orientation of the transducers within the pile.

Significant improvements and advances in instrumentation, data acquisition hardware, and computer technology have been made in recent years. The software applications, however, have greatly fallen behind and have not taken full advantage of the existing technological infrastructure, thereby limiting the effectiveness and potential of the CSL method, as well as other deep foundations integrity testing methods (Chernauskas and Paikowsky, 1999).

A new state-of-the-art CSL testing system, however, has recently been developed that utilizes unique software to take advantage of the new hardware (Amir and Amir, 1998a). This system is called the PISA (Pile Integrity Sonic Analyzer). The PISA is based on a lightweight, portable, pen touch, computer that operates in a Windows graphical environment. This system is easy to use and efficient with regard to its ability to make the collected data available in a real-time manner. The following paper provides the basic background theory on the CSL integrity testing method, a description of the PISA system, and a summary of a recent case history including large size rock socketed drilled shafts, defects detection, and verification.

OVERVIEW OF ULTRASONIC INTEGRITY TESTING METHODS

Cross-Hole Sonic Logging

Cross-Hole Sonic Logging (CSL) is the most common integrity testing method for drilled or cast-in-place foundations. A piezoelectric transducer is used to generate a signal that propagates as a sound (compression) wave within the concrete, while another transducer is used to detect the signal. Each transducer is placed into a vertical PVC or steel tube that has been attached to the reinforcement cage and filled with water prior to the concrete placement. The water acts as a coupling medium between the transducer and the tube. A typical tube arrangement and testing principles are presented in Figure 1.

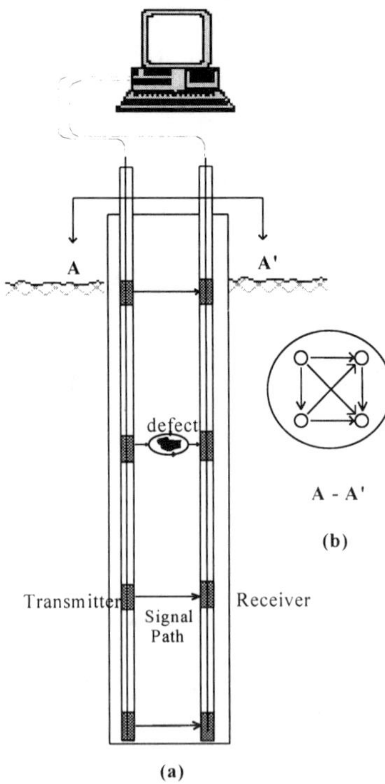

Figure 1 Typical CSL Testing Set-up Showing (a) Transmitter and Receiver at Different Depths, and (b) Plan View of the CSL Tubes with Possible Test Combinations.

The source and receiver transducers are lowered to the bottom of their respective tubes and placed such that they are in the same horizontal plane. The emitter transducer generates a sonic pulse (on the order of 10 pulses per second), which is detected by the receiver in the adjacent tube. The two transducers are simultaneously raised at a rate of around 300 mm/s (1ft/sec) until they reach the top of the drilled shaft. Typically this process is repeated for each possible tube pair combination (perimeter and diagonals). Figure 1b shows the six tube combinations that can be tested (logged) using a configuration of 4 tubes within a drilled shaft. Increased shaft diameter calls for a larger number of tubes, which increases the number of combinations and thereby the resolution of the testing zone.

In homogeneous, good quality concrete, the stress/sound wave speed, C, is typically around 3,800 m/s (12,000 to 13,000 ft/s) and is related to the modulus, E, and bulk density (unit weight, γ, and gravitational acceleration, g) as follows:

$$C = \sqrt{\frac{E \bullet g}{\gamma}} \qquad (1)$$

If for any reason the condition of the concrete is compromised, the wave speed will be reduced relative to that of the "good or sound" concrete value. Figure 2 presents a typical sonic signal for which the propagation time between the transducers is measured. The vertical axis is the signal amplitude (microvolts) and the horizontal axis is the time (microseconds). The point where the amplitude begins to rapidly fluctuate indicates the arrival time of the signal to the receiver (a.k.a. threshold time). Since the distance between the two tubes is known, the wave speed of the concrete between the tubes can be evaluated by the following relationship:

$$C = \frac{t}{L} \tag{2}$$

This is, of course, only a rough estimate, as the picking of the arrival time, t, is not objective and the distance between the tubes, L, is only known at the top of the shaft. The signal arrival times can then be plotted with depth to generate a log for the particular tube combination as presented in Figure 3. In addition to the threshold times, the energy of each signal may also be plotted with depth. This information can be used to compare signals of one zone to another where lower energy and/or later arrival times correspond to a compromised concrete quality and/or defect.

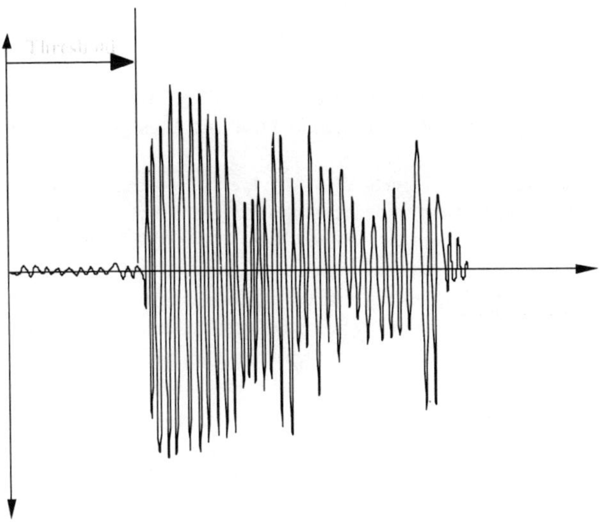

Figure 2 CSL Typical Testing Signal

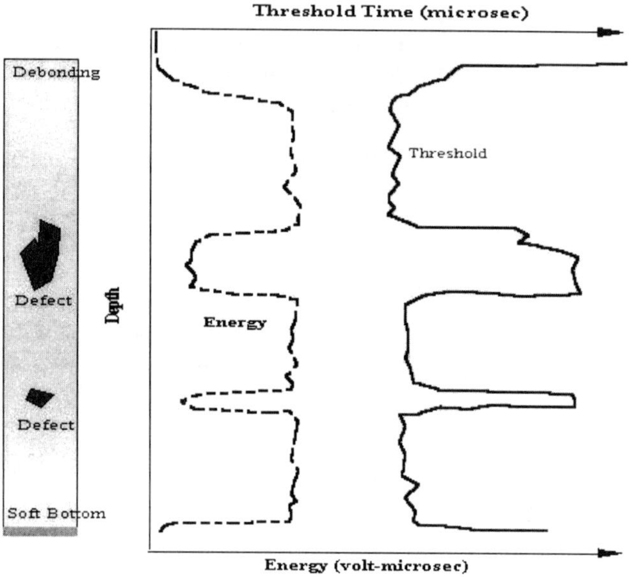

Figure 3 Presentation of CSL Test Results in the Form of Threshold Time and Energy Verse Depth

Advantages to this method include the direct assessment of pile integrity and the ability to position the transducers in different elevations to create more signals, allowing the development of a tomographic presentation of the investigated zone. The limitations of the method include detection of defects only when they exist between the tubes. The testing can be performed only on drilled shafts for which access tubes were installed. Debonding between the tubes and concrete is common if testing occurs long after the concrete placement. Testing in fresh concrete is also difficult as certain zones may cure at a lower rate, creating difficulties in the interpretation of the threshold time and energy. These zones may therefore be interpreted as poor quality concrete.

Single-Hole Sonic Logging

Single-Hole sonic logging (SSL) is a variation of the direct transmission CSL method in which the source and receiver are placed in the same tube and the signal travels in a vertical direction (refer to Figure 4). For drilled shafts and caissons, the method is limited to defects adjacent to the tube and is usually used only when a drilled shaft requires integrity assessment after construction. Due to high coring costs, typically, a single hole is advanced (often down the middle) to the bottom of the shaft or slightly below the depth where a defect is anticipated. It may also be desirable to perform SSL during CSL testing to isolate the location of a defect at a certain depth (i.e. distinguishing whether the defect identified using CSL is adjacent to the tube or in between the tubes).

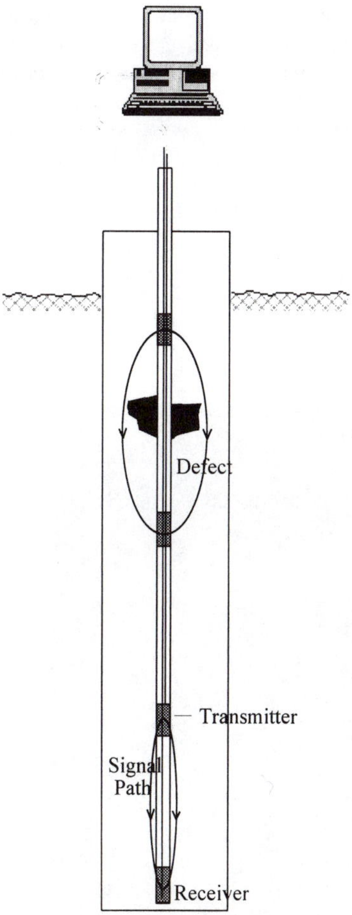

Figure 4 Typical SSL Testing Set-up Showing Transmitter and Receiver at Different Depths.

Recently, however, SSL has been performed within smaller diameter drilled mini-piles and augercast piles (Amir and Amir, 1998b). The use of SSL in these foundation types may become more commonplace in the near future, as research and experience provide insight as to the most efficient vertical placement of the tubes with respect to assessing the lateral integrity. Brettman and Frank (1996) describe a comparison between CSL and SSL tests.

THE PISA CSL/SSL TESTING SYSTEM

General

The PISA (Pile Integrity Sonic Analyzer) is a modular system allowing for adoption, upgrade and incorporation of additional integrity testing technologies. The current integrity testing options available in the PISA include cross-hole sonic logging (CSL) and single-hole sonic logging (SSL) using CHUM (Cross-hole Ultra Sonic Module) and sonic echo (a.k.a. small strain propagation) using the PET (Pile Echo Tester) module. Additional modules are currently under development.

In addition to its modularity, two advantages of the PISA integrity testing system over other systems include its software and portability. The PISA is the only Windows 95/98 based system and is also compatible with Word 2000. The software is updated

ARTHUR LAKES LIBRARY
COLORADO SCHOOL OF MINES
GOLDEN, CO 80401

periodically to incorporate new developments and algorithms that make data collection, interpretation, and report preparation easier and efficient. The PISA is lightweight (only 42.3 N (9-1/2 lb)) and self powered, hence can be easily carried around from shaft to shaft or site to site. This feature is also beneficial for air travel. The system can be also used as a standard laptop, saving the cost and space required for an additional personal computer (PC) when using a dedicated CSL testing system.

Figures 5 through 7 present photographs of the PISA system, including computer and sensors. Figure 8 presents the layout of the pile screen, where one can enter the pile information and select the tube orientation/locations. Selection of the desired tube combinations is accomplished by drawing a line between any two tubes. Figure 9 presents the data collection screen, where real-time graphical presentation of the concrete integrity is provided during testing. If a suspect zone is detected in this stage and the tomography option is enabled, the probes are lowered and raised relative to each other around the suspect zone, to further investigate and delineate the area. The signals can be examined and adjusted by manually picking the points or using preset algorithms to automatically determine the first arrival time (FAT) as shown in Figure 10. Figure 11 represents the typical graphical output for time and energy plots.

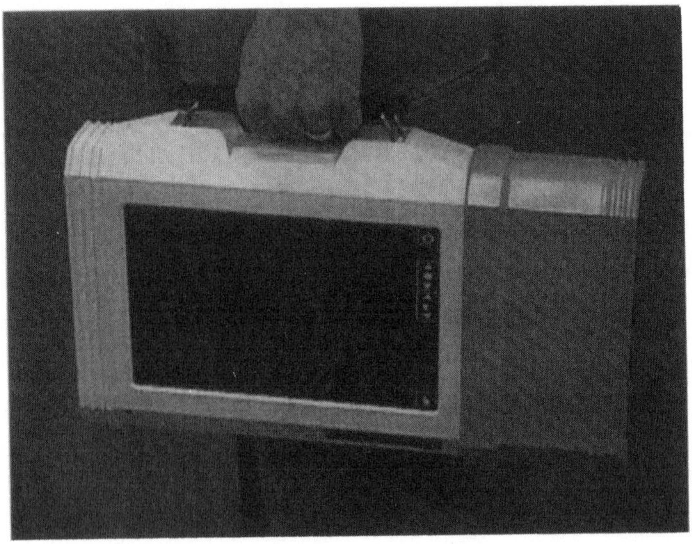

Figure 5 Photograph of the Pile Integrity Sonic Analyzer (PISA)

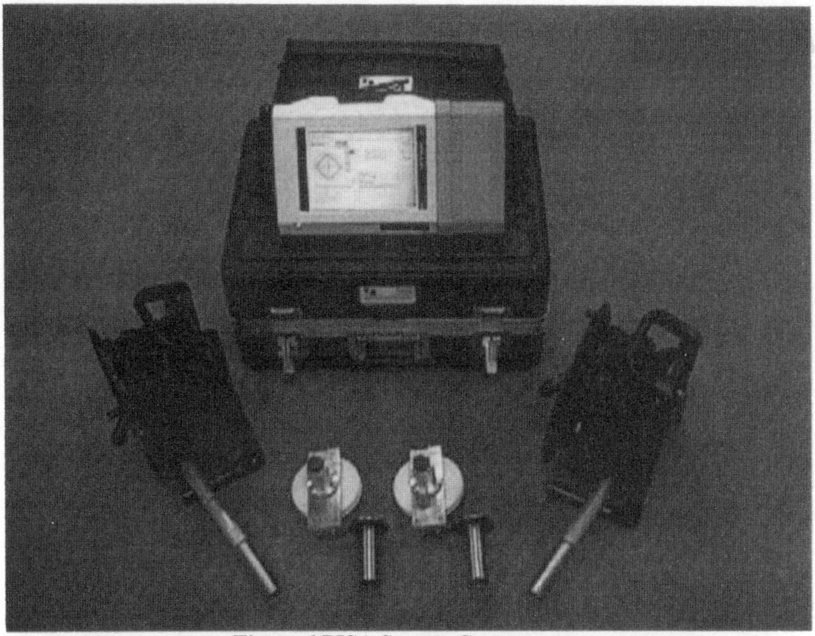

Figure 6 PISA System Components

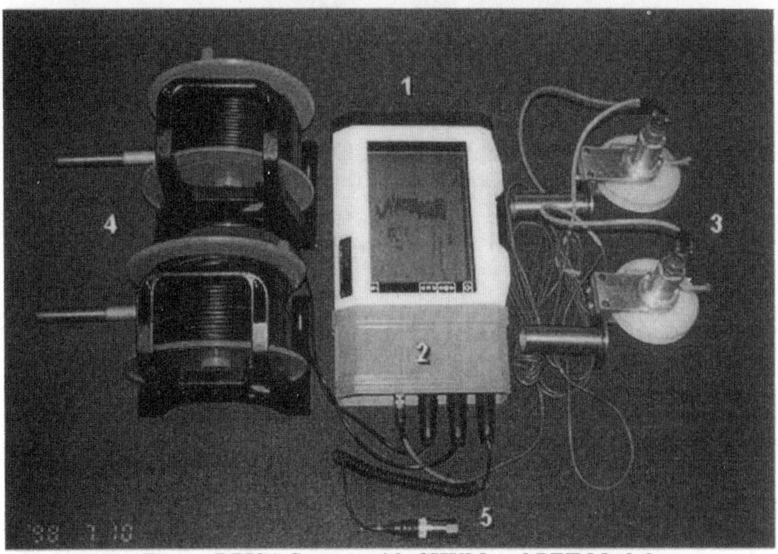

Figure 7 PISA System with CHUM and PET Modules

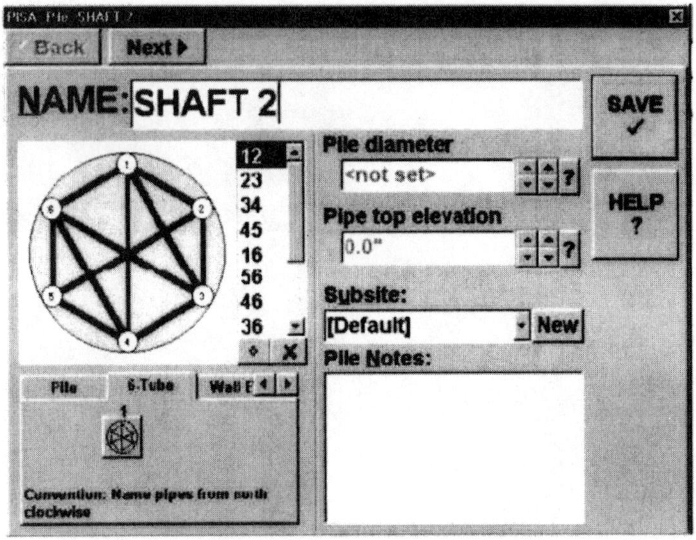

Figure 8 Layout of the Pile Screen

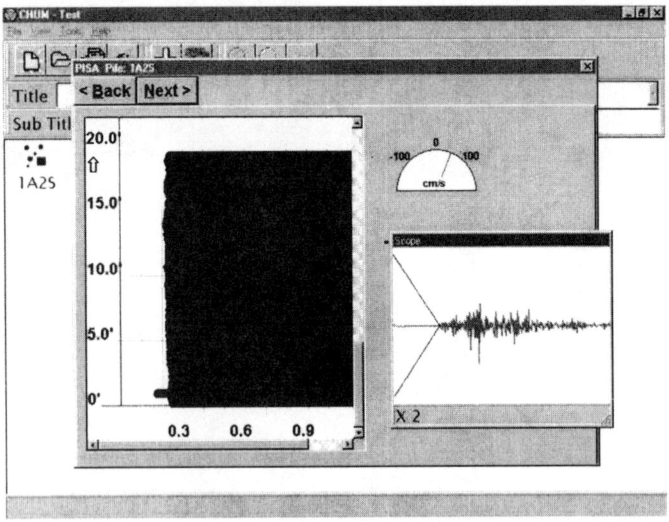

Figure 9 Data Collection Screen

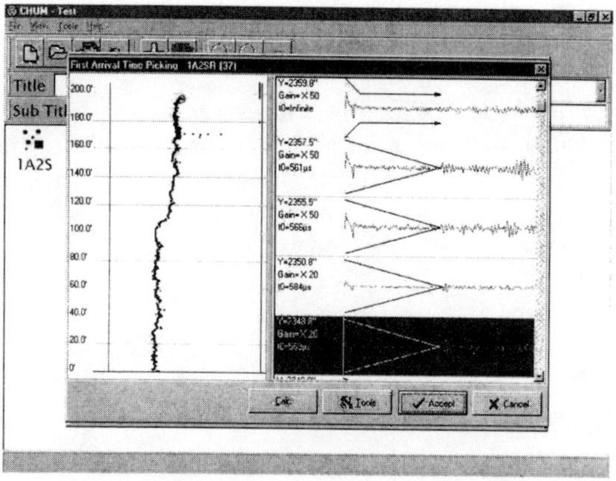

Figure 10 First Arrival Signal Identification Screen

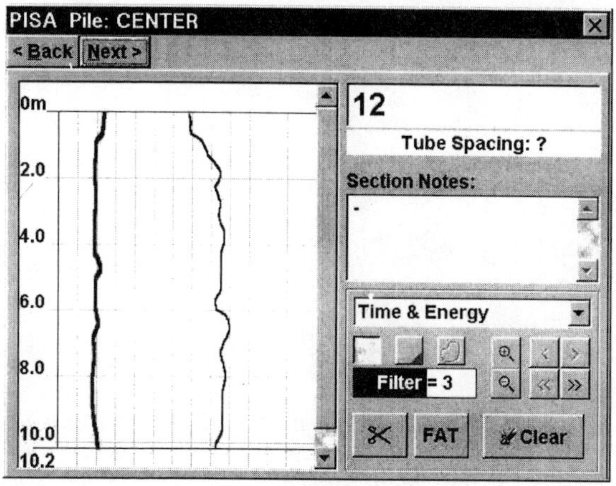

Figure 11 Typical Graphical Output for Time and Energy Plots

CASE HISTORY

CSL testing using the PISA was required for over 100 drilled shafts installed for the support of a major roadway interchange in Boston, Massachusetts. The shafts ranged between 2.1m (7 ft) and 2.7 m (9 ft) in diameter and tapered to 1.2 m (4 ft) to 1.5 m (5 ft) in diameter over the lower portion (15 m (50 ft) to 30 m (100 feet)). The total lengths varied between 36 m (120 ft) and 67 m (220 ft) below ground surface. Various penetrations into rock were required depending on the loading conditions. The shafts were constructed using temporary steel casing to the top of the clay and slurry throughout the remainder of the drilling and concrete placement process. The concrete was placed using a tremie process.

Eight schedule 80 PVC CSL access tubes were attached to the reinforcement cage and placed within the shaft prior to the placement of the concrete. The tubes were filled with water prior to placement in the shaft. CSL testing was performed primarily along the four diagonal tube combinations and four edge tube combinations as shown in Figure 12. Additional testing was performed as needed depending on the results.

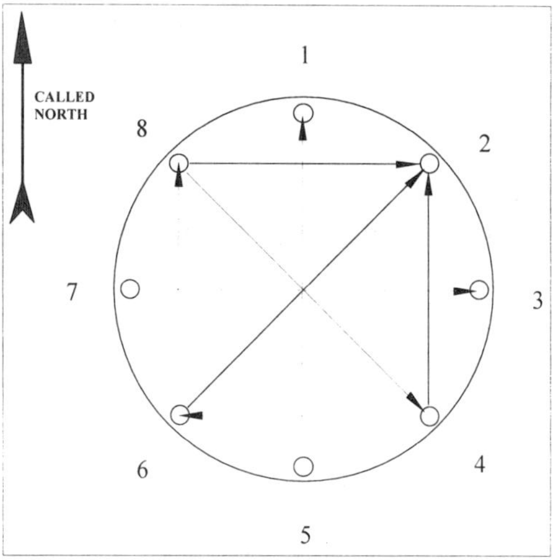

Figure 12 CSL Tube Layout

The CSL testing indicated significant anomalies in four of the first group of shafts installed for phase 1. Typical time and energy plots for the shafts (designated as "1 through 4") showing the various anomalies, are presented in Figures 13 through 16, respectively. The presented data illustrate two soft bottoms (shafts 1 and 2) and two problems in the upper 12.2 m (40 ft) (shafts 3 and 4). Table 1 summarizes the results of the repeated CSL testing performed for all four shafts. Following the testing and outline of anticipated problematic zones, a coring program was undertaken to verify the identified anomalous zones. For shafts 1 and 2, one 10.16-cm (4-in) core was drilled down the center of each shaft into the underlying bedrock. For shafts 3 and 4, multiple cores were taken between 6 m (20 ft) and 9 m (30 ft) as shown in Figures 17 and 18, respectively. The lateral extent of the suspect zones as identified by the CSL testing is presented in the two figures.

Shaft 1 - Tube combinations 4-8

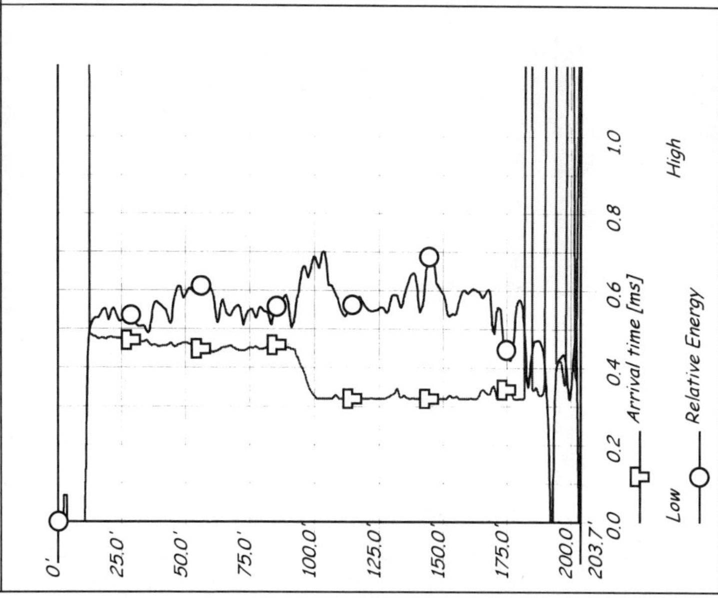

Figure 13 - Shaft 1 - Time and Energy with Depth.

Shaft 2 - Tube combinations 8-4

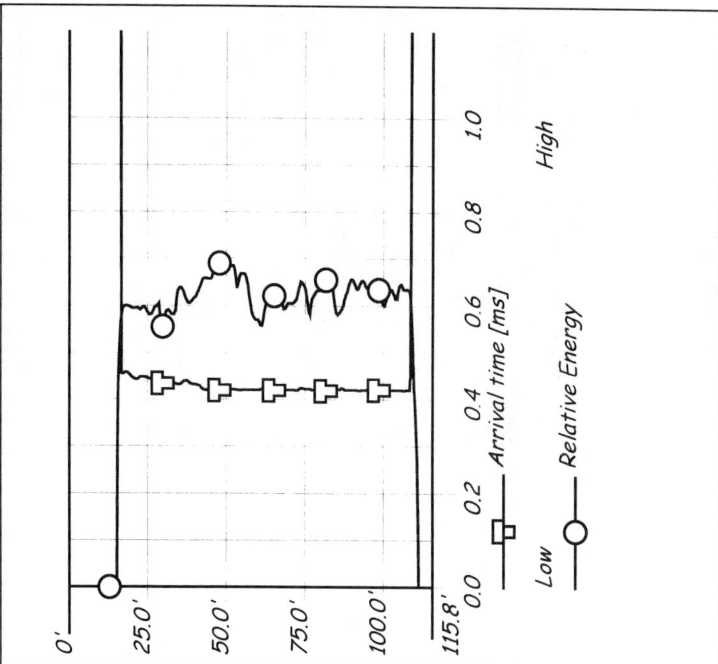

Figure 14 - Shaft 2 - Time and Energy with Depth.

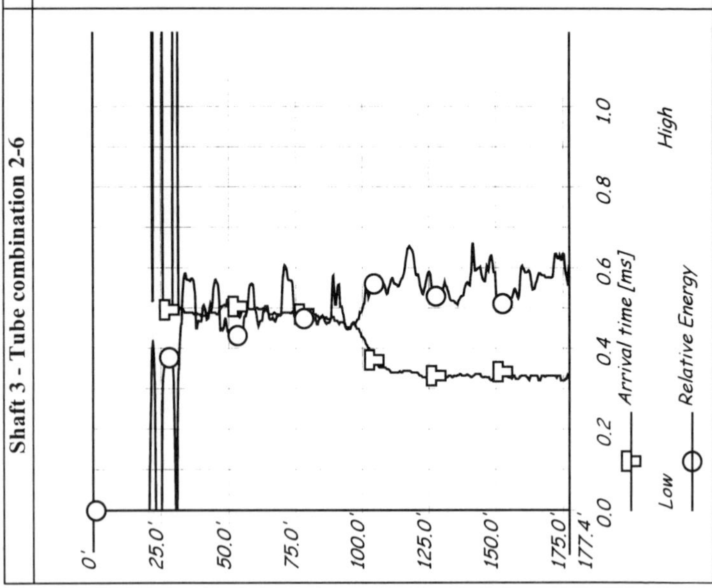

Figure 15 - Shaft 3 - Time and Energy with Depth.

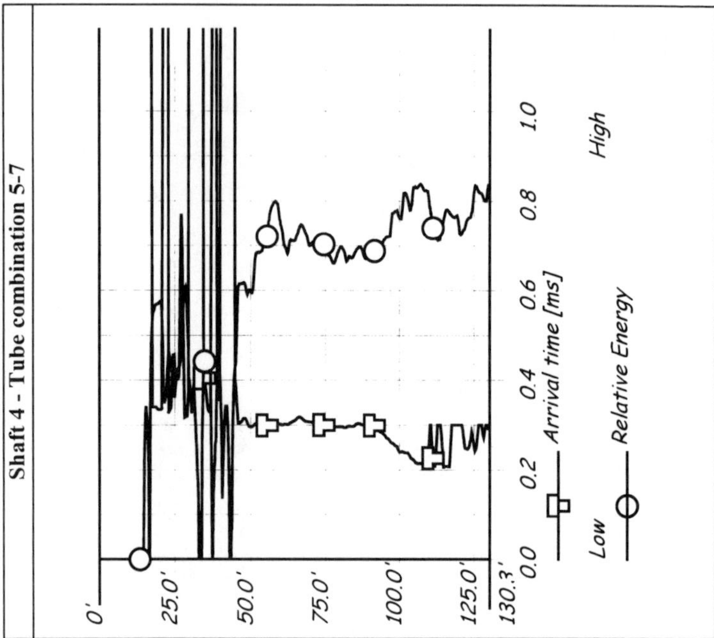

Figure 16 - Shaft 4 - Time and Energy with Depth.

Table 1 Cross-Hole Sonic Logging Test Summary

Shaft	Concrete Age when Tested	Tested[1] Length (feet)	Average[2] Tube Stickup (feet)	Condition[3]
Shaft 1	14 days	197 to 205	10 to 15	Edge Tube combinations: showed weak signal or loss of signal for a 2 to 3' zone in each profile within lower 20 to 30' of shaft (soft bottom). Diagonal Tube combinations: showed frequent weak signal or loss of signal in each profile over lower 20 to 30' of shaft (soft bottom).
	20 days	197 to 204	10 to 15	Diagonal Tube combinations: showed slight improvement over lower 20' of shaft, but still indicated soft bottom in each profile.
	30 days	199 to 203	10 to 15	Diagonal Tube combinations: showed further improvement over lower 20' of shaft, but still indicated soft bottom in each profile.
Shaft 2	9 days	109 to 116	15 to 20	All tested tube combinations indicate soft bottom conditions (lower 2 to 6 feet) except tube combination 1-3.
	30 days	111 to 114	15 to 20	All tested tube combinations indicate soft bottom conditions (lower 6 to 8 feet).
Shaft 3	4 days	177 to 178	15 to 20	Anomaly 25 to 35 feet below top of tubes in most profiles.
	12 days	175 to 178	15 to 20	Anomaly 25 to 30 feet below top of tubes in northwestern quadrant of shaft.
Shaft 4	18 days	130 to 135	15 to 20	Anomaly 20 to 40 feet below top of tubes in most profiles.
	30 days	137 to 139	15 to 20	Anomaly 20 to 40 feet below top of tubes in most profiles.
	44 days	133 to 136	15 to 20	Anomaly 35 to 45 feet below top of tubes in northwestern quadrant of shaft.
	87 days	126 to 133	15 to 20	Anomaly 35 to 45 feet below top of tubes in northwestern quadrant of shaft.

Notes.

1. The tested length represents the average length of tube accessible during CSL testing, including the stickup above the top of the concrete. The tested length may vary for several reasons, including different tube stickups, slack in cables, blockage in tubes, and/or slippage between depth wheel and cable.
2. The average stickup was not the same every time, but varied between the numbers indicated for these tests.
3. The lengths and depths are referenced from the top of the tubes. Subtract the tube stickup to determine the depth in the drilled shaft.

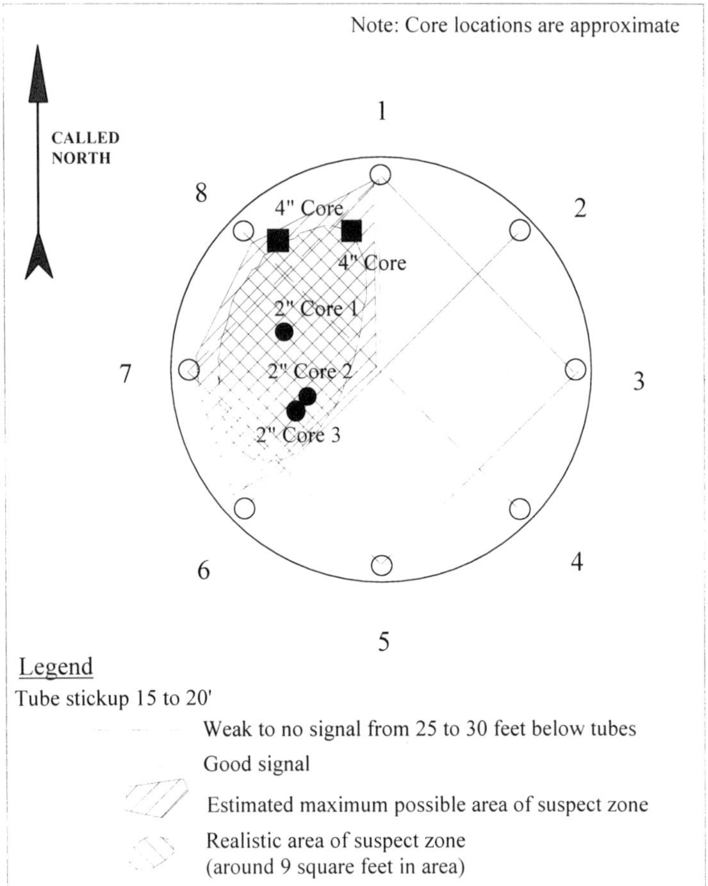

Figure 17 CSL Summary - Shaft 3

The concrete core samples retrieved from the suspect zones in shafts 1 (lower 6.1 m (20 ft)) and 2 (lower 2.4 m (8 ft)) were completely raveled, segregated, and disintegrated suggesting major discontinuity in the lower section of the shaft. The beginning of the defective concrete obtained from the coring coincided closely with the suspect zone identified during the CSL testing. These two shafts are currently undergoing repair procedures to transfer the loads through the defective zone to the underlying rock below.

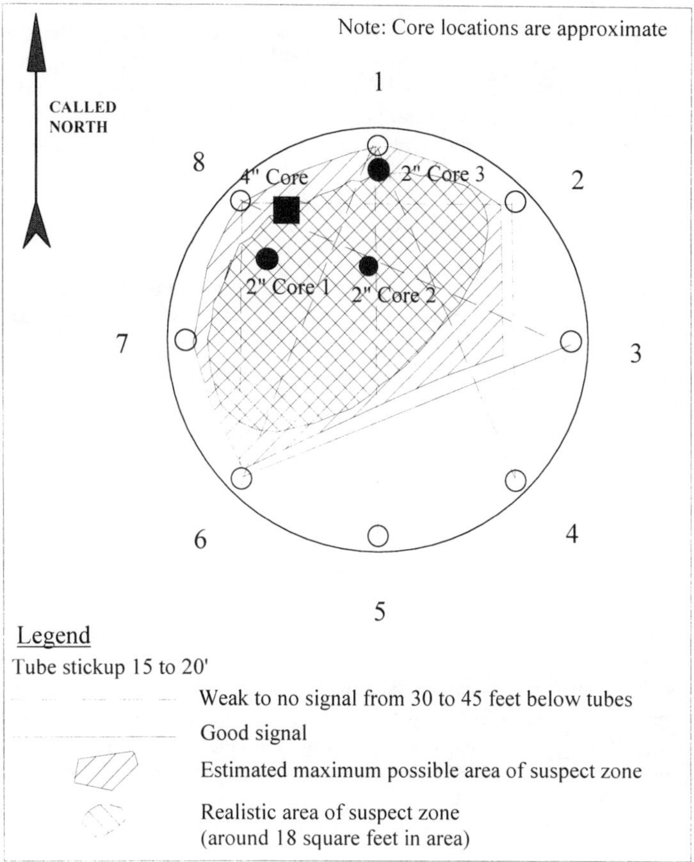

Figure 18 CSL Summary - Shaft 4

The concrete core samples retrieved from the suspect zones in shafts 3 (upper 1.5 to 3 m (5 to10 ft)) and 4 (upper 6.1 to 9.1 m (20 to 30 ft)) indicated mixed results. Moderate to severe segregation was observed in some of the cores for shaft 3. In one core, complete disintegration was observed between 1.5 to 3 m (5 to 10 ft). Some slight to moderate segregation was observed in the core samples retrieved from shaft 4. Compressive strength, Elastic Modulus, and unit weight testing were performed on selected samples from the cores obtained from these two shafts. The compressive strength testing indicated high variability in the values for both shafts. Between 1.5 to 3 m (5 to 10 ft) below the top of shaft 3, the compressive strength was considerably lower (some values were much less than the required 28 day strength) than the areas above and below. Lower compressive strength values were also observed for shaft 4 specimens results below 6.1 m (20 ft). The concrete specimens tested from this shaft exhibited a wide variability in the compressive strength. These depths correspond to the anomalous zones identified during the PISA CSL testing, suggesting compromised concrete condition. Shafts 3 and 4 are still undergoing evaluation with regard to the integrity of the shaft, its structural load carrying ability and subsequent actions if required.

SUMMARY AND CONCLUSIONS

The Pile Integrity Sonic Analyzer (PISA) represents the state-of-the-art equipment in deep foundations integrity testing. The system is lightweight and mobile, comprising of a generic laptop and modular equipment components. The PISA provides real time integrity evaluation and employs common operating systems (e.g. MS Windows), conforming to other requirements (i.e. graphics presentations and word processing). Allowing for easy software updates and hardware features, the PISA represents a new generation of NDT equipment that is better suited for versatile testing demands, advanced analyses and field applications.

A presented case history of detection and configuration of defects in large and complex rock socketed drilled shafts illustrates the importance of non-destructive testing and the success of the examined system. All four detected defects were confirmed to have construction problems. The two deep detected defects were found to be zones of voids and extensive segregated concrete that rendered the drilled shaft as incapable to carry load without a major repair. The shallower zones were found to be associated most likely with a problem in the construction technique. Though currently under investigation, the identified and delineated compromised zone appears to result in a defect that does not decrease the load carrying ability of the shaft below the design load.

APPENDIX 1

REFERENCES

Amir, E.I and Amir, J.M., (1998a), "Testing Piles with Virtual Instruments", Proceedings of the DFI conference, Vienna, Austria, pp. 4.1.1 - 4.1.4.

Amir, E.I. and Amir, J.M., (1998b) "Recent Advances in Ultrasonic Pile Testing" 3rd International Geotechnical Seminar on Bored and Augured piles, Ghent, Belgium.

Baker, C.N., Drumright, E.E., Briaud, J., Mensah-Dwumah, F. and Parikh, G., (1992), "Drilled Shafts for Bridge Foundations", Final Report Under Contract No. DTFH61-88-Z-00040, Federal Highway Administration, February.

Brettman, T. and Frank, M., (1996), "Comparison of Crosshole and Single Hole Sonic Integrity Logging Methods", Proceedings of the 5th International Conference on the Application of Stress Wave Theory to Piles, Orlando, Florida, September, pp. 698-707.

Chernauskas, L.R. and Paikowsky, S.G., (1999), "Deep Foundations Integrity Testing: Techniques and Case Histories", Civil Engineering Practice Spring/Summer 1999, vol. 14, no. 1, Boston Society of Civil Engineers, Section/ASCE, pp. 39-56.

ACKNOWLEDGMENT

The PISA (Pile Integrity Sonic Analyzer) CSL testing system is manufactured by Pile Test.com, Ltd., Herzlia, Israel and distributed in North America by GTR Inc. (Geosciences Testing and Research) of North Chelmsford, MA. GTR personnel carried out and interpreted the described tests. The cooperation of the contractors, consultants and owners associated with the described projects is appreciated. The authors acknowledge the assistance of Ms. Mary Canniff in putting together the manuscript.

SETTLEMENT OF PILE SUPPORTED MAT FOUNDATIONS

Shailendra N. Endley, Ph.D., P.E.[1], Wayne Dunlap, Ph.D., P.E.[2], David Knuckey, P.E.[3], Jeffrey Allen, P.E.[1], Karun Sreerama, Ph.D., P.E.[1]

ABSTRACT

The settlements of four pile supported mat foundations were monitored over a period of 14 years. The mats were supported by 88, 105, 706, and 708 piles. Pile lengths ranged from 17 m to 27.5 m. Load tests were performed on instrumented single piles to determine their load capacity and load distribution along the length. The subsurface conditions at the site consisted of 3m to 4.6 m of dredge soils fill underlain by overconsolidated, stiff to very stiff, clay, loose to medium dense sand and stiff to very stiff, overconsolidated clay. The observed settlements were compared with settlements calculated using four different theoretical methods. Based on the study, a simple correlation is proposed to compute pile group settlements using measured or computed single pile settlements.

INTRODUCTION

The American Rice Facility was constructed around 1985-86 at the Port of Freeport in Freeport, Texas. The project consisted of many concrete silos (up to 40 m high) founded on a system of auger cast pile supported mats. The mats were structurally supported on the auger cast piles. Sustained loads on the mats ranged from approximately 168 kPa to 230 kPa with plan dimensions ranging from 13.7 m by 27.2 m to 25.3 m by 94.5 m. The number of piles supporting each mat ranged from 88 to 708 with lengths ranging from 17 m to 27.5 m. The capacity and load distribution of the piles were determined by performing a series of instrumented pile load tests.

The settlement of four pile-supported mats was monitored over a period of 14 years. The measured settlements are compared with various theoretical methods and conclusions are drawn about the applicability of these methods in predicting pile group settlement.

---

[1] Professional Service Industries, Inc., Houston, TX
[2] Professor, Texas A&M University, College Station, TX
[3] Port of Freeport, Freeport, TX

SITE AND SOIL CONDITIONS

A generalized soil profile is shown on Figure 1. About 3 m to 4.6 m of dredged soil fill from the ship channel was placed at the site approximately 10 years before the construction of this project. The fill was placed on natural soil that existed at about elevation +1.2 m. The consistency of the fill material varied from a very soft to stiff clay to a sand/clay mixture.

Below the dredged fill, a stratum of over-consolidated, stiff plastic clay was encountered and extended to about elevation –4 m. The average undrained shear strength (S_u) of the clay stratum, from unconsolidated undrained tri-axial tests, is about 75 kPa. The plasticity index (PI) is about 40 and the liquid limit (LL) is about 60. Below the clay, a stratum of loose to medium dense sand is encountered and extends to about elevation –11.6 m. The standard penetration test blow counts (SPT – N) ranged from 6 to 21 blows per foot and the percent passing No. 200 sieve ranged from 10 to 27 percent. Below the sand, a stratum stiff to very stiff, over-consolidated clay was encountered with shear strengths ranging from 75 kPa to 129 kPa. The liquid limit (LL) of this material ranged from 63 to 73 with the corresponding plasticity indices (PI) between 40 and 45. This clay stratum extended to elevation –27.5 m, which was the maximum depth of the exploration. Stress-strain relationships measured using consolidated undrained triaxial tests are shown in Figure 2. Detailed subsurface conditions and soil and foundation parameters were presented in Endley et al. (1991).

PROJECT DETAILS

The project consisted of many concrete silos (15.25 m to 47.47 m high) supported on a system of mats. Sustained loads on the mats ranged from approximately 168 kPa to 230 kPa. Due to the presence of random fill (dredge spoil) to depths of 3 m to 4.6 m, it was decided that the mats would be supported on pile foundations. Taking into consideration the subsurface conditions encountered and the presence of a high groundwater table (about Elevation 0), auger cast piles were found to be the most economical foundation option for the project. The diameter auger cast piles was 355 mm. The number of piles supporting each mat ranged from 88 to 708 with lengths of ranging from 16.75 m to 27.5 m. There were four areas of study referred to as the "Rough Rice", "Milled Rice", "Tempering Bins", and "By-Products" structures. The details of each area are as listed in Table 1. Figures 3 and 4 show a plan of the mat foundations and photograph of the site during construction of the mat foundations, respectively.

Table 1 - Project Structure Information

Mat	Plan Dimensions (L x B x D) m	Silo Information (Dia x Ht. - Number)	No. of Piles	Length of Piles (m)	Mat Pressure (kPa)	Pile Load (kN)
Rough Rice	94.5 x 25.3 x 0.91	7.9 x 39.6 - 27	706	27.4	230	841
Milled Rice	39.9 x 31.2 x 0.76	6.4 x 23.1 - 48	708	22.9	210	691
Tempering Bins	13.7 x 27.1 x 0.61	6.4 x 13.2 - 8	105	16.7	168	660
By-Products	20.4 x 20.4 x 0.68	6.4 x 24.3 - 9	88	27.4	175	897

PILE LOAD TESTS

Two pile load tests were performed to determine load capacity, load distribution and settlement characteristics (Endley et al., 1991). The results of the load tests are shown in Figures 5 and 6. The test piles were also instrumented to determine the load distribution along the pile length.

OBSERVED SETTLEMENTS

Settlement readings of mats were measured at various pre-determined locations on the mat foundation over a period of 14 years. Each mat foundation was monitored at 6 to 11 locations, depending on the geometry and accessibility to the location. Figure 3 shows the mat geometry and the monitoring points on each mat. Settlements of the mat foundations were measured by performing leveling survey. The mat elevations at the monitoring points were measured a few days after the mat construction and prior to the completion of silo construction (between September '85 and January '86). The second set of elevation readings were recorded in September 1991 (after approximately 5 years). Third sets of elevation readings were recorded in June 1999 (after approximately 14 years). The settlements observed at these times are shown in Figure 7.

THEORETICAL DETERMINATION OF MAT SETTLEMENTS

Theoretical settlements were calculated for each of the four structures. For the Rough Rice and Milled Rice structures, the load used in the settlement analyses included the dead load as well as the total live (rice) load. For the Tempering Bin structures, dead load plus two-thirds of the live load was used. For the By-Products structure, the dead load plus one-half the live load was used. The amount of live load used in each of the settlement analyses was based on actual loading patterns as indicated by American Rice personnel.

The settlements of the pile groups were calculated based on four different methods. Table 2 shows the comparison of observed and calculated settlements.

Skempton's Method

This is an empirical approach suggested by Skempton (1953) based on the same field observations.

$$\rho_G = \rho_1 \frac{(4B + 9)^2}{(B + 12)^2} \tag{1}$$

where,

ρ_G = Settlement of the group
ρ_1 = Settlement of single pile
B = Width of pile group in feet

The magnitude of ρ_1 was determined from the results of pile load tests performed on two piles, 27.5 m and 16.75 m long as shown on Figures 5 and 6. For rectangular mats, the effective "B" was calculated by taking the square root of the mat area. Because the Milled Rice structure is supported on 24 m long piles, the value of ρ_1 was estimated based on the results of the load tests performed on the 16.75 m and 27.5 m long piles.

Poulos' Method

This method is based on theory of elasticity. The piles are assumed to be freestanding and a rigid pile cap will only influence the behavior of a group by controlling differential settlement between different piles in the group. The settlement by this method is defined as follows:

$$R_s = (R_{25} - R_{16})\ (\sqrt{n} - 5) + R_{25} \tag{2}$$

where,

$$R_s = \frac{\text{Average group settlement}}{\text{Settlement of single pile at same average Load as a pile in the group}}$$

R_{25} = Value of R_s for 25 pile group
R_{16} = Value of R_s for 16 pile group
n = number of piles

The values of R_{25} and R_{16} can be obtained from the published tables (Poulos and Davis, 1980) for these calculations. The value of stiffness factor, k, (ratio of

Young's modulus of pile to soil) was assumed to be 1000. The settlement of a single pile was determined from the two pile load tests previously discussed.

As suggested by "Poulos and Davis", the method assumes that for groups containing more than 16 piles, the group settlement, R_s varies approximately linearly with square root of the number of piles in the group. This procedure was extended up to the group of 708 piles placed for the Milled Rice structure.

Consolidation Settlement of Equivalent Footing

The one dimensional consolidation method was also used to compute the group settlement (Terzaghi and Peck 1967). In this method, the pile group is replaced by an equivalent footing at a depth that is two-thirds the length of the piles. The settlement above this depth is assumed to be negligible. Consolidation tests were performed to determine the C_c and P_c values. The values of equivalent footing pressure exceeded the previous maximum consolidation pressure (P_c) of the clays. Therefore, C_c was used to calculate the settlement.

Two Dimensional Finite Element

The mat and pile foundation system was modeled using finite elements in the software PC-Stran^{tm} (Murphy 1990). The influence of soil on the foundation at each finite element node was modeled using equivalent soil stiffness computed from the measured soil shear strength. The model assumes linearly elastic soil properties. The analysis was based on the long term concrete modulus calculated based on the measured compressive strength of the concrete and the soil Young's modulus (initial slope of the stress-strain curve) calculated based on the measured static stress-strain properties of the soil (Figure 2). The movement of the mat foundation due to the applied loads was calculated from the strains computed from the finite element analysis.

The measured and computed settlements are presented in Table 2 below. The measured settlements in Table 2 represent the range of settlements at various monitored locations on the mat. The time rate of settlements was not considered in the theoretical calculations. The calculated settlements represent the final total settlements and these settlements are compared with the 14-year settlement values.

Table 2 - Comparison of Theoretical and Observed Settlements

Structure (number of piles)	**Settlements (mm)**				
	Observed	**Skempton**	**Consolidation**	**Elastic**	**2-D FEM**
Rough Rice (706)	165 – 275	32.5	468	107.5	234
Milled Rice (708)	138.5 – 187.5	35.3	553.75	161.5	196
Tempering Bins (105)	28 – 30	39	259	52.0	33.75
By-Products (88)	31 - 59	31.25	230	35.3	54.5

CONCLUSIONS

The settlement observations made on four pile supported mat foundations over a period of 14 years indicated that about 80 percent of the settlements occurred during the first 6 years in all the mats. In two of the mats with 105 (Tempering Bins) and 708 (Milled Rice) piles, the mat appears to have stopped settling. However, the settlement seems to be continuing, after 14 years, in the other two mats with 88 (By-Products) and 706 (Rough Rice) piles. The straight line continuation of settlement at location B-7 on By-Products cannot be explained. In general, the settlement measured at the mat corners was less than the settlements measured at the centers. Also, the smaller mats settled relatively uniformly compared to the mats with larger plan dimensions.

The measured mat foundation settlements were compared with calculated settlements. The mat foundation settlements were calculated by four methods. A comparison of the measured and computed settlements indicates that the settlements calculated by the Skempton's method were reasonable for smaller mat sizes and pile groups with 100 piles or less. However, for larger mats with greater than 100 piles, the settlements computed using Skempton's method were significantly less than the measured settlements. The settlement values computed using Skempton's method do not vary significantly with the size of mat.

Settlements calculated using the consolidation theory were significantly greater than the measured settlements, irrespective of the mat size and the number of piles under the mat. The settlements calculated using the consolidation theory were about 2 times the measured value for the larger mats and up to 10 times the measured values for the smaller mats. The settlements computed using the elastic theory of Poulos and Davis (1981) were within the measured ranges, except for the rough rice mat. In the rough rice mat, the calculated settlements were less than the measured. However, the computation of settlement by the elastic theory requires the input of settlement of a single pile, either from calculation or measured during a load test. Some assumptions and extrapolations had to be made in the application of the theory to compute the settlements for the mats by elastic method.

The settlements computed using a simple 2-dimensional finite element model and the measured stress-strain characteristics of the soil appeared to compare reasonably well with the measured settlements. However, based on the nature of the computations and the soil and structure parameters required to be input into the computations, it is not reasonable to expect such computations to be performed in practice, unless simpler software routines along with standard correlations for soil properties are made available.

Figure 8 shows the variation of average observed settlement ratio (ratio of pile group settlement to settlement of single pile) with number of piles in a group. A regression analysis performed on the values indicates that the following relationship predicts the settlement of a pile group, based on measured settlement of a single pile:

$$\frac{\rho_g}{\rho_s} = n^{2/3} \qquad \textbf{(3)}$$

where,

ρ_g = settlement of the pile group
ρ_s = settlement of single pile
n = number of piles in the group

The simple equation also approximately matches results by Poulos's method for group settlements with less than 25 piles at 5-diameter spacing. This equation can be used as a guideline to estimate pile group settlements using measured or computed single pile settlements for essentially friction piles bearing in clays. All other soil properties are reflected in the measured and / or calculated single pile settlements. Further studies are underway to refine the equation.

In general, based on the findings of this study, for pile supported mat foundations in clay soils, it can be assumed that 75 to 80 percent of the settlements would occur in the first 5 to 7 years and 100 percent of the settlements would occur in about 15 years. It appears that measured pile settlements do not correlate well with pile settlements calculated using consolidation theory.

References

Endley, Shailendra, Wayne Dunlap, Jeffrey Allen, and David Knuckey (1991), "Performance of Auger-Cast Piles", Texas Section ASCE Meeting, Fall 1991, South Padre Island, Texas.

Murphy, Joe (1990), "PC-STRAN", Version 4.10, 1984-1990.

Poulos, H.G. and Davis, E.H. (1980), "Pile Foundation Analysis and Design", Series in Geotechnical Engineering, , John Wiley and Sons, Inc., New York, NY.

Skempton, A.W. (1953), Discussion: "Piles and Pile Foundations, Settlement of Pile Foundations", Proceedings of Third International Conference SM&FE, Volume 3, pp. 172.

Terzaghi, K. and Peck, R.B. (1967), "Soil Mechanics in Engineering Practice, 2nd ed., John Wiley and Sons Inc., New York, NY.

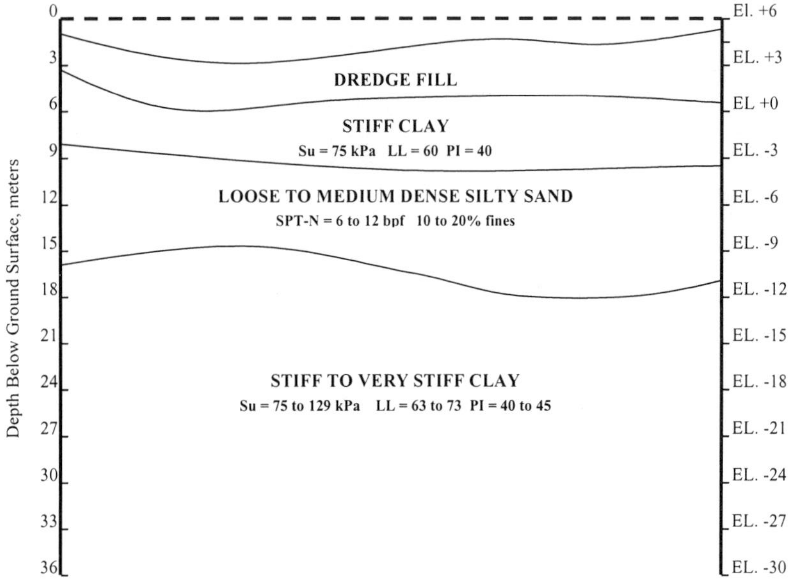

Figure 1 – Subsurface Soil Profile

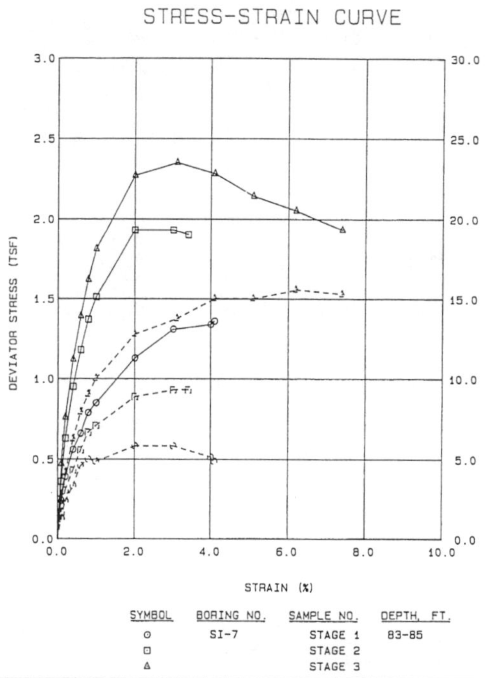

Figure 2 Stress vs. Strain Data of Clays from Tri-axial Tests

Figure 3 Mat Foundation Construction

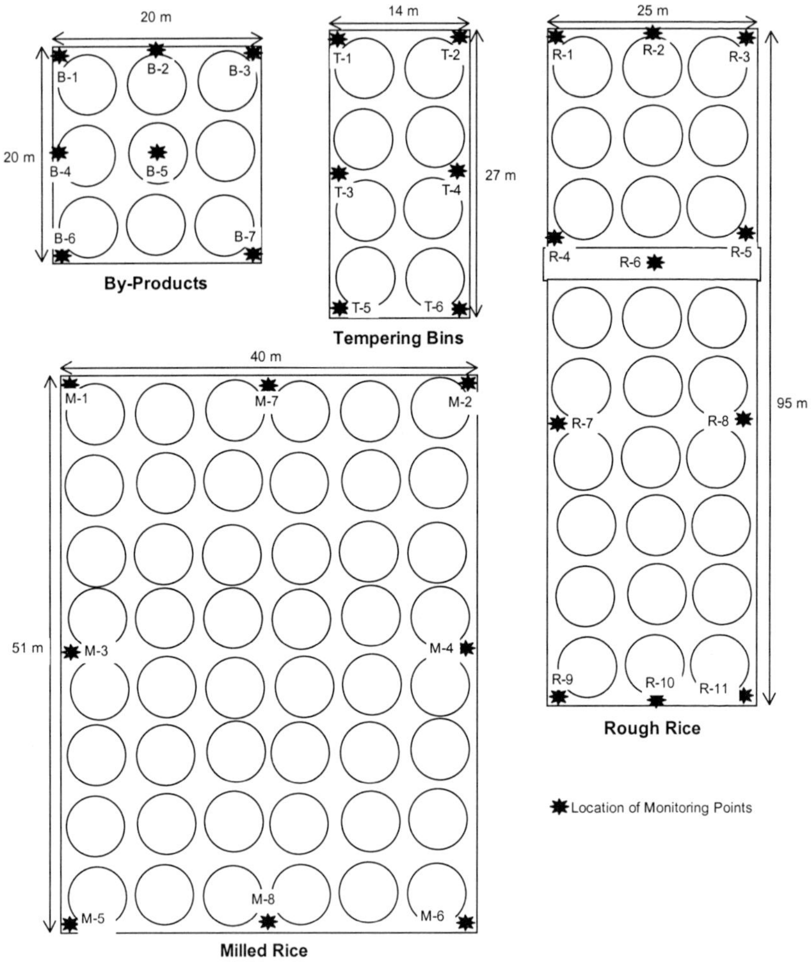

Figure 4 Plan View of The Mats and Silos

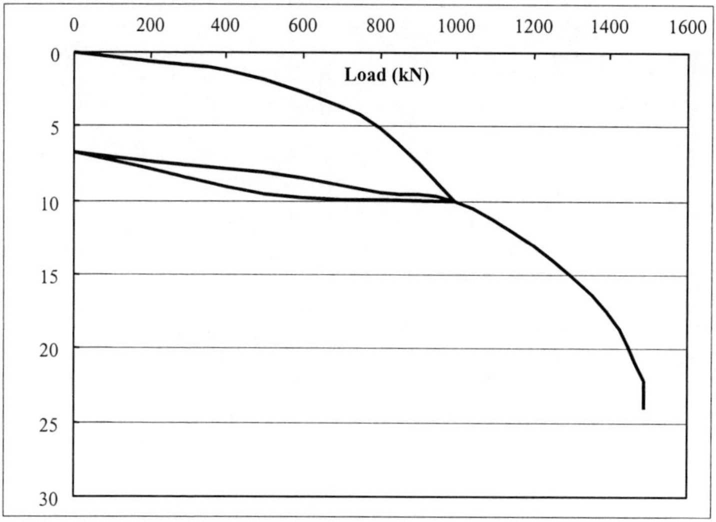

Figure 5 Single Pile Load Test Results
(16.75 meters – Short Pile)

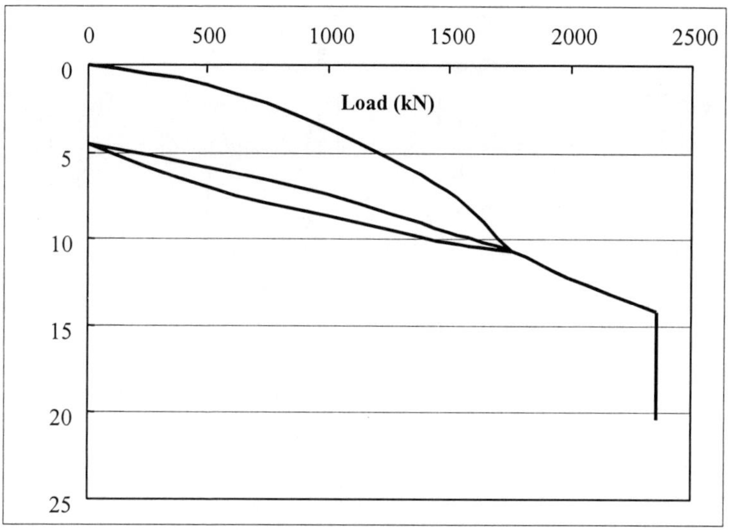

Figure 6 Single Pile Load Test Results
(27.5 meters – Long Pile)

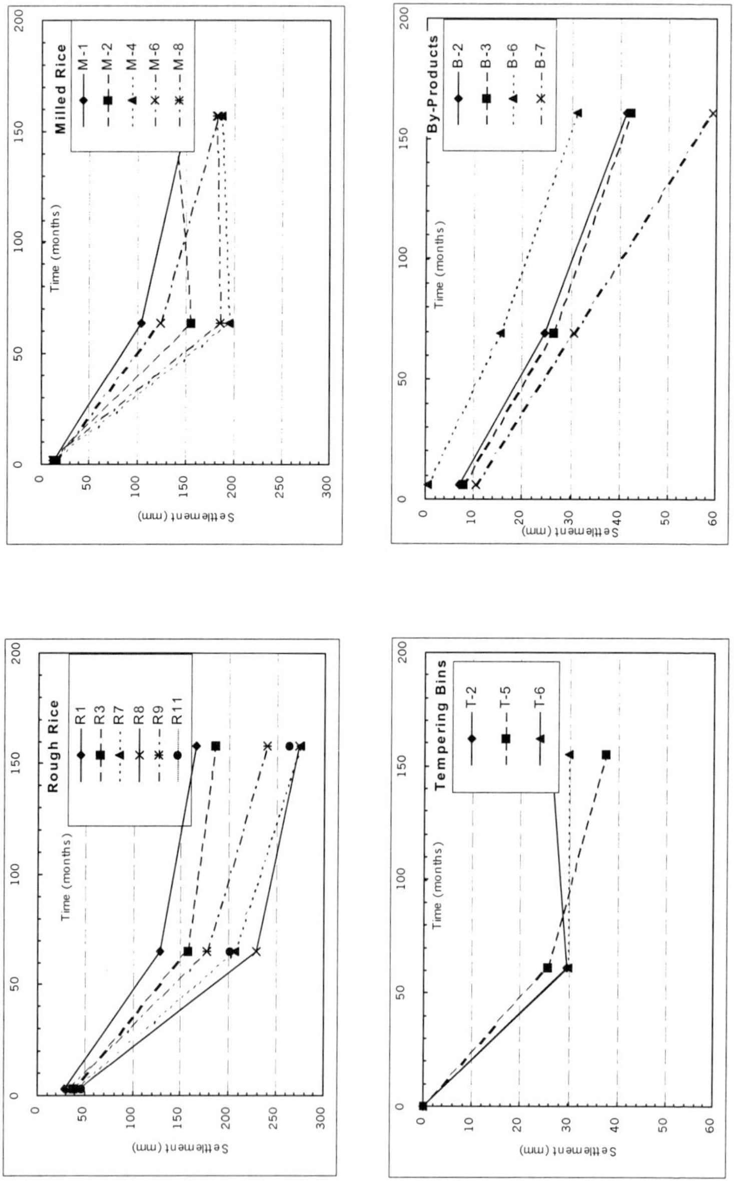

Figure 7 Measured Settlements of Mat Foundations

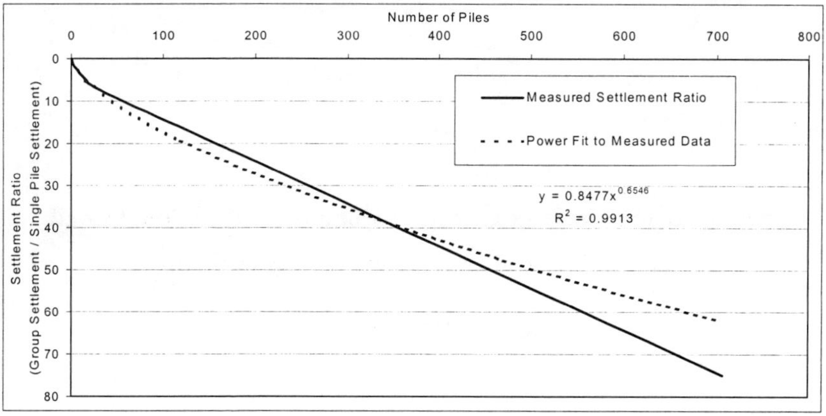

Figure 8 Pile Group Settlement Ratios

Soil Set-up, Variable Concrete Modulus, and Residual Load for Tapered Instrumented Piles in Sand

Bengt H. Fellenius[1], Walter G. Brusey[2], and Frank Pepe[3]

ABSTRACT. Two 20 m long, strain-gage instrumented, closed-toe, 350-mm Monotube piles with a 7 m long bottom section tapering down to a 200-mm diameter at the pile toe were driven into a deposit of silty sand of glacial origin. The piles were concreted and static loading tests were performed—one pile was tested three days after completion of driving and concreting and the other after twenty-three days. The static test established the former pile to 2,000 KN. A maximum load of 2,500 KN was applied to the pile tested after twenty-three days without reaching the ultimate resistance of the pile, confirming the experience of earlier tests at the site that the pile capacity increases with time due to set-up. The actual magnitude of the set-up was not established. Both piles were instrumented with vibrating wire strain gages. The measurements showed that the pile Young's modulus of the composite steel concrete was not a constant but a linear function of the imposed strain. The measured axial strains were used to determine the non-constant Young's modulus of the steel-concrete cross section, enabling calculation of the axial loads in the piles during the test. The test data showed that locked-in load (residual load) acted on the piles before the start of the loading tests. The point of equilibrium (neutral plane) between downward and upward acting residual load forces on the pile was determined to be close to where the pile cross section changed from a straight section to a tapered section. Because of the presence of residual load, the measured loads indicated, falsely, that the soil resistance acting against the tapered section pile was very small. The analysis determined the magnitude and distribution of the actual shaft and toe resistances as well as of the residual load proving that most of the pile capacity was developed in the bottom portion of the pile agreeing well with effective stress principles. In the tapered section, at the maximum applied test load, the resistance was governed by fully mobilized positive shaft resistance and a movement-dependent "toe resistance component". The analysis results suggest that the set-up was due to a stiffening of the soil and not to increase of shaft resistance. The following effective stress soil parameters were established for the two test piles, Piles 2 and 3: beta-coefficient equal to 0.4 and N_t-coefficients of 60 and 75, respectively.

---

[1] Urkkada Technology Ltd., 1010 Polytek St., Unit 6, Ottawa, ON, K1J 9H8

[2] The Port Authority of New York and New Jersey, One World Trade Center, New York, NY 10048

[3] Parsons, Brinckerhoff, Quade, and Douglas Inc., 1 Penn Plaza, New York, NY 10119

INTRODUCTION

Since the early 1990's, the JFK Airport is seeing extensive rebuilding and upgrading. Due to the particular soil conditions at the airport, structures must be supported on pile foundations. The pile of preference at the airport is the Monotube pile, a thin-wall steel pipe with a 350-mm or 450-mm (14 in or 18 in) diameter connected to a 7.6-m (25 ft) lower section that tapers down to a 200-mm (8 in) diameter at the pile toe. The piles were subject to considerable testing at the start of the construction using static and dynamic testing and analysis methods. York et al. (1995) presented results from static loading tests on piles installed to depths ranging from 16 m through 23 m and performed 15 days through 42 days after driving. The capacities established in the static tests were combined with pile capacities obtained from WEAP analysis and Pile Driving Analyzer tests with CAPWAP analysis on driving records from five piles at initial driving and at restrike 49 days after driving. The results indicate that the pile capacity increased on average by approximately 50 percent during the first about three weeks after the driving. In 1994, additional testing was undertaken consisting of static loading tests on two instrumented piles. The results of the 1994 tests are presented in the following.

SOIL CONDITIONS

Details on the soil conditions at the site are available in York et al. (1995). In summary, the site is located adjacent to the Jamaica Bay on the south shore of Long Island, New York, and was developed in the 1940's from an existing marsh by placing sand dredged from the bay bottom to about 4 m height at the subject location. The groundwater table lies at 2.5 m below the fill surface and the pore pressure distribution is considered to be hydrostatic. The fill was placed on the original marsh organic material, which has a thickness at the subject location of about 1 m and consists of soft clay and peat. Below the marsh material, the soil consists of fine to coarse, medium dense to dense, glacial sands to large depth. Fig. 1 shows the results of a cone penetrometer test (CPTu) from the test site (the cone stress is q_t, i. e., it has been corrected for pore pressure on the shoulder). The cone profile identifies the original marsh surface at a depth of 4 m to 5 m and shows that the sand is relatively uniform down to 25 m and contains both dense ($q_t > 10$ MPa), compact (5 MPa $< q_t <$ 10 MPa) layers, as well as, occasionally, loose ($q_t <$ 5 MPa) layers. The pore pressure profile indicates that dilating zones exist at 18 m and at 23 m.

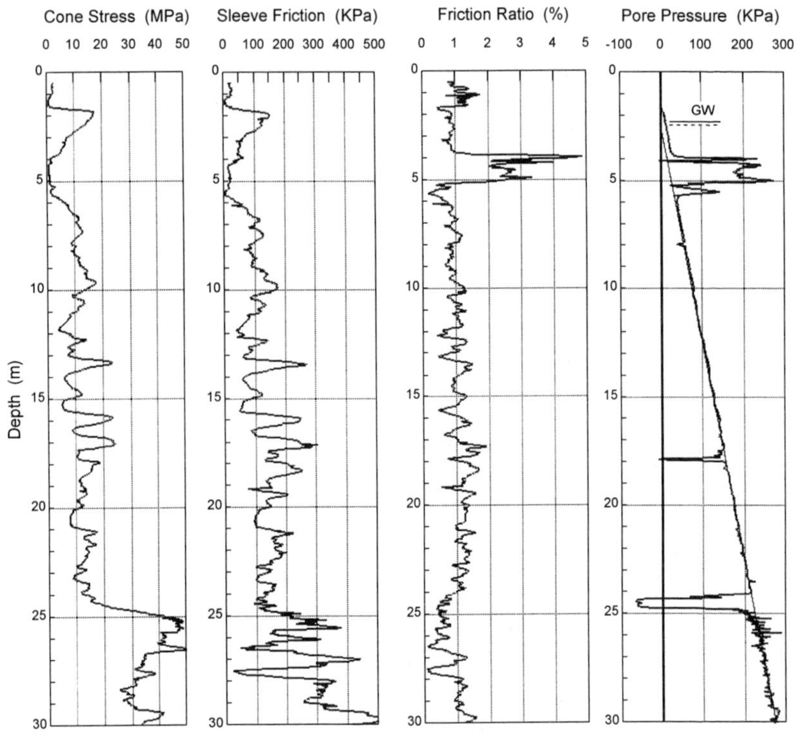

Fig. 1 CPTu Diagram

TEST OBJECTIVES

The objectives of the test are:

1. Confirm the presence of set-up at the site
2. Determine the pile capacity and the load-transfer (the distribution of shaft resistance along the pile and the pile toe resistance)
3. Determine the effect of a linearly variable Young's modulus in the composite (steel and concrete) pile cross section
4. Evaluate the magnitude of residual load

TEST PILES

The test piles, designated Pile 2 and Pile 3, consist of 21 m (70 ft) long, 357 mm (14-in), closed toe Monotube piles with the lower 7.6 m (25 ft) tapered 2.1 percent (0.25 in per foot) to a 203-mm (8-in) toe diameter. The wall thickness was 5.3 mm (0.209 in). The piles were installed by driving about 8 m apart and were filled with a 28 MPa (4,000 psi) concrete immediately after end of driving.

PILE INSTRUMENTATION

The test piles were instrumented with surface-mounted and embedment-mounted vibrating wire strain gages, which were placed at seven depth levels in the piles with two surface-mounted gages and one embedment gage at each level. The distances below the pile head and the depths below the ground surface to the gage levels of are given in Table 1.

The surface-mounted gages, Geokon Model VSM-4000, were attached before installation of the pile at diametrically opposed points and protected by a steel angle welded to the pile. The embedment gage, Geokon Model VSM-4200, was installed after driving, but before concreting, in a 25 mm (1.0 in) diameter pipe with a 3.2 mm (0.125 in) thick wall positioned at the center of the pile. A Geokon GK data-logger and read-out unit was used to record all gage data during the static loading tests.

Gage Level 1 was placed above the ground to serve as reference for determining the elastic modulus of the composite pile, steel and concrete, necessary for converting the strain to stress in the pile.

PILE INSTALLATION

The piles were driven with a Vulcan 010 hammer having a rated energy of 44 KJ (32.5 ft-kips). The penetration resistance at the end of driving was 35 blows/0.3 m for both piles.

Pile 2 met an obstruction at the depth of the original ground surface, later found to be three chunks of concrete. After several attempts to drive through the obstruction, the pile was withdrawn and the obstructions were dug up. The "hole" was then backfilled without compaction and the pile driving was resumed in the same location. In the process, the pile became slightly bent, misaligned by about 3 degrees, and the pile head was curled. As a consequence, some of the strain gages in Pile 2 did not survive the installation and other gages on the pile gave erratic values during the loading test.

TABLE 1 Gage locations in Pile 2 and 3

	Pile 2		Pile 3	
Gage Level	Distance Below Pile Head (m)	Depth Below Ground (m)	Distance Below Pile Head (m)	Depth Below Ground (m)
1	0.5	- -	0.7	- -
Ground surface	1.0	0	0.5	0
2	2.6	1.6	3.0	2.3
3	4.7	3.7	4.9	4.4
4	9.0	8.0	9.2	8.7
5	12.7	11.8	12.9	12.4
Start of taper	13.4	12.4	13.7	12.9
6	16.9	15.9	17.1	16.6
7	20.3	19.3	20.5	20.0
Pile toe	20.9	**20.0**	21.3	**20.5**

STATIC LOADING TEST

The loading tests was by Maintained-Load method with about 62 KN (14 kips) load increments, which were applied at intervals ranging from 6 minutes to 18 minutes. Loads were obtained by jacking against a loaded platform. The loads were applied by a single jack and determined by a separate load cell. Records were obtained by means of the data logger set to scan the gages at 3-minute intervals.

Static axial compression tests were performed three days (Pile 2) and twenty-one days (Pile 3) after the end of driving. Pile 2 was unloaded at the maximum load of 2,200 KN (225 tons), occurring at a movement of 18 mm (0.7 inch) when the load-movement curve reached the offset limit load (Davisson, 1972). Pile 3 was unloaded when the applied load reached 2,500 KN (280 tons), the maximum capacity of the reaction system.

RESULTS AND DATA REDUCTION

Load-Movement Curves

Fig. 2 presents the load-movement curves from the two static loading tests. The two diagrams have been supplemented with the elastic line (marked AE/L) drawn from the Davisson offset limit load.

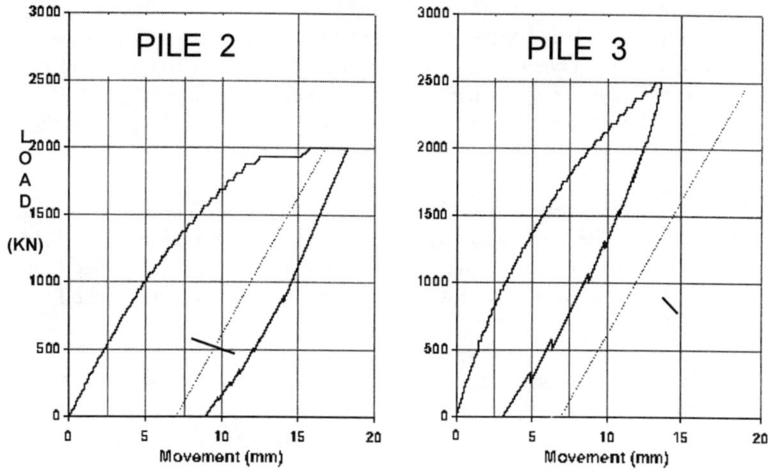

Fig. 2 Pile Head Load-Movement Curves. Piles 2 and 3

Pile 2 demonstrates a less stiff response to the loading than does Pile 3. The ultimate resistance was reached for Pile 2 at the applied load of 2,000 KN, as indicated by the offset elastic line intersecting the load-movement curve. In contrast, no offset-limit intersection occurred for the Pile 3 test curve, which means that the ultimate resistance exceeds the maximum test load of 2,500 KN (280 tons).

The soil conditions and the two test piles are in all respects equal. Therefore, the results of the static tests showing that maximum test load applied to Pile 3 is 25 percent larger than the capacity of Pile 2 indicates that the previously noticed (York et al., 1995) presence of set-up of the sand at the site is valid.

The observed setup is not due to dissipation of pore pressures induced during the driving of the piles (change of effective stress). In the sand at the site, pore pressures induced during the driving dissipate very quickly after the last blow.

Strain Measurements and Elastic Modulus

As mentioned, the initial difficulties in driving Pile 2 resulted in the loss of several of the surface-mounted strain-gages. Also, a few of the surface-mounted strain-gages in Pile 2 did not survive the continued pile driving, which was evidenced by their very erratic reading values. Moreover, while the surface-mounted gages in Pile 3 show readings of similar value, the gages in Pile 2 show differences indicating bending. During both tests, electronic noise affected many individual gage values.

Many of the same-scanning-event gage readings were recorded before, during, or immediately after a load increment was applied to the pile head. Therefore, the time log was carefully reviewed and each reading was placed with its proper load. Readings taken when the load increment was being applied were discarded. In Pile 2, bending caused some gage readings to become obviously erroneous, and these were also discarded as were the values affected by electronic noise.

Only a few of the records from Pile 3 had to be discarded. Fortunately, a large number of readings were taken and this redundancy ensured that reasonably reliable records are available from every load applied to the pile head and from every gage level. However, the accuracy of the Pile 2 test data is lower than that of the Pile 3 data.

Fig. 3 presents a diagram of the applied load versus measured strain. For Pile 2, the strain readings at Levels 1, 2, and 3 are almost identical, whereas the readings at these levels differ for Pile 3. This is due to that only little shaft resistance developed along Pile 2 in the loose backfill.

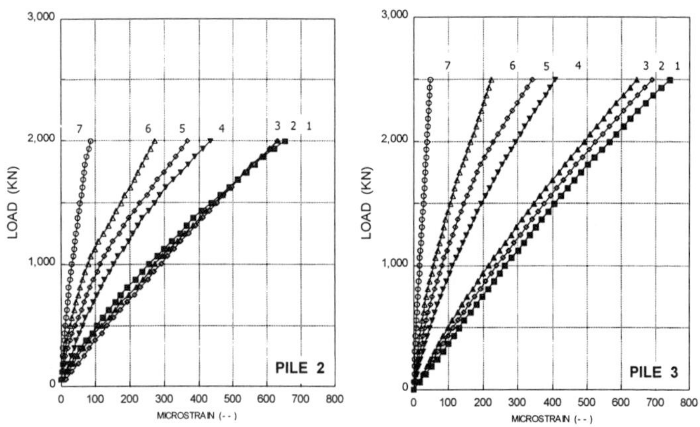

Fig. 3 Strain measured at each gage level. Piles 2 and 3

In contrast to the elastic modulus of steel, the elastic modulus of concrete is not a constant, but a function of the imposed strain. Over the large stress range imposed during a static loading test on a concreted pipe pile, the difference can be substantial between the initial and the final moduli for the composite steel and concrete material. Fellenius (1989) presented a method for the evaluation of the modulus from the actual test data and showed that:

1. For a pile taken as a free-standing column (case of no shaft resistance), the tangent modulus of the composite material is a straight line sloping from a larger modulus value to a smaller.

2. Every measured strain value can be converted to stress by its corresponding strain-dependent secant modulus.

The Fellenius (1989) method for the evaluation is summarized in the following.

The equation for the tangent modulus line is:

$$M = \frac{d\sigma}{d\varepsilon} = A\,\varepsilon + B \qquad (1)$$

which can be integrated to:

$$\sigma = A\,\varepsilon^2 + B\varepsilon \qquad (2)$$

However,

$$\sigma = E_s\,\varepsilon \qquad (3)$$

Therefore,

$$E_s = 0.5\,A\,\varepsilon + B \qquad (4)$$

where
M = tangent modulus
E_s = secant modulus
σ = stress
$d\sigma$ = $(\sigma_{n+1} - \sigma_1)$ = change of stress from one reading to the next
A = slope of the tangent modulus line
ε = measured strain
$d\varepsilon$ = $(\varepsilon_{n+1} - \varepsilon_1)$= change of strain from one reading to the next
B = y-intercept of the tangent modulus line (i.e., initial tangent modulus)

With knowledge of the strain-dependent secant modulus relation, the measured strain values are converted to the stress in the pile at the gage location. The load at the gage is then obtained by multiplying the stress with the pile cross sectional area.

When data reduction is completed, the evaluation of the test data starts by plotting the tangent modulus, that is, for each load increment, values of change of stress over change of strain are plotted versus measured strain. For a gage located near the pile head (in particular, if above the ground surface, e. g., for Gage Location 1 in the two test piles), the modulus calculated for each increment is unaffected by shaft resistance and the calculated tangent modulus is the actual modulus. For gages located further down the pile, the first load increments are substantially reduced by shaft resistance along the pile above the gage location and, therefore, the load change at the gage is smaller than the applied increment of load. Consequently, the modulus values calculated from the first load increments are large. However, as the shaft resistance is being mobilized down the pile, the calculated modulus values become smaller. When all shaft resistance above a gage location is mobilized, the calculated values are the tangent moduli for that gage location.

Presence of shaft resistance above a gage will make the tangent modulus line plot below the modulus line for an equivalent free-standing column. The larger the shaft resistance, the lower the line. However, the slope of the line is unaffected by the amount of shaft resistance above the gage location. The lowering of the line is not normally significant. However, it is a good rule to determine the tangent modulus line by placing one or two gages near the pile head where the strain is unaffected by shaft resistance. An additional reason for having a reference gage level with or even above the ground surface is that such a placement will also eliminate any influence from strain-softening of the shaft resistance. If the shaft resistance exhibits a strain-softening, the calculated modulus values will become smaller, and infer a steeper slope than the true slope of the modulus line. If the softening is not gradual, but suddenly reducing to a more or less constant post-peak value, a spike will appear in the diagram.

Fig. 4 presents the tangent modulus plots for the test piles. The upper diagram pair includes the data from all gage levels with the modulus axis in logarithmic scale (in order to fit all values into the diagram). The lower pair shows the modulus values in a linear scale for the four uppermost gage levels (those in the straight portion of the test piles). The values from the gage levels located in tapered portion of the piles (Levels 5 through 7) are too large to fit into the linear-scale diagram.

Fig. 4 shows that tangent modulus values for the four uppermost gages, Gages 1 through 5, converge to a straight line. The line represents the tangent modulus of the equivalent column representative for a pile unaffected by shaft resistance. The larger resolution lower diagram shows that the tangent modulus line is immediately established for Gage Level 1, very early for Gage Levels 2 and 3, and somewhat later for Gage Level 4.

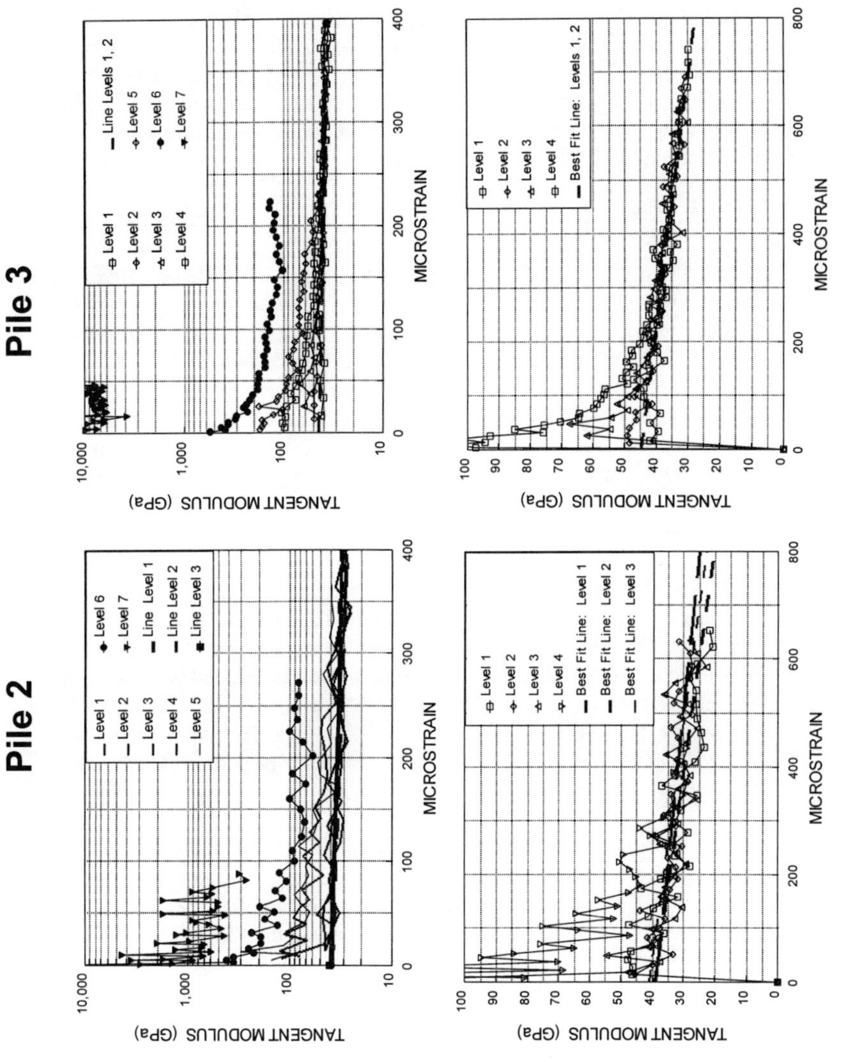

Fig. 4 Tangent modulus diagram. Piles 2 and 3

In contrast to the higher up gage levels, Gage Levels 6 and 7 do not show modulus values that converge to a modulus line. Even at the end of the test, the tangent modulus values are much larger than those for the straight portion of the piles. It is obvious, therefore, that the actual incremental increase of load at the Levels 6 and 7 is always much smaller than the load increment applied to the pile head. This indicates that the resistance along tapered portion of the piles continues to increase with increasing load. This observation will be addressed later in this paper.

Linear regression of the slope of the Pile 3 tangent-modulus line indicates that the initial tangent modulus is 44.8 GPa (the constant "B" in Eqs. 1 through 4). The slope of the line (coefficient "A" in Eqs. 1 through 4) is -0.021 GPa per microstrain ($\mu\ \varepsilon$). The resulting secant moduli are 40.5 GPa, 36.3 GPa, 32.0 GPa, and 27.7 GPa at strain values of 200 $\mu\ \varepsilon$, 400 $\mu\ \varepsilon$, 600 $\mu\ \varepsilon$, and 800 $\mu\ \varepsilon$, respectively. The Pile 2 strain values do not have the same accuracy, as mentioned, and the diagram shows that the three upper levels indicate slightly different tangent modulus lines. The average of the three lines indicates an initial tangent modulus of 40 GPa and the average slope of the line is about the same as found for Pile 3. The approximately 10 percent difference in modulus between Piles 2 and 3 is consistent with the fact that the concrete had cured for only three days at the time of Pile 2 static test as opposed to 21 days for the Pile 3 test.

The following illustrates the importance of establishing the strain dependency of the modulus: At the applied load of 2,400 KN, Pile 3 Gage Level 3 located at a depth of 4.9 m registered a strain of 625 $\mu\ \varepsilon$. At the same load, Gage Level 5 at a depth of 12.1 m registered a strain of 217 $\mu\ \varepsilon$. The strain values correspond to stress levels of 21.9 MPa and 9.6 MPa, respectively. If the 36 GPa average constant modulus had been used, the stress levels would have become 22.5 MPa and 7.8 MPa. That is, the shaft resistance acting between the two levels would have been determined with an about 10 percent to 20 percent error.

The modulus values for the composite section at Gage Levels 1 through 3 (Pile 2) and Gage Levels 1 and 2 (Pile 3) were correlated to the amount of concrete and steel in the pile in order to determine the concrete moduli. The latter moduli were then used in combination with the steel modulus and the slightly different steel area and concrete areas at Gage Levels 6 through 7 in the tapered section to determine the proportional modulus relation for these gage levels.

Load Distribution

The evaluated coefficient "A" and constant "B" are combined in Eq. 4 to determine the secant modulus relation. The relation is used with the strain data to calculate the stress and load at each gage for each load applied to the pile head.

Fig 5 shows the strain gage readings transferred to load and plotted against depth to give the load distribution in Piles 2 and 3 as evaluated from the measurements during the static loading test.

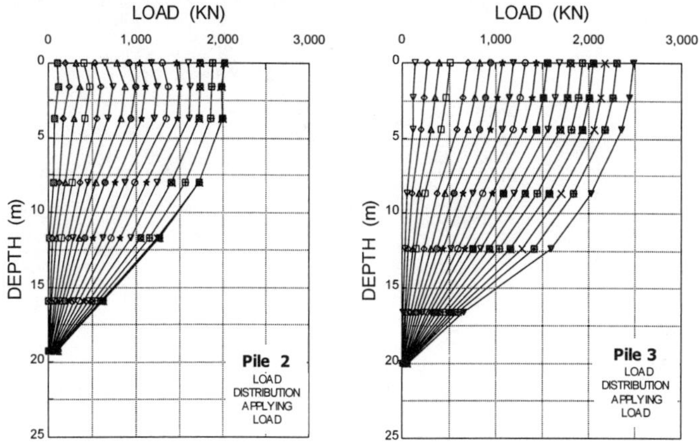

Fig. 5 Load distribution for each load applied to the pile head. Piles 2 and 3

Along the straight shaft portion of the two piles, the calculated distributions of load in the pile are almost identical. This is demonstrated in Fig. 6, where the load distribution curve of Pile 2 showing the values for the gages in the straight portion of the pile has been shifted over (translated) to Pile 3. This means that no set-up of shaft resistance developed along the straight length of the piles .

In contrast, the evaluated distributions within the tapered portion differ for the piles. For Pile 2, the resistance is smaller as indicated by its steeper load distribution curve. Obviously, the 25 percent larger resistance shown for Pile 3 as opposed to the capacity of Pile 2 is obtained along the tapered length of the pile. This means that the increase of capacity (set-up) determined for the time period of three through twenty-three days is not due to increase of shaft resistance expressed by means of the beta coefficient. Instead, the results indicate that the set-up occurred along the taper section and as a consequence of increased soil stiffness during the three-week wait between the two static loading tests, expressed by means of the toe bearing coefficient.

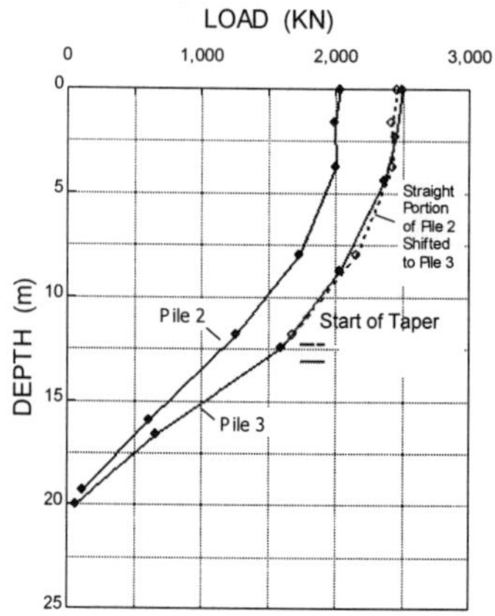

Fig. 6 Load distribution. Comparison between Piles 2 and 3

Residual Load and Soil Parameters

The measured load distributions are not the true load distributions in the piles, as the piles are subjected to residual load. The residual load is generated by fully mobilized negative skin friction along an upper portion of the pile in equilibrium with positive shaft resistance along a lower portion and toe resistance. For the two test piles, the point of equilibrium—the neutral plane—appears to be located shortly below the bottom of the straight portion of the pile, i. e., within the tapered portion of the pile.

Where a straight-shaft pile is subjected to residual load, above the neutral plane, the true load distribution is located at half the distance to the vertical through the pile head load, and is easily constructed (e. g., Altaee et al., 1992). However, below the neutral plane and in a tapered portion, no similarly easy construction is possible. Here, the true load distribution was determined in a series of trial and error using static analysis based on effective stress principles. The analysis was performed using the UniPile program (Fellenius and Goudreault, 1997; details are available on Internet address www.unisoftltd.com).

First, the true shaft resistance along the straight shaft portion of the piles down to Gage Level 5 was determined and employed in a back analysis using effective stress to determine the beta coefficients, which gave β equal to 0.40—same for both piles.

The conical shape makes the effective stress analysis less direct. It conceivably increases the resistance along the pile shaft as opposed to straight piles. Nordlund (1963) suggested a taper modification factor to use to increase the unit shaft resistance in sand for conical piles. The Nordlund correction factor is a function of the taper angle and the soil friction angle. The taper angle of 1° (0.25-inch/foot) of the subject piles in a sand with a 33° to 37° friction angle would give a Nordlund correction factor ranging from about 2 through about 5.

Fellenius (1996) proposed to use a more direct calculation method consisting of dividing up the soil into sub-layers of some thickness and at the bottom of each such sub-layer project the donut-shaped diameter change. The projected area is treated as an extra pile toe, which analysis is similar to the analysis of a step-taper pile. The shaft resistance is calculated using the mean diameter of the pile over the same "stepped" length. The shaft resistance over each such particular length consists of the sum of the shaft resistance and the extra-toe resistance.

In the continued back-analysis of the test results, the beta-coefficient established for the sand above gage level 5 was assumed applicable also to the soil along the tapered portion of the piles. The assumption is valid considering the uniformity with depth of the soil at the site evident in the CPTu diagram (Fig. 1) and the observation that no set-up developed along the straight length of the piles.

At Gage Levels 5, 6, and 7, the analysis is controlled by the condition that the measured loads are equal to the calculated values of residual load minus the calculated values of true loads. This leaves the analysis of the full length pile with only the toe bearing coefficient, N_t, unknown (applicable to the donut projection of the taper). Calculations were made at every 0.3 m length of the pile and resulted in the following soil parameters for the two test piles: beta-coefficient equal to 0.4 and N_t-coefficients of 60 and 75 for Piles 2 and 3, respectively.

The results of the analysis is presented in Fig. 7 for both Piles 2 and 3, showing the measured and true load distributions, and the difference between these, which is the residual load acting on the pile at the start of the static loading test. The distributions shown are for the maximum test load applied to the piles, which for Pile 2 is the pile capacity defined by the offset limit.

The strain measurements do not include strain caused by the residual load. However, the 'residual strain' could be added and a new modulus line be determined. This refinement would have negligible effect on the modulus line determined from Gage Levels 1 through 3 used for determining the tangent and secant moduli. At Gage Levels 4 and 5, the strain due to residual load is small in relation to the strain measured at the end of the test, which means that the residual load has negligible effect. In contrast, in the middle of the tapered length, Gage Levels 6, the load determined from the strain measured at the end of the test and the calculated residual load before the start of the test are about equal. At the load of 2,500 KN applied to Pile 3, for example, the measured strain is 224 microstrain resulting in a "measured load" of 656 KN. Considering that the measured strain developed when the pile was preloaded by the residual load and that the secant modulus decreases with the strain, the "measured load" is overestimated. On the assumption that the residual strain is equal to the measured strain, 224 microstrain would result in a measured load reduced to 620 KN. Applying this value to the analysis results in a 10 KN to 15 KN reduction in the true load and a similar increase in the residual load with respect to the first determined values. One could go through a few additional loops of calculation toward the end results of a slightly smaller measured load, a slightly larger residual load and a moderately affected true load. However, the precision gained would not warrant the substantial effort required to correct for the residual load..

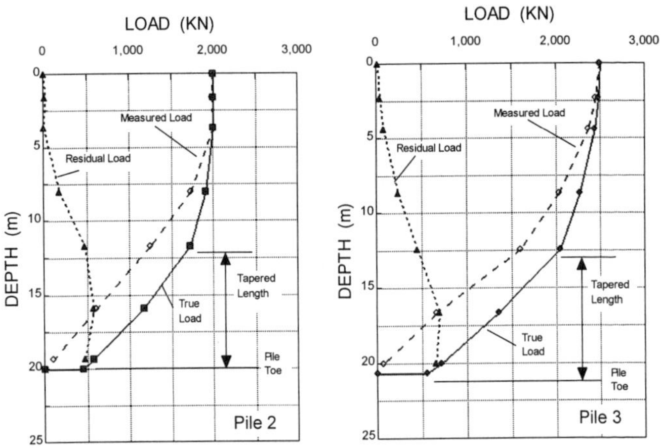

Fig. 7 Distribution of measured load, true load, and residual load. Piles 2 and 3.

CONCLUSIONS

The maximum load applied to Pile 3, tested twenty-three days after installation was 25 percent larger than the capacity of Pile 2 tested after three days. This confirms the experience of earlier tests at the site that the pile capacity increases over time due to set-up. Because the maximum load applied was smaller than the ultimate resistance, the actual magnitude of the set-up was not established.

Because all driving-induced pore pressures have dissipated before the static loading test on Pile 2, the observed set-up is not due to change of effective stress.

The measured strains were used to calculate the tangent and secant moduli for the steel-concrete cross section, enabling calculation of the loads in the piles during the test—the measured loads.

The point of equilibrium (neutral plane) between downward and upward acting residual load forces on the pile was determined to be close to where the pile cross section changed from a straight section to a tapered section.

Because of the presence of residual load, the measured loads indicated, falsely, that the soil resistance acting against the tapered section of the piles was very small. The analysis determined the magnitude and distribution of the actual shaft and toe resistances as well as of the residual load proving that most of the pile capacity was developed in the bottom portion of the pile agreeing well with effective stress principles. In the tapered section, at the maximum applied test load, the resistance was governed by fully mobilized positive shaft resistance and a movement-dependent "toe resistance component". The analysis results suggest that the set-up was due to a stiffening of the soil and not to increase of shaft resistance.

The following effective stress soil parameters were established for the two test piles: beta-coefficient equal to 0.4 and N_t-coefficients of 60 and 75 for Piles 2 and 3, respectively. Note, the N_t-coefficients represent the resistance mobilized at the respective toe movements.

APPENDIX I — REFERENCES

Altaee, A., Fellenius, B. H., and Evgin, E., 1992. Axial load transfer for piles in sand. I: Tests on an instrumented precast pile. Canadian Geotechnical Journal, Vol. 29, No. 1, pp. 11 - 20.

Davisson, M. T., 1972. High capacity piles. Proceedings of Lecture Series on Innovations in Foundation Construction, American Society of Civil Engineers, ASCE, Illinois Section, Chicago, March 22, pp. 81 - 112.

Fellenius, B. H., 1989. Tangent modulus of piles determined from strain data. The American Society of Civil Engineers, ASCE, Geotechnical Engineering Division, 1989 Foundation Congress, Edited by F. H. Kulhawy, Vol. 1, pp. 500 - 510.

Fellenius, B. H., 1996. Basics of foundation design. BiTech Publishers, Richmond, BC, 134 p.

Fellenius, B. H. and Goudreault, P. A., 1997. UniPile Version 3.0 for Windows, User Manual, Unisoft Ltd., Ottawa, 64 p.

Nordlund, R. L., 1963. Bearing capacity of piles in cohesionless soils. American Society of Civil Engineers, ASCE, Journal of Soil Mechanics and Foundation Engineering, Vol. 89, SM3, pp. 1 - 35.

York, D. L., Brusey, W. G., Clemente, F. M., and Law, S. K., 1995. Set-up and relaxation in glacial sand. American Society of Civil Engineers, ASCE Journal of Geotechnical Engineering, Vol. 120, No. 9, pp. 1498 - 1513.

APPENDIX II — SYMBOLS USED

M = tangent modulus

E_s = secant modulus

σ = stress

$d\sigma$ = $(\sigma_{n+1} - \sigma_1)$ = change of stress from one reading to the next

A = slope of the tangent modulus line

ε = measured strain

$d\varepsilon$ = $(\varepsilon_{n+1} - \varepsilon_1)$= change of strain from one reading to the next

B = y-intercept of the tangent modulus line (i.e., initial tangent modulus)

DEEP FOUNDATION CAPACITY – WHAT IS IT?

George G. Goble[1], M.ASCE and Mohamad H. Hussein[2], M.ASCE

Abstract: As foundation design becomes more structured by codes the question of axial strength definition becomes more important. Most deep foundation elements, when tested to soil failure, do not exhibit a clearly defined failure load. But the design procedure requires that a specific ultimate capacity be taken from the load test results. The load test result is also affected by the procedure used in load testing and particularly by the loading rate. Loading procedures are discussed and suggestions are made for standardization of these procedures. Failure load definition is also discussed and some commonly used methods are reviewed. It is suggested that more consideration should be given to the requirements of the structural engineer and that load testing and load test result evaluation methods should be standardized.

INTRODUCTION

Capacity is usually the basic consideration in the design of deep foundations. But, what exactly do we mean by capacity? The basic function that is usually assigned to a deep foundation is to transfer a load to deeper, stronger soil layers or in a few cases to limit and control settlements in soft soils. We presume that if the foundation has inadequate strength it will displace, probably slowly, into the ground. However, to make design decisions we must have a precise definition of inadequate strength and thus we imply that the foundation will be adequate until some load level is exceeded at which point it will fail. The nature of structural design is such that this approach is necessary. However, the determination of strength is not so simple and clear cut.

Structural design was converted to Load and Resistance Factor Design (LRFD) in several design codes beginning in the early 1960's. This approach to design has developed into a procedure where failure modes are considered separately from

---

[1]Principal, GRL and Associates, Inc., 5398 Manhattan Circle, Suite 100, Boulder, CO 80303
[2]Partner, GRL and Associates, Inc., 8008 South Orange Avenue, Orlando, FL 32809

serviceability limits and in some countries it has become known as Limit States Design. However, until recently LRFD was not used for the design of any geotechnical facilities in the United States. Today, LRFD is used in foundation design in the design codes of AASHTO, ACI, AISC, and PDCA in the United States. Design codes based on LRFD are also available in Canada, Australia, and the European Community.

When LRFD is used it is implied that well-defined methods are available to determine strength. In the case of deep foundations, there are two failure modes due to axial loads, structural failure and soil failure. Structural failure can be reliably predicted with currently available methods for strength prediction from structural design methods. This paper is primarily concerned with the definition of failure as limited by soil failure.

Of course, axial strength is not the only design limitation. Other considerations include lateral behavior, settlement, downdrag, etc. These considerations may be considered to be strength limitations or they may be performance limitations that do not include failure but only some type of unsatisfactory performance. The latter type of limitation is referred to as a serviceability limit state. In this paper, limit states other than the axial strength due to the soil failure condition will not be considered.

Deep foundation design, installation, and quality control is unusual in that it is possible to statically test to failure and if the failure is a soil failure, certainly the most common case, the pile can still be used in service. The load is applied to induce some substantial displacement and when removed a portion of the displacement is recovered. If a still greater load is applied the load-displacement curve will approximately follow the previous unloading curve and then may continue to higher loads and displacements. An example of this type of load test curve with multiple load cycles is shown in Figure 1. In other cases, the load displacement curve may be nearly plastic as illustrated in Figure 2.

TESTING PROCEDURE

Given the conditions of Figure 1, how should the load test be performed? Thirty years ago the most common load test procedure was to impose an increment of axial load and hold it for an extended period of time after displacement stopped (or became less than a specified level). The load increments applied were a fairly large fraction of the design load. Due to the long hold periods the time required to perform the test was measured in days. This required 24 hour monitoring of the test operations and, thus, the cost of the test was increased as was the possibility of an error in the test performance. When field tests must be monitored 24 hours a day it can become difficult to staff the test and it may be necessary to use less well-qualified test personnel. The performance of a good quality load test requires engineers and technicians that are well trained in the performance of the test.

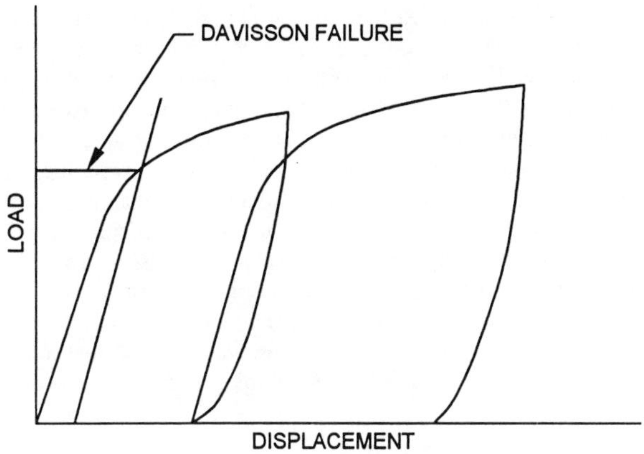

Figure 1 Load Test on Primarily End Bearing Pile in Sand with Two Loading Cycles.

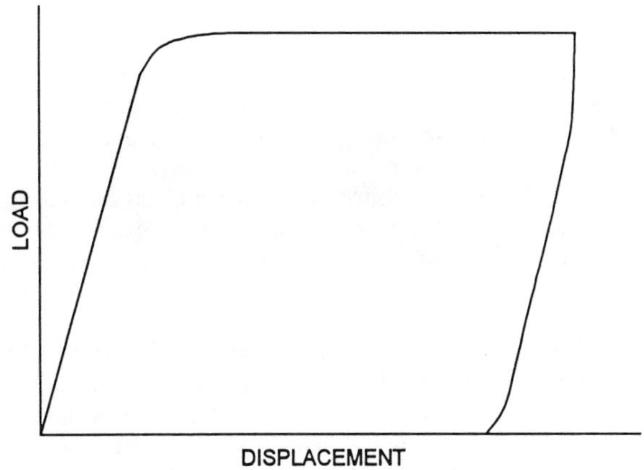

Figure 2 Load Test on Friction Pile.

If a large increment of load is applied at widely spaced time intervals there is a tendency for failure to occur at the level of load application. A particular load level is carried. The next level is applied rapidly and the load level is reached followed immediately by gradual pile failure. This behavior may bias the test result.

During the early days of the research project at Case Western Reserve University (Goble et al 1975) Constant Rate of Penetration Tests (CRP) were used extensively. In this test procedure, a particular, continuous rate of displacement is selected and imposed at the pile top. Such an approach is attractive, particularly to the structural engineer since it is similar to procedures used for testing structural elements in the laboratory. Rate of displacement for pile testing is not easily selected and if the rate is fairly fast the capacity can depend on the rate with increased capacities at higher loading rates. A CRP test has the additional disadvantage that it is difficult to explain to typical field personnel that must execute it. Then test specifications appear in which unnecessary complexity in testing procedures are specified. For example, specifications have been used in which a non-pulsating pump is required. During the Case Project it was found that such refinements were unimportant. This should not be surprising since if this type of loading were critical deep foundations would be sensitive to vibration and such considerations would have to be considered in design.

Modern practice in the United States has moved to a test procedure where the load is applied in smaller, closely spaced intervals. ASTM D1143 includes the specification of a Quick Load Test Method in which the load is applied in increments of 10 to 15% of the proposed design load and held for a 2-1/2 minute time interval. With this approach a test is usually completed in one to two hours. Such a test procedure is easily defined for the field personnel and it will probably be applied correctly.

In the recent past, it has become more common in Western Europe to perform static load tests where the loading rate is controlled, frequently with feedback controlled systems. It is argued that such a system type is required to obtain meaningful results. This topic has become of great interest among European foundation engineers. Because of the difference in pile stiffness the selection of a loading rate is complex. If soil failure is limiting it would be desirable to use a reasonably constant rate of loading of the soil. A very flexible pile will load the soil more slowly than a very stiff pile for a given rate of pile top displacement.

The selection of loading rate is important because time dependent effects can influence the definition of pile capacity if widely varying loading rates are used. An example of this problem is shown in Figure 3. First, a maintained load test was performed on this driven pile located in West Palm Beach, Florida (Goble and Likins 1973). The 18-inch prestressed concrete pile was driven through silica sand to a blow count of 40 blows per foot at a depth of 35 feet. After completion of the maintain load test a CRP test was performed with a rate of 0.02 inches per minute. This rate was taken from experience testing steel piles that was considered to be successful. Figure 3 shows the large difference in the failure load between the two tests. This problem seems to be more serious accentuated in this case where the pile is lightly loaded and, therefore, relatively stiff compared to the soil. In addition, it is quite short and this accentuates the problem.

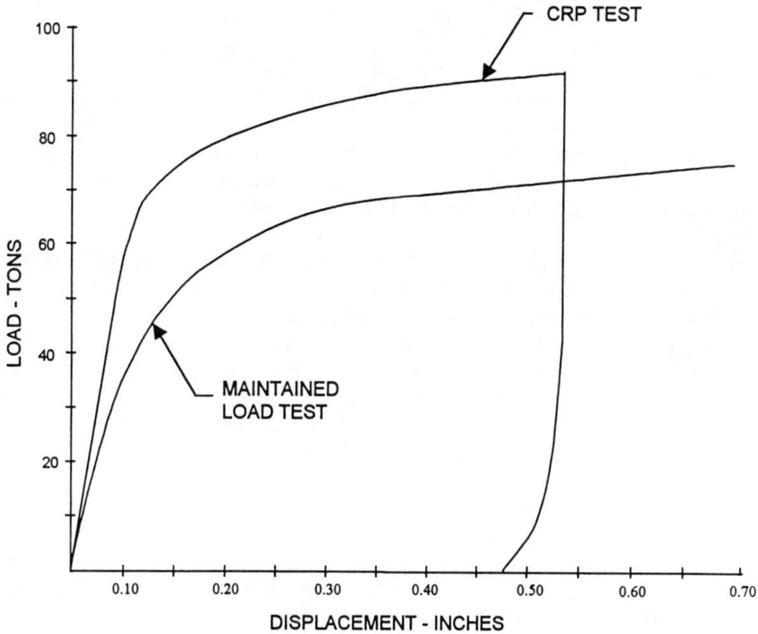

Figure 3 Static Load Test Results for 18 Inch Square Concrete Pile by Two Test Methods.

The failure load was reached in about five minutes in the CRP test with this loading rate. The use of the ASTM Quick Test procedure can provide a simple approximate control of loading rate. Since load increments are selected as fractions of the design load the time to failure will be approximately constant for a given factor of safety. In the example of Figure 3, the time to failure would have been on the order of 45 minutes or more. Had the test have been performed at that rate the load test curve would have probably more nearly followed the path of the maintained load test.

It would be useful to perform some load tests on a single pile at a variety of rates to try to quantify the loading rate where the strength is no longer strongly sensitive to rate. If such tests were performed for a variety of soils it would give an idea of the sensitivity of the test result to loading rate.

Longer holding times for the loads have been popular. The goal of such test methods has been to obtain a measure of long term behavior. However, a test of two or three days will not measure settlement behavior since a much longer loading period is required. It is important and advantageous to use the same loading procedure for all tests to make the comparison of results easier.

What can cause problems in pile load testing? The load must be accurately measured. This implies that a load cell should be used. The load cell must be treated as a sensitive measuring instrument and it must be re-calibrated frequently. Today the use of load cells has become quite common. These load cells are usually of good quality and can be expected to give good results. The use of jack pressure has been shown to under-measure capacity due to jack friction (Fellenius et al, 1983).

Displacement must be carefully and accurately measured with the gages supported on unyielding supports that are unaffected by the motion of the test pile. Displacement measurements are sometimes quite inaccurate due to non-rigid supports for the displacement measurement transducer, slack in the displacement system and other problems. Inaccurate displacements can produce large changes in the defined failure load for most of the evaluation procedures. Errors in either load or displacement measurements can seriously effect the defined capacity. A static load test requires care in its performance if it is to produce accurate results.

Perhaps the most serious problem in all of static load testing is the failure to continue the loading to failure. A distressing percentage of all pile load tests are performed to only twice the design load. If failure does not occur the test is deemed a success and production pile installation begins. It is entirely possible that the pile failure load may be several times the required ultimate load. Less stringent installation criteria could be used in such cases. All load tests should be carried to the highest load possible.

LOAD TEST EVALUATION

Some load test curves will have a sharply defined failure such as illustrated in Figure 2 and the failure load is easily defined. The example load test curve of Figure 1 does not have a well-defined failure load and this is the more common case. However, it is necessary that a defined failure load be obtained from the test data. A large number of different failure criteria have been proposed for determining the "ultimate" capacity of a deep foundation element. In fact, in most cases, a true ultimate capacity does not exist. Particularly for deep foundations where there is substantial toe capacity, the load test will exhibit considerable strain hardening.

As a graphic example, the Michigan State Highway Commission (1965) reported that one static load test was carried to a displacement of about two feet. At the end of this large displacement the load had approximately doubled beyond the point on the curve where the stiffness showed a large decrease. This characteristic becomes more pronounced as the deep foundation element increases in diameter. In some large drilled shafts with tips in sand the test load displacement will be strongly rounded with a very large displacement required to even approach something like ultimate load. It is important that a larger view be taken in the evaluation of the load test result and maybe one that is more in agreement with the evaluation of other construction materials such as steel or concrete.

It is important that the requirements of the structural engineer be considered since the performance of the foundation is often limited by the requirements of the structure. In past and current practice there is usually limited communication between the structural and geotechnical engineer and often the one has little understanding of the requirements of the other. Consider two examples, pile support in the design of a commercial building and a freight-handling wharf. In the case of the commercial building, the ratio of dead to live load on the foundation will be large. On the other hand, the wharf dead load will be very small, so the forces designing the piles will consist mostly of live load. Thus, the commercial building load can be determined quite accurately and a smaller factor of safety is justified. (Remember that part of the factor of safety belongs to the structural engineer.) For the wharf, the design load will be almost entirely live load and it can be highly variable. A larger factor of safety is appropriate. In the case of the commercial building, the tolerable settlement may be quite small while the dead load on the wharf will be small producing little settlement and, in addition, the structure may be quite settlement tolerant.

Test Evaluation Methods

The Offset Method (also referred to as the Davisson Method) defines failure as the intersection of the elastic stiffness of the pile (PL/AE) drawn through an offset on the abscissa that depends on the pile diameter. The method is mostly, but not completely, independent of judgement since the elastic modulus of the pile material, particularly for drilled shafts may not be easily established. Since the offset is defined by the pile diameter the capacity is dependent on the pile diameter.

The Diameter Ratio Method is particularly popular in Europe. The failure load is defined as the load associated with a pile top set of 10% of pile diameter. This method yields relatively high failure loads particularly for large diameter piles. It does not make much sense for the structure that larger displacements are tolerated when the foundation pile diameter is increased. Obviously, this method favors large pile diameters.

The Chin Method assumes that – at least near the failure load – pile top deflection and the quantity pile top deflection divided by pile load fall on a straight line and that therefore the load-set curve is a hyperbola. The inverse of the straight line defines the failure load. In many cases failure is extrapolated using this method. Probably, among the most commonly used methods in the world, the Chin method produces the highest capacity value of any method for a given load set curve. It is frequently used in the Far East.

Kulhawy, (1997) has suggested that the load test curve be defined by two linear portions at the beginning and at the end with a curved portion between. The portion between the two straight lines is defined by the points and the slopes at the junctions between the straight and curved portions. This recommendation was developed for use with drilled shafts and Kulhawy reports that available drilled shaft data fits this description very well. Rather predictable relationships between the transition points are reported based on available data.

The Fixed Total Set Method is proposed as a means of being more responsive to the requirements of the structural engineer. The pile capacity is defined by three points on the load test curve. With this data, the structural engineer could specify an appropriate displacement where the capacity is defined. For example, failure may be defined at a pile top set of 25 mm. This method is quite simple and reasonable. The absolute value at which capacity is to be taken may be variable or several values may be picked from the load test curve as an indication of elasticity and plasticity of the load set curve. (For example, loads may be picked at pile top sets of 20, 40 and 80 mm. A plunging pile would then be characterized as 900/1000/1020 kN while a displacement pile in granular soil could have a test load result of 800/1100/1250 kN.) This method would form a basis for a rational way in which the interaction of structure and foundation can be calculated. To the authors knowledge, this method of characterizing pile capacity with several values has not been utilized to date.

CONCLUSIONS

Pile design is continually being refined and improved and pile design capacities are increasing. As the process is refined it is appropriate that methods for performing and evaluating the results of static load tests should become more standardized. Some load testing and evaluation methods have been summarized and discussed critically. As a minimum, some limits should be placed on loading rates to limit the effect of time dependent deformations. Selected methods of load testing must also be selected with consideration of the practical problems that arise in the field.

The method that is used for evaluating the load test result must also be standardized. A major consideration in selecting the evaluation method is the requirements of the structural engineer.

APPENDIX-1

REFERENCES

ASTM 1143-81 (Reapproved 1994), Standard Test Method for Piles Under Static Axial Compression Load, ASTM 100 Barr Harbor Drive, West Conshohocken, PA 19428.

Fellenius, B.H., A. J. O'Brien, R. E. Riker, and G. T. Tracy, (1983), Dynamic Monitoring and Conventional Pile Testing Procedures, Proceedings, Symposium on Dynamic Measurements of Piles and Piers, ASCE Convention, Philadelphia, PA.

Goble, G. G. and G. E. Likins, 1973, Static and Dynamic Pile Test in West Palm Beach, Florida, Report to the Florida Department of Transportation and The Federal Highway Administration, Submitted by Case Western Reserve University, Cleveland, Ohio.

Goble, G. G., F. Rausche, and G. E. Likins, 1975, Bearing Capacity of Piles from Dynamic Measurements – Final Report, Prepared for the Ohio Department of Transportation and the Federal Highway Administration, Case Western Reserve University, Cleveland, Ohio.

Kulhawy, F. H. and A. Hirany, 1997, What is Foundation Failure, Proceedings of the International Conference on Foundation Failures, The Institution of Civil Engineers, Singapore.

Michigan State Highway Commission, 1965, A Performance Investigation of Pile Driving Hammers and Piles, Lansing, Michigan.

PERFORMANCE EVALUATION OF AN INSTRUMENTED TEST PILE CLUSTER

Edward L. Hajduk[1] and Samuel G. Paikowsky[1]

ABSTRACT: As part of a research project investigating time dependent pile capacity gain, a heavily instrumented test pile cluster was installed at a bridge reconstruction site in Newbury, Massachusetts. The test pile cluster consisted of (i) three-instrumented test piles, (ii) a surrounding ground piezometer field, and (iii) a complex data acquisition array to monitor the instrumentation. An extensive testing program, which included various static, dynamic, and Statnamic tests, was carried out over a lengthy time period to determine the pile capacity with time.

Electrical resistance, piezo-electric, piezo-resistive, and vibrating wire instruments were used throughout the cluster to record acceleration, displacement, pressure, and strain within the test pile cluster. Primary test pile instrumentation consisted of an alternating pattern of piezometers and strain gages. Auxiliary instrumentation within the piles included accelerometers and total pressure gages. The ground piezometer field consisted of a total of 10 vibrating wire pressure transducers installed within normally consolidated and silty sand soil layers. Four separate data acquisition systems were assembled into an on-site data acquisition array to monitor the instrumentation throughout the testing period.

This paper details the maximum loading conditions that each individual test pile was subjected to over the course of the testing program and evaluates the performance of the individual gage models and designs, instrument types, and data acquisition systems. Conclusions are drawn regarding (i) the reliability of the various gages and gage types, (ii) the performance of similar gage types, (iii) the selected measurement frequencies for each gage type, and (iv) the performance of the data acquisition array during the harsh conditions of driving and testing over time.

[1]Graduate Research Assistant and Professor, respectively, Geotechnical Engineering Research Laboratory, Department of Civil and Environmental Engineering, University of Massachusetts Lowell, Lowell, MA 01854.

INTRODUCTION

A full-scale instrumented test pile cluster was installed and tested at the University of Massachusetts – Lowell (UML) Geotechnical Engineering Research (Newbury) Site in Newbury, Massachusetts as part of a long-term research investigation of pile capacity gain of driven piles with time. The Newbury Site was located adjacent to a bridge reconstruction site on Route 1 on the Newbury – Newburyport, Massachusetts border. This site was used previously by the University of Massachusetts - Lowell for helical anchor and Multiple Deployment Model Pile (MDMP) testing (Paikowsky and Peach, 1995 and Paikowsky and Hart, 1998). In order to aid in the analysis of these tests and the instrumented test pile data, Paikowsky and Chen (1998) completed a comprehensive study of the engineering parameters of the subsurface at the site based on extensive field and laboratory testing. A locus plan of the site is presented in Figure 1.

The test pile cluster was comprised of three instrumented test piles, a surrounding ground piezometer field, an on-site data acquisition array with remote data retrieval capabilities, and an extensive testing schedule. The instrumented test piles consisted of one 32.4cm (12.75in) diameter x 1.3cm (0.5in) wall thickness x 31.4m (103ft) long closed ended steel pipe end-bearing pile (Test Pile #1 or TP#1), a 32.4cm (12.75in) diameter x 24.4m (80ft) long closed ended steel pipe friction pile (TP#2), and one 35.6cm (14in) square x 24.4m (80ft) pre-stressed concrete friction pile (TP#3). The ground piezometer field, which consisted of 7 piezometers within a normally consolidated clay layer and 3 piezometers within a silty sand layer, was installed around the planned test pile locations to monitor excess pore pressure buildup and dissipation due to test pile installation. The on-site data acquisition array of four separate data measurement and control systems was located within a storage trailer that was placed adjacent to the test pile and ground piezometer locations. Paikowsky and Hajduk (1999) described the design and construction of the test pile cluster in detail. A plan view of the test pile cluster at the Newbury Site is presented in Figure 2. A cross-section of the site, showing the layouts of the test piles and the ground piezometer field relative to the location of TP#1 and the soil profile, is presented in Figure 3. Additional instrumentation, consisting of soil accelerometers and total pressure cells, was also installed at the Newbury Site in a joint research project with GeoDelft of Delft, Holland. The description of this instrumentation is beyond the scope of this paper and will be addressed in future publications (Hajduk et al., 2000).

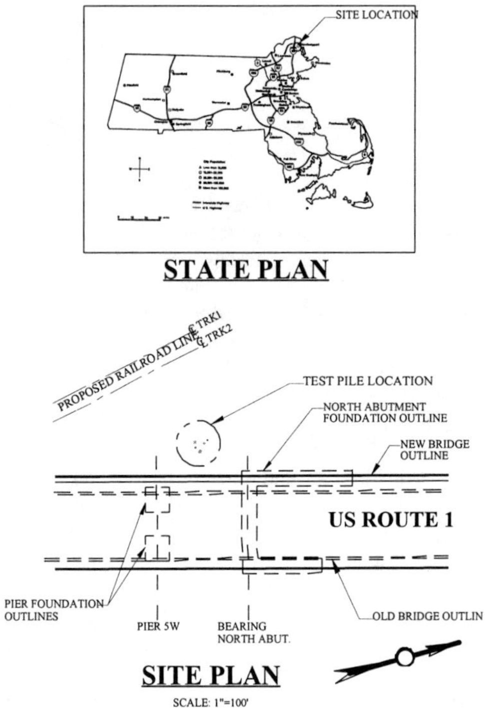

Figure 1. University of Massachusetts Lowell's Geotechnical Engineering Testing (Newbury) Site (after Paikowsky and Hajduk, 1999).

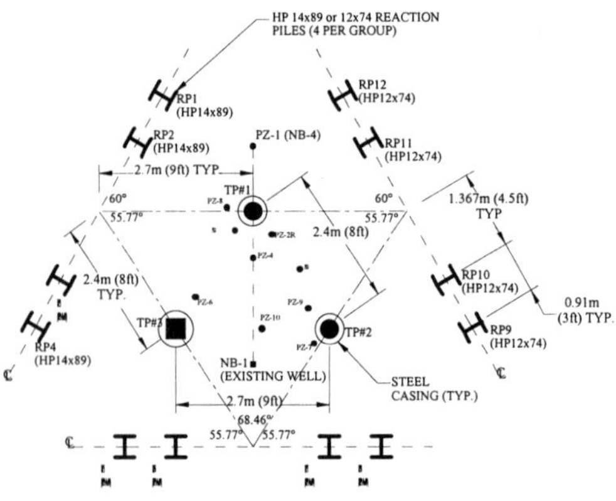

Figure 2. Newbury Site Plan View (Paikowsky and Hajduk, 1999).

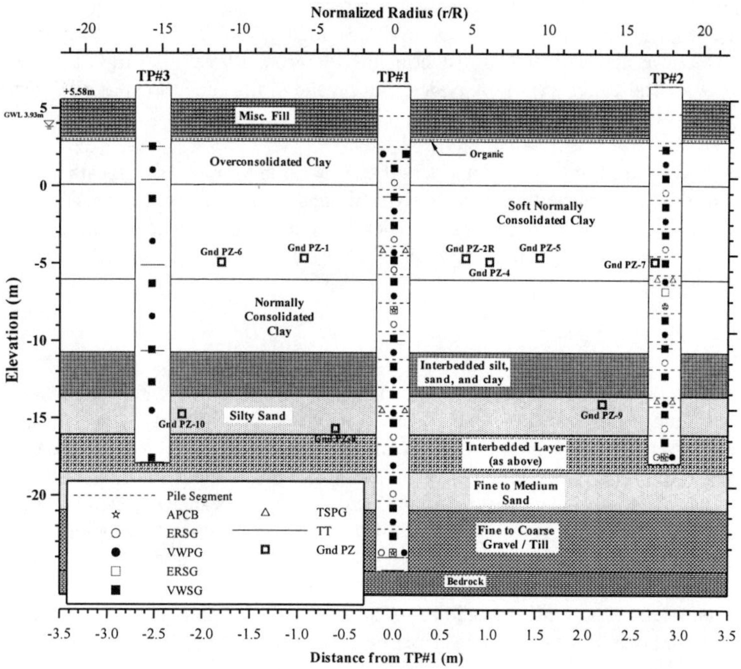

Figure 3. Newbury Site Soil Profile with Test Pile Layouts.

The soil stratigraphy of the Newbury Site is typical of the conditions found in the Boston area. The general soil profile at the pile testing location (from ground surface downward) consists of the following soil strata: 8 feet (2.4 meters) of granular fill composed of very dense, brown sand and gravel intermixed with frequent concrete fragments, overlies a thin layer (approximately 1 foot (0.3 meters)) of highly compressible organic silt and peat. In order to protect the test piles, these layers were excavated and cased with steel pipe sections. Below the fill and organics is an approximately 45 feet (13.7 meters) thick deposit of a marine clay, known as Boston Blue Clay. The clay consists of approximately 9 feet (2.7 meters) of very stiff to medium stiff, overconsolidated layer (crust), over 20 feet (6.1 meters) of soft normally consolidated clay and 16 feet (4.9 meters) normally consolidated clay. An interbedded deposit of silt, sand, and clay approximately 9.5 feet (2.9 meters) thick underlies the clay. Below this interbedded deposit is a layer of silty sand approximately 8 feet (2.4 meters) thick. Another interbedded deposit of silt, sand, and clay approximately 7.5 feet (2.3 meters) thick underlies the silty sand. Below this interbedded deposit is a layer approximately 8 feet (2.4 meters) thick of medium dense to dense, fine to medium sand. Underlying the fine to medium sand is a dense glacial till consisting of medium dense to dense, fine to coarse sand and gravel, with traces of silt and rock fragments. Good quality, unweathered bedrock was encountered at approximately 100 feet below the ground level (elevation 81.7 feet) (Paikowsky and Chen, 1998). A typical soil profile of the site at the test pile cluster is presented in Figure 2.

Over the course of a testing period that lasted approximately 2 years, the test pile cluster was subjected to repeated dynamic and static tension, compression, and Osterberg Cell load tests of various procedures and load rates. In addition, Test Piles #2 and #3 were subjected to incremental Statnamic load tests at the end of the test period.

INSTRUMENTATION

A total of nine different instrument types were installed within the test pile cluster to measure various parameters associated with pile capacity gain with time. A summary of the various instruments and the associated measured parameters is presented in Table 1. Primary instrumentation for the test piles consisted of vibrating wire piezometers and strain gages. Electrical resistance piezometers and strain gages were also installed within the two steel pipe piles to supplement the vibrating wire gages. The layout of each pile consisted of an alternating pattern of piezometers and strain gages, with each piezometer located halfway between two-strain gage sets. This pattern would allow for correlation of pore pressure dissipation and average capacity gain along discrete pile segments.

Table 1. Summary of Measurement Parameters, Instruments, and Instrument Types for the Instrumented Test Piles (after Paikowsky and Hajduk, 1999).

Parameter	**Measuring Instrument**	**Gage Type**	**Instrument Abbreviation**
Excess Pore Pressure Buildup and Dissipation	Piezometers	Electrical Resistance	ERPG
		Vibrating Wire	VWPG
Total Radial Pressure	Liquid Total Soil Pressure Gages	Pressure measurement w/vibrating wire piezometer	TSPG
Dynamic Pile Testing – Interior Stress Wave Measurements	Strain Gages	Electrical Resistance	ERSG
	Accelerometers	Piezo-resistive	APCB
Load Distribution Along Pile During Static Testing	Strain Gages	Vibrating Wire	VWSG
		Electrical Resistance	ERSG
	Tell-Tales	Slender Rod	TT
Telltale Displacement	Displacement Transducers.	Electrical Resistance	LVDT
Pile Temperature	Thermistor	Electrical Resistance	TH

Paikowsky and Hajduk (1999) explain in detail (i) the rationale behind the instrument type selection, (ii) the selection of the individual gage models, and (iii) the various gages, gage assemblies, and installation of these assemblies into the test pile cluster. A summary of the selected gage models for the various instrumentation types installed within the test pile cluster and the corresponding number of gages per pile is listed in Table 2.

Table 2. Test Pile Cluster Instrumentation Summary.

Gage Type	**APCB**	**ERPG**	**ERSG**	**TSPG**	**VWPG**	**VWSG**	
Model	PDA Piezo-resistive	Kistler Type 4100	MG LWK-06-250W-34	Geokon Model 4500H	Geokon Model 4500	Geokon Model VSM-4200	Geokon Model VCE-4200
Test Pile #1	2	7	2	4	12	28	0
Test Pile #2	2	6	2	4	8	20	0
Test Pile #3	0	0	0	0	4	0	6
Gnd PZ	0	0	0	0	10	0	0
Total	**4**	**13**	**4**	**8**	**34**	**48**	**6**

In addition to the above-mentioned instrumentation, an Osterberg Load Cell was installed at the base of Test Pile #1 and the SmartPile System was installed in Test Pile #3. The performance of these systems is not evaluated in this paper.

Instrumentation Identification

An instrumentation identification system was developed to provide classification of the test pile, relative location, orientation, and instrument type for each gage in the test pile cluster. The identification system also assisted in the development and installation of data acquisition systems. The instrumentation identification system used the following nomenclature:

X-Y-Z-Q

Where:

X = Test Pile Number
Y = Segment Number (for TP#1 & 2) or Gage Number (for TP#3)
Z = Gage.
Q = Orientation within the pile (TP#1 & 2 only)

X may be 1, 2, or 3, depending on the test pile. **Y** may be segments 1 through 16 (Test Pile #1), segments 1 through 12 (for Test Pile #2), or gage 1 through 6 (Test Pile #3). The variables for **Z** are listed in Table 1. Variable **Q** may be A or B, according to the orientation of the instrument within the test pile and is discussed in detail by Paikowsky and Hajduk (1999). Two examples of the instrumentation identification system are the following:

1-6-VWSG-A and **2-10-TSPG-B**

The first identification is of vibrating wire strain gage (VWSG) A of segment 6 in Test Pile #1. The second identification is that of the total soil pressure gage (TSPG) B of Test Pile #2 in segment 10. A TH located in front of the gage identification would specify the thermistor of that gage. The ground piezometers were labeled as **Gnd PZ**, with the specific number of the gage following PZ. For example, ground piezometer 3 was labeled as **Gnd PZ-3**.

DATA ACQUISTION ARRAY

Four separate data acquisition systems, assembled into an onsite data acquisition array, were used to monitor the various gages used within the test pile cluster. The core of this array consisted of three separate data acquisition systems, with each system assigned to monitor one of the three primary gage types (vibrating wire, electrical resistance, and dynamic instrumentation). This arrangement was chosen to simplify the setup and design of the data acquisition array and to provide backup measurements should one system fail. The fourth data acquisition system provided flexibility to the array by (i) recording non-essential instruments during pile testing, thereby freeing resources from the core systems and (ii) recording various instruments prior to core data acquisition installation.

The primary data acquisition system used throughout the testing was the Micro-10 (Model 8020) Datalogger, assembled and marketed by Geokon, Inc. of Lebanon, NH. Although the Micro-10 is capable of recording the measurements of a variety of instrumentation, it was only used to monitor the vibrating wire gages and their thermistors. A total of 2 Micro-10 Dataloggers, labeled ADAS-1 and ADAS-2, each with two attached Model 8032 multiplexers, were used during the project. Each Micro-10 was equipped with a modem to allow for remote data retrieval.

A CR9000 Measurement and Control System, manufactured by Campbell Scientific of Logon, UT, was used to record the majority of electrical resistance instrumentation used in the test pile cluster. This instrumentation consisted of the electrical resistance pressure gages (ERPG) and strain gages (ERSG), the linear variable displacement transducers (LVDT), and the thermistors of various gages. The CR9000 is a 16 bit, modular, portable, multiprocessor data acquisition system that can sample at rates up to 100 K readings per second.

Proprietary software developed by Campbell Scientific, named PC9000, was used to control and communicate with the CR9000. PC9000 is a Windows based program that provides program generation, data graphics and analysis, and data retrieval. A modified BASIC programming language, entitled CR BASIC, was used to program the CR9000 for a variety of data processing and analysis routines.

A Hewlett Packard Data Acquisition System (HP DAS) was used at various times during the testing program to record the static load test instrumentation and thermistors. The HP DAS was primarily used to monitor the LVDT's and electrical resistance load cell of a static load test conducted on Test Pile #3 and to record a limited number of thermistors before and after the driving of Test Pile #1. The visual programming language HP VEE was used to trigger the scanning multi-meter, set the number of channels to be monitored and sampling frequency, display the real time data to the computer screen, and store the data to the hard drive with a time stamp (Paikowsky and Hart, 1998).

The final data acquisition system of the instrumented test pile cluster was a Pile Driving Analyzer (PDA). The Pile Driving Analyzer is a dynamic pile testing data acquisition and analysis system developed by Pile Dynamics Incorporated of Cleveland, Ohio. For the test pile cluster, the PDA was used to measure the exterior and interior dynamic response of the tested pile during driving, restrike, and Statnamic tests. This included the interior piezo-resistive accelerometers (APCB). The PDA used during the Test Pile Program consisted of a 486SLC 25MHz processor with 8MB RAM and a 240MB hard drive. A summary of the various data acquisition systems and their assigned instruments is presented in Table 3.

Table 3. Test Pile Instrumentation Summary.

Gage Type	APCB	ERPG	ERSG	TH	TT (LVDT)	TSPG	VWPG	VWSG
DAS	PDA	CR9000	CR9000	Micro-10 CR9000 HP DAS	CR9000 HP DAS	Micro-10	Micro-10	Micro-10

Data Acquisition Array Configurations

The four data acquisition systems listed above were incorporated into an array that allowed for flexible configuration and measurement frequency settings. In addition, the system had limited capability to record additional instrumentation installed at the site (pile top load cells, ground accelerometers, dilatometers, miniature piezocones). The array was arranged into three major configurations during the test pile-testing program. Each configuration corresponded to the installation of a specific test pile. Figure 8 shows the schematics for the various configurations of the data acquisition array. Paikowsky and Hajduk (1999) detail the various data acquisition array configurations.

Measurement Frequencies

In order to adequately follow the phenomena of pore pressure buildup and dissipation and pile capacity gain, minimum measurement frequencies were established for each gage and testing event. The recording capabilities of the data acquisition systems and experience of the University of Massachusetts – Lowell in investigating time dependent pile capacity gain with the Multiple Deployment Model Pile (MDMP) (Paikowsky and Hart, 1998) influenced the selection of these recording frequencies. A summary of these frequencies related to testing activity is provided in Tables 4.

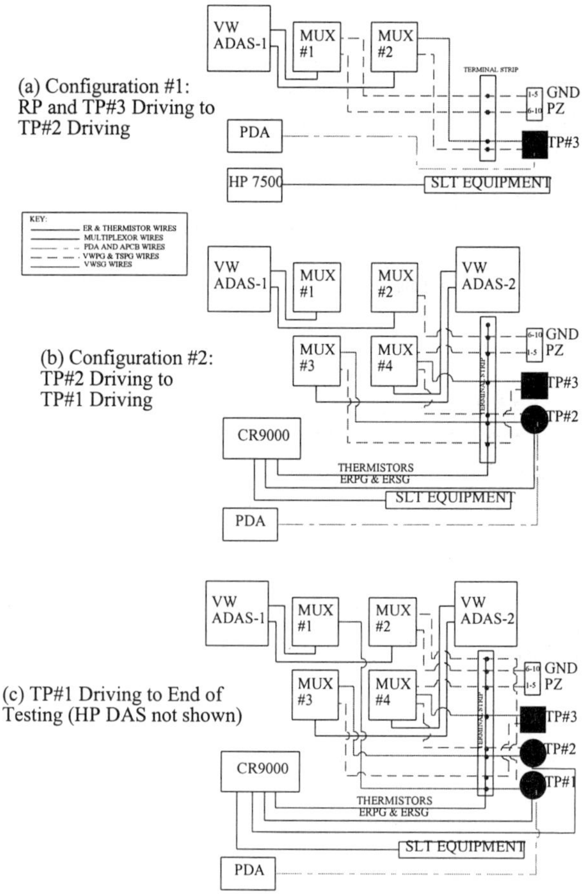

Figure 8. Data Acquisition Array Configurations (Paikowsky and Hajduk, 1999).

Table 4. Frequency of Measurements for Various Test Pile Cluster Instruments (after Paikowsky and Hajduk, 1999).

Test Pile	Gage Type	Testing Activity				
		During Driving and Restrikes (+ 12 hrs)	During SLT (+ 12 hrs)	EOD to 1st Restrike or SLT	Between Restrikes and SLT (before EOD + 7 days)	Between Restrikes and SLT (after EOD + 7 days
Tested Pile	APCB	10000 Hz	N/A	N/A	N/A	N/A
	ERPG	1 Hz	1 Hz	1 Hz	1rdg/15min	1rdg/hour
	ERSG	10000 Hz[1]	1 Hz	1rdg/15min	1rdg/15min	1rdg/hour
	VWPG	1rdg/30ec	1rdg/30ec	1rdg/30ec	1rdg/15min	1rdg/hour
	VWSG	1rdg/30ec	1rdg/30ec	1rdg/15min	1rdg/15min	1rdg/hour
	TSPG	1rdg/30ec	1rdg/30ec	1rdg/30ec	1rdg/15min	1rdg/hour
Adjacent Pile(s) & Gnd PZ Field	APCB	N/A	N/A	N/A	N/A	N/A
	ERPG	1 Hz	1 Hz	1 Hz	1rdg/15min	1rdg/hour
	ERSG	1rdg/15min	1rdg/15min	1rdg/15min	1rdg/15min	1rdg/hour
	Gnd PZ	1rdg/30sec	1rdg/30ec	1rdg/30ec	1rdg/15min	1rdg/hour
	VWPG	1rdg/30ec	1rdg/30ec	1rdg/30ec	1rdg/15min	1rdg/hour
	VWSG	1rdg/15min	1rdg/15min	1rdg/15min	1rdg/15min	1rdg/hour
	TSPG	1rdg/30ec	1rdg/30ec	1rdg/30ec	1rdg/15min	1rdg/hour

[1] During Blows. Between blows only 1 Hz between blows.

EOD = End of driving, SLT = Static Load Test

TESTING CONDITIONS

The instrumentation within the test pile cluster was subjected to a variety of harsh conditions during the testing program. Three different types of tests caused these conditions: dynamic, static, and kinetic (i.e. Statnamic). Each test directly produced an assortment of accelerations, strains, and stresses on the test piles. The shearing and remolding of the surrounding soil caused by these events produced changes in (i) the lateral soil pressure on the pile and (ii) the pore pressures adjacent to the test piles and surrounding ground piezometer field. In addition to these conditions, each pile top was subjected to extreme seasonal temperature variations during the testing period.

An identification system was developed to identify the test number and type for each loading event. This identification system would provide a quick reference that indicated the test pile number, test type, and test number. The following nomenclature was used to identify the testing:

TP#A – B - C

Where:

A = Test Pile Number

B = Test Type
C = Test Type Number

A may be 1, 2, or 3, depending on the test pile. The variables for **B** are listed in Table 5. Examples of typical test identification are the following:

TP#2SLT1 or TP#3R5

The first identification is of Static Load Test #1 of Test Pile #2. The second identification is of Restrike #5 for Test Pile #3.

Table 5. Test Pile Load Testing Abbreviations.

Test Type	Abbreviation
End of Driving	EOD
Static Load Test (Axial Compression)	SLT
Static Load Test (Axial Tension)	SLT(T)
Dynamic Restrike	R
Osterberg Load Test	OST

The dynamic testing consisted of pile installation (driving) and dynamic restrikes. Dynamic restrikes were performed in accordance with standards of the Massachusetts Highway Department (1995). Up to four different static load test methods were conducted on the test piles: short duration, slow maintained, static cyclic, and Osterberg (Test Pile #1 only). All static load tests were conducted as axial compression with the exception of one tension test on Test Pile #2 (TP#2SLT3T) and the Osterberg Load tests. The slow maintained and short duration tests were conducted according to the standards of the Massachusetts Highway Department (1995). The static-cyclic tests were conducted according to criteria developed in a concurrent research project presented by Paikowsky et al. (1999).

The maximum forces, accelerations, and total set for each test pile were determined from the individual data of each separate test. Tables 6 through 8 shows these values for the individual tests for Test Piles #1 through #3, respectively.

Table 6. Maximum Load Testing Conditions of Test Pile #1.

Test	EOD	OST1	OST2	OST3	OST5	OST6	OST7	OST8	OST9	SLT10	SLT11	SLT12
Date	5/29/97	5/30/97	5/30/97	6/5/97	6/13/97	7/11/97	10/6/97	12/11/97	1/11/98	1/13/98	1/14/98-1/15/98	1/20/98
Maximum Load: (kN)	2473	246	285	381	444	589	672	719	456	615	4024	3567
Maximum Acceleration (g):	920	N/A	N/A	N/A	N/A	N/A	N/A	N/A	N/A	N/A	N/A	N/A
Number of Blows:	575	N/A	N/A	N/A	N/A	N/A	N/A	N/A	N/A	N/A	N/A	N/A
Total Set (mm):	N/A	26.2	10.0	11.5	5.6	12.5	10.3	16.7	30.9	40.1	36.6	1.1

Table 7. Maximum Load Testing Conditions of Test Pile #2.

Test	EOD	R1	SLT1	R2	SLT2	SLT3 (T)	R3	SLT4	SLT5	R4	SLT6	SLT7	SLT8	SLT9	STN
Date	2/24/97	2/25/97	2/28/97	3/4/97	3/12/97	3/14/97	3/20/97	3/24/97	5/26/97	5/29/97	10/3/97	10/22/97 10/24/97	10/27/97	10/28/97	4/23/98
Maximum Load: (kN)	2438	5222	623	5138	717	433	2789	720	729	3430	Unk.[1]	853	815	826	1044
Maximum Acc. (g):	1031	928	N/A	1056	N/A	N/A	1231	N/A	N/A	699	N/A	N/A	N/A	N/A	4
Number of Blows:	307	14	N/A	6	N/A	N/A	17	N/A	N/A	5	N/A	N/A	N/A	N/A	4
Total Set (mm):	N/A	216	14.5	64	29	19	91	28.4	30.9	67	5.5	32.3	20.7	21.4	6

[1] Load was not recorded (incomplete test)

Table 8. Maximum Load Testing Conditions of Test Pile #3.

Test	**EOD**	**R1**	**R2**	**R3**	**SLT1**	**R4**	**R5**	**R6**	**SLT2**	**R7**	**SLT3**	**SLT4**	**SLT5**	**STN**
Date	12/13/96	12/14/96	12/17/96	12/27/96	1/14/97	1/15/97	3/4/97	3/20/97	3/26/97	5/29/97	12/2/97-12/4/97	12/4/97-12/5/97	12/5/97	4/23/98
Maximum Load: (kN)	3358	2624	2660	2553	924	2931	3265	3140	1219	4764	1347	1217	1258	2303
Maximum Acc. (g):	218	118	101	140	N/A		217	222	N/A	426	N/A	N/A	N/A	7
Number of Blows:	484	11	6	9	N/A	5	4	9	N/A	6	N/A	N/A	N/A	6
Total Set (mm):	N/A	52	52	52	56	24	18	15	29	24	27	21	50	110

INSTRUMENTATION PERFORMANCE

Performance evaluation of any project is a subjective process that can be influenced by (i) the chosen assessment areas and (ii) the definition of success and failure. Due to the magnitude of instrumentation within this project, it was decided to limit the presented performance evaluation to three areas: (i) instrument durability, (ii) comparisons between similar instrument types, and (iii) measurement frequencies. By limiting the evaluation to these three areas, qualitative and quantitative assessments could be made of individual gages, the various gage types, and the project as a whole.

Instrument Durability

Instrument durability is crucial for a project with the extended testing period. It was therefore important to include an assessment of instrument dependability in the evaluation of test pile performance. Resilience of the instrumentation within the test pile cluster was assessed in terms of overall durability, i.e. the ability to withstand the entire testing period, and performance durability, the ability to take data during actual test events.

The assessment of overall durability was made with a comparison between the number of instruments that worked at the start and end of the testing period for each major test project component and the project as a whole. Instrumentation was broken down into the selected gage types previously described and listed in Table 1. The percentage of working instruments at the end of the testing, labeled as % Durable, was computed to allow comparisons between the major instrument types. The results of this analysis are presented in Table 9. The results of Table 9 show that the majority of vibrating wire instrumentation could withstand the harsh conditions throughout the testing period. Failure of the various vibrating wire instruments did not align with specific testing events.

The majority of electrical resistance gages did not withstand the testing program. Since failure of all the electrical resistance strain gages and piezo-resistive accelerometers did not occur during specific tests, their failures were attributed to the intrusion of water into the two steel pipe piles. The cause of the electrical resistance strain gages malfunctioning during static load testing before complete failure was not discerned. Only the electrical resistance piezometers continued to function throughout the project duration.

Reviewing measurements over the course of the testing period assessed performance durability for each gage. The review of the individual gages showed that, when functioning, the piezo-resistive accelerometers and electrical resistance piezometers and strain gages performed to expectations. This review also showed that several vibrating wire gages experienced sudden measurement changes and erratic readings during and between specific testing events.

Table 9. Test Pile #1 Instrumentation Durability Summary.

Test Pile	Stage	Gage Type					
		APCB	**ERPG**	**ERSG**	**TSPG**	**VWPG**	**VWSG**
TP#1	Start of Testing	2	7	4	4	12	28
	End of Testing	0	7	0	4	11	28
	% Durable	**0**	**100**	**4**	**100**	**92**	**100**
TP#2	Start of Testing	2	6	4	4	8	20
	End of Testing	0	5	0	3	7	20
	% Durable	**0**	**83**	**0**	**75**	**88**	**100**
TP#3	Start of Testing	0	0	0	0	4	6
	End of Testing	0	0	0	0	4	6
	% Durable	**N/A**	**N/A**	**N/A**	**N/A**	**100**	**100**
Gnd PZ	Start of Testing	0	0	0	0	10	0
	End of Testing	0	0	0	0	10[1]	0
	% Durable	**N/A**	**N/A**	**N/A**	**N/A**	**90**	**N/A**
TOTAL	Start of Testing	4	13	8	8	34	54
	End of Testing	0	12	0	7	31[1]	54
	% Durable	**0**	**92**	**0**	**88**	**91**	**100**

[1] Gnd PZ-2 was replaced during the testing period.

Sudden measurement changes were experienced by all three vibrating wire gages: piezometers, total soil pressure gages, and strain gages. Each specific jump occurred during dynamic testing events, either during driving and/or during restrikes. An example of a sudden measurement jump is shown in Figure 9. During the first restrike of Test Pile #2 (TP#2R1), gage 2-6-VWPG-B experienced a sudden jump in pore pressure of approximately 84kPa (12psi), then continued with pore pressure dissipation. Comparison with the measurement of the adjacent gage 2-6-VWPG-A showed that this jump was unrealistic. The sudden measurement increase was subtracted from the subsequent measurements (see area marked with arrow in Figure 9). Comparison with the adjacent gage showed similar pore pressure dissipation measurements after the correction, thereby confirming the assumption that the sudden increase in pore pressure was unrealistic. A total of 4 out of 96 vibrating wire gages experienced significant measurement jumps during dynamic testing.

The sudden changes in the vibrating wire measurements were attributed to relaxation of the gage wire tension caused by dynamic loading. This relaxation, also referred to as slippage, causes a shift in the zero reading (i.e. zero drift) of the gage. Since vibrating wire gages convert the change in wire tension to an applied measurement, such as pressure, a change in the original or zero reading will cause a subsequent change in the measurement.

A number of vibrating wire gages also experienced erratic readings during testing events and over the testing period. An example of erratic gage reading experienced during testing is shown in Figure 10. In this example, the measurements of gage 3-4-VWSG-A are shown compared to the pile top load with time over the length of testing for TP#3 SLT1. The readings show both small and large fluctuations during both loading and constant load periods. Upon removal of these inconsistent readings, a clear measurement can be obtained (see Figure 10). Although minor erratic measurements over the testing period were recorded by the majority of vibrating wire instrumentation, 4 gages experienced continuous and significant measurement variations over the course of testing.

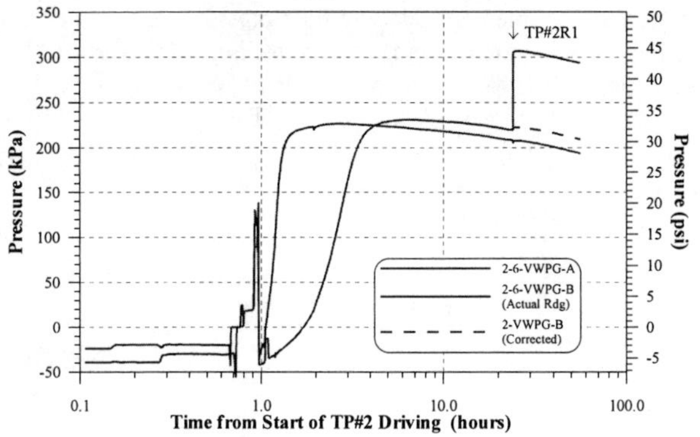

Figure 9. Example of Vibrating Wire Measurement Jump during Dynamic Testing.

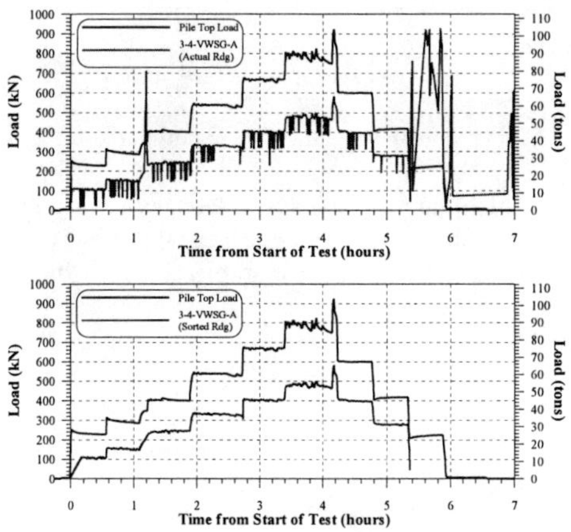

Figure 10. Example of Erratic Vibrating Wire Measurements.

Instrument Comparisons

Comparison of records from two similar instrumentation types would allow for performance assessment between the two major gage types used throughout the project, electrical resistance and vibrating wire. These two gage types were both used to measure pore pressure and pile capacity along the lengths of the two steel pipe piles, TP#1 and #2.

Comparisons between the two gages type measurements with time showed that each gage was capable of measuring pore pressure buildup and dissipation. A typical example of this comparison is shown in Figure 11. Figure 11 shows nearly identical pore pressure measurements with time for gages 1-6-ERSG-A and 1-6-VWPG-A after the driving of Test Pile #1. The shift in pore pressure buildup between 1 and 100 hours from the start of driving is associated with the locations of the gages on the pile, pile installation, and porous stone saturation and is not attributed to a difference between the gages.

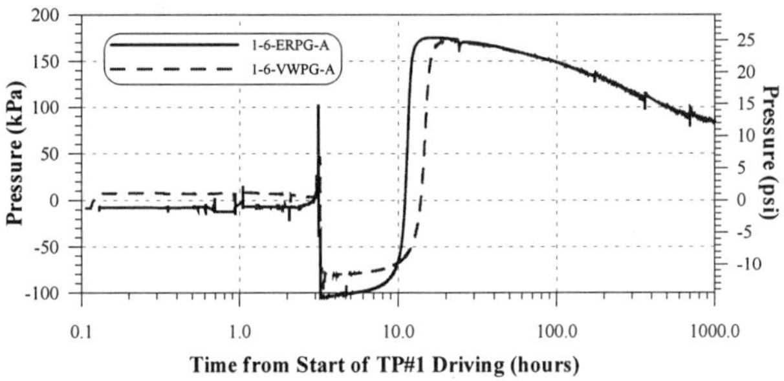

Figure 11. Typical ERPG and VWPG comparison measurements.

Examination of Figure 11 notes a difference in measurements between the two gages at approximately 3 hours from the start of driving. At this time, it appears that the ERPG records a higher pressure before each gage experiences a sudden drop in pressure. By plotting the same gage readings with depth, a better understanding of the measurement difference is observed. Figure 12 shows that during penetration into the overconsolidated clay layer, each gage experiences a loss in pressure. Before this loss of pressure occurs, a slight pressure buildup was registered by each gage. The ERPG, which is being recorded at a higher frequency than the VWPG, records more data points of the buildup. The lower recording frequency of the VWPG misses the majority of this buildup. Examination of similar comparisons between the two gage types for the installation of each of the steel pipe piles reveals comparable patterns of measurement.

A comparison of the performance of the electrical resistance and vibrating wire strain gages could not be conducted since there were no concurrent measurements involving the two gage types. The lack of simultaneous readings was a result of the failure of the ERSG's during static load testing and the inability of the VWSG's to record measurements during dynamic events.

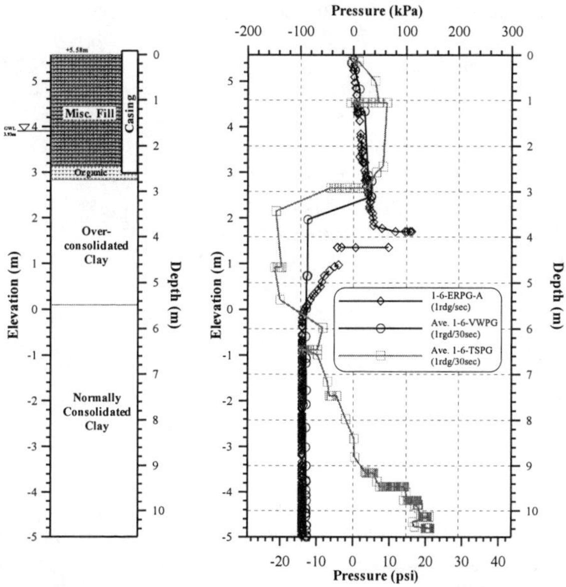

Figure 12. Typical ERPG, TSPG, and VWPG measurements with depth through the overconsolidated clay layer.

Measurement Frequencies

The measurements from each gage were examined to determine if the selected sampling frequencies for the various testing events (see Table 4) adequately monitored the desired phenomena. For the majority of test pile cluster instrumentation, the selected recording rates were adequate to measure the desired phenomena. However, rapid total and pore pressure buildup along the pile and in the ground piezometer field, caused by pile installation and other dynamic events, were not adequately recorded by the corresponding vibrating wire instrumentation. For example, comparison between the readings of electrical and vibrating wire piezometer types revealed that the reading rate of the vibrating wire instrumentation was not adequate to record the buildup of pore pressure during pile penetration into the overconsolidated layer (see Figure 12). Comparison of ERPG to adjacent TSPG recordings, also shown in Figure 12, revealed that the same problem existed for the total soil pressure measurements.

Piezometer Assembly Evaluation

Examination of the pile piezometers based on durability and comparative analysis also produced an evaluation of the piezometer assemblies used with the test piles. A consistent loss of pore pressure was observed for all test pile piezometers during instrument penetration into the overconsolidated soil layer. A substantial delay was also observed in the pile piezometers before an increase in pore pressure was recorded.

To determine the cause of this delay in pore pressure measurements, additional comparisons were made between adjacent pore and total pressure measurements. Typical comparisons between the ERPG, TSPG, and VWPG measurements are shown in Figures 12 and 13 with depth and time, respectively. In Figure 13, an initial decrease in pressure is observed in both gage types, a substantial time lag occurs before the pile piezometers register an increase while the total pressure cell records an almost immediate pressure increase. The pressure vs. depth plots in Figure 12 show that the total pressure cells begin to register pressure increases within the normally consolidated layers while the piezometers remain constant.

Based on these observations, it was concluded that a loss of saturation occurred within the piezometer assemblies during penetration into the overconsolidated clay, most likely due to dilation of the clay during shear caused by the installation. The only common element of both piezometer assembly designs was the size and material type of the porous stones (see Paikowsky and Hajduk, 1999). It was therefore assumed that the permeability of these porous stones was too high to prevent a loss of saturation due to dilation of the overconsolidated clay.

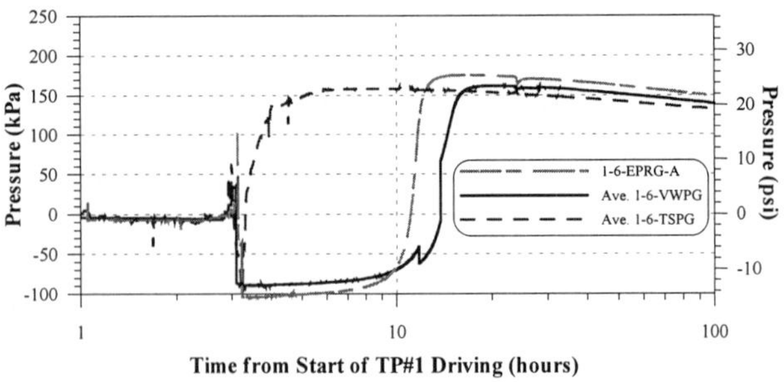

Figure 13. Typical ERPG, TSPG, and VWPG measurements with time.

Data Acquisition Array

The data acquisition array was able to provide near continuous measurements of the different gages employed throughout the 17 months of testing during the project. Although supplied with constant power and equipped with battery backups, several significant data gaps that ranged from several hours to 5 days were experienced during the testing period due to power loss at the site. Individual DAS battery backups were not adequate to operate the various components of the array during extended blackout periods. To prevent extended periods of time with no instrument recordings, the download of data from the array was adjusted during the testing program.

Although housed in a trailer adjacent to the test site, the data acquisition array was subjected to extreme temperature variations over the course of the 17 month testing program. These temperature variations did not affect the array performance.

Piezometer measurement comparisons revealed that the VWPG and TSPG did not effectively record the complete pore and total pressure buildup along the pile during pile installation. This limitation was not due to the performance of the array but rather due to the recording limitation of vibrating wire gages, which have a maximum rate of approximately 1reading per gage per second. In order for the vibrating wire piezometers to have a recording frequency comparable to the electrical resistance piezometers, a Micro-10 would have to be assigned to each VWPG. However, the addition of more Micro-10 dataloggers to the array would not be economically or practically feasible for the project.

The propriety software of each component data acquisition system used within the array provided sufficient flexibility to change instrument-recording frequencies according to the testing event. The finite memory capabilities of the individual data acquisition systems limited the capabilities of the array to store large quantities gathered at high measurement frequencies. Required downloads on data where then conducted at various intervals to prevent data loss. These intervals were dependent on recording frequency and array component. The finite amount of memory also affected the ability of certain components to perform multiple recording tasks. For example, the CR9000, which recorded the electrical resistance instrumentation, could not simultaneously record the ERSG's during driving and provide continuous monitoring of the ERPG's in the configuration used during the project. The ERPG's were therefore not recorded during actual pile driving, but during the periods between blows. This produced time gaps in the EPRG measurements. Examples of these time gaps as they relate to pile penetration are shown in Figure 12.

Aside from the loss of data from power outages and system limitations, several large blocks of data were lost due to operator inexperience and error. A data acquisition array of the size and complexity used for the instrumented test pile cluster requires a trained engineer to set recording frequencies, synchronize time measurements, and download the acquired data.

Measurement coordination between components of the data acquisition array was achieved through use of absolute date/time measurements and synchronization of the component time clocks.

CONCLUSIONS

The performance evaluation of the instrumented test pile cluster presented in this paper yielded several conclusions:

1. The vibrating wire instrumentation proved to be more durable to the testing and site conditions than the piezo-resistive and electrical resistance gages used throughout the project. All the piezo-resistive accelerometers stopped functioning during the testing period, which resulted in a failure rate of 100% (4 out of 4 gages). Of the electrical resistance components, all electrical resistance strain gages failed during the duration of testing, resulting in an overall failure rate of 43% (9 out of 21) for these gages. In comparison, only 4 out of 96 (4%) vibrating wire gages ceased recording over the testing duration.
2. An additional 8 out of the 96 vibrating wire gages suffered from significant erratic readings and/or zero drifts at various times over the course of testing. Two of these gages that experienced zero drift also experienced complete failure. When combined with the gages that failed, a total of 10 out of 96 vibrating wire (10%) did not function to expectations. This percentage was considered an excellent rate of success for this gage type.
3. The overall instrument failure rate for the instrumentation within the test pile cluster was 17% (21 out of 121 gages). This failure rate considered the gages that ceased operating (17 gages) and the erratic vibrating wire gages that did not completely fail (4 gages). The failure rate was considered excessive even for a project of this size and complexity.
4. Comparison of the vibrating wire and electrical resistance piezometers showed that the vibrating wire gages did not sample at a rate high enough to accurately capture the buildup of pore pressure in the overconsolidated layer along the piles during driving. This limitation was attributed to limitations of the gage type and not the data acquisition array.
5. Comparison of the adjacent pore and total pressure measurements showed that the porous stones used in the piezometer assemblies did not have a permeability that could prevent loss of saturation during driving through the overconsolidated clay layer. Therefore, there was a lack in gage response until the system could become saturated and take relevant measurements.

6. The data acquisition array performed adequately during the test pile project and was not affected by the site conditions. Data gaps were attributed to extended power outages and operator error.

Overall, the performance of the instrumented test pile cluster was considered to be outstanding when taking into account the length of test period and the amount and variety of testing involved. Lessons learned in this evaluation should be applied in the reuse of the two steel pipe piles.

ACKNOWLEDGEMENTS

The presented results are a part of a long-term research project supported by the Massachusetts Highway Department (MHD) and assisted by the Federal Highway Administration (FWHA). The authors would like to specifically acknowledge the assistance of Mr. Nabil Hourani and Mr. John Pettis of the MHD and Mr. Al DiMillio and Mr. Carl Ealy of the FHWA.

Also acknowledged are the support and cooperation of all the individual and companies associated with this research project, in particular:

Pile Driving Contractor: P.A. Frisco, Inc., Ms. Paula Frisco, President
Geotechnical Design Firm: Shannon and Wilson, Inc
Mr. Alex Feldman, Project Manager
Mr. Gerard Buechel, Senior Associate
Mr. Jim Panton, Project Engineer
Instrumentation Firm: Geokon, Incorporated
Mr. John McRae, Vice President
Mr. Alex Neuwirt, Electrical Engineer
Mr. Ken Alger, Machinist
Concrete Pile Donator: Mr. Charlie Guild of AM&FC.
UML Geotechnical Engineering Research Lab:
Mr. Gary Howe, Director of Laboratories
Mr. Leo Hart, Christopher Palmer, John Chen, Steve Tien, Mike Bachand, and Ms. Elena Demeris.
University of South Florida (Statnamic Testing):
Dr. A. Gray Mullins
Carl Lewis

APPENDIX -I References

Campbell Scientific, 1995, "CR9000 Measurement and Control System Instruction Manual", Campbell Scientific Inc., Logan, Utah.

Hajduk, E.L., 2000. "Theoretical and Full Scale Field Testing Examination of Pile Capacity Gain with Time", Doctor of Engineering Dissertation to be submitted to the Department of Civil and Environmental Engineering, University of Massachusetts at Lowell, Lowell, Massachusetts.

Hajduk, E.L. Paikowsky, S.G., Hölscher, P, and van der Pool, T., 2000. "Accelerations of a Driven Pile and the Surrounding Soil", paper under review for publication in the Proceedings of the 6th International Conference on the Application of Stress-Wave Theory to Piles, San Pao, Brazil, 2000.

Hewlett Packard, 1990. HP 75000 Series B/C, 5½-Digit Multimeter HPE1326B/E1411B User's Manual, Manual Number: E1326-90003.

Hewlett Packard, 1991. HP 75000 Series B, Mainframes HP E1300A and E1301AUser's Manual, Manual Number: E1300-90005.

Hewlett Packard, 1993. HP 75000 Series B, 16-Channel Relay Multiplexer Module HP E1343A/44A/45A/47A User's Manual, Manual Number: E1345-90003.

Hewlett Packard, 1994. HP 75000 Series B/C, 4-Channel D/A Converter Module HPE1328A User's Manual, Manual Number: E1328-90003.

Massachusetts Highway Department, 1995. "Standard Specifications for Highways and Bridges".

Model 8020 (Micro-10) Multi-Channel Datalogger Instruction Manual, 1996. Geokon, Inc., Lebanon, New Hampshire.

Model 8032 16/32 Channel Multiplexer Instruction Manual, 1996. Geokon, Inc., Lebanon, New Hampshire.

Paikowsky, S.G., and Chen, Y.L., 1998, "Field and Laboratory Study of the Physical Characteristics and Engineering Parameters of the Subsurface at the Newbury Bridge Site", Research Report submitted to the Massachusetts Highway Department, Geotechnical Engineering Section, Boston, Massachusetts.

Paikowsky, S.G., and Hajduk, E.L., 1999. "Design and Construction of an Instrumented Test Pile Cluster", Research Report submitted to the Massachusetts Highway Department, Geotechnical Section, Boston, Massachusetts.

Paikowsky, S.G., and Hart, L. J., 1998. "Development and Field Testing of Multiple Deployment Model Pile (MDMP)", Research Report submitted to the Massachusetts Highway Department, Geotechnical Section, Boston, Massachusetts.

Paikowsky, S.G., Operstein, V., and Bachand, M.B., 1999. "Express Method of Pile Load Testing by Static Cyclic Loading", Research Report submitted to the Massachusetts Highway Department, Geotechnical Section, Boston, Massachusetts.

Paikowsky, S.G. and Peach, D. L., 1995. "A Mechanized Anchorage System for Quick Development Structures", Research Report submitted to The United States Army Natick RD&E Center, Natick, MA.

PDA Manual, 1995. Pile Dynamics, Inc., Model PAK, Cleveland, Ohio.

VISUAL OBSERVATION AND MEASUREMENT OF AERIAL STRESS DISTRIBUTION UNDER A RIGID STRIP FOOTING

Samuel G. Paikowsky[1], Christopher J. Palmer[1], and Albert F. Dimillio[2]

ABSTRACT

Recent advances in tactile sensor technology have produced very thin sensors allowing the simultaneous measurement of normal pressure at multiple points. Such sensors were used under the base of a laboratory model of a rigid strip footing (76.2mm x 609.6mm).

The developed contact stresses are peaked along the centerline and lower along the edges. This distribution, visually observed for the first time, matches well the anticipated theoretical distribution and is quite different than the even distribution routinely assumed in design.

The insight obtained by the observed contact stress distribution can assist in a better geotechnical and structural economical design of shallow foundations.

BACKGROUND

A successful and economic design of a shallow foundation requires the knowledge of the maximum bearing pressure of the soil under the footing. Design methods are based on Bearing Capacity theories assuming that the contact pressures beneath a footing are uniform. In the case of eccentric loading conditions, the design methods are simplified and a linear approximation of the actual stress distribution is used. In reality, however, the distribution of contact stresses is neither constant nor linear.

---

[1] Professor and formerly a Graduate Research Assistant, (respectively), Geotechnical Engineering Research Laboratory, Department of Civil and Environmental Engineering, University of Massachusetts - Lowell, 1 University Avenue, Lowell, MA 01854 Email - Samuel_Paikowsky@uml.edu

[2] Highway Research Engineer, Federal Highway Administration HNR-30 Construction, Maintenance and Environmental Design Division, 6300 Georgetown Pike, McLean, VA 22101 Email – Al.Dimillio@fhwa.dot.gov

Elastic analysis (Borowika, 1936, 1938) and measurements (Schultze, 1961 and Barden, 1962) indicate that the stress distribution beneath symmetrically loaded footings is not uniform (see Bowles, 1996).

Terzaghi and Peck (1948) described the problem of the distribution of contact pressure on the base of uniformly loaded rigid footing, in the following way:

"Since the settlement of the base of a perfectly rigid footing is by necessity uniform, the distribution of the load required to produce uniform settlement of the loaded area. If the sub-grade consists of a perfectly elastic material, of clay, or of sand containing thick layers of soft clay, a uniformly loaded area assumes the shape of a shallow bowl or trough. In order to obtain uniform settlement it would be necessary to shift part of the load from the center of the loaded area toward the edges. Hence, the contact pressure on the base of a rigid footing resting on such a sub-grade increases from the center of the base toward the rim **(see Figure 1).** *On the other hand, if a uniformly loaded area is underlain by sand, the settlement is greater at the edges than at the center. Uniform settlement can be obtained only by distributing the load so that its intensity decreases from a maximum at the center to a minimum at the rim. Hence, the distribution of the contact pressure on the base of a rigid footing on sand has the same characteristics."* **(see Figure 2)**

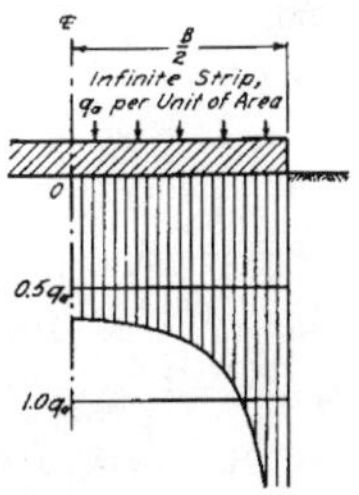

Figure 1: Distribution of contact pressure on base of uniformly loaded rigid footing of very great length, resting on perfect elastic, homogeneous and isotropic subgrade (Terzaghi and Peck, 1948)

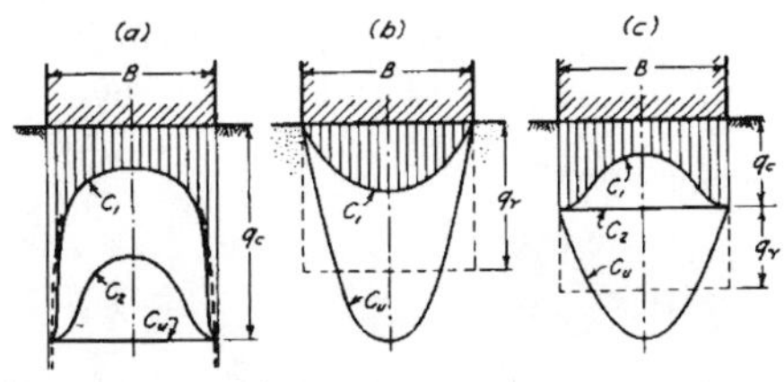

Figure 2: Distribution of contact pressure on base of smooth rigid footing supported by (a) real, elastic material; (b) cohesionless sand; (c) soil having intermediate characteristics. Curves C_u refer to contact pressure when footing is loaded to ultimate value. (Terzaghi and Peck, 1948)

The relationship depicted in **Figure 1** is based on the theory of elasticity and suggests stresses at the edges approaching infinity (in theory) but limited to the yield stress, (in practice) as shown in **Figure 2a**. The state of stress depicted in **Figure 2** represents the conditions expected under a rigid footing on cohesionless soils. An ideal flexible footing on the other hand would develop uniform stresses while experiencing uneven settlement. In practice, the building of a rigid foundation is feasible while testing of flexible foundations encounters the problem of a 'flexible' loading system, which is not easily achievable.

A revolutionary technology enabling the measurement and presentation of normal stress distribution over an area in real time was recently developed. A flexible, grid-based, tactile pressure sensor allows pressure to be measured in up to 2288 (51 x 44) sensing locations. The overall shape and size of the measured area can vary with sensors up to 427 by 488 mm in size. Pressure ranges are possible up to 172 MPa (25 ksi). The system was first developed for dental purposes and has been used in other medical and mechanical applications as well. The original examination and application of the system to granular materials was presented by Paikowsky and Hajduk (1997). Further technological developments continue to evolve and some are utilized in the present study.

OBJECTIVE

The objective of the study presented in this paper was to investigate the possibility of applying the tactile technology to examine the stress distribution beneath shallow foundations. The placement of tactile sensors on the bearing surface of the foundation allows the direct measurement of the contact pressures (stress distribution) in real time during the foundation loading.

SCOPE

The present preliminary research focuses on tests conducted in the laboratory on a model rigid strip footing, examining the feasibility of such tests, the adequacy of the measuring systems and the potential of the study.

TEST SETUP

Overview

The test setup includes three mechanical and four data acquisition components. The mechanical components are: (i) the testing container and granular material, (ii) the load application system, and (iii) the model foundation. The four data acquisition systems (with three separate PC's), measure: (i) contact stress distribution, (ii) applied load, the displacement response of the foundation and the granular material at the surface, (iii) digital imaging system to monitor detailed exterior soil displacement, and (iv) general observation using a regular video camera. **Figure 3** presents a schematic of the test setup components and **Figure 4** shows a cross-section of the foundation and the related setup.

The Mechanical Components

Testing Container and Granular Material

A 63.5 x 116.8 x 94 cm deep (25 x 46 x 37 inch) box was constructed of 0.5 inch PLEXIGLASS® to fit into an INSTRON testing machine being used for load application. Ottawa sand has been used in all tests and was selected for its standardization, grain size and grain shape. Its uniform particle size and round grain

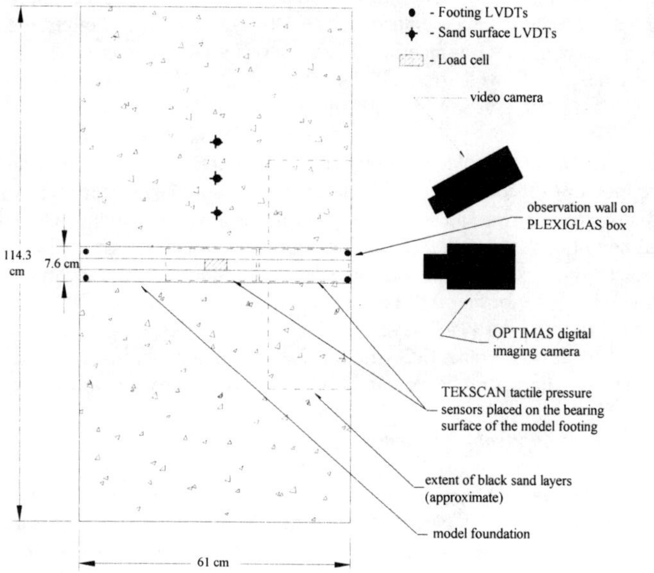

Figure 3 Schematic of model foundation test setup.

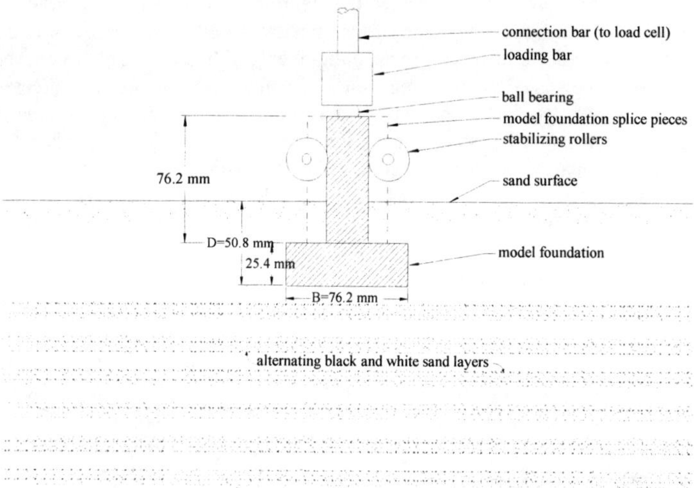

Figure 4 Profile view of model foundation setup (LVDT's not shown).

shape make it ideal for use with the tactile sensor, providing non-puncturing edges and approximately equal contacts per sensel. This sand has been used extensively in the Geotechnical Engineering Research Laboratory at the University of Massachusetts Lowell (e.g. Paikowsky et al., 1995) and has the following characteristics: 99.7%SiO_2, D_{50}=0.50mm, C_c=1.08, C_u=1.40

The sand was placed through a pullivation (raining) process, through a set of wire meshes (45° to each other). The sand was poured evenly from a constant height and at a constant rate to achieve a uniform density. At 10 cm (3.94 in) below the final height of the sample, the surface was leveled and a 1-cm (0.39-in.) thick layer of black (dyed) sand was rained and leveled along the observation side of the box below where the footing will be placed. A 2-cm high regular Ottawa sand was then rained on top of the black layer and so on. This process produced a total of five black sand layers. The final surface is leveled for footing placement. **Figure 4** presents the alternating black and white sand layers below the model foundation.

Load Application System

The load is applied to the footing in the form of a constant rate of displacement using an electro-mechanical INSTRON universal testing instrument Model No. 1127. The INSTRON has an integral load cell attached to the center of the crosshead (which is positioned between two vertical tracks) capable of measuring loads to 250 kN (56202 lbs.). In order to allow for a clear view of the failure plane beneath the foundation, the container had to be placed off-center in relation to the testing machine. This required the design of a special load application system consisting of a horizontal arm fabricated from aluminum channel stock, bolted to the INSTRON load cell. The underside of the arm had an adjustable plate with a ball-type connection for moment free exact alignment with the model. The vertical loading system consisted of three elements: a connecting rod from a direct shear machine, a 4.448 kN (1000-pound) load cell, and the model loading bar. The loading bar connects to the model foundation through two ball bearings placed 304.8 mm (12 inches) apart. The bearings provide the foundation with a moment free load application and the 304.8mm (12-inch) spacing allows for uniform moments along the model itself, (acting as a beam).

The Model Foundation

The size of the model foundation was dictated by the size of the testing container and the extent of the expected failure zone (to prevent boundary effects). The maximum length of the model was governed by the interior width of the box to be 60.69-cm (24 inches), as illustrated in **Figure 3**. The failure zone calculations were based on Terzaghi's (1943) and Hansen's (1970) log spiral failure mode and Rankine's wedges, determining the maximum width of the foundation to be 7.62-cm (3 inches) thereby resulting in a length to width ratio of eight (L/B=8) characteristic of a strip footing. The foundation is comprised of a bearing plate and a wall plate constructed of 25.4 x 76.2 mm (1 x 3-inch) aluminum stock.

The model was built as two separate pieces to allow the sensors to be placed through the bearing plate in the middle of the footing. The wall plates were attached to the bearing plates with a total of eleven bolts and the two sections were connected using 8 bolts and two coupling 12.7-mm (0.5-inch) plates. **Figure 5** presents the design of the model foundation and a photograph. The two circles with crosshairs on the top edge of the wall plate are ball bearing seats for the load application system.

The model is aligned with the cross section of the footing against the transparent (observation) wall of the testing box (see **Figures 3 and 4**). Placement of pressure sensors on the shorter 203.2-mm (8-inch) section as well as the center of the footing is accomplished through the slot between the two coupling plates. The footing design allows the use of a third sensor. This sensor is placed on the outer end of the longer model section. The sensor is wrapped around the end of the foundation and the connections are passed through a recessed channel between the model and the wall of the testing container, (see **Figure 5b**). The photograph in **Figure 6** presents the footing with an attachment of two sensors covering two 203.2 mm (8-inch) sections.

The Data Acquisition Systems

Contact Stress Distributions and Tactile Pressure Sensors

TEKSCAN tactile pressure sensors are constructed of two polymeric sheets with pressure sensitive semi-conductive ink printed on one side of each sheet. The sheets are then laminated together with the printed sides facing each other. Usually, the ink is printed in rows on one sheet and columns on the other. When the two sheets are pressed together a grid of sensing areas is formed. A "handle" which is clapped to the end of the sensor and connected to a data acquisition card installed in a PC that reads the sensor. The computer scans the sensor sequentially by rows and columns and records the change of resistance in each of the sensels. This resistance change is then converted into Raw Sensor Data (RSD) units, which are then correlated to a pressure using a calibration process. Each individual sensor has its own unique properties and must be calibrated separately. For additional details and procedures, see Paikowsky and Hajduk (1997).

The 9800 TEKSCAN sensor was selected for the model foundation tests due to its size, sensel dimensions, and excellent performance in the calibration process. The sensing area of the 9800 measures 76.2 x 203.2 x 0.1 mm thick (3 x 8 x 0.00393 inches) and consists of 6 columns and 16 rows of sensels for a total of 96 sensing locations. Each sensel measures 9 x 11 mm and has an effective sensing area 12.7 x 12.7 mm (161.3 mm^2). Images of two 9800 sensors mounted at the bottom of the foundation are shown in **Figure 6**.

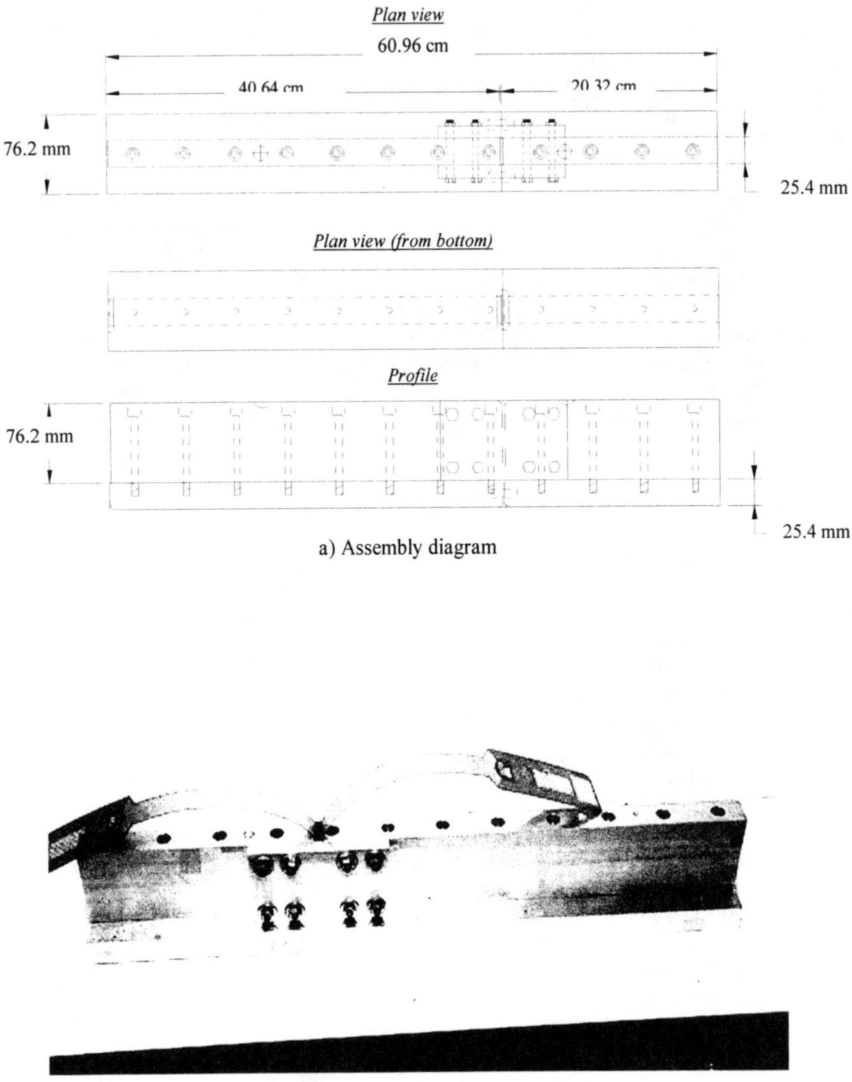

a) Assembly diagram

b) Photograph of model foundation

Figure 5 Model rigid strip footing.

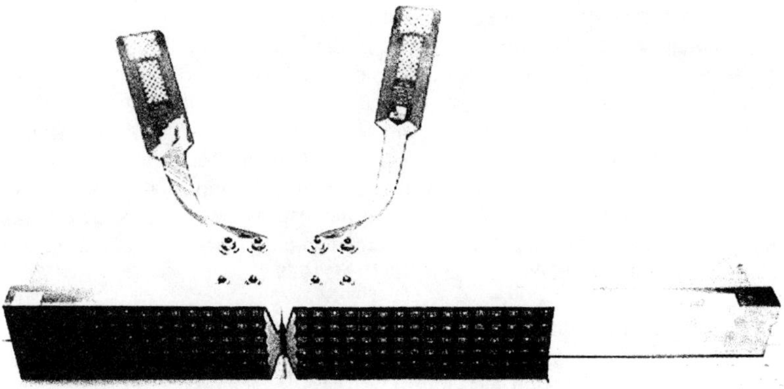

Figure 6 Tactile sensors attached to the base of the strip footing, arrangement for model test #2 and #3.

Foundation and Soil Instrumentation for Monitoring Total Load and Displacement

The footing was instrumented both before and after placement in the testing box. The TEKSCAN pressure sensors must be attached prior to the placement of the model in the container (see **Figure 6**). Following the placement of the model an additional layer of overburden sand was rained around the footing to provide embedment depth, and roller guides were bolted to the side of the box to ensure linear vertical displacement and prevent tilting of the foundation during the test (see **Figure 4**).

Displacements and applied load were monitored by a Hewlett – Packard, HP7500, VEE Data Acquisition System, (DAS). Four linear voltage displacement transducers (LVDTs) were attached to the corners of the bearing plate, and an additional three LVDTs were placed on the sand surface in the middle of the footing at distances 76.2, 152.4, 228.6 mm away from the edge of the bearing plate as shown in **Figure 3**. All LVDTs are connected to a frame attached to a reference frame and are independent of movements of the testing box. A 4.448 kN (1000 lb.) load cell connected to the loading bar was used to measure the total load applied to the foundation.

Digital Imaging System

High resolution Kodak Megaplus Model 4.2 (MP 4.2) CCD camera, Bit Flow imaging board, Data Raptor – VL frame grabber (single slot local BUS board) and OPTIMAS digital imaging processing and analysis software were used to measure the displacement of the sand beneath the footing. The digital images have a resolution of 2000 pixels x 2000 pixels and are taken at a frequency of about one frame every 15 seconds. For the size of the image taken in these tests, the resolution is high enough to identify individual grains of the Ottawa sand. A grid was placed over the transparent observation wall of the box to provide reference points for both the video and OPTIMAS cameras prior to the test. The imaging camera was set-up normal to the box surface and at the average height of the expected displacement field. After the test, the imaging system is calibrated using points of known coordinates (e.g. the grid) in the initial frame. Points, or individual grains of sand, can then be tracked in subsequent frames and the displacements can be determined. Details of motion monitoring techniques utilizing image processing and analysis system are presented by Paikowsky and Xi (1997, 1999).

Video Photography

An analog color 8mm CCD camera was used to monitor and record the entire process, and provide, via a monitor, visual information of the overall foundation response during the test.

TESTING PROCEDURE

Following the placement of the footing, the overburden sand rained, the data acquisition systems are connected and stabilized and the test can be started. Initial image frames and zero-load data bursts are taken by both the TEKSCAN and the HP DAS systems. The data acquisition systems are then setup for the test and double-checked. Next, the INSTRON testing machine is set for a constant displacement rate of 1 mm per minute. The INSTRON displacement is started and the tests proceeds and followed up via a strip chart graphing the load-displacement relations. The test proceeds well into the post peak or to total displacement of at least 25.4-mm (1 inch).

CALCULATED BEARING CAPACITY

The bearing capacity and ultimate load were calculated using Terzaghi's, Meyerhof's, Hansen's, and Vesic's bearing capacity equations. The ultimate loads and contact pressures for the model at an embedment depth of 25.4-mm (1-inch) ranged from 2127 to 3239N (478 to 728 lbs.) and 45.52 to 69.66kPa (6.6 to 10.1 psi). At a depth of 50.8 mm (2 inch), the calculated ultimate load ranged from 3.34 to 4.45kN (750 to 1000 lbs.). These calculations allowed the mechanical design of the loading system connections and the determination of the measurement ranges of the monitoring system (e.g. load cells) and the tactile sensors.

TEST RESULTS

Overview

Four model foundation tests have been conducted to date. The first test did not have alignment rollers and as a result, the foundation tilted and failed typically of eccentric loading. New equipment was acquired and used for Test 4. Three sensors were mounted under the entire bearing area of the footing. Test 4 complex results were not yet analyzed and as such, this manuscript presents the results for the second and third tests. The major difference between the two tests were the sand density influencing the foundation behavior, stress distribution, and the conditions next to the boundaries. Two 9800 tactile sensors were used in both tests 2 and 3. One sensor measured the stress distribution on the 203.2mm (8 inch) segment of the footing adjacent to the observation wall and another was placed under the middle 203.2 mm (8 inch) segment of the footing such that two thirds of the footing's contact stresses were observed.

The Foundation's Load vs. Displacement Response

In both tests, the load and displacement of the foundation were measured and the typical load - displacement curves were plotted. **Figures 7a and 7b** show the load vs. displacement relationship for tests 2 and 3 respectively. Ultimate failure in the second test occurs at a displacement of 10 mm and a corresponding load of 2.50 kN. The shape of the curve suggests that the model underwent a local shear failure. In the third test, the foundation failed at a displacement of 16.1 mm and a corresponding load of 5.09 kN. The shape of the curve for Test 3 suggests a general shear failure. Post failure stick-slip behavior is seen in both tests.

The sand densities for the second and third tests were 16.51kN/m^3 (105 pcf) and 17.61kN/m^3 (112 pcf) respectively. Although small, this difference is significant in relation to the sand's uniformity and hence the density variation effects on the behavior of the model foundation can be clearly seen. In Test 2, the load increased almost linearly with the displacement until failure and then continued to gain capacity with additional large displacements. In Test 3, the load increases linearly with a greater stiffness. After failure the load decreases rapidly to a residual value, indicative of a dense sand strain softening behavior.

Figures 8 and 9 show the detailed displacement (of the various LVDTs) vs. load for Tests 2 and 3 respectively. The detailed load – displacement behavior allows for evaluation of the tilting of the foundation. It can be seen that the foundation tilted along its short axis in Test 2, with a differential settlement of approximately 2-mm from one end to the other. In Test 3, only slight tilting occurred along the long axis.

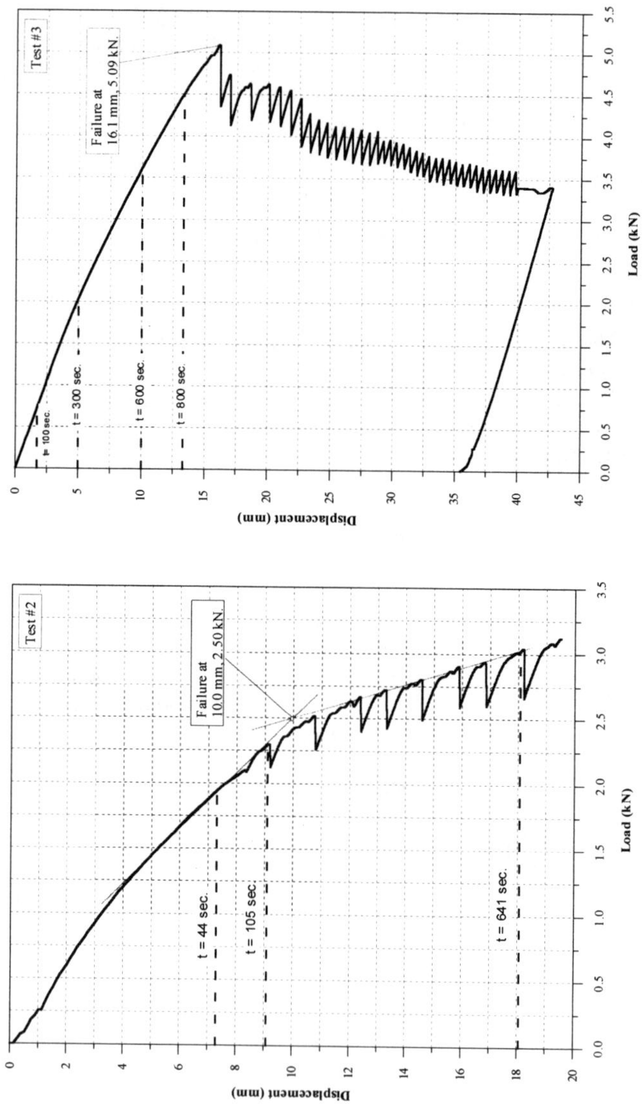

(a) Load vs. displacement plot for Test #2.

(b) Load vs. displacement plot for Test #3.

Figure 7 Load vs. displacement for Test 2 and Test 3

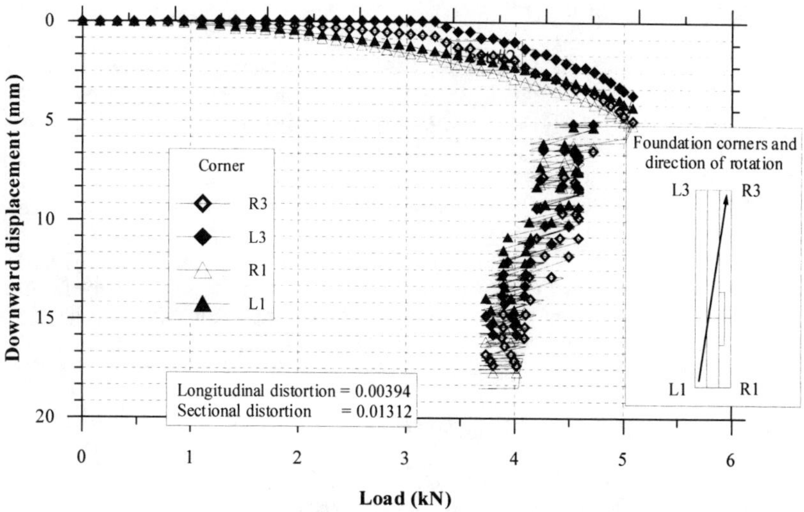

Figure 8 Load vs. displacement curves for the footing LVDTs for Test #2.

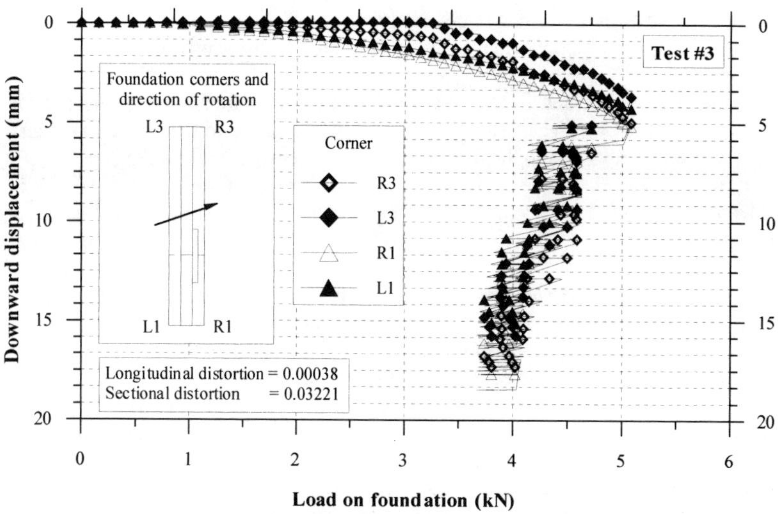

Figure 9 Load vs. displacement curves for the footing LVDTs for Test #3.

Soil's Response at the Surface

The behavior of the sand surface adjacent to the footing also differed between tests 2 and 3. **Figures 10 and 11** show load vs. displacement for the three LVDTs mounted on the sand surface (see **Figure 3** for details). In interpreting the information presented in **Figures 10** and **11**, one needs to correlate the data to **Figures 8** and **9**. The 0 (zero) displacements of the sand surface represent initial loading, permanent displacements represent the final condition of the soil at the surface (e.g. about 2.5 mm in Test 3). Only slight movement of LVDT S3 can be seen in Test 2 while all the surface LVDTs moved between 2 and 3 mm in Test 3. These differences are also associated with the variation in the sand density for the two tests and the resulting mode of failure. The general shear failure in the sand exhibit larger surface deformation (e.g. heave) while local shear failure result in little or no surface deformations.

Aerial Stress Distribution vs. Load Cell Measurements

The data collected from the tactile sensors is stored in a movie file and can only be viewed using the TEKSCAN software. The data can then be converted into a graph file and the sensor response vs. time is displayed. Similarly, the data from the movie file can be converted into a frame file that allows the analysis of an individual data frame. Both the graph and the frame files can then be exported as data files and imported into a spreadsheet program. The appropriate calibration for the sensor is applied and the actual pressure measured by the sensor is obtained.

Three frames were selected for analysis of Test 2 and four frames were selected for the analysis of Test 3. The locations of these frames relative to the applied load and time (i.e. in direct relations to the displacement) are shown as part of **Figures 7a and 7b** for tests 2 and 3, respectively. The post failure data analysis of Test 3 is problematic due the to stick-slip behavior and the unloading conditions and hence is not presented.

The average bearing pressure can be calculated from the load cell data for each frame, providing an independent evaluation of the aerial stress measurement. **Figure 12** shows the average cross-sectional stress distributions and the average bearing pressures from both the tactile sensors and the load cell for the three selected frames of Test 2. **Figure 13** shows the same for the four selected frames of Test 3. Recall that in Test 2 the foundation tilted lengthwise about its short axis (and not to the side) as detailed in **Figure 8**. The stress profiles shown in frames 1 through 3 of Test 2 indicate a symmetric stress distribution, confirming that side-to-side tilting about the long axis did not take place. The stress profiles shown in frames 1-4 of Test 3 suggest the opposite. In frame 3 (t=600 sec.), the stress is higher toward the right side of the footing. This is also seen in frame 4 indicating that the footing is tilting toward the side with higher stresses. The foundation LVDT data for Test 3 confirms tilting in the clockwise direction.

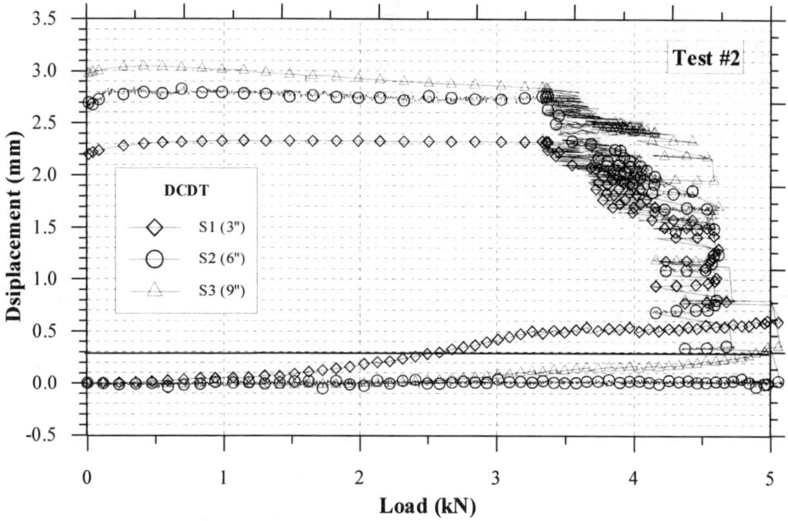

Figure 10 Load vs. displacement curves for LVDTs on the sand surface for Test #2.

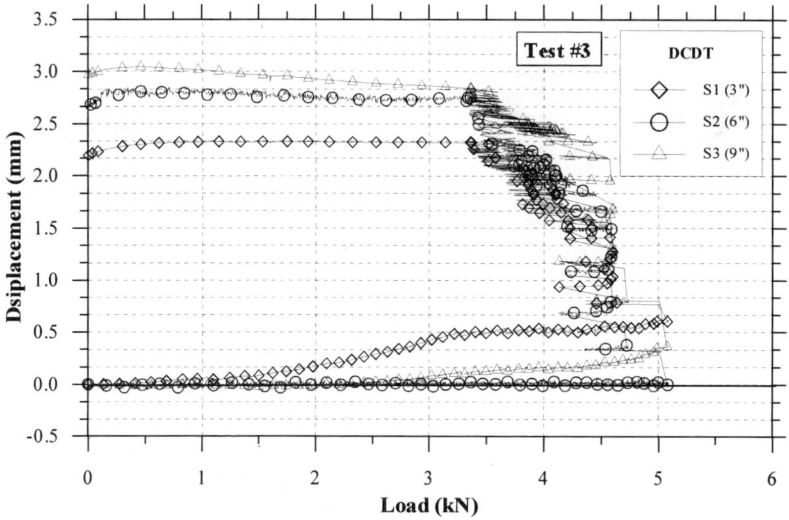

Figure 11 Load vs. displacement curves for LVDTs on the sand surface for Test #3.

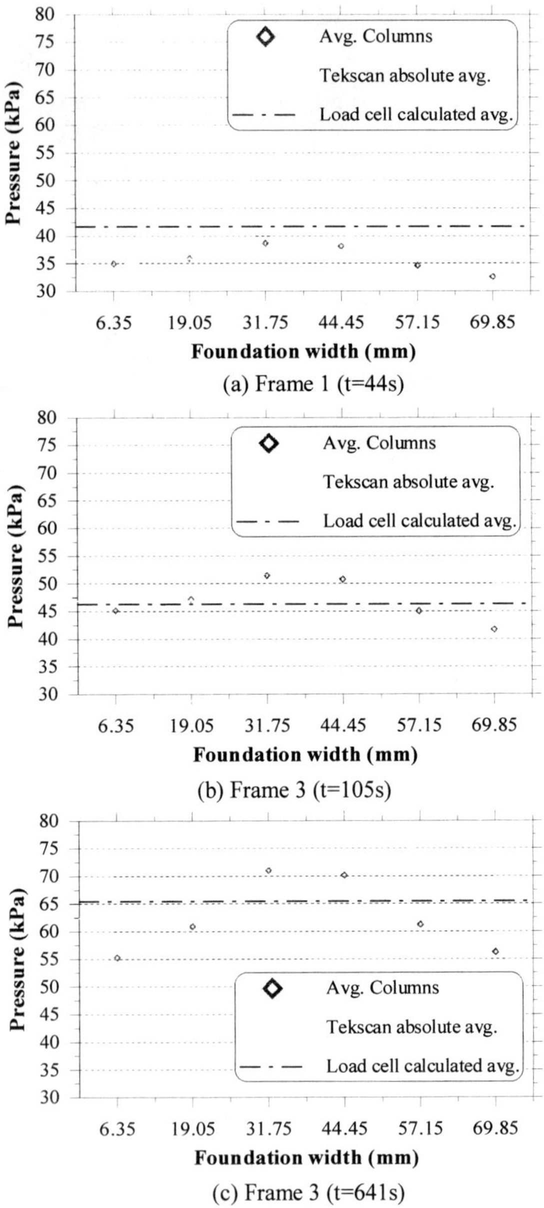

Figure 12 Average cross-sectional stress distributions and the average bearing pressures from TEKSCAN sensors and the load cell for the three frames analyzed of

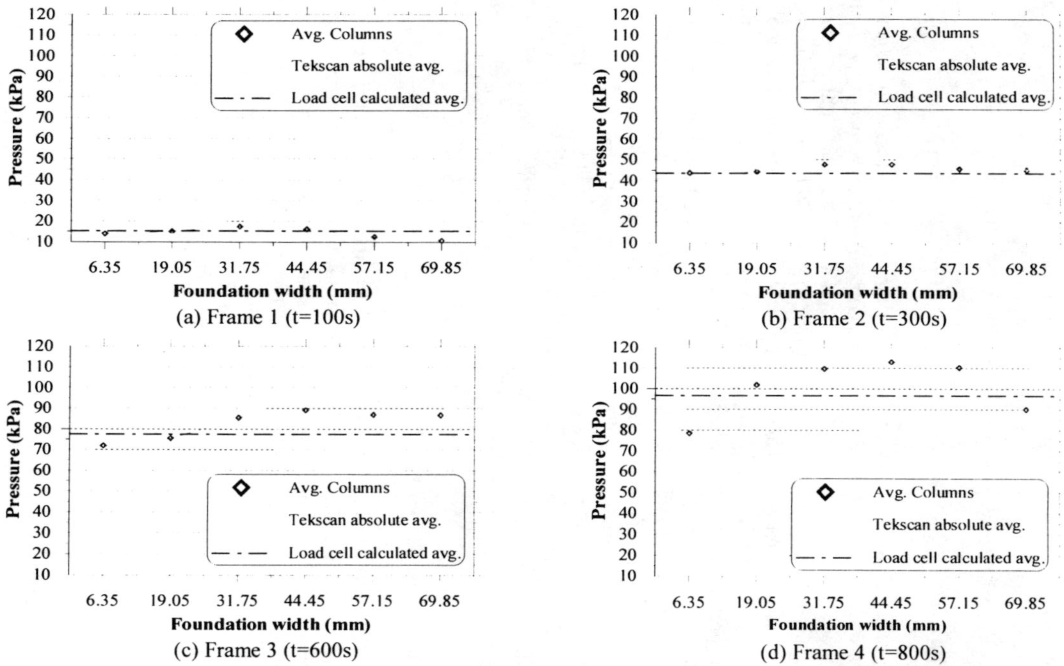

Figure 13 Average cross-sectional stress distributions and the average bearing pressures from TEKSCAN sensors and the load cell for the three frames analyzed of Test #3.

The absolute average of the tactile sensors, as marked in **Figures 12** and **13**, is the average pressure at all sensing locations. This number, when compared to the calculated average from the load cell, provides an independent assessment of the accuracy of the tactile sensors. In general, a very good agreement between the two measurements. For Test 3, a maximum difference occurs in frame 3 at only + 6%. In Test 2, the maximum difference occurs early in the test and is shown in frame 1. This large (relatively) difference of approximately + 14%, is due to the small magnitude of the loads acting on the foundation for this frame. As a larger load is placed on the footing, the percent error decreases as shown in frames 2 and 3 of Test 2.

Actual 2D and 3D Stress Distributions

Figures 14 and 15 show two and three dimensional stress distributions for the peak contact stresses, i.e. frames 3 and 4 in Tests 2 and 3, respectively. Notice the localized high stresses adjacent to the boundary in Test 2 and the higher stresses in the central region of the model in Test 3. Due to the low density of the sand, the stresses at the observation wall (length = 0) are lower in test 2 than in the center of the footing. As the footing displaced downwards, the sand is moved and densified. At the wall, lateral strains are prevented in one direction, resulting in higher contact pressures. Non-uniform density of the sand may also contribute to this condition. The stresses for Test 3 are lower near the observation wall indicating that the condition at the boundary seems to be opposite to those of Test 2. This was caused in part by the increased density of the sand (as the lateral spreading was limited normal to the wall) and by a space between the end of the footing and the observation wall. Sand trapped between the wall of the box and the footing and caused increased friction. This end friction on the foundation reduces the contact stresses under the footing at the boundary and detected by the measurements presented in **Figure 15**.

SUMMARY AND CONCLUSIONS

The presented model tests are unique due to the presence of the tactile sensors. For the first time, the actual stress distribution at multiple discrete points below a rigid strip footing could be recorded and analyzed. Due to the sensors small thickness (0.1 mm), the disturbance in the stress behavior in the granular material is minimal. The results of the two model foundation tests presented in this paper lead to the following conclusions:

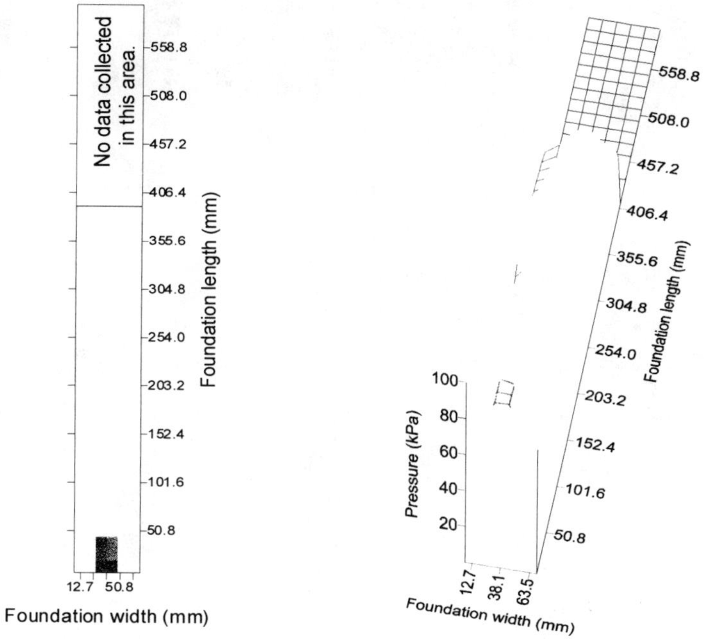

a: 2-dimensional image

b: 3-dimensional

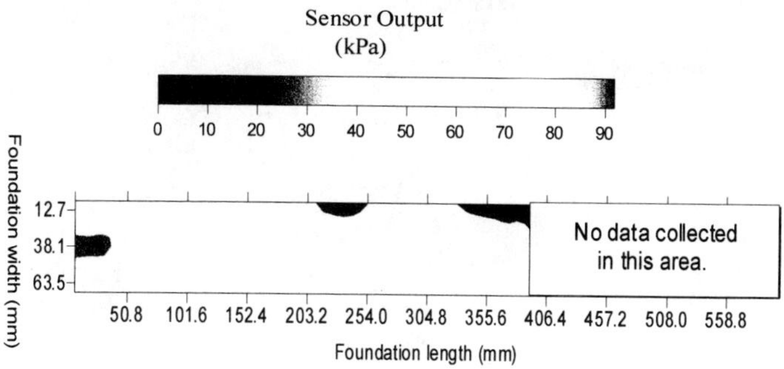

c: 2-dimensional

Figure 14 Two and three dimensional presentation of the stress distribution in frame 3 of Test #2.

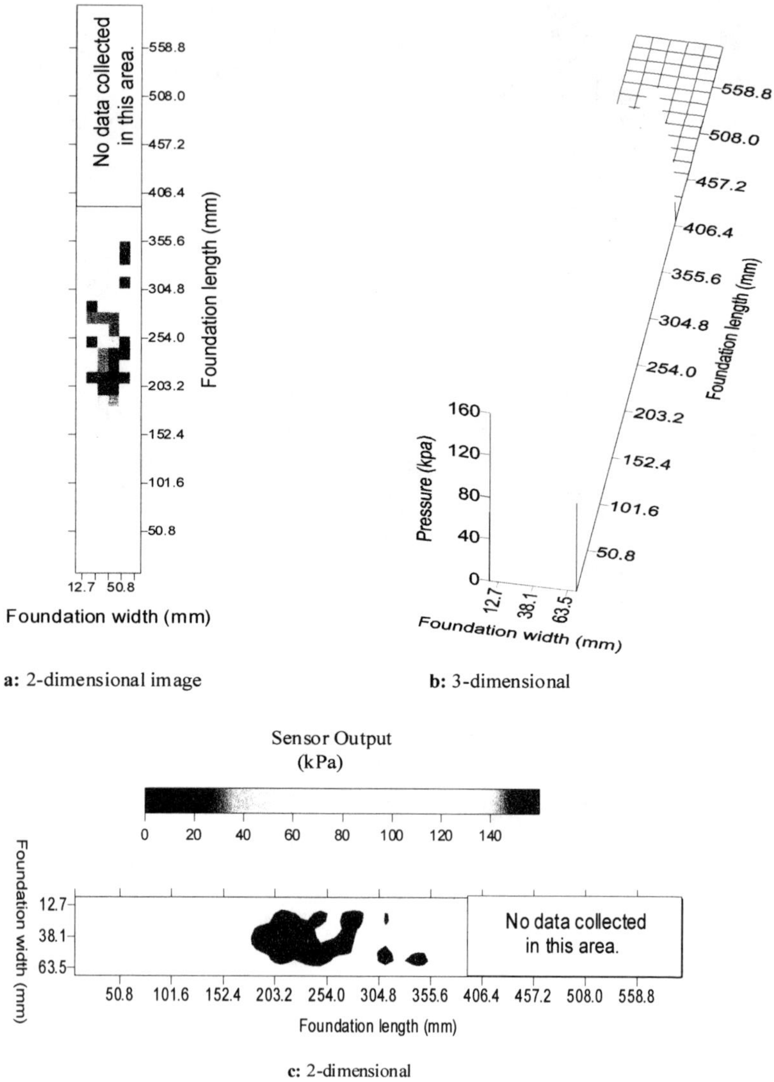

a: 2-dimensional image

b: 3-dimensional

c: 2-dimensional

Figure 15 Two and three dimensional presentation of the stress distribution in frame 4 of Test #3.

1. The application of the tactile sensing technology to geo-materials is revolutionary and very promising. The presented preliminary tests revealed visual and quantitative data that could not have even being imagined only short time ago. The stress distributions measured accurately by the sensors reflect those predicted by foundation and stress distribution theories. The general accuracy of the sensors was shown to be on the average within ± 10% (although higher errors were seen at very low pressures). These values need, however, to be reviewed in light of the very large variation in stresses between one zone under the foundation to another and hence reflect very high accuracy.

2. The variations in the stress distributions due to the influence of the boundaries and sand density can be easily and clearly measured and quantified for the first time. The sand density greatly affects the distribution of the contact stresses on the bearing surface of the footing and therefore the load-displacement behavior. The effect of the boundaries was also seen to vary greatly with only a slight change in sand density. A relatively small variation in the sand density caused a different failure mode in each test. Sand raining tests using the Ottawa sand determined that lower raining rates result in denser sand. Sample uniformity, however, is troublesome to predict due to the difficulty in controlling the density of a uniform round material and installation procedures (e.g. leveling of the sand) required for the placement of the black sand layers.

3. The 9800 type sensors used in the present study have large sensing areas relative to the particle size of the sand. The large sensels are very suitable for stress distribution measurements within granular materials. The use of sensors with larger number of sensels will result in a higher resolution when used with a larger foundation.

4. The digital imaging system was utilized in both tests but the results have not been analyzed yet. Difficulties exist in identifying representative displacements along the sand/wall interface (where the displacements are observed) and accurate point selection for frame analysis. The initial points on the transparent grid can be used for initial calibration of the camera but tracking the displacement of individual particles proved difficult. The installation of a greased rubber membrane between the sand and the wall of the box with a printed grid (on the membrane) is the preferable solution. The grid intersections could be easily tracked by the high resolution CCD but the application of the technique requires effort beyond that possible under the current preliminary study.

5. The limited tilting of the foundation in one direction during all tests suggests that it is caused by an eccentric load problem in the loading system. The deflection of the cantilever-loading arm (bolted to the load cell of the Instron loading machine) seem to be the source of the introduced eccentricity. As a result, the load is applied to the right of center, causing a clockwise rotation in all tests. Stiffening of the loading arm may be required to eliminate foundation tilting.

REFERENCES

Atkinson, J.,1993. *The Mechanics of Soils and Foundations.* McGraw Hill.

Barden, L. 1962. *Distribution of Contact Pressure under Foundations.* Geotechnique, Vol. 18, no. 1, March, pp.1-24.

Borowicka, H. 1936. *Influence of Rigidity of a Circular Foundation Slab on the Distribution of Pressures over the Contact Surface.* Proc. 1st International Conference of Soil Mechanics. Cambridge. Vol. 2, pp. 144-149.

Borowicka, H. 1938. *The Distribution of Pressure under a Uniformly Loaded Elastic Strip Resting on Elastic-Isotropic Ground.* 2nd Cong. Int. Assoc. Bridge and Structural Eng.. Berlin. Final Report

Bowles, J. E., 1996, *Foundation Analysis and Design*, Fifth Edition, McGraw Hill.

Das, B. M. 1990. *Principles of Foundation Engineering,* 3rd edition. PWS-KENT Publishing Company, Boston MA.

Das, B. M. 1990. *Principles of Geotechnical Engineering,* 2nd edition. PWS-KENT Publishing Company, Boston MA.

Fang, H. Y. 1991. *Foundation Engineering Handbook,* 4th edition. Van Nostrand Reinhold, New York.

Hansen J.B., 1970, "A Revised and Extended Formula for Bearing Capacity", Danish geotechnical Institute, Bulletin 28.

Lambe, T. W., and Whitman, R. V. 1969. *Soil Mechanics*, Wiley, New York.

Paikowsky, S.G. and Hajduk E.L., 1997 "*Calibration and Use of Grid-Based Tactile Pressure Sensors in Granular Material,*" Geotechnical Testing Journal, GTJODJ, Vol. 20, No. 2, June, pp. 218-241.

Paikowsky, S.G., and Xi, F., 1977, "*Photoelastic Quantitative Study of the Behavior of Granular Material with Application to the Problem of Friction Along an Interface*", Research Report submitted to the Air Force Office of Scientific Research, September 1997.

Paikowsky, S.G. and Xi, F., 1999, *Particle Motion Tracking Utilizing High Resolution Digital CCD Camera*, in print, ASTM Geotechnical Testing Journal.

Paikowsky, S.G., Player C. M., and Connors P.J., June, 1995, "*A Dual Interface Apparatus for Testing Friction of Soil Along Solid Surfaces*", ASTM,

Geotechnical Testing Journal, Vol. 18 No. 2, pp. 168-193

Schultze, E. 1961. *Distribution of Stress beneath a Rigid Foundation.* 5th ICSMFE. Vol. 1, pp. 807-813.

Tatsuoka, F., Molenkamp, F., Torii, T., and Hino T. 1984. "*Behaior of Lubricated Layers of Platens in Element Tests,*" *Soils and Fondations.* Vol. 24, No.1, JSSMFE, Tokyo, Japan. pp. 113-128.

Terzaghi K., 1943, "*Theoretical Soil Mechanics*", John Wiley and Sons

Terzaghi K. and Peck R. B., 1948, "*Soil Mechanics in Engineering Practice*", John Wiley and Sons.

Terzaghi, K., Peck, R. B., and Mesri, G. 1996. *Soil Mechanics in Engineering Practice.* 3rd edition, John Wiley and Sons, Inc., New York.

ACKNOWLEDGEMENT

An initial support of the presented research was provided by the Federal Highway Administration (FHWA), by the National Science Foundation (NSF), NYI Grant MSS-5398090, and the University of Massachusetts Lowell. Continuation of the research is possible through the support of the NSF Grant No. CMS-9908612 with a complementary support by the Tekscan Corp., Boston, MA.

EFFECT OF DIFFERENT SURFACE FINISH METHODS APPLIED TO BEARING SOIL LAYER ON PERFORMANCE OF HIGHWAY BRIDGE SPREAD FOOTING

By S. M. Sargand,[1] Member ASCE, T. Masada,[2] Associate Member ASCE, and R. L. Engle.[3]

ABSTRACT: Shallow foundation has a potential to be a successful alternative to deep foundations at many highway bridge construction sites. To further encourage their uses, comprehensive performance data must be compiled and made available to the practicing civil engineers. A research team at Ohio University has completed a major study on the field and laboratory performance of highway bridge spread footing foundations for the Ohio Department of Transportation and the Federal Highway Administration. Majority of the spread footings placed over the bearing soil layers were instrumented with sensors to measure their overall performance. At one of the five bridge construction sites involved in the study, the upper zone of the bearing soil layer received a few different treatments due to localized unique subsurface conditions and changes in payment agreement between the Contractor and the State. This paper presents detailed field performance of the spread footings at this site, which rested on bearing soil layers with different finish methods. Although the amount of data obtained in the study was still limited, the field gave a preliminary indication that the surface finish method might not have much influence on the field performance of these spread footings.

---

[1] Russ Professor of Civil Engineering, Ohio Research Institute for Transportation and Environment, Ohio University, Athens, OH 45701. E-mail: sargand@bobcat.ent.ohiou.edu.

[2] Assistant Professor, Department of Civil Engineering, Ohio University, Athens, OH 45701.

[3] Assistant Administrator, Office of Structural Engineering, Ohio Department of Transportation, Columbus, OH 43223.

INTRODUCTION

Shallow foundations have been utilized to support various types of civil engineering structures over the years. However, one type of structures to which the shallow foundations have not seen a much widespread application across the U.S. is the highway bridges. Traditionally, bridge/foundation engineers take a conservative approach in designing highway bridge foundations. They usually specify deep foundations unless a rare site condition (ex. a very shallow bedrock or a highly consolidated, stiff soils) exists. This is because their confidence level in the shallow foundation is less than that in the deep foundations. Spread footing can be a viable alternative to the deep foundations, provided that subsurface conditions are reasonable. Since the cost associated with the spread footing is 30 to 80% less (Amar et al. 1984), spread footing should be considered seriously at almost all bridge construction sites. To further encourage their use, comprehensive performance data must be compiled from successful case histories and made available to the practicing civil engineers.

A research team at Ohio University has completed a multi-year study on the field and laboratory performance of highway bridge spread footing foundations in Ohio. Detailed information on the study can be found in a report by Sargand et al. (1997). At one of the five bridge construction sites involved in the study, the upper zone of the bearing soil layer received different treatments due to localized unique subsurface conditions and changes in payment agreement between the Contractor and the State. Majority of the spread footings placed over the bearing soil layers were instrumented with sensors to measure overall and differential settlement, contact pressure distribution, and tilting of abutment walls/pier columns. This paper presents detailed field performance of the spread footings at this site, which rested on bearing soil layers with different finish methods.

SITE AND BRIDGE CONSTRUCTION

General Descriptions

The bridge construction site is located on the west side of Cincinnati, Ohio. Soils are originally of glacial moraine, and the bedrock is composed of alternating strata of limestone and shales. Ten boreholes were placed within the bridge construction area to depths ranging from 12.2 to 29.0 m (40 to 95 ft.) below the ground surface. According to the boring log data, the most common soil type was a silty clay soil classified as A-7-6, A-6-a, or A-6-b. In the rear abutment area, granular soils classified as A-2-7, A-1-b were found right above the bedrock. The depth to bedrock (below the bottom of footings) was not uniform. It was 2.8 m (9.3 ft.) at the rear abutment, 2.7 m (8.7 ft.) at Pier 1, 5.1 m (16.7 ft.) at Pier 2, 10.1 m (33 ft.) at Pier 3, 24.2 m (79.5 ft.) at Pier 5, and 23.0 m (75.5. ft.) at the forward abutment. No groundwater table was detected in boreholes within the influence zone of the footings. The Standard Penetration Test (SPT) data are represented by the resistance N-values (blows per ft. penetration) and summarized in Table 1.

Table 1: Standard Penetration Test (SPT) Results

Bore Hole I.D. (1)	Borehole Location Information (2)	SPT-N Value @ Depth (m) below Footing Base: 0.0 (3)	0.3 (4)	1.2 (5)	2.7 (6)	3.4 (7)	4.6 (8)	5.8 (9)	7.3 (10)	9.1 (11)
H-1	Near R.A.	10	56	90	100+			100+	100+	
H-2	Near Pier 1.	12	89		100+					
H-3	Near Pier 2.	9			66	30	85	100+		
H-4	Near Pier 3.	31		20						100+
H-6	Near Pier 5.	20			8		35			
H-7	Near Pier 6.	22	15		8	31				
H-8	Near Pier 7.	11			14	77			37	
H-9	Near Pier 8.	21	13	6		16				
H-10	Near F.A.	12		6	30					40

[Note] "R.A." = Rear Abutment; and "F.A." = Forward Abutment.

The purpose of the construction project was to replace an aging nine-span bridge structure, built in the 1930's, with a new one. The new bridge also had nine spans and was a composite-deck type. It had a deck width of 16.5 m (54 ft.) and a length of 204 m (670 ft.). The longest and shortest spans were 26.3 m and 20.4 m (86.2 ft. and 66.9 ft.), respectively. The bridge was supported by two abutments and eight piers. Some basic information on the new bridge structure are summarized below:

- Footing: Width = 3.4 to 4.6 m (11 to 15 ft.)
  Thickness = 0.9 to 1.0 m (3 to 3.25 ft.)
  Combined area = 337.5 m^2 (3,633 ft^2.)

- Spans: Rear abut. to Pier 1 (span 1) = 26.0 m (85.4 ft.)
  Pier 1 to Pier 2 (span 2) = 26.3 m (86.2 ft.)
  Pier 2 to Pier 3 (span 3) = 26.3 m (86.2 ft.)
  Pier 3 to Pier 4 (span 4) = 20.3 m (66.5 ft.)
  Pier 4 to Pier 5 (span 5) = 20.4 m (66.9 ft.)
  Pier 5 to Pier 6 (span 6) = 20.4 m (66.9 ft.)
  Pier 6 to Pier 7 (span 7) = 20.4 m (66.9 ft.)
  Pier 7 to Pier 8 (span 8) = 20.4 m (66.9 ft.)
  Pier 8 to Forw. abut. (Span 9) = 22.6 m (74.0 ft.)

- Superstructure: Concrete box beams = 175.1 kN/m (12 kips/ft.)
  Concrete deck = 58.4 kN/m (4 kips/ft.)

- Projected ADT = 47,190 (year 2011).

Each of these foundations was constructed in two phases. Construction work was divided into four staged so that the traffic could be maintained throughout the construction:

Stage 1 ----- Construction of entire Phase I section of the new bridge next to the old bridge on the south side; traffic was routed through the old bridge.

Stage 2 ----- Demolition of the old bridge; traffic was routed through the Phase I section of the new bridge.

Stage 3 ----- Construction of the remaining (Phase II) section of the new bridge; traffic was still routed through the Phase I section.

Stage 4 ----- Completion of Phase II section of the new bridge; two phases were combined to open the entire new bridge to the traffic.

Construction work commenced in the middle of March, 1994 and continued until July 1, 1995, during which time the Phase I section of the new bridge was opened to the traffic on Aug. 6, 1994. Figure 1 shows pictures of the old and new bridge structures. Figure 3.b shows a general view of the new bridge structure, taken at the end of Stage 2.

Finish Methods Applied to Bearing Soil Layers

During the Phase I footing construction, the bearing soil layers received different finish methods prior to concreting the footing. This was due to localized unique subsurface conditions and changes in payment agreement between the Contractor and the State. Figure 2 presents this data through simplified drawings.

The top 75 mm (3 inches) of the soil was replaced with compacted crushed limestone (meeting the Ohio DOT #305 gradation specifications) layer in the footing construction areas of Pier 8 and the forward abutment. While excavating for Pier 3 and 5 footings, masonry foundations of unknown age were discovered. A decision was made at each of these locations to remove the concrete block entirely and place unreinforced concrete pad up to the bottom elevation of the footing and then to construct the new footing on top. The plain size of the unreinforced concrete pad placed at each location was larger than the corresponding footing size. A similar situation was encountered in the rear abutment construction zone, and the same solution was practiced. However, the size and shape of the unreinforced concrete pad were identical to the footing above (see a photograph shown in Figure 3.a). In the Pier 2 footing area, a small pocket of highly organic soil was detected within its northern half section. This material was removed, and concrete was placed to fill the void. In the Pier 1 footing construction area, crushed limestone was used to fill a similar small void left by removing highly organic soil (the State decided not to pay for the concrete used to fill these depressions). All of the other footings at Piers 4, 6, and 7 were placed directly on top of the in situ soil layer that was compacted with no special finishing methods applied.

Field Instrumentation Methods

In the study described in this paper, sensor selections and field instrumentations were designed to measure three performance parameters (settlement, contact pressure, rotation) of the spread footing. Figure 4 presents the basic instrumentation scheme. Automatic level instrument (Topcon AT-F2) was utilized to measure vertical movements of the settlement points. Vibrating-wire pressure cells (Model 4800E by geokon), with a 690 kPa (100 psi) range and a 0.7 kPa (0.1 psi) sensitivity, were installed to record the contact pressure between the footing and bearing soil layer. These pressure cells were calibrated inside a special chamber prior to the field installation. Each pressure cell was placed against a 75-mm (3-in.) thick fine sand lense to assure a uniform pressure distribution over its sensing disk face. An accelerometer- based tilt sensor (Model 50344 by Slope Indicator), with a ± 30° range and a 0.003° sensitivity, was used to monitor rotational movement of abutment walls and pier columns. Overall instrumentation plan is summarized in Table 2. At a few selected pier foundations, tilting of the pier columns was monitored in the longitudinal as well as transverse directions.

Table 2: Summary of Field Instrumentation Plan

Spread Footing I.D. (1)	No. of Settlement Monitoring Points (2)	No. of Contact Pressure Cells (3)	Monitoring Stations for Tilting: Longit. (4)	Transv. (5)
R.A.	3	3 (heel, key, toe)	1 (center of wall)	N/A
Pier 1	5 (corners, center)	4 (near corners)	2 (outer columns)	2
Pier 3	5 (corners, center)	4 (near corners)	1 (center of wall)	0
Pier 4	1 (center)	0	0	0
Pier 5	1 (center)	0	0	0
Pier 6	5 (corners, center)	2 (each side)	2 (outer columns)	0
Pier 7	1 (center)	0	0	0
Pier 8	5 (corners, center)	4 (near corners)	2 (outer columns)	2
F.A.	3	3 (heel, key, toe)	1 (center of wall)	N/A

[Notes] "R.A." = Rear Abutment; "F.A." = Forward Abutment; "Longit." = Longitudinal Direction; "Transv." = Transverse Direction; "N/A" = Not Applicable.

FIELD PERFORMANCE OF SPREAD FOOTINGS

The foundation movements were monitored until about 5 months after the completion of the new bridge structure. Table 3 summarizes the field performance data.

Table 3: Summary of Field Performance Data

Spread Footing I.D. (1)	Settlement (mm): Total (2)	Post-deck (3)	Ave. Contact Pressure (kPa) (4)	Tilting of Wall/Column (°) Longit. (5)	Transv. (6)
R.A.	35	13	287 to 412	0.00 to 0.10	N/A
Pier 1	20	5	268 to 383	-0.05 to 0.00	0.00 to 0.10
Pier 3	14	7	134 to 192	-0.08 to 0.08	N/A
Pier 4	15	3	N/A	N/A	N/A
Pier 5	5	1	N/A	N/A	N/A
Pier 6	10	2	172 to 239	-0.05 to 0.00	N/A
Pier 7	11	2	N/A	N/A	N/A
Pier 8	14	1	57 to 86	-0.10 to -0.05	-0.05 to 0.10
F. A.	13	3	N/A	0.00 to -0.05	N/A
Average-1	15.2	4.1	184 to 262	-0.05 to 0.02	-0.025 to 0.10
Average-2	12.0	2.3	(Average values are not computed due to a lack of data).		
Average-3	16.8	5.0			
Average-4	11.5	3.0			

[Notes] Average-1 = average values computed among all of the spread footings.
Average-2 = average values computed among the spread footings which rested directly on in-situ soil (Piers 4, 6, and 7).
Average-3 = average values computed among the spread footings which rested on bearing soil layer with special finish treatments (Abutments, Piers 1, 3, 5, and 8).
Average-4 = average values computed among the spread footings which rested on bearing soil layer with special finish treatments, excluding the footings at rear abutment and Pier 1 where the dead load was larger (Forward abutment; Piers 3, 5, and 8).

According to the table, average settlements were below 38 mm (1.5 in.) in all cases. Rear abutment footing experienced the largest settlement of 35 mm, and Pier 1 footing experienced the second largest settlement of 20 mm. This may be understandable, since Spans 1 and 2 were the two longest spans. Post-deck settlement typically constituted 7 % to 50% (average 25%) of the total settlement. Average maximum contact pressure measured under the footings was between 184 and 262 kPa (26.7 to 38 psi). The tiltmeter data showed that the pier structure could rotate over a small angle both in the longitudinal and transverse directions.

Four different average values were computed to make a comparison among the settlement data. The definitions of these averages are given below Table 3. In calculating the average among the footings that rested on bearing soil layer with special finish methods applied, the footings at rear abutment and Pier 1 were first included (to obtain Average-3) and then excluded (to obtain Average-4), because of the fact that they were under heavier loads. A comparison among these average settlement values provides a preliminary indication that the surface finish method might not have much influence on the field performance of the spread footings (see Average-2 and Average-4). Due to a lack of data, average values could not be calculated for the remaining performance data.

SUMMARY AND CONCLUSIONS

The field study conducted by the authors demonstrated overall that the spread footings on both cohesionless and cohesive soils could support the highway bridge structures successfully, provided that the subsurface conditions are relatively favorable (i.e., free of unsuitable materials, no high groundwater table, well-consolidated state). None of the footings settled was more than 51 mm (2 inches). No significant differential settlement was observed anywhere at the site. Maximum contact pressure was measured to be less than 250 kPa (2.6 tsf). And, the degree of tilting remained well within $\pm$ 0.3°for the abutment walls and pier columns.

In this paper, field performance of highway bridge spread footings, resting on bearing soil layers that had received different surface finish methods, was discussed. Although the amount of data obtained in the study was still limited, the field data provided a preliminary indication that the surface finish method might not have much influence on the field performance of the spread footings.

ACKNOWLEDGEMENT/DISCLAIMER

Authors are grateful to the Federal Highway Administration (FHWA) and the Ohio Department of Transportation (Ohio DOT) for proving financial support for this study. The contents of this paper do not necessarily reflect the official views and/or policies/guidelines of the Ohio DOT.

APPENDIX-I. REFERENCES

Amar, S., Baguelin, F., and Canepa, Y. (1984). "Etude experimentale du comportement des fondations superficielles." Annales de l'ITBTP, 427, Series Soils et Fondations, 189, Paris, France, Sept.

Sargand, S. M., Hazen, G. A. (1997). "Field and Laboratory Performance Evaluation of Spread Footings." FHWA/OH-98/017, Final Report to the Ohio Department of Transportation and the Federal Highway Administration, Ohio University, Athens, OH, Dec., 234+ pp.

Sargand, S. M., and Masada, T. (1999). "Spread Footing Foundation for Highway Bridge Applications." Journal of Geotechnical and Geoenvironmental Engineering, Vol. 125, No. 5, American Society of Civil Engineers, May, pp. 373-382.

(a) Old Bridge Structure (partially demolished)

Figure 1. Photographs of Old and New Bridge Structures

(b) New Bridge Structure (before its opening)

Figure 1. Photographs of Old and New Bridge Structures

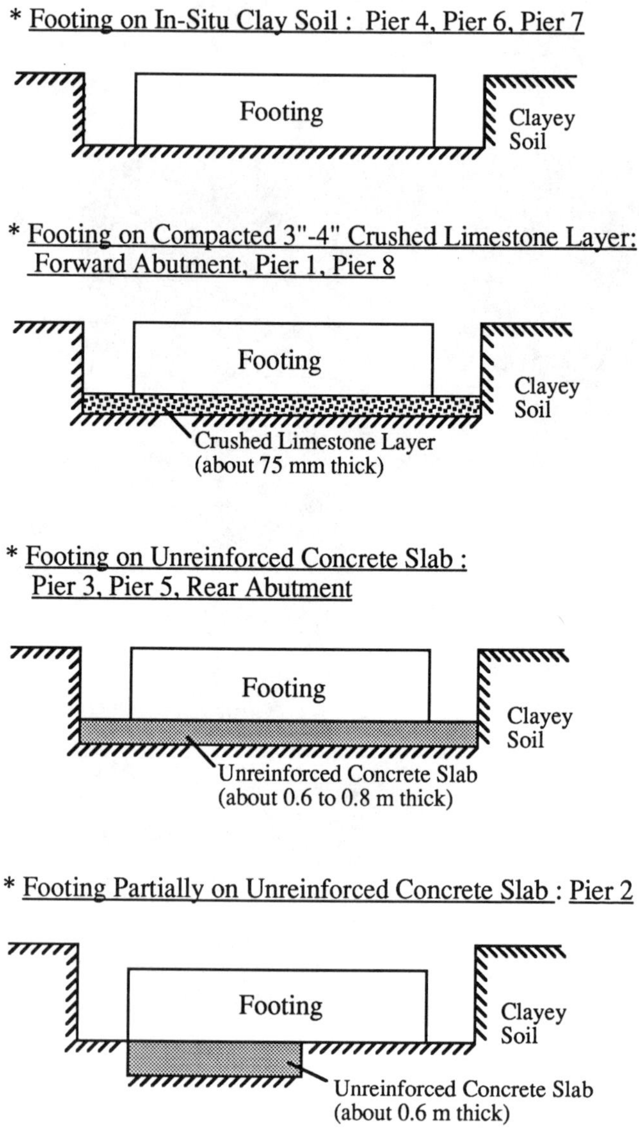

Figure 2. Different Practices Used in Preparing Bearing Layer for Phase I Foundations

(a) General View of Rear Abutment Construction (Phase I)

Figure 3. Photographs Taken During Construction of New Bridge Structure

(b) General View of New Bridge (Phase I) at End of Stage-2 Construction

Figure 3. Photographs Taken During Construction of New Bridge Structure

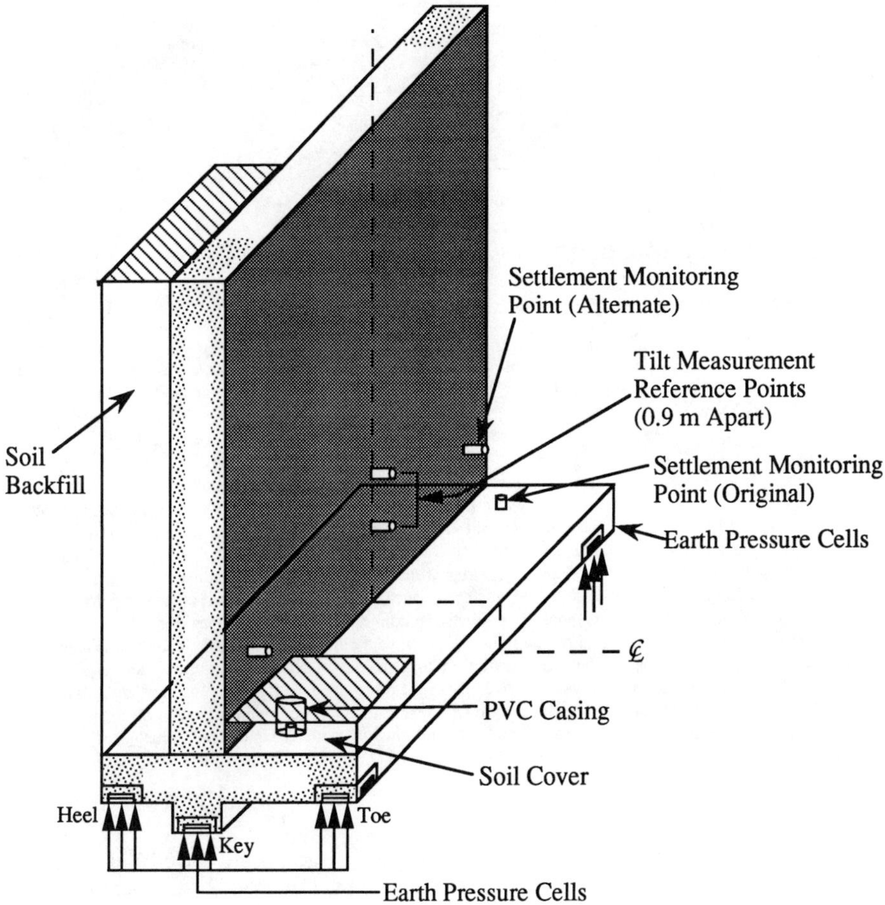

Figure 4. Basic Instrumentation Schemes for Spread Footing for Highway Bridge Abutment

LATERAL LOAD PREDICTION AND TESTING OF 3.05M DIA. SHAFTS

Trevor D. Smith[1] (Member), Risheng Park[2] (Ass. Member) and Richard Hannan[3] (Member)

ABSTRACT

Four 3.05m diameter permanent steel-cased shafts were laterally load-tested over water in the Columbia River, downstream of Bonneville Dam, Oregon. The site investigation into the subsurface gravels, sandy gravels and near mudline silts included pre-bored pressuremeter tests. Design life lateral loads arrive from the pile supported concrete fish flume superstructure being submerged during periods of high river flows, producing large repeat loads to the shafts. The industry standard LPILE code was employed with *user defined* P-y curves constructed from Pressuremeter data to predict the first cycle load deflection response. Predicted mudline deflection & slopes were close to the maximum structurally tolerated by the flume, 25m above the mudline. Full-scale lateral instrumented load testing on the production shafts validated the analysis and P-y curves up to the maximum 2.22MN. A power law model was used to evaluate the repeat load cycles in the load test and establish the long term accumulated cyclic deflection. Optional foundation improvements were also examined, including a 12.2m diameter concrete Drag Collar designed from the Pressuremeter data.

INTRODUCTION

Bonneville Dam is a large hydroelectric scheme generating 558 MW of power and situated 55 km. east of Portland, Oregon, on the Columbia River. To improve the survival rate of juvenile fish hatching upstream a $38M Juvenile Fish Bypass System (JBS) is nearing completion to carry fish around the 2 powerhouses, 2 navigation locks and the spillways. The JBS safely transports fish 2km downstream

[1] Professor of Civil Engineering, Portland State University, PO Box 751, Portland, OR 97207-0751, trevor@eas.pdx.edu

[2] Staff Geotechnical Engineer, L. R. Squier Assoc. , 4260 Galewood Street, Lake Oswego, OR 97035, piaor@squier.com

[3] Senior Geotechnical Engineer, US Army Corps of Engineers, Portland District, Portland, OR 97201, Richard.W.Hannan@nwp01.usace.army.mil

to be discharged from a concrete flume above swift water currents. To maintain minimum elevations above the waterline the square concrete flume has twin Outfalls over water; one for normal water levels (L) and a second for high river, flood, conditions (H). These flood stages, based on spring snow and variable power generation demands fully submerge the low flume seasonally below water elevations and provide high hydraulic forces. These forces can be classified as repeat, one way, cyclic monotonic loads, always in the downstream direction and normal to the flume axis. The high flume may also see cyclic loads with 100year frequency floods, but much less often. Flume invert elevations above the river mudline are approximately 25m to the high (H) flume and 20m to the low (L) flume. To support each of the 125m elevated flume lengths over water 3 large shafts of 3.05m diameter were chosen, carrying flume spans of 35m on each H & L flume, Figure 1.

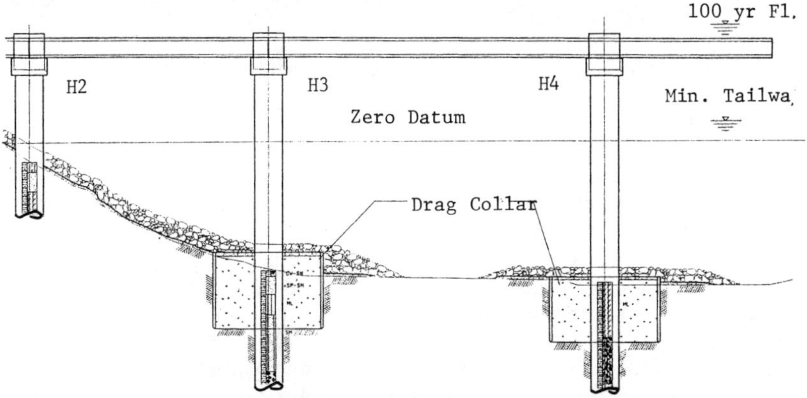

Figure1: Elevation of the High Outfall Flume at Scale 1:250

Generalized subsurface conditions comprise alternating layers of coarse grained medium to dense gravels, sandy gravels, and some sand, below a near surface soft silt zone up to 6m in thickness. This silt zone at shafts H4 and L4 has a depth of twice the pile diameter and dominates the lateral load performance. With the predominance of gravels and sandy gravels a permanent steel casing 25mm thick was vibro driven and then chiseled into bedrock at depth between 27m and up to 35 m below river bottom. After removing the soil plug a rebar cage was lowered and concrete tremied up to the flume elevation.

Tolerable structural flume deflections are of the order 0.75 m. With the 20m to 25m stickup the mudline slope, as well as mudline deflection, contribute to flume deflections. Limited in-water work periods, predator fish, environmental concerns and hydraulic considerations prevented most accepted design approaches to stiffen the deflection performance of the shafts. Predicted shaft and resulting flume deflections necessitated consideration of a variety of construction alternatives. The selected option for construction bids, was a 12.2 m diameter mass concrete 'Drag Collar' set into the surface silts and mobilizing frictional drag by bearing on the gravels. The purpose of the four full-scale instrumented load tests was 3 fold:

1) Validate the PMT based LPILE predictions for first cycle loadings.
2) To gain unload-reload cyclic load and creep data for evaluation of accumulated deflection over a 50 years service life for the flumes.
3) To determine if the concrete Drag Collars were required to be installed by the contractor at an additional $1M project cost.

The load testing took place during an approved 'in-water' work period from a purpose built driven pipe pile supported trestle in May 1998[4].

OFFSHORE INVESTIGATION AND SHAFT CONSTRUCTION

Generalized Conditions

The subsurface investigation was conducted in 2 phases. Phase 1 comprised two onshore exploratory boreholes using an open ended Becker hammer drill at the location of H1 and L1 shafts. One of these boreholes also included SPT and pre-bored pressuremeter testing (PMT) by mud rotary techniques through the Becker casing. These tests confirmed the feasibility of PMT for all the offshore shafts at H2 to H4 and L2 to L4 as the prime investigative tool to characterize the gravels, sandy gravels and silts. The Phase 2 offshore investigation was undertaken from a spud located barge, with further anchorage in swift current provided by a Columbia river tug during exceptionally heavy river flows in December 1996. The testing program included 20 PMT tests with occasional SPT tests below the PMT test zone, when practical, in the silts. Some limited CPT testing in artesian pressure bearing sands from 10m to 16m, encountered at H3, and diamond coring in the dense gravels above the sedimentary bedrock were also made.

Conditions along both the H and L flume alignments were similar with silts, interbedded sands, gravelly sand and sandy gravel. Each shaft location had a dedicated borehole in which PMTs and SPTs were taken. Table 1 indicates the variety of soil conditions offshore. It can be seen most shafts would encounter near surface silts and loose to medium sands during installation. A permanently installed 25mm thick steel casing vibro-driving was selected to advance the casing through the coarser gravels and into bedrock.

[4] This construction trestle was partially destroyed by severe flood conditions occurring 2 weeks *after* the conclusion of the load testing program

L3		L4	
Depth (m)	Soil Conditions	Depth (m)	Soil Conditions
0-3	Stiff Silt	0-5.5	Silt and Sandy Silt
3-7.3	Dense Gravely Sand	5.5-9.8	Med-Dense Gravel
7.3-11.5	Very Dense Gravel	9.8-14.6	Loose-Med Sand
11.5-16	Medium Sand	14.6-22.8	Dense Sandy Gravel

H3		H4	
Depth (m)	Soil Conditions	Depth (m)	Soil Conditions
0-3.1	Fine Sand	0-5.8	Soft Silt
3.1-4.9	Stiff Silt	5.8-11.3	Very Dense Gravel
4.9-10.6	Gravelly Sand	11.3-15.9	Fine Sands
10.6-14.3	V.Dense Gravel		

Table 1 Summary of Soil Conditions at the Load Tested Shafts

Offshore Pressuremeter Testing

A TEXAM hydraulic control unit was used operating the long NX probe of 1850 cc deflated volume. The use of the PMT was directed toward the upper 18m of all deposits, which controls lateral response of the shafts. Drilling with NJ rod and the circulation of drilling mud was protected from the 5.5m/s river currents inside a 203mm diameter conductor pipe set below the river bottom. From a total of 18 prepared test sections 5 collapsed under artesian pressure in the clean sands and some occasional squeezing of the hole was evident. Consistent with all insitu testing in coarse gravels the PMT does have limitations in achieving the maximum success rate of highest quality data in gravely soil. Table 2 illustrates the range of PMT properties; net limit pressure P_L*, initial modulus, E_o, and reload modulus, E_R, for the offshore boreholes in all soil encountered. By following procedure B of ASTM D4719-87 no creep pressures are measured. The overall success for the tests, conducted over 22m of fast moving water, in these difficult conditions and aggressive soils was remarkable, and provided quality soil characterization for P-y work.

With the exception of the near surface silts all soils were revealed as showing fully drained behavior. As a way of assessing test quality in these material a maximum 3 star (***) rating is used, and shown in Table 2.

Test No.	Depth Below Mudline (m)	Initial Modulus Eo (kPa)	Net Limit Pres. Pl* (kPa)	Ratio Eo/Pl*	Reload Modulus Er (kPa)	Soil Type
3 ***	4.6	5506	512	10.7	23078	Silt
6 ***	3.2	5506	613	9.0	28250	Silt
7 ***	5.5	6560	670	9.8	20396	Silt
9 **	7.4	>4980	991	5.0	12310	Sandy Silt
10no star	9.75	>10500	----	----	-----	Silty Gravel
11 **	7.0	9528	1110	8.6	20636	Gravely Sand
12 **	8.23	16758	2050	8.2	72970	Sandy Gravel
15 ***	8.96	9145	1039	8.8	53201	Sandy Gravel
16no star	11.28	>1700	-----	-----	>7600	Loose Gravel
19 ***	16.03	4596	411	11.2	16087	Silt
20 ***	11.43	8427	527	16	25376	Sandy Silt
22 **	16.67	16280	1603	10.1	33611	Gravel
25 ***	21.31	5315	723	7.4	34760	Sand

NOTE: Test numbers not listed are either calibration tests, or tests which were not be attempted due to heave

Table 2: Summary of Offshore Pressuremeter Tests

The full *** rating is given to the highest quality PMT test curve which contains a well defined modulus, sufficient plastic behaviour to approach a limit pressure, $P_{L,}$ and shows no adverse effects from borehole preparation.

DESIGN LOADING AND P-Y METHODOLOGY

Predicting the resistance to lateral loading for a shaft is a complex, non-linear, soil-structure interaction problem, which has received considerable attention in the last 20 years. Significant parameters include the magnitude, duration and frequency of the loads, shaft stiffness and geometry, and the soil layering and properties. Since the low flume (L) is submerged during projected frequent high river elevations and the high flume (H) by flood conditions, repeat cycle downstream loads, of low frequency, are expected. In addition, load from river currents to the submerged portion of the shafts above the riverbed and combined loads resulting from ice, floating debris, wind and seismic must all be considered. Tolerable flume joint lateral movements must be restrained to 0.76m differential to the river shore. The foundation type was restricted to the single large shafts to minimize adverse hydraulic effects, such as eddying, and avoid providing predator fish habitat that threaten the juveniles. Due to these considerations, the non-linear LPILE code was selected using PMT based P-y procedures.

P-Y Procedures

The method chosen as most suitable to capture these large shafts was the Briaud/Smith method (Smith 1987, Briaud 1992). These procedures take account of installation disturbance, soil layering, pile stiffness, and the depth of reduced soil resistance, Dc, effects. The method calls for construction of both the 'front' pressure (Q-y) and side shear (F-y) mobilization curves, which are added for the P-y curve, after slope and disturbance effects are included, Figure 2.

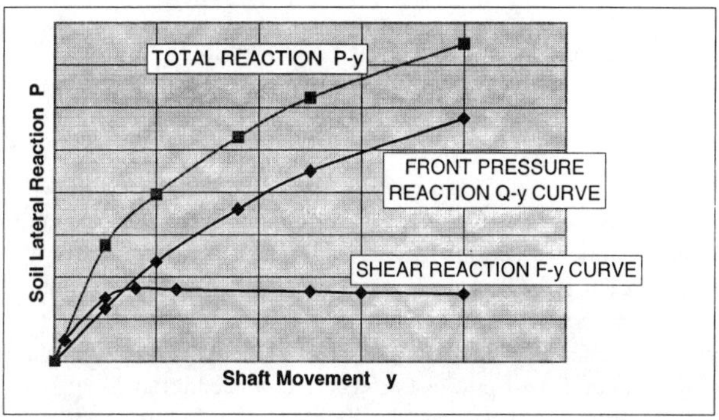

Figure 2 Front Pressure and Side Shear Components of the P-y Curve

Use of the SPT together with density descriptions was made as a correlation tool to complete a full set of PMT curves for each soil at each location. Tolerable movements at the mudline would be restricted to less than 4% of the shaft diameter. The depth of reduced resistance, Dc, varied between 1 and 1.8 pile diameters in depth, and controls the important near surface curves. The Briaud/Smith procedure calls for use of the initial PMT modulus on zero displacement piles to capture the lack of any driving benefits. Most empirical procedures are derived from load tests in sand on *closed end* full displacement piles. Closed end piles would feel the beneficial increase in horizontal stress and modulus from higher confining pressures and soil densification. These conditions would not apply at the Bonneville tests due to the vibrodriven installation of the open permanent casing and possible liquefaction of clean granular deposits, including gravels. The choice of the appropriate early strain soil modulus, Es, is critical to predict movement of these shafts. Most theoretical work supports the conclusion that initial Es values are *not* functions of shaft diameter.

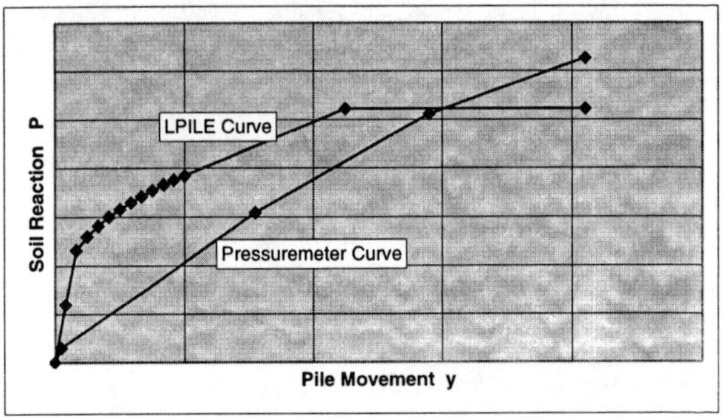

Figure 3 P-y Curves at 1.5 m Depth for H3 Shaft

A typical P-y curve at a depth of 1.5m (1/3 way down the Dc effect) at the H3 shaft is shown in Figure 3 which compares the PMT curve and LPILE subroutine generated curve for medium sand with ϕ= 36 degrees and k= 16.3 MN/m^3. The SPT blow count at this depth was 24. Using the PMT based curves prediction of shaft riverbed, and flume, deflections were made for all offshore shafts by the structural engineers with LPILE. These deflections were close to the maximum tolerated and raised concerns about the flume movement and the accumulated effects of repeat cycles over the structure's 50-year life. The geotechnical design team assembled a database of 58 lateral load pile tests in the technical literature, of which 24 were in granular soils and contained cyclic loading. A preliminary relationship was established between the initial first cycle deflection, rebound, and repeat load deflection increase for movements less than 5% of the shaft diameter.

The decision to proceed with a full-scale proof load test on the production shafts was based on the need to provide the following:

- Confirmation that first cycle deflections were correctly predicted by LPILE, and that the diameter effect are properly represented with PMT P-Y procedure.
- An understanding of pile behaviour under repeat, low frequency, lateral cyclic loads and high moments.
- Evaluation of possible creep under constant load.
- Provide a basis to activate, or delete, the large proposed 'Drag Collar' improvement option contained in the contract.

LOAD TESTING PROCEDURES AND INSTRUMENTATION

Test Setup

Of the six shafts vibro-driven in the river, four had significant 'stick up' heights and predicted deflections which were of concern. These were H3, H4, L3 and L4. Testing could be achieved by passing a multistrand steel cable through steel sleeves for the shaft pairs H3 to H4, and L3 to L4, which could then be jacked toward each other for reaction, thus placing the cable in tension, Figure 4.

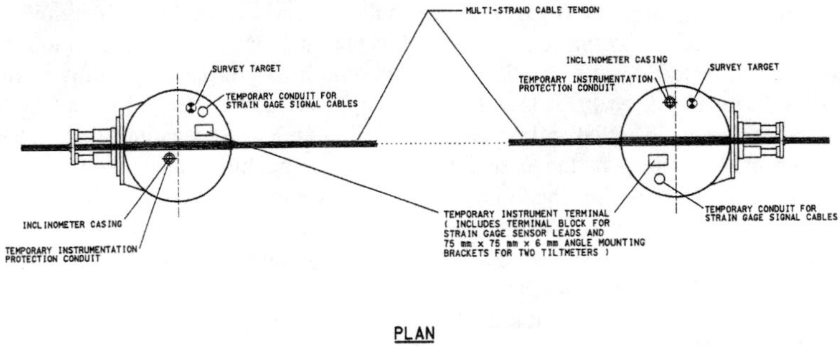

Figure 4: Plan of Load Testing Configuration in Tension

The load would be applied perpendicular to the river current and in line with the proposed flume. This prevented any shaft preset and soil stress in the future downstream current loading direction. Thus, a total of 4 load tests would be conducted on offshore shafts of diameters much larger than currently reported in the literature.

The shaft placement and load-testing requirement called for the contractor to install a temporary support trestle clear of the fluctuating river elevations. This 105 m x 22 m trestle was timber decked and provided access to all construction plant. Eighty 0.91m diameter steel pipe piles supported the trestle. The 3.05 m production shafts were concrete filled up to the final design elevation with additional stiffening into the extended casing portion used for load application via the cable. An anchor slab was constructed around the 200mm diameter steel sleeve and stiffened the shaft where the rams reacted externally against the shaft.

A 40m long, 19 strand, multi-cable tendon with 4.0 MN capacity was hung to twin 1.8MN jacks at the shaft, giving a 3.6 MN load capacity to each shaft. The anchor head assembly comprised bearing plates, hydraulic jacks, yoke, load cell and anchor head. Each jack had a maximum throw of 0.61m, which provided for a total cable extension of 1.22m. At the peak anticipated load the cable stretch in tension would consume 260 mm leaving a total combined shaft deflection, at the point of load application, of 0.96m. An LPILE prediction for this loading geometry configuration indicated this amount of deflection at the cable point would generate of the order 40mm to 70mm shaft deflections at the riverbed.

Instrumentation

A wide array of instrumentation was present on each of the 4 shafts. These comprised tiltmeters, strain gages, indinometers, survey points and load cells. Geokan model 6350 vibrating wire tiltmeters were installed at the shaft top, in pairs, to measure tilt in the loading direction and the orthogonal direction. Thirty-six Geokon vibrating wire strain gages were installed on H4 and L4, and thirty-two on H3 and L3. These were placed in sets of 4 at the quadrant points around the rebar cage and protected during installation. Inclinometers were used to monitor shaft deflected shape after each load, and unload, application. On H3 and L3 a SINCO 70mm casing was installed and on H4 and L4 a 85mm casing was installed. These were grouted in place at the base into NX sized core holes drilled through the concrete and extended into bedrock. (Two survey points on each shaft provided additional data and a check to the inclinometers on the top of shaft.) Finally, a vibrating wire load cell was placed in line with the anchor head and yoke at the North end of each cable tendon, at H3 and L3.

An 80 channel Geomation 2380 automated data acquisition/reduction system monitored strain gages, tilt meters and load cell every 2 minutes. Four 24-pair direct burial type signal cables were used to connect instrumentation to the Geomation 2380. Frequency of monitoring the survey points was 10 to 15 minutes. The Contractor's personnel operated the hydraulic load system for applying jack pressure.

Test Procedure

A number of limiting 'red flag' (stop test) criteria were agreed upon by the team. The maximum allowable peak mudline deflection of 100mm, or, a maximum allowable permanent deflection of 50mm, controlled the allowable load level. Test loads governed by the jacks could not exceed 3.6 MN. By way of comparison peak equivalent 100-year flood loads were estimated at 3.1 MN for shaft H4 and 2.67 MN at L4.

Each test load was initiated at 222 kN and increased in increments of 444 kN. A minimum unload condition of 222 kN was maintained for the entire test to ensure no slippage in the stressing system. The typical data acquisition sequence followed:

- Load application, shaft deformation electronically tracked till stabilization reached (½ to 1 hour).
- Hydraulic load and survey data gathered
- Inclinometer readings on both piers, (1 hour)
- Inclinometer data downloaded from SINCO Datamate and processed by Digipro®

- Tiltmeter, load cell and strain gage data extracted
- A second set of hydraulic loads & survey data entered to establish any shaft 'drift'

To assist field crews in control of the test and allow data evaluation by the design team a 'decision tree' was employed to establish the proper course of action after each load step. The full series of tests conducted on the H shafts lasted 9 days and, following 7 days of demobilization and transfer, a further 9 days to the L shafts. Typically, the loading sequence varied between 2 hours to 6 hours with a creep portion of 24 hours built into the L tests at about the 1.78 MN load.

Table 3 presents the unrestrained 'stick up' height from riverbed to the loading cable, load levels and number of cycles for all shafts.

SHAFT: STICK-UP HEIGHT (m) H3:23.8, H4:24.5, L3:24.6, L4:24.2

Peak Load (MN)	.445	.890	1.34	1.78	2.22
H3 & H4 # of cycles	2	4	5	3	2
L3 & L4 # of cycles	6	6	5	1*	2

*24 hour creep test.

Table 3 Unload-Reload Test Summary

. Each unload cycle was taken down to the baseline minimum of 222kN prior to reloading up to the next selected load level.

LOAD TEST RESULTS AND DESIGN SERVICE LIFE

Results

Figures 5 and 6 show the mudline deflection, measured by the inclinometer against applied load, recorded by the load cell, for L and H shafts respectively. The plots confirm nonlinearity is present at all these comparatively low load levels for such large shafts. This confirms that nonlinear P-y based LPILE application is appropriate even with peak-measured deflections less than 3½% of shaft diameter. The LPILE reported predictions are based on subroutines in the software and follow the procedures given by Reese et. al. (1974) for submerged granular soils. It does appear the vibro installation procedure for these shafts, with zero displacement, leaves these sands and gravels in a loose state following possible partial liquefaction. This seriously damages the small strain modulus and produces *higher* deflections.

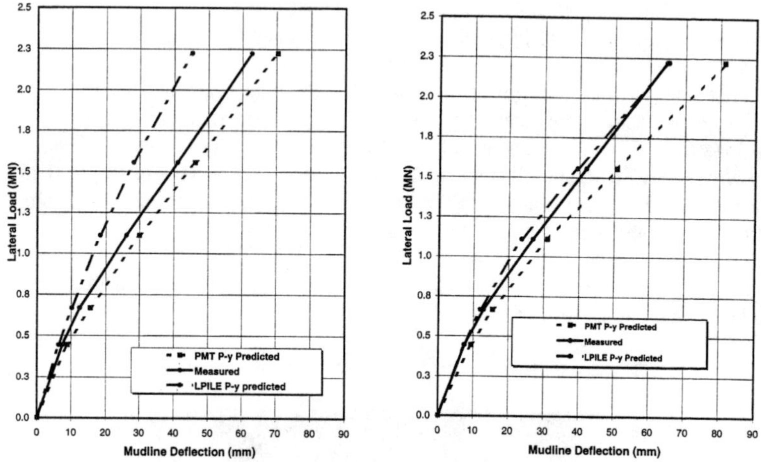

Figure 5: Measured and Predicted Load Deflection for L3 and L4

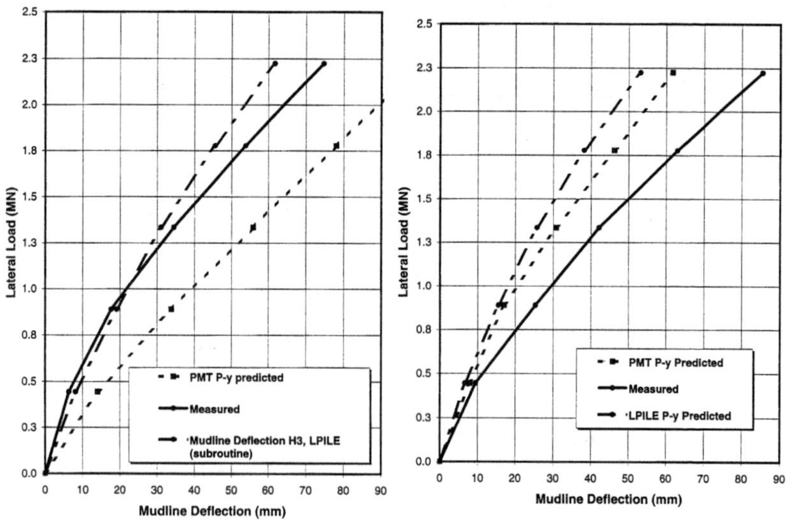

Figure 6: Measured and Predicted Load Deflection for H3 and H4

Figures 5 and 6 also show the predicted response from both P-y methods. The PMT procedures for P-y construction are shown to have done very well in predicting the response, with acceptable small conservatism. Taken together the LPILE subroutines and PMT based methods have quite well 'bracketed' the shafts response in 3 of the 4 load tests. It should be noted that at the time of shaft installation H4 encountered a large boulder, which required rock chisels to break up and facilitate removal from inside the open casing. This is believed to have further loosed the gravels and produced higher deflections under load. All other shafts were vibro-driven continuously without interruption.

Design Service Life

To model the effects to peak deflection from repeat load flood cycles the popular power law relationship below was used:

$$y(N) = y(1) \times N^{\alpha} \quad \text{Eq. (1)}$$

Using historic hydrologic data from Bonneville Dam, a statistical model was developed by the Portland District to represent the expected variations of river elevation from flood and electric power generation requirements. These variations were converted by the design team to shaft load levels and the number of load cycles projected over a 50-year design life. These are presented in Table 4 for both H and L shafts. The higher number of loads and cycles at the L shafts are caused by the frequent submerged conditions for the low elevation flume.

L3 and L4		H3 and H4	
Design Load (MN)	No. of Cycles	Design Load MN	No. of Cycles
0.756	2660	0.267	40
1.023	362	0.556	49
1.29	72	0.734	29
1.82	144	1.0	35
2.36	26	1.334	42
2.713	2	1.779	4

Table 4: High (H) and (L) Shaft Loads and Cycles for 50-year life

Back calculated α values for each of the 4 shafts load tested, and subjected to the number of cycles presented earlier, ranged from 0.07 to 0.1. From the site-specific α values, and equation 1, the mudline and flume deflection could be calculated for the load ranges given in Table 4. Predicted deflection at L3, L4, H3 and H4 were 45%, 64%, 30%, and 82% respectively of the maximum allowable using the actual load test result (not LPILE predictions), and equation 1. Thus total deflections for all shafts over 50 years were less than the maximum allowable, which if exceeded would have rendered the flume inoperable.

The effect of the boulder and subsequent disturbance at H4 is seen in the highest percentage of allowable deflection consumed by the expected load cycles over 50 years. All shafts have acceptable factors of safety. As a result of the 24-hour constant load test at 1.78 MN for L3 & L4, creep was not judged to be of concern. At the present, the instrumentation continues to be monitored to track the actual performance of the shafts.

SUMMARY

From a review of the literature the Authors believe these load tests set a new record in diameter for load testing under horizontal load. As a result of the site investigation and full-scale offshore cyclic load testing to these 3.05m diameter shafts a number of conclusions can be advanced.

- These large diameter, zero displacement shafts, installed by vibro techniques lower the pile lateral soil modulus. Great care should be exercised in constructing P-y curves, and some judgement applied to selecting the initial modulus of subgrade reaction. Backcalculated constant of subgrade reaction values at these shafts are around 2.3 MN/m^3 for the 2.22 MN load.
- Successful high quality pre-bored pressuremeter tests can be made offshore in silt/sand/gravel deposits.
- Pressuremeter based P-y procedures were successfully employed and satisfactorily accommodated the pile diameter effect.
- Calibration of predictive models by full scale testing permitted design life deflection issues to be resolved. The cost saving to the Corps of Engineers from deleting the Drag Collar retrofit option amounted to $1M.
- The need for effective teamwork between owner-structural and geotechnical consultants is of paramount importance, particularly on soil/structure issues. The Portland District Corps of Engineers technical staff commented on the excellent teamwork demonstrated in this project.
- Consideration should be given to ensuring resources are available to monitor large foundation throughout the design service life. Thus future data analyses can confirm the long term performance is on target to ensure a full service life, or if flood load produce excessive deflection remedial measures can be planned well in advance.

APPENDIX I : REFERENCES

Briaud, J.L. (1992) The Pressuremeter, Balkema Books, pp 144-151

Reese, L., Cox, W. R. and Koop, F.D. (1974) "Analysis of Laterally Loaded Piles in Sand", OTC Paper 2312, Proc., 7th Offshore Technology Conference, Houston, TX

Smith, T.D. (1987) "Friction Mobilization F-y Curves for Laterally Loaded Piles from the Pressuremeter", Proc., Prediction and Performance in Geotechnical Engineering, Calgary, Canada. pp 89-96

APPENDIX II : SYMBOLS

$y(1)$ = Measured deflection on the first (initial) cycle.

$y(N)$ = Deflection after N cycles

N = Number of cycles

α = Power exponent

Es= Pile Soil Modulus

RAMMED AGGREGATE PIERS DEFEAT 75-FOOT LONG DRIVEN PILES

Kord J. Wissmann[1], Associate Member, ASCE, Nathaniel S. Fox[2], Associate Member, ASCE, and John P. Martin[3], Member, ASCE

ABSTRACT: A six-story parking garage was designed to be supported by 23 m (75 foot) long steel pipe piles driven through loose silts and silty sands. Rammed Aggregate Piers with shaft lengths ranging between 2.1 m and 2.7 m (7 feet and 9 feet) were proposed and accepted as an alternative to the driven piles. In this system, highly densified aggregate piers are incorporated within the upper weak and compressible soils resulting in a composite bearing material that is substantially stiffer than the unimproved soil, and on which shallow foundations can be supported with tolerable settlements. Methods for analyzing and predicting settlements of footings supported on aggregate pier-reinforced soils are discussed. Design parameters were verified by the performance of aggregate pier modulus load tests. Total settlements of footings were estimated to be less than 4 cm (1-1/2 inches). Measured footing settlements were less than 1.3 cm (0.5 inch). The implementation of the aggregate pier system saved the project owner $187,000.

INTRODUCTION

The new six-story Justice Center parking garage in Washington County, Oregon was supposed to be supported by 23 m (75 foot) long steel pipe piles driven through loose silts and silty sands. Instead, the parking deck was designed using high bearing pressure spread footings supported by soil reinforced with 2.1 m to 2.7 m (7 foot to 9 foot) long *Rammed Aggregate Piers™*. The project owner saved $185,000 in foundation costs, a value that is greater than 50% of the bid price for

---

[1] Chief Engineer, Geopier Foundation Co., Inc., 515 Sunrise Drive, Blacksburg, VA 24060, kordw@earthlink.net

[2] President, Geopier Foundation Co., Inc., 11421 East Aster Drive, Scottsdale, AZ 85259, geopier6@aol.com

[3] Vice President, Geopier Foundation Company – Northwest, 3234 Southeast 39th Ave., Portland, OR 97202, johnm@geopier.com

driven piling. The performance of the foundation system may seem astonishing – foundation settlements have been less than 1.3 cm (0.5 inch). This paper describes how such a dramatic change in foundation concepts is possible, the engineering mechanics that govern the behavior of Rammed Aggregate Piers, the results of aggregate pier load tests performed for the project, and the settlement performance of the completed structure.

BACKGROUND

The six-story parking garage, built by the Hoffman Construction Company, is a cast-in-place reinforced concrete structure with a ductile frame design (Figure 1). Column loads were determined by KPFF structural engineers to range between 180 kN and 8000 kN (40 kips and 1,800 kips).

FIGURE 1: PHOTOGRAPH OF COMPLETED PARKING GARAGE

Geotechnical Subsurface Conditions

Figure 2 presents profiles of the soil boring and Cone Penetration Test (CPT) sounding that were performed at the parking garage site. The explorations indicate the following generalized stratigraphy:

- **Loose silty fine sand and sandy silt** extends from below the surficial pavement to a depth of about 6.1 m (20 feet). Standard Penetration Test (SPT) N-values ranged between 6 and 8 (average of 7). CPT tip resistances generally ranged between 1.4 MN/m^2 and 5.7 MN/m^2 (15 tons per square foot (tsf) and 60 tsf) with an approximate average of 2.9 MN/m^2 (30 tsf). Laboratory test results indicate an average dry density of 1420 kg/m^3 (89 pcf) and an average soil moisture content of 33 % for this layer.
- **Medium-dense silty fine sand** extends from a depth of about 6.1 m (20 feet) to depths greater than 15.2 m (50 feet) below grade. SPT N-values ranged between 11 and 31 (average of 17). CPT tip resistances generally ranged between 3.8 MN/m^2 and 15.2 MN/m^2 (40 tsf and 150 tsf) with an approximate average of 9.5 MN/m^2 (100 tsf).

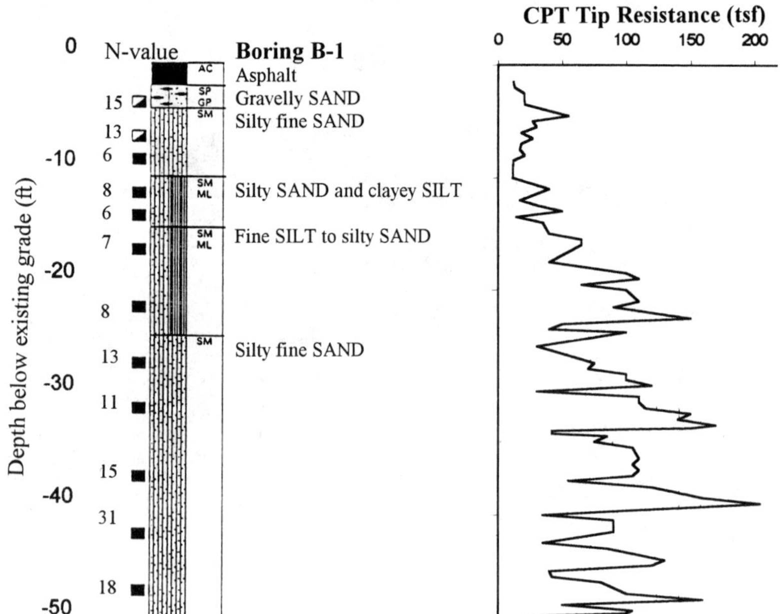

FIGURE 2: SUBSURFACE CONDITIONS

Foundation Challenges

The design of parking garage foundations posed a significant challenge because of the heavy column loads combined with the weak and compressible soil conditions. Conventional shallow spread footings were deemed not to be feasible because of the likelihood of excessive settlements that would exceed the settlement criterion of 3.8 cm (1.5 inches). Consequently 23 m (75 foot) long 30.5 cm (12 inch) diameter concrete-filled driven steel pipe were specified. The low-bid cost of the pipe piles was $360,000.

An Innovative Proposal

As a value-engineering alternative to the deep foundations, Rammed Aggregate Piers were proposed for construction by *Geopier™* Foundation Company – Northwest. The patented system is designed to improve the subsurface soil conditions and allow the use of high-bearing pressure shallow spread footings for foundation support. The proposal included providing design calculations, performing two modulus load tests to confirm assumed design parameters, installing 521 production piers for foundation support, and recommending foundation settlement monitoring. The total cost of the design-build Rammed Aggregate Pier system was $175,000. The project owner accepted the proposal on the basis of technical merit and the large foundation cost savings exceeding 50% of the cost of the driven piles.

CONSTRUCTION

Prior to the construction of the Justice Center parking garage, Rammed Aggregate Piers were used to support structures at approximately 45 project sites in 11 different states including 4 project sites in Oregon. Generalized construction methods are described by Lawton and Fox (1994) and Lawton et al. (1994).

The aggregate piers installed at the Justice Center parking garage were constructed by drilling 84 cm (33 inch) diameter holes to depths ranging between 2.1 m and 2.7 m (7 feet and 9 feet) below the footing bottoms, placing controlled lifts of aggregate stone within the cavities, and compacting the aggregate using a specially designed high-energy impact tamper (Figure 3). The first lift consists of clean stone and is forced into the soil thus forming a bottom bulb below the excavated shaft. The bottom bulb effectively extends the design length of the aggregate pier element by one pier diameter. The piers were completed by placing additional one-foot thick lifts of aggregate over the bottom bulb and densifying the aggregate with the beveled tamper.

The piers were designed and installed to cover approximately 35% of the gross area of the overlying footing elements. High bearing pressure spread footings, with an allowable composite bearing pressure of 270 kN/m^2 (5,600 psf), were

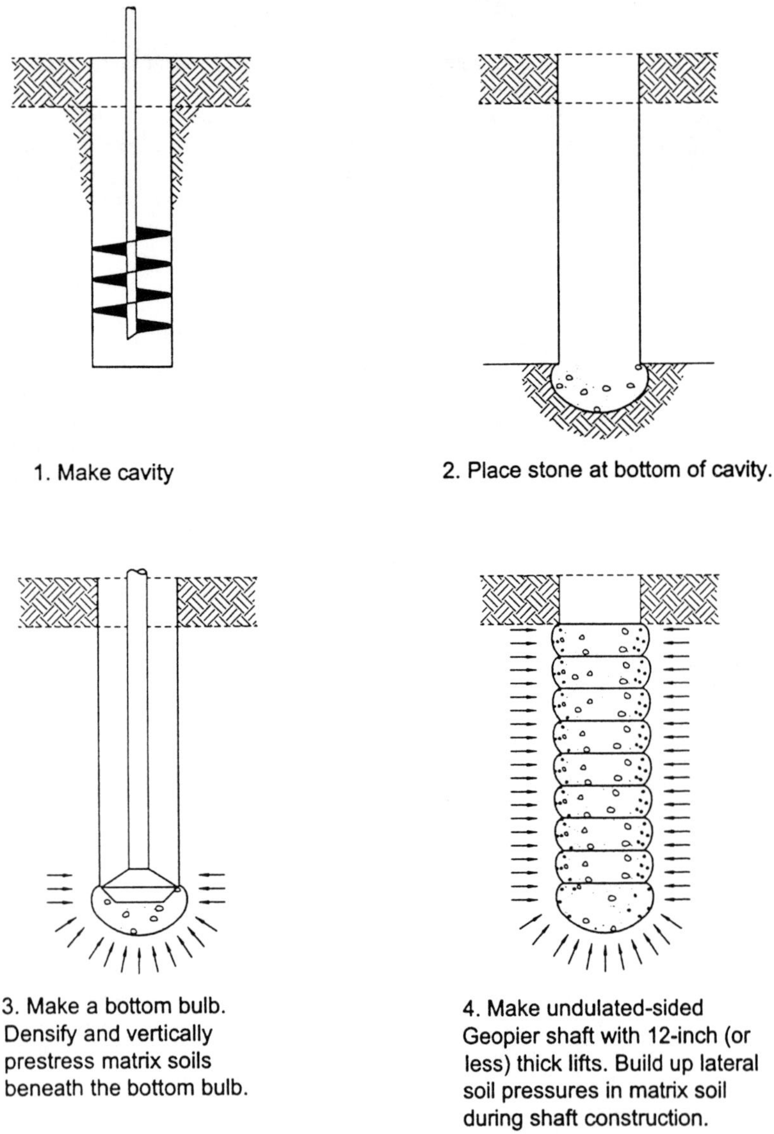

FIGURE 3: RAMMED AGGREGATE PIER CONSTRUCTION

constructed directly over the aggregate piers installed at the parking garage site. The smallest square footings were 2.3 m (7.5 feet) wide and were supported by 3 aggregate piers. The largest square footings were 6.7 m (22 feet) wide and were supported by 25 aggregate piers.

As shown on Figure 2, the 2.1 m to 2.7 m (7 foot to 9 foot) long aggregate piers do not extend completely through the Layer 1 silts and are not considered to be end-bearing elements. Rather, the piers are designed to improve the overall stiffness of the subsurface soils at depths in which footing-induced stresses are the highest in order to limit long-term foundation settlements to the design criterion.

During densification, the beveled shape of the tamper forces stone laterally into the sidewall of the excavated cavity. This action increases the lateral stress in the matrix soil thus providing additional stiffening.

MECHANICS

The aggregate piers greatly increase the bearing capacity of the reinforced soil matrix and significantly reduce foundation settlements. Settlements calculations are performed to estimate the compression of the zone of matrix soil reinforced by the aggregate piers (upper zone) and to estimate the compression of the zone of soil that is subject to footing stresses and is located below the tips of the piers (lower zone).

Upper zone calculation procedures are based on a spring analogy (Lawton and Fox 1994 and Lawton et al. 1994) as described in the following:

1. The footing is assumed to be perfectly rigid relative to the foundation materials. Thus, the stresses applied to the composite foundation materials depend on their relative stiffnesses (R_s) and area coverage. The total downward force (Q) on the footing, which may be expressed as the product of composite stress (q) and footing area (A), is resisted by a total upward resisting force in the rammed aggregate piers (Q_g) and soil (Q_s) materials:

   $$Q = q\,A = Q_g + Qs = q_g\,A_g + q_s\,A_s\ , \tag{1}$$

   where q_g is the stress at the top of the aggregate pier elements, A_g is the area of the pier elements below the footing, q_s is the vertical stress on the matrix soil below the footing, and A_s is the area of the matrix soil in contact with the bottom of the footing.

2. Because the footing is essentially rigid compared to the bearing materials, the settlement of the pier will equal the settlement of the matrix soil. The settlement of the foundation (s) can be written in terms of aggregate pier

stress and aggregate stiffness modulus (k_g) or in terms of the matrix soil stress and matrix soil stiffness modulus (k_s):

$$s = q_g / k_g = q_s / k_s \quad . \tag{2}$$

3. Equation 2 can be rewritten to express the matrix soil stress in terms of the aggregate pier stress and the ratio of the pier and matrix soil modulus values (R_s):

$$q_s = q_g (k_s / k_g) = q_g / (k_g / k_s) = q_g / R_s \, . \tag{3}$$

4. Combining Equations 1 and 3 and defining area ratio (R_a) as the ratio of A_g to A:

$$q = \{q_g A_g / A + q_g A_s /(A R_s)\} = [q_g R_a + q_g (1 - R_a) / R_s\} = \\ = \{q_g [R_a + 1/R_s - R_a/R_s\} = \{q_g [R_a R_s + 1 - R_a] / R_s\} \, . \tag{4}$$

5. Rewriting q_g in terms of q:

$$q_g = \{q R_s / [R_a R_s + 1 - R_a]\} \, . \tag{5}$$

6. Upper-zone settlements are computed using Equations 2 and 5 which depend on the applied composite footing stress, the relative stiffness of the aggregate pier and soil materials, the area ratio of the aggregate pier elements, and the aggregate pier stiffness modulus.

Estimates of settlements in the lower zone material below the bottom of the aggregate pier bulb are computed using conventional geotechnical settlement analysis procedures well described in the literature (Terzaghi and Peck 1967) combined with soil elastic modulus values interpreted from the results of the penetration tests. The analysis includes the assumption that the lower zone footing-induced stress may be estimated using solutions for a footing supported by an elastic half-space. This assumption is believed to be conservative because the presence of the piers results in a more efficient stress transfer with depth below the footing bottoms than the stress transfer that occurs for conventional spread footings. Example design calculations are shown on Figure 4.

DESIGN PARAMETER VERIFICATION: MODULUS LOAD TESTS

To verify the assumed modulus values used for the aggregate piers, two full-scale aggregate pier modulus tests were conducted prior to construction. The tests were performed by placing circular steel plates over the full cross-sectional area of an installed aggregate pier element and then applying pressure in gradual increments. The maximum applied stress corresponded to 150% of the design stress computed at the top of the aggregate pier elements.

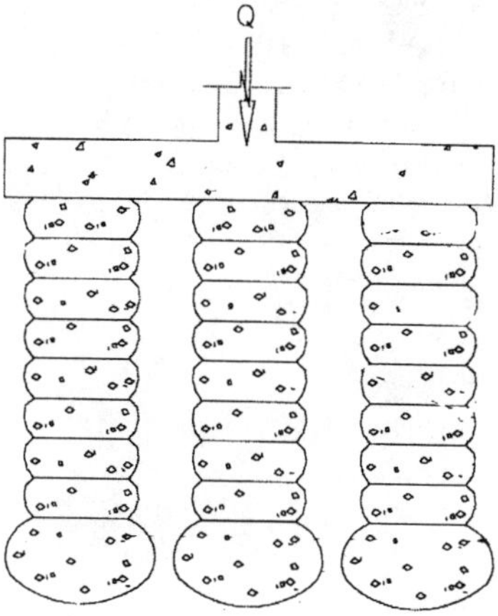

$Q = 375$ kips

Footing width, $B = 9.5$ ft.

$q = Q/B^2 = 4160$ psf.

No. piers, $N = 5$

Pier diameter, $d = 33$ inches.

$R_a = A_g/A = 0.33$.

Design pier stiffness modulus, $kg = 240$ pci.

Assumed soil stiffness modulus, $ks = 20.8$ pci.

Stiffness ratio, $R_s = k_g / k_s = 11.5$.

Settlement Calculations

Aggregate pier stress, $q_g = q\, R_s / \ (R_s\, R_a - R_a + 1) = 11{,}495$ psf.

Upper-zone settlement $= q_g / k_g = 11{,}495$ psf / 240 pci = <u>0.33 inch</u>.

Thickness of upper zone = 7 feet + 33 inches = 9.8 feet.

Thickness of compressible lower zone, $H_{lz} = 2\,B - 9.8$ feet = 9.2 feet.

Ratio of depth to center of lower zone to footing width, $z/B = 14.4$ ft / 9.5 ft = 1.5.

Westergard influence factor for z/B, $I_\sigma = 0.12$.

Estimated lower zone modulus of elasticity, $E = 150$ ksf

Lower-zone settlement $= (q\, I_\sigma\, H_{lz} / E)$ = <u>0.37 inch.</u>

Total settlement = 0.33 inch + 0.37 inch = <u>0.7 inch.</u>

FIGURE 4: DESIGN CALCULATIONS

Test results indicate that top of pier deflections ranged between 0.5 cm (0.20 inches) and 0.81 cm (0.32 inches) for 2.7 m (9 foot) long and 1.8 m (6 foot) long piers, respectively, when subjected to 100% of the pier design stress (Figure 5). These values were used to compute aggregate pier modulus values ranging between 65.4 MN/m^3 (241 pounds per cubic inch) and 104 MN/m^3 (385 pci) in accordance with Equation 2. The measured values confirmed and exceeded the design-level modulus value of 65.1 MN/m^3 (240 pci).

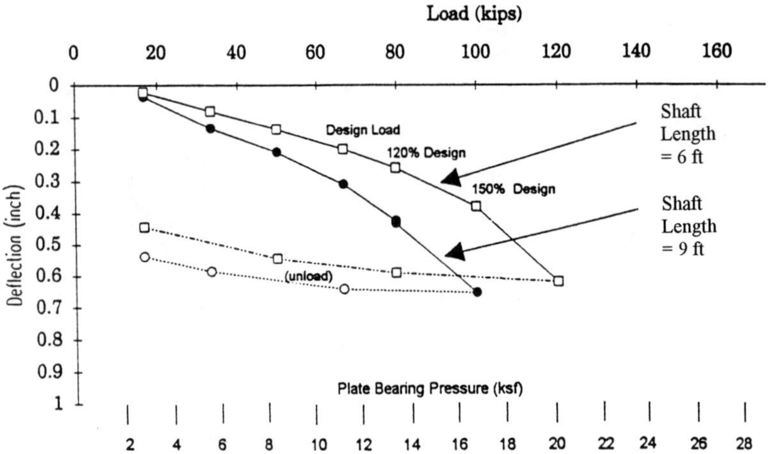

FIGURE 5: MODULUS LOAD TEST RESULTS

FOUNDATION SETTLEMENTS

The results of settlement measurements are presented on Figures 6 and 7. Measurements were taken on the top of the concrete footings overlying the aggregate piers. The measurements indicate that less than about 0.25 cm (0.1 inches) of settlement occurred per each completed floor of construction. Measurements are available for the first five floors of the parking garage. Measurements are not available for the last floor of construction because the project owner decided to halt construction after the fifth level and to add the sixth level at a later date. Total settlements are estimated to be on the order of 1.27 cm (0.5 inches) for the instrumented columns when loaded by all 6 floors of the parking garage with major column loads ranging from 2,700 kN to 8,000 kN (600 kips to 1,800 kips).

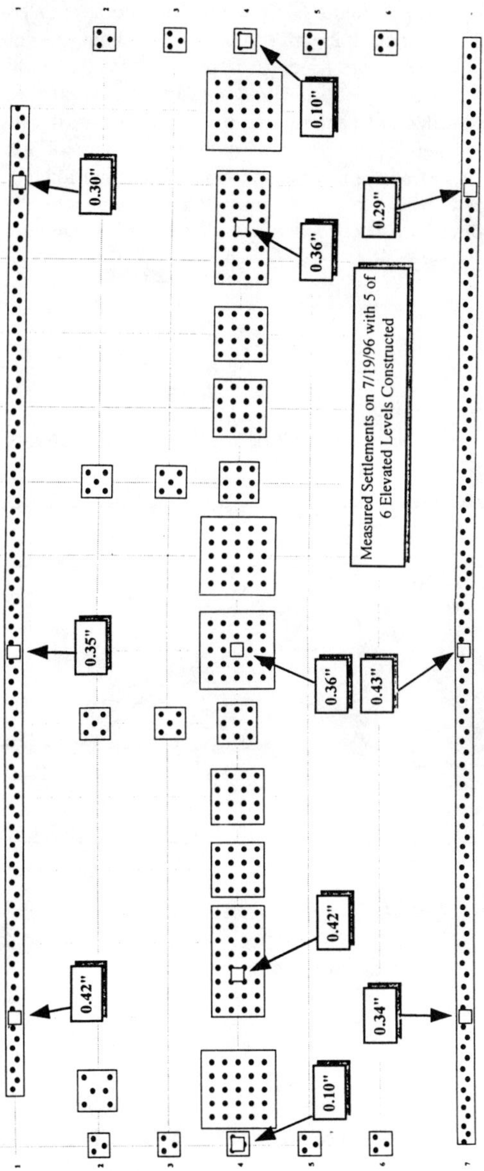

FIGURE 6: FOUNDATION FOOTPRINT

The spatial distribution of settlements after the completion of five floors of construction is shown on Figure 6. It is interesting to note that small, lightly-loaded footings settled only about 0.25 cm (0.1 inches) or about one-fourth as much as the adjacent heavily-loaded large footings despite the fact that footing-bottom stresses are comparable. While this observation appears to confirm the notion that footing settlement magnitudes are proportional to footing width, it calls into question the commonly held belief that a large, heavily-loaded footing will have a great influence on the settlements of adjacent footings. One explanation for this observation is that the stress regime below footings supported by aggregate piers is so markedly changed that conventional concepts of foundation load spreading are not applicable.

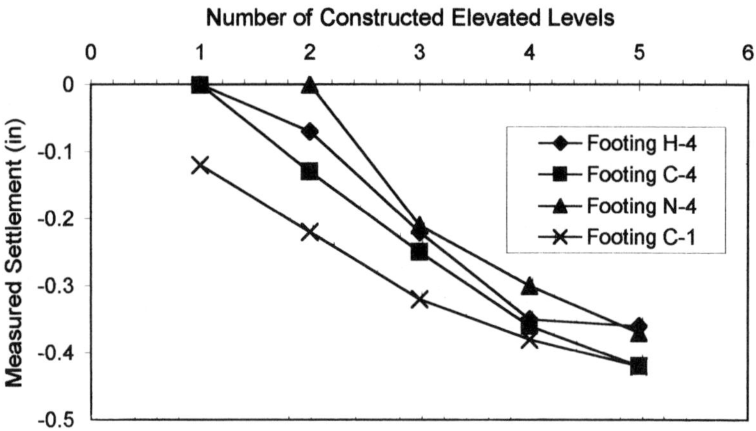

Note: Column loads at instrumented footings ranged between 600 kips and 1200 kips after construction of 5 elevated levels

FIGURE 7: SETTLEMENT MEASUREMENTS

SUMMARY AND CONCLUSIONS

The Washington County Justice Center Parking Garage is supported by an innovative intermediate foundation system consisting of high capacity spread footings supported by Rammed Aggregate Piers. This foundation system was selected because of sound engineering mechanics used for design and because of the significant cost savings available to the owner. The following itemized statements summarize foundation construction:

1. The 2.1 m to 2.7 m (7 foot to 9 foot) long aggregate piers replaced steel piles designed to be driven to 23 m (75 feet) below grade.
2. The aggregate piers do not extend to the underlying medium-dense silty sand layer but, rather, are terminated within the upper silt stratum.
3. Spread and strip footings placed above the rammed aggregate piers were designed for a composite bearing pressure of 270 kN/m^2 (5,600 psf).
4. Assumed design parameters were verified by field modulus tests.
5. The settlements of footings supported by the aggregate pier elements have been measured to be less than 2.5 cm (1 inch).
6. The aggregate piers facilitated a 50% savings in foundation costs to the project owner.

To date approximately 70 major structures are supported by Rammed Aggregate Piers in the Pacific Northwest; aggregate piers nationwide support more than 250 structures (Fox 1999).

APPENDIX-I. REFERENCES

Dames and Moore (1995). Geotechnical investigation report for Washington County Justice Center, 1995.

Fox, N.S. (1999). Personal communication, Geopier Foundation Company, Inc. project summary list.

Lawton, E. C., and N. S. Fox (1994). "Settlement of structures supported on marginal or inadequate soils stiffened with short aggregate piers." *Geotechnical Specialty Publication No. 40: Vertical and Horizontal Deformations of Foundations and Embankments*, ASCE, 2, 962-974.

Lawton, E. C., N. S. Fox, and R. L. Handy (1994). "Control of settlement and uplift of structures using short aggregate piers." *In-Situ Deep Soil Improvement,* Proc. ASCE National Convention, Atlanta, Georgia, 121-132.

Terzaghi, K., and R.B. Peck (1967). *Soil Mechanics in Engineering Practice,* John Wiley and Sons, New York, New York.

APPENDIX-II. SYMBOLS USED

A = Gross footing area.
A_g = Footing area supported by aggregate piers.
A_s = Footing area supported by matrix soil.

k_g	=	Stiffness modulus of aggregate pier.
k_s	=	Stiffness modulus of matrix soil.
Q	=	Total downward force on footing.
Q_g	=	Resisting force of aggregate pier.
Q_s	=	Resisting force carried by matrix soil surrounding aggregate pier.
q	=	Composite bearing pressure at base of footing.
q_g	=	Stress applied to top of aggregate pier.
q_s	=	Stress applied to matrix soil surrounding aggregate pier.
R_a	=	Ratio of cross-sectional area of aggregate piers to gross footing area.
R_s	=	Ratio of relative stiffness of aggregate pier and matrix soil.
s	=	Footing settlement.

Performance and Evaluation of Fills, Embankments, and Slopes

MONITORING THE CUT-OFF WALL PERFORMANCE OF UPPER HUIA DAM

Alaa S. Ahmed-Zeki[1], Nicholas Logan[2], Wallace McQuarrie[3], and Raveen Jaduram[4]

ABSTRACT: Upper Huia concrete gravity dam was constructed in Auckland-New Zealand, 70 years ago. The dam was not designed using present standards for uplift, although a concrete cut-off wall was constructed as a means to reduce the uplift pressures beneath the dam. This paper presents a description of the installation of vibrating wire piezometers to continually monitor the effectiveness of the foundation cut-off wall. The piezometer readings showed that the cut-off wall was providing the reduction in pore water pressures required for maintaining the stability of the dam.

BACKGROUND

Upper Huia Dam is a concrete gravity dam with a vertical upstream face and no curvature. It has a maximum height of 36.6 meters and a crest length of 166 meters, creating a lake having a storage capacity of about 2.44×10^6 m^3. Construction took place during the period 1926-1929. Located in the Waitakere Ranges west of Auckland (see Figure 1), it is one of Watercare Services Limited's 10 water supply dams in the Auckland region.

The dam is serviced by an uncontrolled overflow spillway consisting of seven weirs. It has a capacity of passing a flow with 1:100 recurrence, but the Probable Maximum Flood has been estimated to result in about 0.4 meters of overtopping.

[1]Geotechnical Engineer-Design Services (aahmed-zeki@water.co.nz), [3]Senior Systems Engineer, [4]Water Planning Manager, Watercare Services Limited, Private Bag 92 802, Penrose, Auckland-New Zealand.
[2]Instrumentation Specialist/Geotechnical, Opus International Consultants Limited-Central Laboratories, P. O. Box 30 845, Lower Hutt-New Zealand.

A cut-off wall of about 0.9 m width was excavated beneath the upstream face, and along the entire length, of the dam. Old rails were reported to have been vertically placed at about 1 meter centers, at the cut-off wall/dam interface to improve the bond (tensile strength) of concrete. The wall was designed to be approximately 3 meters deep, but was extended to about 6.1 meters at the deepest section of the dam to intercept two water bearing seams encountered during excavation. The cut-off wall was extended laterally into the abutments to form wingwalls.

Safety Evaluation reviews undertaken for Upper Huia dam in 1990, 1991 (Ref. 4 and 5) and 1998 (Ref. 6) included structural analyses evaluating the behavior of the dam under different loading conditions. Both studies note that even though the dam's design does not comply with the present standards for uplift, the upstream concrete cut-off wall was constructed as a means of reducing seepage pressures beneath the dam. These reviews emphasized the significance of the cut-off wall on the overall stability of the dam and recommended the implementation of a monitoring program.

GEOLOGY OF THE DAM SITE

The geology of the strata underlying Upper Huia dam has been classified (Refs. 2, 3, 4, and 5) as breccia/conglomerete, interbedded with tuff. These are volcanogenic sediments known as Piha and Nihotupu Formations.

Piha Formation consists of indistinct beds of brown-grey, poorly sorted, gravel to boulder breccia-conglomerate consisting of angular to sub-rounded, very fine to medium grained basaltic andesite clasts in a sandy matrix of similar material. Fractures are commonly extremely widely spaced. At the dam site, this formation grades into, or interfingers with, Nihotupu Formation.

Nihotupu Formation consists of finer grained sediments of yellow-grey to grey, poorly sorted sandstones/silstones (tuff) containing angular to sub-rounded grains of basaltic andesite. Fractures are typically widely spaced, while bedding thickness varies between thin to very thick, and is also lensoidal making beds discontinuous over relatively short distances.

The rock on which the dam was founded ranges from moderately weathered to unweathered.

BRIEF REVIEW OF STABILITY ANALYSIS

Evaluations of the dam stability were undertaken in accordance with the US Bureau of Reclamation and Federal Energy Regulatory Commission guidelines. Stresses within the dam body will not be discussed here, although the performed analyses have shown that the internal stresses should cause no concern under all loading conditions. The focus of this paper is on the stability of the dam against overturning moments and sliding.

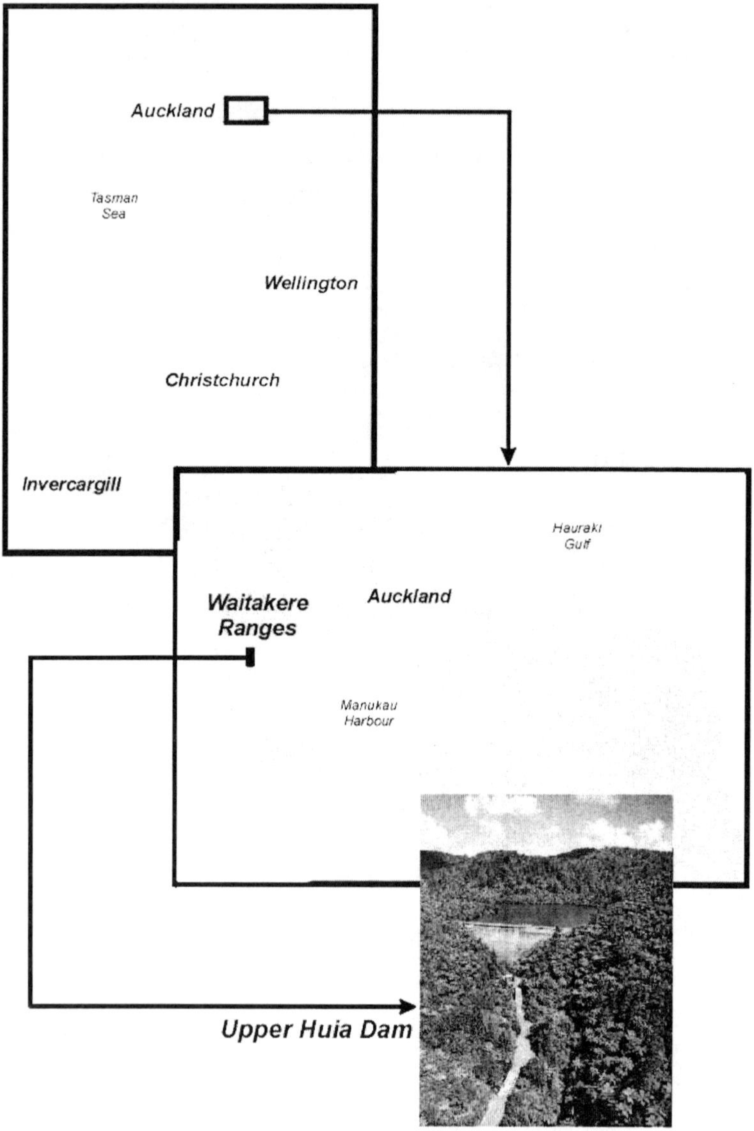

Figure 1 Location plan

As mentioned above, stability analyses were undertaken in 1990, 1991 (Refs. 4 and 5) and 1998 (Ref. 6). Usual, unusual, and extreme loading conditions were considered. The usual loading corresponds to the normal operating condition with the lake at the spillway's crest elevation (Top Water Level, TWL) at 166.22mRL, plus the dam's dead weight, and the uplift pressures. The unusual loading condition combines a reservoir head at Probable Maximum Flood (PMF) level of 167.83mRL, the dam's dead weight, and the uplift pressures. The extreme loading case adds the effects of ground shaking due to the Maximum Credible Earthquake and the arising hydrodynamic reservoir load, to the usual loading case.

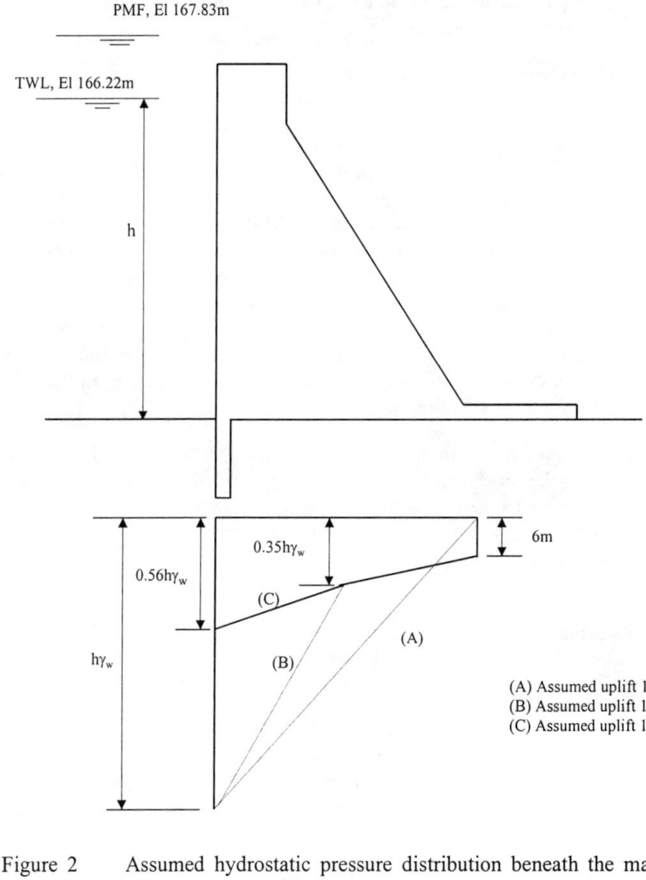

Figure 2 Assumed hydrostatic pressure distribution beneath the maximum cross section of the dam (schematic) (Ref. 6)

The critical factor in these stability analyses was the estimation of the uplift pressures beneath the dam. Based on the piezometric data gathered through the years till 1998, the assumed uplift load in the 1990 and 1991 studies, which took into account the effects of a full reservoir head at the heel of the dam, was considered overly conservative (Ref. 6). The study undertaken in 1998 adopted reduced uplift pressures as observed from existing pneumatic piezometers' readings for the period 1991-1998 (i.e. considers an effective cut-off wall) as presented in Figure 2 for the maximum dam cross-section. For the cross-sections through the right and left abutments, a reservoir head of 0.63 $h\gamma_w$ (γ_w is the unit weight of water) was assumed at the heel of the dam, while decreasing linearly to 4.8 mH_2O (hydraulic head in meters) at the toe of the right abutment section, and down to zero at the toe of the left abutment section. In essence, the study (and further communications) established that catastrophic failure will not be a concern so long as the uplift pressures at the upstream heel of the dam do not reach 75% of full reservoir head.

DESIGN CONCEPTS

Type of Piezometer

Deterioration of the cut-off wall, due to aging or as a result of unusual and extreme events, may lead to a reduction in its effectiveness, jeopardizing the dam's stability due to rising uplift pressures. This issue required a piezometer type with a rapid response which allows a timely reaction from the operation staff. As well as being suitable for a long term application, the vibrating wire (vw) type piezometer was chosen in preference to a strain gage resistance type or the pneumatic type; the three demonstrate a short hydrodynamic time lag. The failure rate of the pneumatic piezometers has been high at Watercare's dams, although some have been satisfactorily functioning for more than five years. The automation requirement also eliminated the pneumatic type.

Three prominent instrumentation manufacturers were invited to provide quotations and the final selection was governed by the total costs. Slope Indicator Company (Sinco) of Seattle, Washington, was chosen. Piezometers similar to those selected have been installed in Western Australia inside boreholes beneath concrete dam(s) and have been functioning satisfactorily for more than 11 years.

Borehole type vibrating wire piezometers were purchased with a range of -3.5 to +35 meters of water head and a 60-micron sintered stainless steel filter tip.

Future Replacement and Calibration

The piezometers were installed inside NQ cored boreholes, vertically pre-drilled through the dam body into the foundation rock. To allow for future replacement and calibration capabilities, it was decided that the four boreholes remain ungrouted, and that they be sealed with inflatable packers.

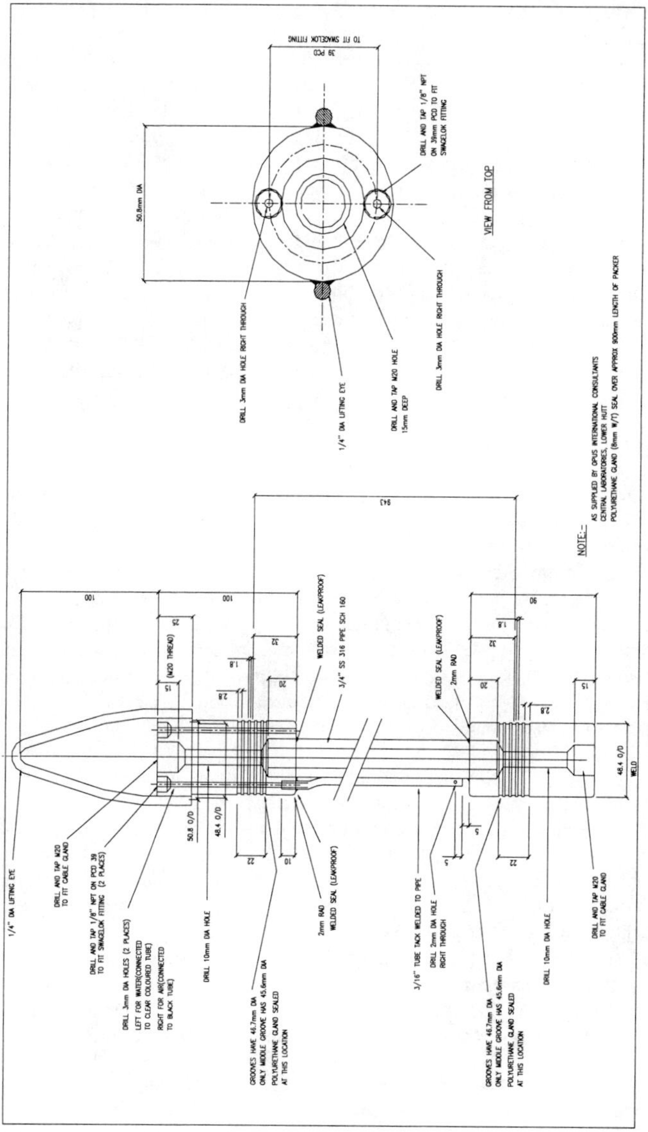

Figure 3 Inflatable packer details

The use of inflatable packers generates a sealed system that ensures that the piezometers respond to a minimal volume change of seepage water, reflecting the corresponding change in pore pressures immediately beneath the dam.

Stainless steel (SS) packers incorporating a polyurethane gland (see Figure 3) were installed immediately above the concrete dam/rock interface as shown in Figures 4 and 5. The piezometer's cable passes through a watertight gland in the center of the packer to the top of the borehole. A 4mm stainless steel wire was attached to the packer for future retrieval. Two 3/16 inch tubes come out of the packer (water inlet and air outlet) for the inflation process and terminate at the top of the borehole with SS shut-off valves. A SS "hat-like" cover bolted to the concrete seals the borehole, where the SS wire is attached to a ring welded to the underside of this cover. The piezometer's cable eventually connects to the Moscad receiver/transmitter box on the dam crest. Within the Moscad PLC (Programming Logic Controller), a program scales the piezometer analog readings (as well as other instrumentation available on site) and prepares the data for radio transmission back to the Control Room. The Moscad PLC sends the data back when polled.

Automated Data Acquisition

To attain the objective of the project, continuous and effective real time surveillance that allows around-the-clock data tracking without on site-monitoring was established through connection to the Company's Supervisory Control and Data Acquisition (SCADA) system. For compatibility with Watercare's SCADA, the cut-off wall monitoring system required a computerized data logging system and instrumentation with 4-20 mA input/output capabilities to automatically compile data from the piezometers. To obtain signal modification from a frequency to a 4-20 mA current, VWD500 interface units, supplied by Sinco, were used which were customized to match the selected piezometers and their ranges.

The VWD500 contains circuitry to provide transducer excitation, signal amplification, filtering and detection. A micro-computer chip co-ordinates all functions, calculates and scales output to 4-20 mA (full scale) current loop. Each vw piezometer is fed into a VWD500 interface. The VWD500 drives and reads a piezometer and processes the read signal to produce a 4-20mA current proportional to the pressure on the piezometer. The current output of each VWD500 is fed into a separate analog input of a Moscad Telemetry PLC. The Moscad sends the piezometer readings back to the Control Room by way of digital radio. The program in the Moscad also detects if a fault occurs between the VWD500 and the Moscad. If a piezometer's replacement becomes necessary, a reprogramming service on the VWD500s' micro-chip will be required.

A 12 volt solar power source is available at the dam site. Consequently, the manufacturer's on-site power source was not required, and due to the automation requirement, nor was a local data logger. However, the idea of having easy to use Liquid Crystal Displays (LCD) on site which can read 4-20 mA signals was favored and implemented. The on-site LCDs can display readings in

mH_2O whenever required with the convenience of a push button, although not corrected for atmospheric pressure changes.

Accuracy of Readings

For a closed system such as the installed vw piezometers, atmospheric pressure compensation is required to improve accuracy. In an attempt to avoid the necessity to purchase and incorporate a fifth piezometer, the possibility of using a "vented type" piezometer was thoroughly explored but discarded for fear of jeopardizing the long term reliability of the foundation piezometers. The incorporation of a fifth, low-pressure piezometer was thus necessary. A Sinco vibrating strip (vs) piezometer was chosen with a range of -1.4 to +14 meters of water head.

Although the chosen piezometers have temperature sensors, early readings have shown, and agreed upon by the manufacturer, that the variations due to temperature changes were trivial.

The installed system which required the use of interface cards/convertors to modify the piezometer signals, is still expected to produce an accuracy of better than ±1% of full scale. The readings obtained and compared to those read by Sinco's VWP Indicator (portable read-out unit) show that this is currently being achieved.

Lightning Strikes

To protect the VWD500s and the associated system's electronics from lightning strike-induced power surges, 30V varisters have been connected from each leg of each piezometer down to earth. It is worth mentioning that the regional area of the dam has had no reported bad experiences with lightning strikes.

PROJECT IMPLEMENTATION

Drilling

Four NQ size boreholes were drilled; two at the abutments and two in the supply tunnel (which houses the discharge pipe) as shown in Figures 4 and 5. Two small wire line drilling rigs were used due to access restrictions (no vehicle access to the dam crest and space limitation in the supply tunnel), transported to the site using a helicopter.

The two small rigs used for the job performed full core recovery drilling which allowed the depth of the dam/rock interface to be established with sufficient accuracy (also rock coring resulted in a fair face of the dam concrete for the assured installation of the inflatable packers). The aim was to terminate drilling as close as possible to the dam base, but within a rock zone representative of the general rock condition, specifically, in terms of the coefficient of permeability. Installing the piezometers at deeper elevations would not provide useful information on the effectiveness of the cut-off wall in reducing the pore water pressures immediately beneath the dam.

Other major issues considered prior to the commencement of drilling were:

- Inclined drilling from the downstream face of the dam, but this was later discarded as no savings were expected, besides facing the risk of drilling close to and towards the cut-off wall.
- Sand backfilling around the vw piezometer was considered but discarded for ease of replacement in case of failure or for future calibration purposes.
- Wash boring until the final few meters close to the expected depth of the dam/rock interface was considered to speed up drilling, but was not possible, as an appropriate sized rig could not be conveniently transported to the site by helicopter.
- The likelihood of encountering artesian pressures while drilling in the supply tunnel was considered and accounted for using blow-out preventer equipment.
- Verticality of drilling (specifically for the two boreholes drilled from the dam crest) to achieve the required location to place the piezometers, which is close to and behind the concrete cut-off wall beneath the dam.
- Grouting of the boreholes within the foundation rock for long term stability, and re-drilling the next day, extending slightly deeper than the depth reached the previous day, to expose ungrouted rock, with full core recovery that ensures the quality of grouting. Eventually, the good quality of drilled rock proved that this was not required.
- Insertion (with a capability of future retrieval) of a perforated stainless steel, or a slotted PVC pipe for the purpose of supporting the borehole walls within the foundation rock. Again, the good quality of drilled rock proved that this was also not required.

Installation

The piezometers and packers installation works were undertaken in a manner similar to what Dunnicliff (1982) described as "instrumentation specialist contracting with the owner." Connection to SCADA was performed by Watercare's staff.

At each borehole location, the piezometer cable was passed through the packer and the piezometers' depth below the packer was set out so that the sensor diaphragm was 180 mm below the base of the packer. Following attachment of

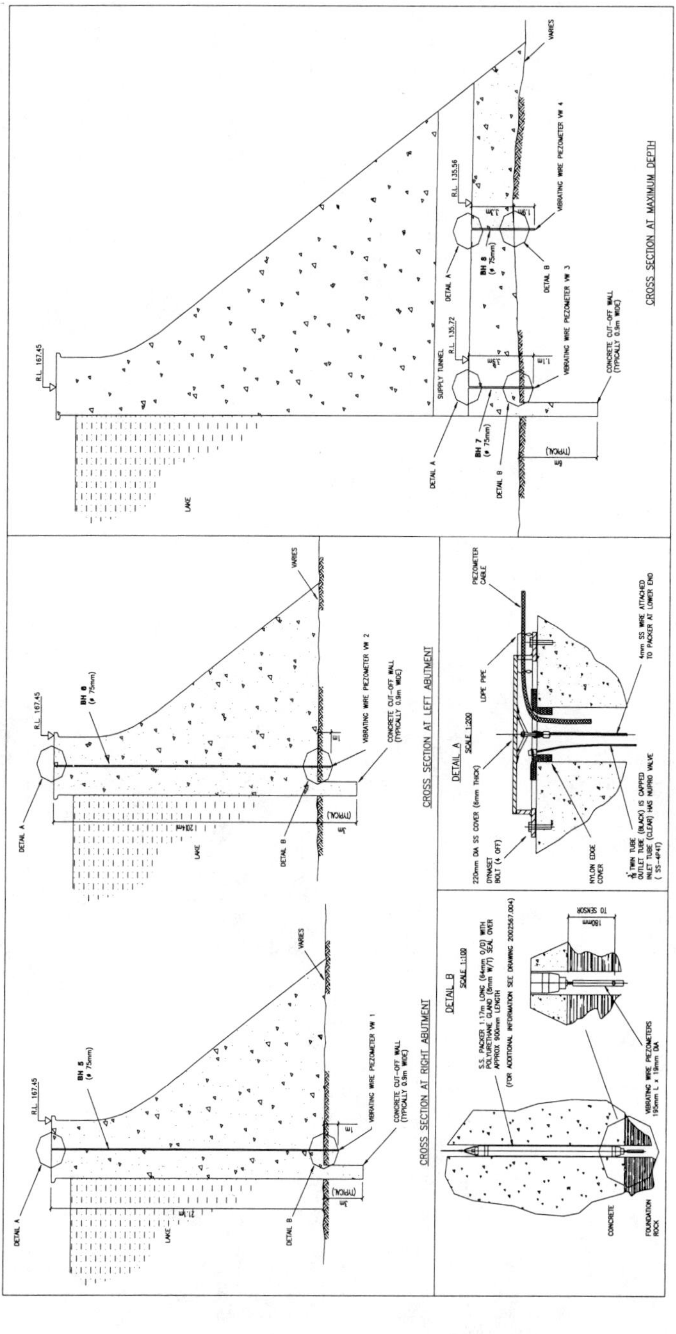

Figure 4 Dam cross-sections showing installation details

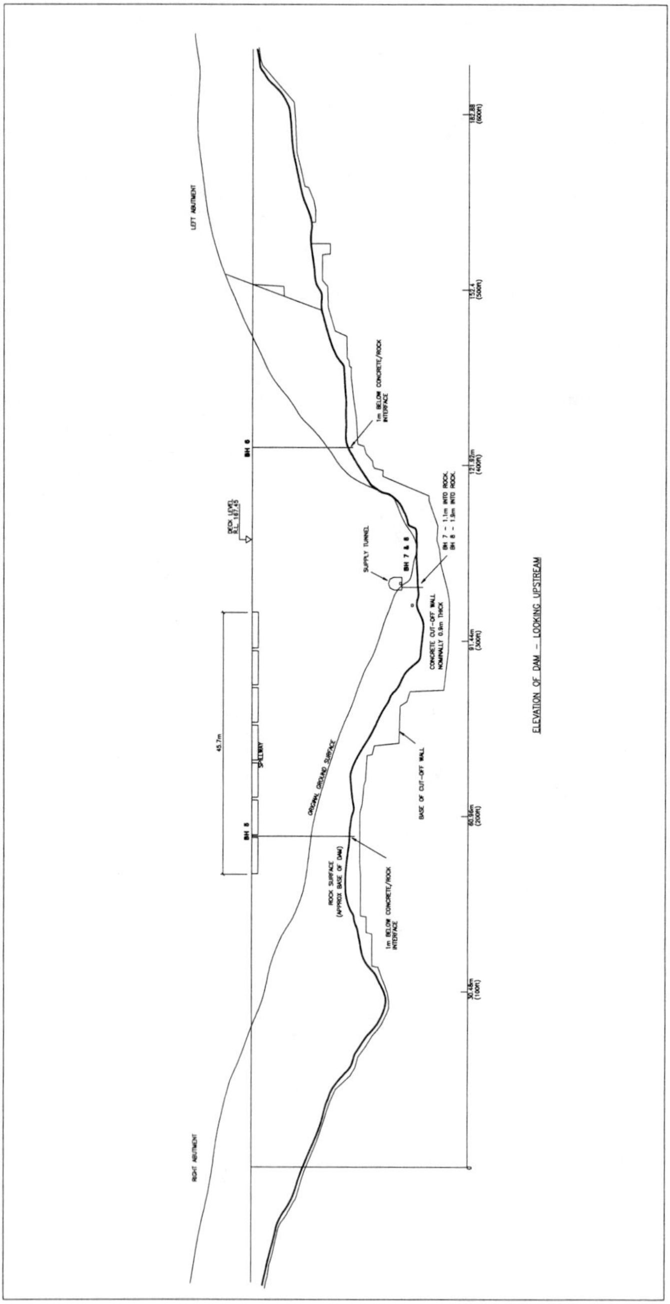

Figure 5 Longitudinal profile showing the locations of the vw piezometers' boreholes

the hydraulic 3/16 inch tubing to the packer, the packer/piezometer assembly was then lowered down the borehole with the aid of the SS wire. Inflation then took place using de-aired water, and while monitoring the rise in pressure in stages and the associated change in volume, pressure was taken to more than 2.5 times the maximum reservoir head. The VWP Indicator was used to record the initial piezometers' readings and water temperatures.

The installation of the interface cards (VWD500s) and connection to the SCADA system were undertaken afterwards. Also the on-site LCD indicators were installed which now provide piezometric heads in mH_2O and, although not corrected for atmospheric pressure changes, they still provide reasonably indicative readings of how the foundation seepage pressures are fluctuating.

PLC programming was carried out on the Moscad unit, so that the power supply for the 4-20 mA is switched on by the RTU every 60 minutes; readings are obtained on the hour, allowed to settle, and the 4-20 mA output is read. Once a reading is obtained, the power supply is then switched off. The readings on the Intouch software (Graphics Display) at the Control Room were calibrated by simulating a minimum piezometer output and a maximum piezometer output. Eventually the piezometer readings were displayed at the Control Room, allowing comparison against those measured using Sinco's read-out unit, and the readings from the on-site displays.

FINDINGS AND BASELINE DATA

Good quality rock was encountered during drilling. It is likely that the four drilled boreholes will not dry out as long as the reservoir remains operational, judging by the fact that they were at least half full with water while the lake was about 52% full. This is confirmed by the readings of the existing pneumatic piezometers for the 1991-1998. Thus, a standard type filter tip for these piezometers is considered appropriate.

Observations and in-situ permeability testing ensured that water carrying joints were intercepted and there was no need for further drilling. The aim was to have the piezometers as close as possible to the dam/rock interface. The coefficient of permeability values collected from falling head tests (as well as rising head tests in the supply tunnel boreholes) ranged between 1.9×10^{-8} and 1.3×10^{-7} m/sec. These values compare reasonably well with the values measured in 1991 (Ref. 5).

Prior to the piezometer and packer installation, it was observed, while carrying out permeability testing, that the groundwater at the left (east) abutment influenced the water level (experienced an occasional head rise) in Borehole 6 (BH6), which would be an indicator of the existence of a crack through the dam body. Another indication of a crack existence, probably at nearly 152 mRL, was observed by a change in the characteristics of the head decrease versus time relationship from the falling head test. This crack is believed to have affected the observed value of the Basic Time Lag (T) and to a lesser degree the coefficient of permeability; the latter being somewhat compensated as testing continued below the crack location. The calculated T value is more realistic than the measured value. Coefficient of permeability K_h versus Basic Time Lag T plots in Figure 7 demonstrates this observation.

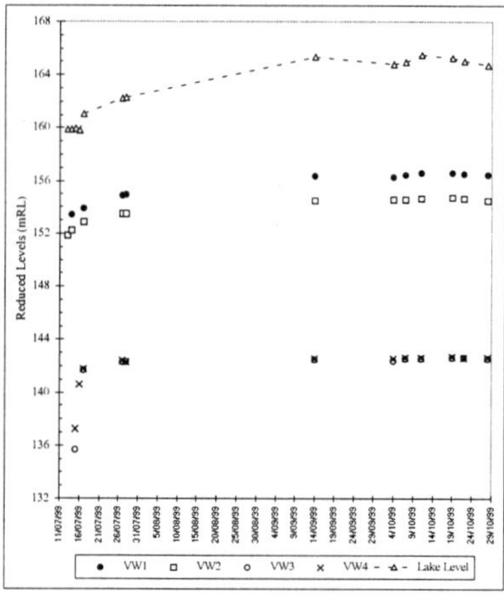

Figure 6 Early piezometer readings following installation

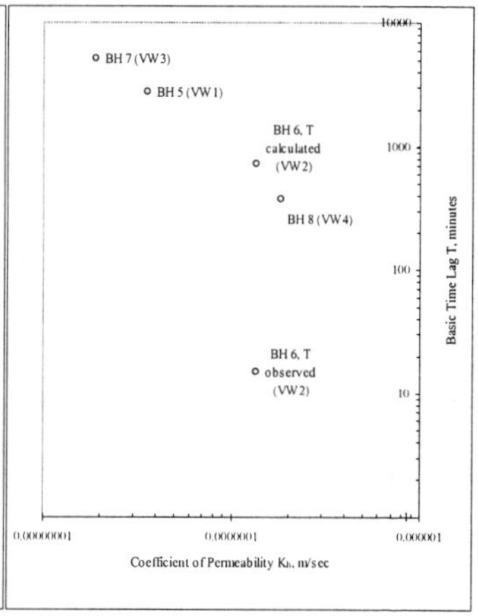

Figure 7 Coefficient of Permeability versus Basic Time Lag

As expected, artesian flow was encountered in BH7 and BH8 drilled through the supply tunnel in the valley section. The rise in water level from the NQ sized holes was 2.8 mm per minute and 34 mm per minute in BH7 and BH8 respectively. Measurements of the water level using a dip meter have shown that reductions in uplift pressure caused by the cut-off wall were about 46 and 68% in BH5 and BH6 (right and left abutments) respectively. These reduction figures compare well with the piezometer readings afterwards.

Close comparison of the piezometers' readings against the water level measured using a dip meter in the abutment boreholes, was influenced by the lake level fluctuation, in addition to the time needed for the borehole water level to equalize.

DATA COLLECTION AND CUT-OFF WALL PERFORMANCE

Following installation, first and subsequent snap shot readings were undertaken using the portable read-out for each piezometer. Figure 6 shows how the early readings stabilized and then maintained a close response to lake level fluctuations.

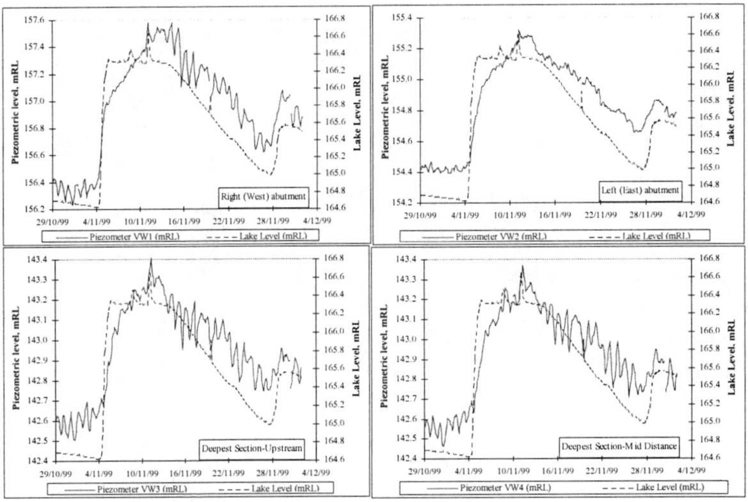

Figure 8 Piezometer readings versus lake level (Data from SCADA for a 2-hour interval)

Figure 8 presents the plots of the piezometers' readings against lake level changes using automatically collected data (from SCADA) for a period of about five weeks. Data is for a 2-hour interval. The lake level rose above top water level (TWL) of 166.22 mRL during this period, and the dam site received a significantly high record of rainfall, reaching a total of more than 300 mm. All piezometers show very close response to lake level variations. Within the valley section, piezometers VW3 and VW4 showed head fluctuations of about 200 mm. VW1 at the right abutment showed a similar observation but to a lesser magnitude.

VW3 and VW4 showed very similar magnitudes of piezometeric head, reflecting a pressure distribution that deviates from that of a flow net. This is apparently due to the presence of preferential seepage paths within the rock mass. The four piezometers' readings show that the uplift pressures are essentially well below those that could trigger an overturnning or a sliding instability.

Piezometeric heads at the locations of VW3 and VW4 obtained from a seepage model against a lake level at TWL (Figure 9) at the locations of the two piezometers were approximately 15 and 13.5 mH_2O respectively, assuming an effective cut-off wall. The corresponding values with no cut-off wall, rise to 29.5 and 19.5 mH_2O.

As part of the surveillance and emergency plans, procedures are currently being prepared regarding the time to reach a safe lake level should the uplift pressures reach alarm levels. The existing capacity of discharge is 1.8 m^3 per sec,

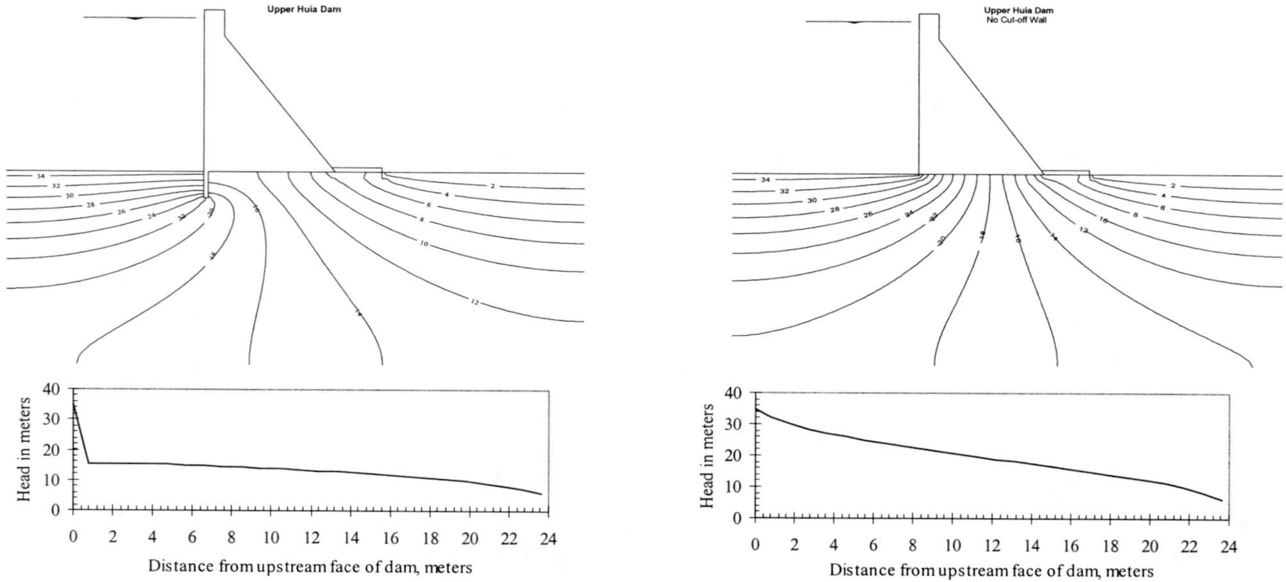

Figure 9 Flow net with and without a cut-off wall, and the corresponding pore pressure distributions beneath the dam (maximum cross-section)

and to draw down the reservoir level, for example, from 166.22 to 165 mRL would take 37 hours.

Seepage monitoring and deformation surveys are also being conducted to assist in ensuring the satisfactory performance of the dam.

CONCLUDING REMARKS

Upper Huia dam is one of Watercare's oldest dams which was designed and constructed in the late 1920s with criteria that do not comply with present day practice regarding uplift. Critical locations were chosen beneath the dam where four vibrating wire piezometers were installed to provide continuous monitoring of the performance of the cut-off wall. Should a reduction in the effectiveness of the wall be observed by these instruments, prompt action can be undertaken. Alternatives for remedial works (e.g. drainage, anchorage) will then be sought. Currently, the vibrating wire piezometers are showing that the uplift pressures are within the limits whereby the stability of the dam is maintained under all loading conditions.

ACKNOWLEDGEMENTS

Watercare Services Limited is thanked for the approval to publish this paper. The assistance of Barbara Pendrey, Alan Brooks, and Ian Moses in draughting is most appreciated. Fred Tapp undertook the design and supervision of the electrical part of the job and software installation. Alan Kennard pointed out errors in the manuscript.

Appendix-I. References

Dunnicliff, J. (1982) "Geotechnical Instrumentation for Monitoring Field Performance", publication no. 89, Transportation Research Board, Washington, D.C., April, 1982.

Kermode, L. O. (1988) "Helensville-Waitakere", Infomap 290, Sheet Q10/11, 1:100 000, Department of Lands and Survey, Wellington, New Zealand.

Kermode, L. O. (1992) "Geology of the Auckland Urban Area", Scale 1:50 000, Institute of Geological and Nuclear Sciences, Lower Hutt-New Zealand.

Riley Consultants Limited (1990) "Upper Huia Dam Safety Evaluation, Stage 1", Auckland, New Zealand, October 1990.

Riley Consultants Limited (1991) "Upper Huia Dam, Stage 2 Evaluation of Safety", Auckland, New Zealand, May 1991.

Woodward-Clyde (NZ) Limited (1998) "Upper Huia Dam Safety Evaluation", Auckland, New Zealand, May 1998.

DESIGN, CONSTRUCTION, AND PERFORMANCE OF ROCK TEST FILLS AT SATURN CORPORATION

Don W. Dotson,[1] James Wm. Martin,[2] and Robert W. O'Connor,[3] Members, ASCE

ABSTRACT: In the mid-1980s, Saturn Corporation built manufacturing facilities of over 604,000 m^2 (6,500,000 ft^2) on a 8620 ha (1300 acre) site in Spring Hill, Tennessee for a total construction cost of over $3.5 billion. Although the facilities were constructed on gently sloping terrain, site grading requirements mandated cuts of up to 9 m (30 ft.) and fills of up to 15 m (50 ft.). Design criteria required that the subgrade under the entire plant be of a uniform consistency so that equipment could be relocated anywhere within the structure without concern for poor performance of the slab and subgrade. Saturn authorized a geotechnical study that included the construction of test fills to establish engineering and construction parameters for the construction of the engineered fill. During the construction of the test fills, the source of the shot rock fill, the lift thickness, the moisture content, and type of compaction equipment were varied while the unit weight of the fill was measured and plate load tests were conducted. This paper presents the results of the experimental work conducted at Saturn to establish the engineering and construction parameters of the fill used in the construction.

INTRODUCTION

In the mid-1980s, Saturn Corporation built manufacturing facilities of over 604,000 m^2 (6,500,000 ft^2) on a 8620 ha (1300 acre) site in Spring Hill, Tennessee at a total construction cost of over $3.5 billion. Although the facilities were constructed on gently sloping terrain, site grading requirements mandated cuts of up to 9 m (30 ft.) and fills of up to 15 m (50 ft.). Design criteria required that the subgrade under the entire plant be of a uniform consistency so that equipment could be relocated anywhere within the structure without concern for poor performance of the slab and subgrade. Saturn authorized a geotechnical study that included the construction of

[1]Senior Geotechnical Engineer, Ogden Environmental and Energy Services Company, Inc. 3800 Ezell Road, Suite 100, Nashville, Tennessee 37211. Adjunct Professor, Department of Civil Engineering, Tennessee State University, Nashville, Tennessee, 37209.
[2]Vice-President, Ogden Environmental and Energy Services Company, Inc. 3800 Ezell Road, Suite 100, Nashville, Tennessee 37211.
[3]Project Manager, Gannett Fleming, Inc. 11818 Rock Landing Drive, Suite 101, Newport News, Virginia 23606.

test fills to establish engineering and construction parameters for the construction of the engineered fill.

A total of eleven test fills were constructed at two locations on the site. The test fills were constructed with limestone and shaly limestone of the Bigby-Cannon Formation. Unconfined compressive strengths from representative specimens obtained from that formation averaged approximately 55 MPa (8000 psi). The rock borrow was excavated by blasting from borrow sources located near the northeast and east-central portions of the site. The shot rock used to construct test fills A through H was obtained from the northeast borrow source and included a well-graded range of particle sizes from rock flour to 76 cm (30 in.) in maximum dimension. Fills I, J, and K were constructed with material obtained from the east-central borrow source, which consists of rock high in shale content. As a result of the higher shale content, these three test fills contain considerably more plastic fines generated by breakdown of the shale fraction during compaction. Although gradation distributions of representative samples of fill material were not determined, the fills that were constructed with the shale limestone were estimated to contain up to about thirty percent fines (minus No. 200 sieve).

Prior to placement of fill, the overburden was completely removed from the test fill areas. In order to accurately control the thickness of each lift, a leveling course of shot rock was placed within areas where steps in the bedrock surface occurred. Each test fill was constructed of three 90 cm (36 in.) thick lifts placed directly on the clean bedrock surface or above a leveling course which was compacted with eleven passes of an Ingersoll-Rand SP-60DD 18 t (20 ton) vibratory roller. The test fills were placed as indicated in Table 1 using a variety of compaction efforts as applied by a Caterpillar 825 Wedgefoot 32 t (35 ton) compactor and/or the SP-60DD.

Table 1 Fill Placement Procedures

Site	Test Fill	Compactive Effort	Water
I	A	No compaction other than applied by a Caterpillar D-8 Dozer during placement	None
I	B	5 passes with Ingersoll-Rand SP-60DD	None
I	C	8 passes with Ingersoll-Rand SP-60DD	None
I	D	11 passes with Ingersoll-Rand SP-60DD	None
I	E	5 passes with Cat 825 Wedgefoot	None
I	F	2 passes with Cat 825 Wedgefoot	None
I	G	2 passes with Cat 825 followed by 3 passes with Ingersoll-Rand SP-60DD	None
I	H	2 passes with Cat 825 followed by 5 passes with Ingersoll-Rand SP-60DD	None
II	I	11 passes with Ingersoll-Rand SP-60DD	23-35 l/m^3 (4-6 gal/ cy)/lift
II	J	5 passes with Ingersoll-Rand SP-60DD	23-35 l/m^3 (4-6 gal/ cy)/lift
II	K	5 passes with Ingersoll-Rand SP-60DD	None

Each of the eleven test fills was load-tested in general accordance with ASTM D1194 (*Standard Test Method for Bearing Capacity of Soil for Static Load on Spread Footings*). Four Dywidag threaded reaction anchors were installed into the underlying bedrock at each of the test locations. A reaction beam was attached to the threaded reaction anchors and four circular steel bearing plates ranging from 36 to 96 cm (14 to 38 in.) in diameter were stacked concentrically on the prepared subgrade. A 181 t (200 ton) hydraulic loading jack was placed between the reaction beam and the base plates. Loads were applied to the subgrade in 7 t (15 kip) increments. Three dial gages mounted to an independent reference beam were used to measure deflection induced by the applied load. The deflection readings were recorded and averaged for each load increment. Each load was stabilized to a rate of 0.025 mm (0.001 in.) of deflection per minute before additional load was applied. Ten of the eleven tests were continued until at least 13 mm (0.5 in.) of settlement occurred. The loading test on Fill F was terminated when one of the reaction anchors failed in tension. Deflection readings were also recorded during the unloading cycle of each test in order to evaluate elastic rebound of the subgrade.

The effect of water on the performance of the rock fill was evaluated in two ways; first, by placing the fill with water and, second, by flooding the fill surface while load was being applied to the subgrade. In order to evaluate the effect of water, Test Fills I and J were constructed using the same compaction criteria as Fills B and D except that 23-35 l/m^3 (4-6 gal/cy) of water were applied prior to compaction of each lift. Fills I and J were constructed with shot rock from a somewhat more shaly borrow source than Fills B and D. Therefore, for comparison purposes, Fill K was constructed using the same shaly limestone and the same compaction criteria except that no water was applied. However, because Test Fills I, J, and K were all compacted differently, the effect of water content was not strictly isolated as the independent variable.

To assess the effect of introducing water to the subgrade during the loading cycle, the subgrade was inundated at Test Fills G, I, J, and K at a predetermined applied bearing pressure. The additional deformation related to flooding was recorded at each of these locations until the rate of deformation stabilized.

Following completion of load testing, in-place density tests were performed at Test Fills B, C, F, G, and J. In addition, three in-place density tests were performed outside of the test fill areas (in an area called the Canyon) to evaluate any difference in density caused by compacting the fill with about 127 l/m^3 (22 gal/cy) of water applied at the surface of each lift. The in-place density tests were performed by excavating and weighing about 3 m^3 (4 cy) of material. Because of the size of the particles involved and the desire for accuracy, this amount was the largest that could be handled with available, on-site equipment. The excavation was then lined with an impervious membrane and filled with a measured quantity of water to determine the volume of the excavation.

TEST RESULTS

Plate Loading Test

Load versus deflection graphs for each of the eleven test fills are shown in Figures 1 through 3. In general, the load-deflection curves indicate that the subgrade was dramatically improved by a few passes of the SP-60DD while only a modest reduction in settlement results from more than five passes. Additional passes with compaction equipment, extend the elastic ranges of deformation to higher stress levels (Figure 4). The elastic limits shown by this figure consist of the initial, tangent portion of the load-deflection curve. The limits of elastic deformation for each test, when compared with the corresponding stress level, provides criteria for selecting the compaction effort required to provide an adequate margin of safety against excessive settlement of loaded footings. For instance, when a relatively crystalline limestone fill containing few plastic fines is compacted with five or more passes of the SP-60DD, essentially all of the deformation under the influence of the applied total load of 575 kN/m^2 (12 ksf) will be within the elastic range. For a shaly limestone, eight passes with the roller were required to accomplish a comparable effect. If a lesser compactive effort is applied to the fill, loads in excess of the elastic limit will produce greater magnitudes of permanent, plastic deformation at a rate with increases at successive loads.

The slope of the initial, tangent portion of the each curve is the subgrade reaction modulus (k) which is a measure of the elastic stiffness of the subgrade expressed in kilonewtons per square meter per meter (pounds per square inch per inch) of deflection. Because the maximum design bearing pressure of footings is typically within the straight line range of deformation, subgrade reaction moduli are considered to be a good indicator of the relative effect of compactive effort. Figure 5 shows that subgrade reaction moduli increase linearly with additional passes applied by the SP-60DD. Where six or fewer passes are applied by the SP-60DD, the resulting subgrade reaction modulus is essentially the same as the modulus produced by substituting the first two passes of the SP-60DD with two passes of the 825 Wedgefoot Roller. However, where more than six passes are applied, the SP-60DD alone is more effective with respect to increasing the subgrade reaction modulus than combined passes with the Wedgefoot Roller followed by the SP-60DD. In addition, the lowest subgrade reaction modulus 52 MPa/m^3 (190 pci) was recorded for Test Fill E which received five passes with the 825 Wedgefoot Roller. That result does not conform to the trends established by the other test data, which suggest that the Wedgefoot may randomly provide inadequate compaction at depth within individual lifts due to difference in the fill material.

The data indicates that the initial passes of both types of rollers are generally effective in compacting the upper surface of the fill. In some types of material, such as shaly limestone, the 825 Wedgefoot may be more effective than the vibratory roller owing to the ability to crush rock particles into voids within the rock mass. As the fill becomes more dense at the surface, the influence of vibratory compaction becomes increasingly more effective than the Wedgefoot compactor because the dense surficial interval of rock fill propagates the dynamic energy of the vibratory roller deeper into the fill resulting in further densification at depth.

Water Placement

The subgrade at Fills I, J, K, and G was flooded during the loading cycle and the additional settlement related to flooding was recorded (Figure 3). This test procedure was implemented in an attempt to simulate the long-term influence of infiltrating rain water or percolating ground water within the fill. Prior to flooding the subgrade, Fills I, J, and K were loaded in incremental steps to a bearing pressure of 546 kN/m2 (11.4 ksf) and the corresponding deflection was stabilized to a rate of 0.025 mm (0.001 in.) per minute. Additional settlement related to flooding the subgrade ranged from 1 to 1.8 mm (0.04 to 0.07 in.). Fill G was loaded in incremental steps to a bearing pressure of 2188 kN/m^2 (45.7 ksf) and the rate of deflection was stabilized. Settlement under the influence of that load continued while the subgrade was flooded and did not stabilize until flooding was suspended. Net settlement caused by the flooding at Fill G was 10.2 mm (0.40 in.) after about one and one-half hours.

The test results indicate that settlement following the application of water will reach a limiting value when the applied loads are within the elastic range (e.g., Test Fills I, J, and K); however, when the applied loads are well beyond the elastic limit (e.g., Test Fill G), water introduced to the subgrade apparently serves to lubricate interparticle contacts causing a continuation of settlement. The magnitude of settlement caused by flooding depends on several variables, including the stress regimes, the thickness of fill, the depth of water infiltration into or through the fill, the degree of interlocking between individual rock pieces, the gradation of the rock (particularly with respect to percentage of fines), and the percent voids within the fill. Because of the number of variable factors which could govern settlement following inundation of the fill, an accurate estimate of the magnitude of deformation related to a specific load is difficult to ascertain. In general terms, however, it is probable that the potential for settlement related to infiltrating rain water runoff or ground water flows can be minimized by placing fills with large quantities of water so as to achieve the densest possible compacted state. For this project, an average rate of application of 104 l/m^3 (18 gal/cy) of water per lift was used for construction of production fills. In theory, optimum moisture and density could be determined with as few as three moisture-density tests and should be considered on large rock-fill projects.

As evidenced by the near identical fit of the straight line portion of the load-deflection data from Test Fills J and K (Figure 3), placement of fill with only 23 to 35 l/m3 (4 to 6 gal/cy) of water per lift does not improve the subgrade with respect to settlement. However, when the fill is placed with approximately 104 l/m^3 (22 gal/cy) of water per lift, the in-place density is substantially increased (see Table 2) which presumably relates to a proportional decrease in settlement potential.

Despite the good correlation of data obtained from this study, the findings are subject to several variables specifically related to the borrow source, including gradation, percent fines, and angularity and distribution of individual rock pieces within the fill. Among those potential sources of variation, the presence of rock fines and decomposed shale is considered the most likely to increase the settlement potential of the fill. Comparison of plate load test results (performed on-site, but not presented in this paper) for Test Fills B and D with I, J, and K provides evidence that settlement under the influence of an 383 to 575 kN/m^2 (8 ksf to 12 ksf) load could be on the order of 25% higher than predicted at locations where the fill is constructed with shaly limestone instead of the more crystalline limestone which is prevalent on the site.

In-Place Density Determination

Measured densities for rock fill placed without water and at different compactive efforts, range from 20.8 to 22.0 kN/m^3 (132.3 pcf to 140.1 pcf) with an average of 21.4 kN/m^3 (136.1 pcf). To measure the effect of larger quantities of water, an additional fill (called the Canyon Fill) was placed and flooded with about 127 l/m^3 (22 gal/cy) of water per lift and compacted with eight passes of the SP-60DD per lift. At the Canyon Fill, in-place densities range from 23.7 to 24.3 kN/m^3 (151.1 to 154.8 pcf) with an average of 24.1 kN/m^3 (153.4 pcf). Wetting the fill with this larger volume of water per lift during placement resulted in an average increase in density of 13%. This is attributed to rock dust fines being flushed in to the interstices of the rock mass and retention of moisture within the fill. Although most of the density testing which was performed to evaluate the effect of wetting the fill during placement was within areas that were flooded with approximately 22 gallons of water per cubic yard, lesser amounts of water may have a comparable effect. Only one density test (Fill J) was performed on a fill which was placed with less than 127 l/m^3 (22 gal/cy). At that location, 23 to 35 l/m3 (4 to 6 gal/cy) of water per lift was used and the in-place density was 22.5 kN/m^3 (143.1 pcf) or 5% higher than the average density of Fills B, C, F, and G which were placed dry. Because the fill compacted with 23 to 35 l/m3 (4 to 6 gal/cy) is composed of shaly limestone with more fines than the dry fills, the higher density is most likely related to material differences and water retention rather than the method of placement.

Measurable increases in in-place density were also effected by increased compactive effort; however, the trends developed on the basis of four tests where different compactive efforts were used are insufficient to develop a reliable relationship. Nevertheless, in-place density testing is considered a viable method of quality control when combined with the observation of trained technical personnel during field placement. Individual in-place density test results are summarized by Table 2.

Table 2 Summary of In-Place Density Results

Location	Compactive Effort	Water During Placement l/m^3/lift (Gal/cy/lift)	Density kN/m^3 (pcf)
Fill B	5 passes with SP-60DD	None	20.8 (132.3)
Fill C	8 passes with SP-60DD	None	22.0 (140.1)
Fill F	2 passes with 825 Wedgefoot	None	21.2 (134.7)
Fill G	2 passes with 825 Wedgefoot 3 passes with SP-60DD	None	21.6 (137.2)
Canyon	8 passes with SP-60DD	127 (22)	24.3 (154.3)
Canyon	8 passes with SP-60DD	127 (22)	24.3 (154.8)
Canyon	8 passes with SP-60DD	127 (22)	23.8 (151.1)
Fill J	5 passes with SP-60DD	23-34.5 (4-6)	22.5 (143.1)

SUMMARY

Based on the tests results, a minimum of eight passes per lift of an SP-60DD vibratory roller were used to compact the rock fill over the entire project. Using this compaction criterion and the results of the plate loading tests, relationships were developed showing expected settlement versus width of square footing for various bearing pressures. Ultimately, all footings were actually designed based on an allowable bearing capacity of 380 kN/m^2 (8 ksf). Owing to the settlement induced by inundation of the subgrade when the bearing plates were loaded to about 546 kN/m^2 (11.4 ksf), the fill was thoroughly saturated where placed above deep depressions in the bedrock or beneath exterior wall footings where the subgrade would be exposed to surface water infiltration. In general, the rate of application was about 104 l/m^3 (18 gal/cy) of water per lift; however, the requirements for watering the fills were adjusted in the field depending upon the gradation and percentage of rock fines within the borrow. Since construction, the rock fill underlying many of the major manufacturing facilities at Saturn has continued to perform well.

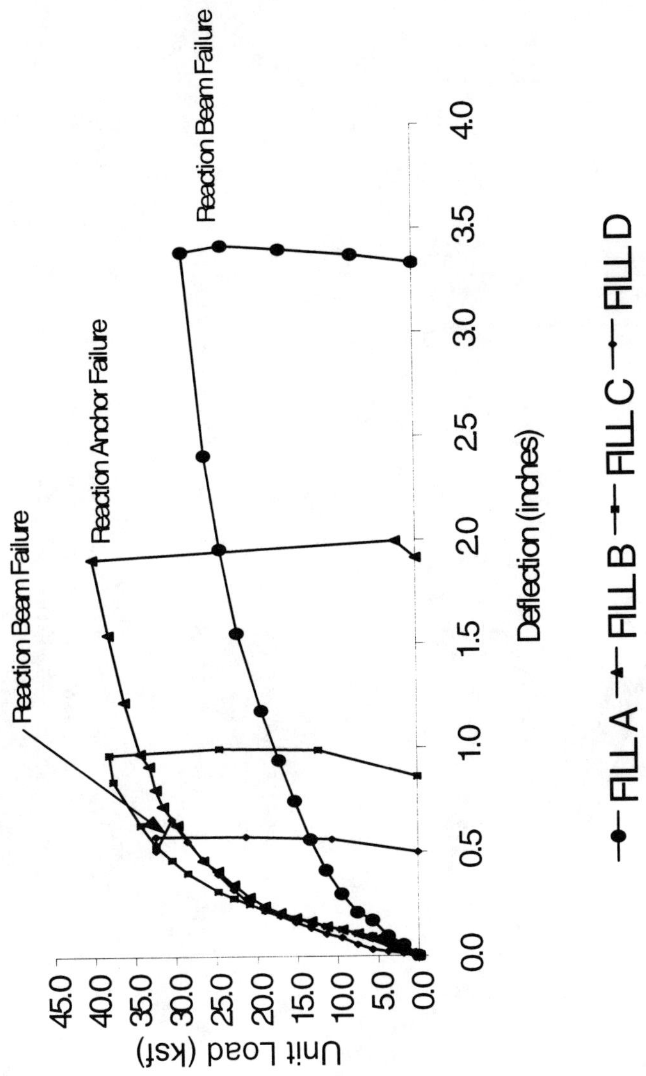

Figure 1
Deflection vs. Applied Load for Test Fills A – D
(See Table 1 for Placement Procedures)
(1 kN/m^2 = 20.885x10^{-3} ksf, 1 inch = 2.54 cm)

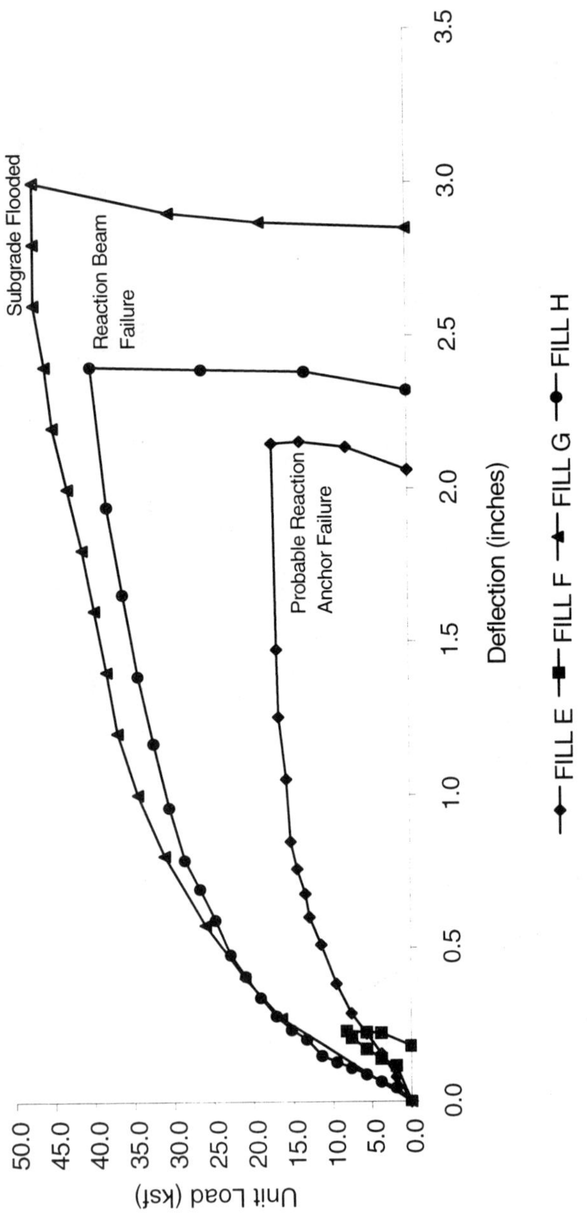

Figure 2
Deflection vs. Applied Load for Test Fills E – H
(See Table 1 for Placement Procedures)
(1 kN/m^2 = 20.885x10^{-3} ksf, 1 inch = 2.54 cm)

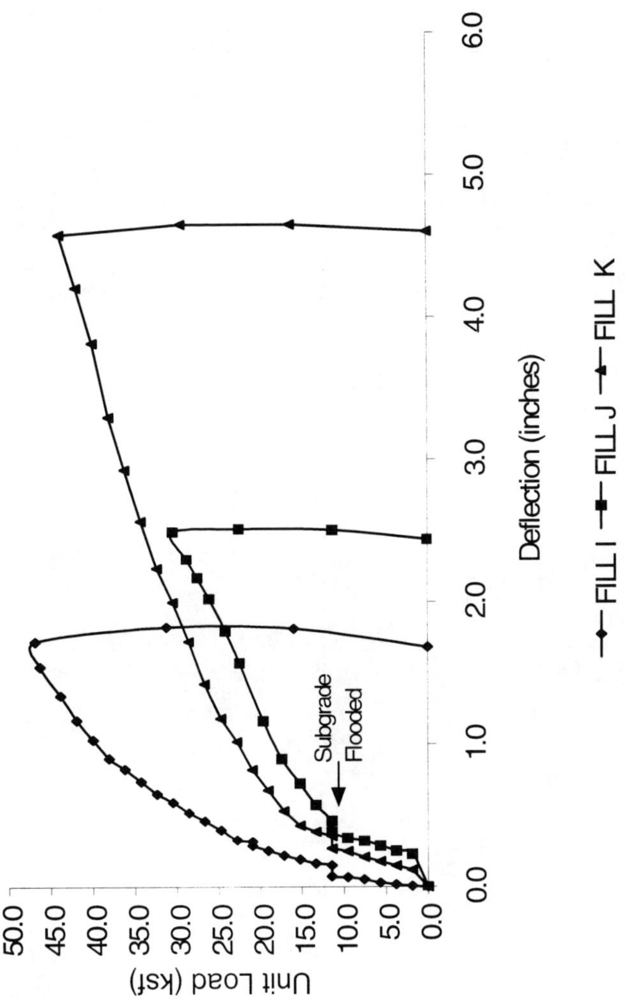

Figure 3
Deflection vs. Applied Load for Test Fills I – K
(See Table 1 for Placement Procedures)
(1 kN/m^2 = 20.885x10^{-3} ksf, 1 inch = 2.54 cm, 1 gpm = 3.785 liters/minute)

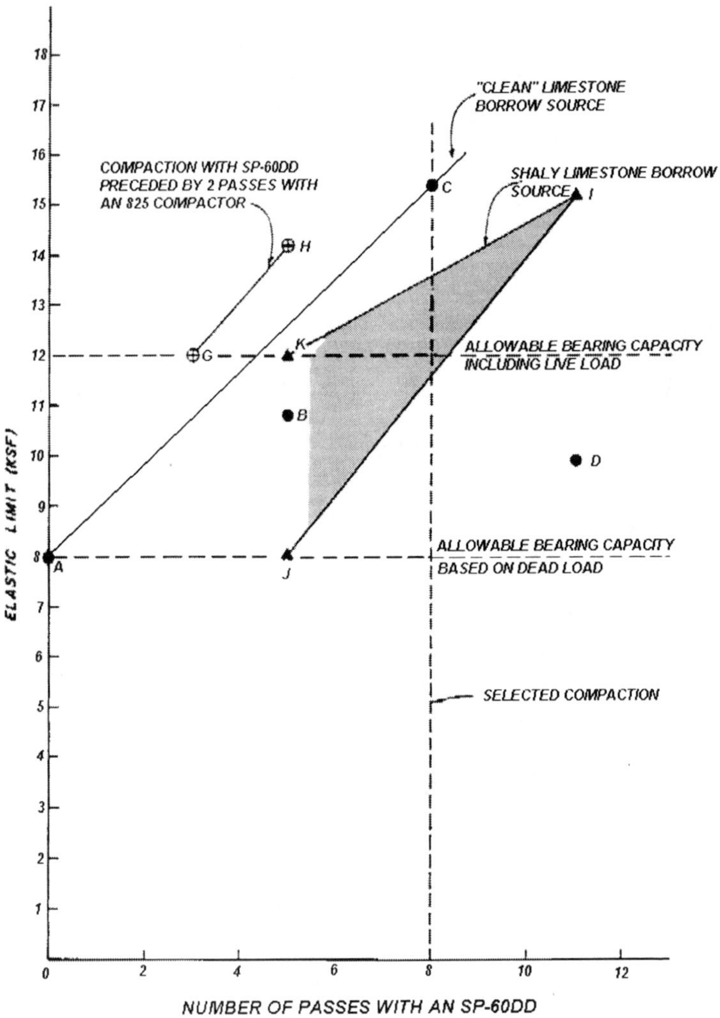

Figure 4
Variation of Elastic Limit with Compactive Effort
(See Table 1 for Placement Procedures)
($1\ kN/m^2 = 20.885 \times 10^{-3}$ ksf)

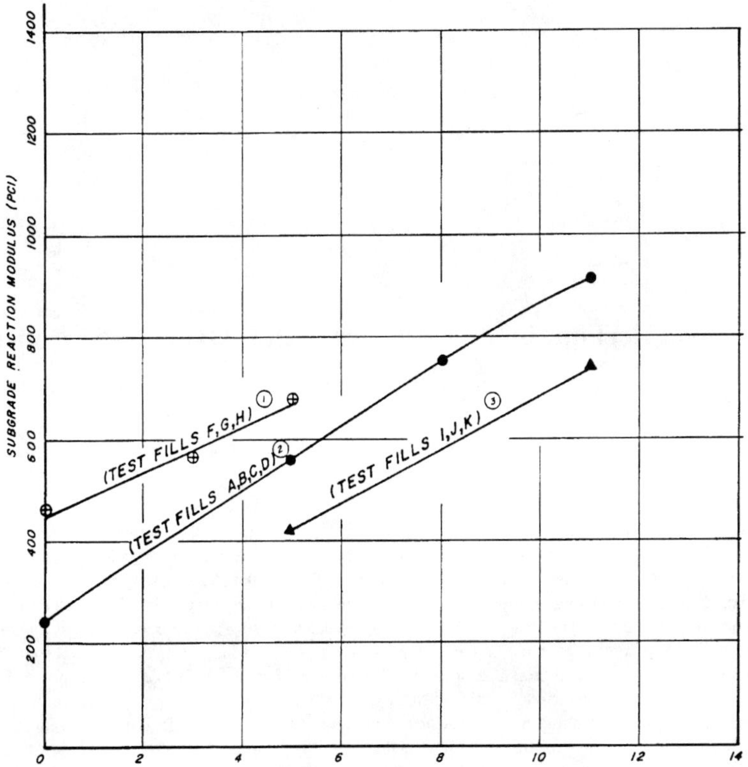

Number of Passes with an SP-60DD

Figure 5
Variation of subgrade Reaction Modulus with Compactive Effort
(See Table 1 for Placement Procedures)
(1 kN/m^3 = 0.003682 pci)

Bureau of Reclamation Experience with Instrumented Rolled Earth Fill Dams

Jeffrey A Farrar M. ASCE[1]

ABSTRACT: This report summarizes the work of Dr. G.P. Gould who studied instrumented performance of Earth Dams during a 3 year stay at the Bureau of Reclamation. Settlement and pore pressures were measured in over 20 earth dams from a period of the 1930s through the 1950s. Detailed measurements were made possible by using multiple cross arm settlement columns and hydraulic pieziometer arrays. Compression records indicated most of the consolidation occurs during construction, and that for soils compacted dry of optimum, most of the compression occurs by reduction of air voids. From the compression data, typical compressions indices were correlated to Unified Soil Classification System soil types. Three phase media theory was used to develop equations to predict pore air and pore water pressure development in compacted soils during compression. Compressible soils compacted wetter than 0.6 % of dry of optimum moisture, exhibited undesirably high pore water pressure increases during construction. On dams where soils were compacted to an average 1.5 to 0.5 % dry of optimum, there were no significant excess pore pressures. This study and several related studies performed during the 1950s at Reclamation constitute some of the most important contributions to dam engineering. By using this information, earth dam designers can confidently design earth dams, and gradually, instrumentation has been reduced in new earth dam construction.

INTRODUCTION

This report summarizes a comprehensive study of compression and pore pressure behavior of over 20 rolled earthfill dams constructed by the U.S. Bureau of Reclamation. This study was summarized by Dr. James P. Gould (deceased of Musser

---

[1] U.S. Department of the Interior, Bureau of Reclamation, Earth Sciences and Research Laboratory, PO Box 25007, D-8340, Denver CO 80225, jfarrar@do.usbr.gov.

Rutledge Engineers) in the 1950s (Gould, 1954, 1959). The study is probably one of the largest compilations of dam performance by an agency. This work is cited in some dam engineering texts (Sherard, et al.1963). These types of compilations have added to engineering experience in dam construction.

By using results of extensively instrumented dams, the observational method was applied. Engineers learned the effects of compaction and water content on the performance of the structures. By learning from this behavior we can avoid undesirable deformation, cracking, and excess pore pressure generation during construction. Today such extensive instrumentation is not required.

The dams were constructed from 1930s to the 1950s. The impervious sections of these dams were compacted with 12 passes of a sheeps foot roller in 150 mm compacted lifts. The sheepsfoot rollers were loaded to 2 tons per linear foot of roller, with tooth pressures ranging from 1750 kPa in earlier dams and to 2919 kPa on newer larger dams. A wide variety of soils were used, but there were not enough examples of construction of very fine, high plasticity soils.

Optimum moisture and maximum density values were determined according to the "Standard Proctor" test (ASTM D-698). The degree of compaction specified was generally an average of 100% of standard effort. Compaction control was based on the minus #4 control fraction. Compaction control techniques varied somewhat. Some of the earlier structures used needle pressure monitoring of the fill as preferred by Proctor. But in most all cases the compaction curves were also monitored. The Rapid Method of control (ASTM D 5080) was used in later construction. Experience began to indicate undesirably high pore pressures developed with soils compacted near optimum moisture. Limiting moisture studies were performed using consolidation tests to target compaction moisture of about 1 to 2 % dry of optimum. Extensive in-place density and measurements were made. Typical testing frequency was one test per 2,000 m^3 of fill.

Standard field laboratory testing was performed as part of the construction. Permeability and settlement characteristics of the fill were documented using record tests samples taken at 30,000 m^3 intervals. These tests were performed by the field laboratories during winter slow down of construction. The permeability and settlement tests were performed using constant head tests on 200 mm diameter specimens (USBR 5600, 1990). The loads applied were in proportion of the fill height or standard load of 700 kPa. This data, coupled with the construction monitoring would aid in determining the cause of embankment responses measured.

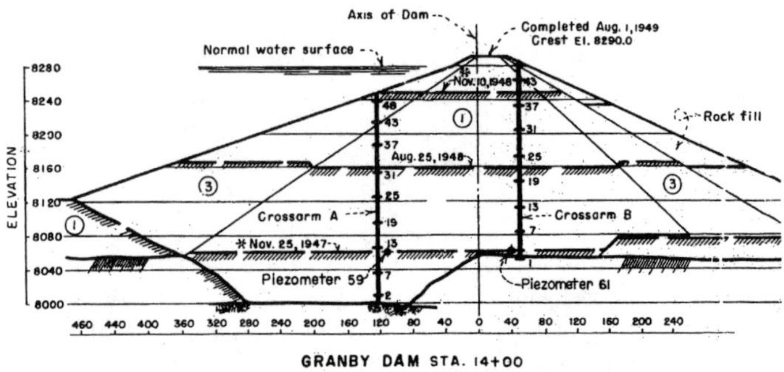

Figure 1 - Example of Crossarm installation at Granby Dam (1m = 3.28ft)

INSTRUMENTATION SYSTEMS

To monitor compression, cross arm settlement columns were installed in one to six locations in 25 dams. An example of the cross arm installation in Granby Dam is shown on Figure 1. Horizontal cross arms, are connected to a system of telescoping riser pipes. The arms and pipes are added as construction progresses. The arms are located at 1.5 m vertical intervals. The vertical settlement was measured to 3 mm using a torpedo measuring device. The details of this instrumentation can be found in Reclamation's Embankment Instrumentation Manual (1987).

Pore pressures were measured by an array of hydraulic pieziometers in 26 earth dams. These arrays were placed in instrumented sections of the dam and several zones were monitored. The hydraulic gauges use water filled tubing hooked to bourden tube gages in a control house. These devices were capable of measuring negative pressures, up to cavitation, in the fill. Information on the construction of these devices can also be found in Reclamation's Embankment Instrumentation Manual (1987).

COMPRESSION BEHAVIOR

The compression data show that the majority of compression occurs during the construction of the embankment. Figure 2 shows a plot compression of Deer Creek Dam during and after construction. Even in dams where high pore pressures developed, the post-construction settlement is not large. Post construction settlement amounted to less than 0.2% in the first 3 years, and less than 0.4% for periods of up to 14 years. Even though compaction moisture as low as 3% dry of optimum were used in construction, there was no significant settlement upon wetting recorded for any of the dams.

Compression data confirmed that the compression occurred in the central portion of the dam in a vertical direction. Lateral displacements were not significant. Gould compiled average values of measured compression and plotted them as a function of the weight of the overlying fill. After subtracting pore pressures, he obtained curves of vertical compression versus vertical effective stress.

Figure 3 shows that the compression curves approximate straight lines on log plots. Compression at approximately 700kPa of all materials range from 1 to 4 %. Table 1 summarizes the compression and compaction characteristics of the dams. The data show that the compression is governed by compaction water content and soil properties, and not a function of the total density attained.

Compression behavior of the soils in the dams are compared against Unified Soil Classification soil type (ASTM D 2497) on Table 1. This data shows that uniform materials such as silty soils (GM, SM, ML) had low degree of compression, while plastic fine grained soils (CL, SC) were much more compressible. Coarse grained soils with plastic fines (SC) were moderately compressible. For Lean Clay soils, compression at 700 kPa ranged from 2.8 to 4.6 % and the variation was primarily due to compaction water content, with wet soils showing higher degrees of compression. This collection of data is often used to design camber height for new dams . Table 2 is a summary by Sherard, et al. (1963) of compression of soil groups under low and high embankment loading. If the soils contain gravels, the compression is reduced by the amount of gravel in the fill. Consolidometer tests performed on the minus # 4 fraction can be used to design for compression after a simple correction for gravels content. Users of these design methods are cautioned that the data is based on typical Reclamation practice of strict construction control and use of proper construction equipment.

PORE PRESSURE BEHAVIOR

In order to understand the behavior of partially saturated compacted soils, compression theories were developed considering both pore air and pore water pressure development. Soils compacted wet of optimum can develop high construction pore pressures, which sometimes can reduce the factor of safety to the lowest level in the life

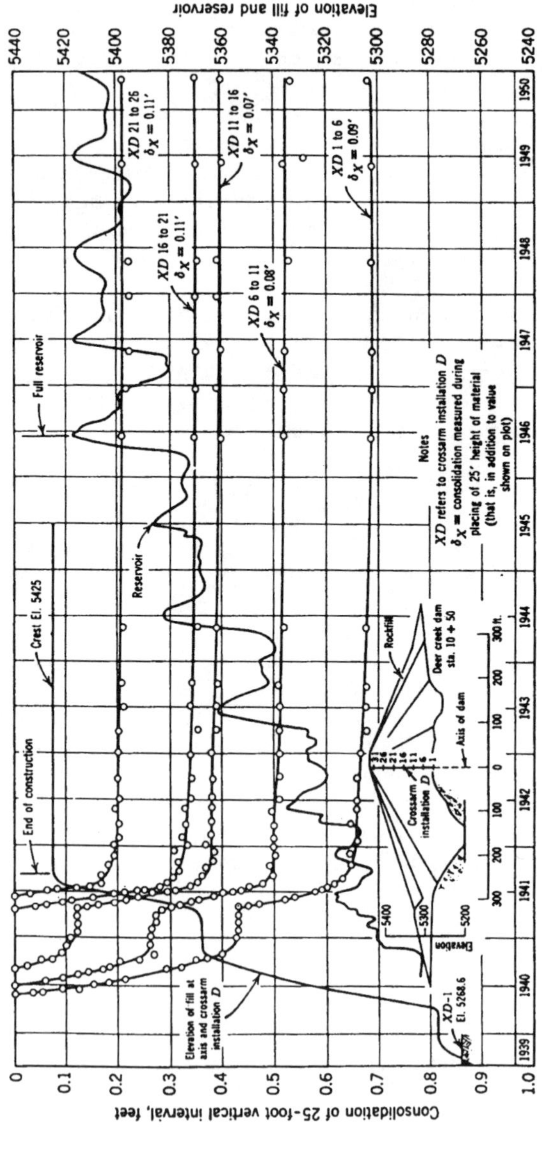

Figure 2 - Example of Compression Behavior at Deer Creek Dam (1m = 3.28ft).

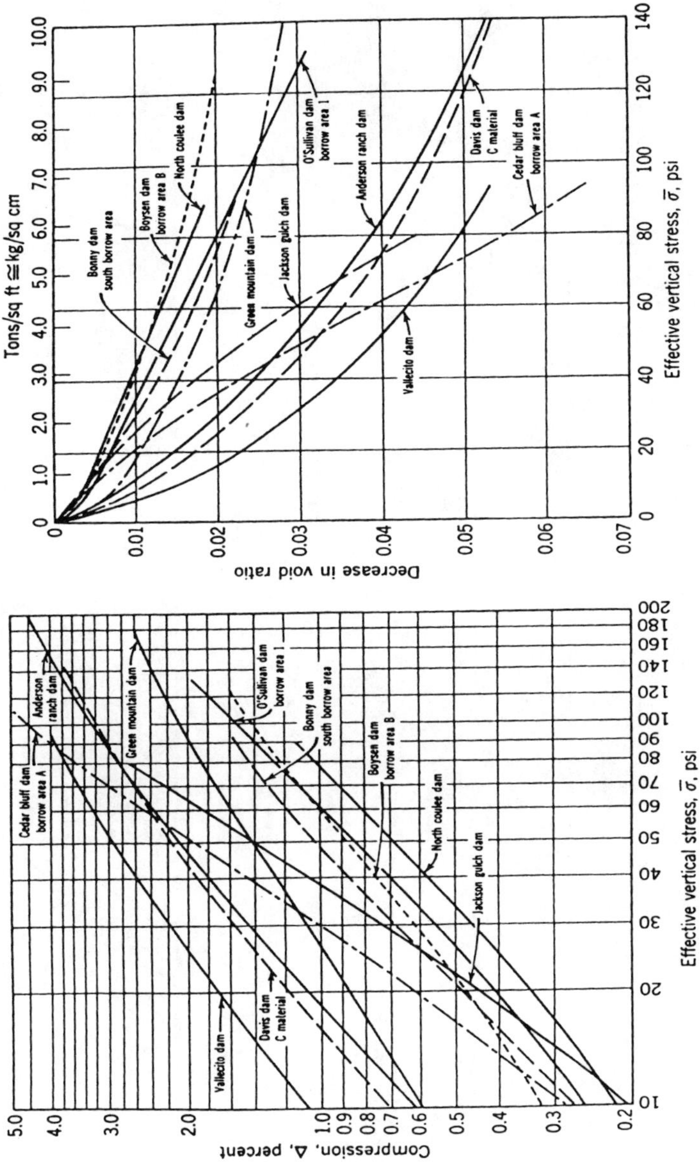

Figure 3 - Typical measured compression curves (a) Logarithmic plot (b) Arithmetic plot (1psi = 6.9 kPa)

Table 1 - Summary of Measured Compression of Impervious Sections of USBR Dams After Construction (1psi = 6.9 kPa, 1 lb/ft^3 = 16.01 kg/m^3).

SOIL TYPE	CLASSIFI-CATION SYMBOL	DAM NAME (BORROW AREA)	AVERAGE MEASURED EMBANKMENT COMPRESSION (VERTICAL)		AVERAGE SOIL PROPERTIES					
			at 10 p.s.i. (%)[2]	*at 100p.s.i. (%)*[2]	*Dry Density (Fill) -#4 mtl. (lb/ft.^{3})*[3]	*Water Content (Fill) -#4 mtl. (lb/ft.^{3})*[3]	*Opt. Water Content (Lab) (%)*[1]	*Rock Content (>#4 Sieve) (%)*	*Liquid Limit (%)*	*Plas-ticity Index (%)*
Silty										
gravel	GM-SM	Davis (A & B)	0.2	1.0	122.0	9.8	11.5	45	23	2
and	GM-GP	Boysen (B-1)	0.3	0.9	121.6	10.3	10.4	50	18	2
coarse	GM-SM	Boysen (B)	0.3	1.4	122.5	10.8	11.4	39	20	5
silty	GM-SM	Caballo	0.3	1.1	117.9	11.6	12.8	37	--	--
sand	GM-SM	Cachuma	0.3	1.1	112.7	13.7	14.3	37	25	6
Fine	SM	Davis (B)	0.5	1.8	118.1	11.9	12.2	14	21	1
silty	SM	Long Lake (1)	0.3	1.6	107.2	14.6	17.1	0	20	1
sand	SM	Bonny (S.A.)	0.3	1.6	113.4	12.7	13.9	0	21	2
and	SM	O'Sullivan (1 &	0.3	1.6	107.6	15.5	17.1	0	22	NP
silt	ML	7)	0.4	2.1	111.6	13.2	14.5	0	20	1
of low	ML	Enders	0.4	1.9	106.9	15.4	16.2	0	21	2
plasticity	ML	Bonny (N.A.)	0.2	1.3	101.6	15.7	18.2	0	22	2
		No. Coulee								
Clayey	SC-GC	Cachuma	0.3	2.0	110.9	14.2	15.3	36	-	-
gravel	SC-SM	Heart Butte (NE)	0.4	2.6	112.0	15.1	16.3	4	26	6
and	SC-SM	Boysen (K)	0.4	1.9	121.8	11.3	11.8	27	21	5
coarse	SC	Anderson Ranch	0.6	3.3	121.2	12.5	13.0	10	28	12
clayey	SC	Granby	0.8	2.4	126.8	9.6	10.9	27	22	8
sand of	SC	Green Mtn.	0.6	2.1	132.2	8.4	8.9	35	21	8
low	SC	Shadehill	0.4	2.0	122.4	11.4	12.4	30	25	10
plasticity	CL-GC	Deer Creek	0.4	2.6	110.6	15.1	15.5	30	26	10
	CL-ML	Fresno	0.7	3.6	119.1	12.2	12.4	3	20	5
Clay	CL-ML	Long Lake (LC)	0.2	2.9	100.0	18.5	20.5	0	24	4
of low	CL-SC	Vallecito	1.1	4.2	123.2	12.2	11.5	16	-	-
to	CL	Cedar Bluff (C)	0.2	2.8	112.7	12.9	13.9	0	31	14
medium	CL	Horsetooth	0.2	3.0	112.5	11.6	13.9	2	26	12
plasticity	CL	Medicine Ck. (A)	0.3	3.5	104.5	16.8	17.6	0	30	8
	CL	Heart Butte	0.5	3.8	115.3	13.5	14.5	2	24	10
	CL	(S.W.)	0.2	3.7	100.7	18.9	21.4	5	40	21
	CL	Jackson Gulch(1)	0.3	4.0	113.5	13.4	14.9	6	27	11
		Carter Lake								

[1] Standard Proctor compaction test. [2] Effective stress (pore pressures subtracted). [3] Measured by control test during construction.

Table 2. Relationship between Embankment Compression and Soil Type as Determined from Measurements on USBR Dams (Sherard, et al., 1963).

EMBANKMENT SOIL TYPE	APPROXIMATE RANGE OF MEASURED COMPRESSION (%) at 70 kPa.	at 700 kPa
1. Silty gravel and coarse silty sand (GM and SM)	0.2-0.3	0.9-1.4
2. Fine silty sand and silt of low plasticity (SM-ML)	0.2-0.5	1.3-2.1
3. Clayey sands and gravels (GC-SC)	0.3-0.8	1.9-3.3
4. Clay of low to medium plasticity (CL and CL-ML)	0.2-1.1	2.8-4.2

of the structure. The attempt was made to compact the soils dry of optimum, such that most of the compression during construction occurs in the air void structure. For soils compacted dry of optimum, Hilf and Gould derived equations that could predict the compression of air during construction and predict continued pore water pressure increase as fill progressed. The equations were based on Terzaghi consolidation theory with some of the following assumptions;

- The individual soil grains and the water in the voids are incompressible.
- Deformation from weight of superimposed fill is in vertical strain without lateral deformation.
- The decrease in total volume is caused by compression of air and dissolving of air into the water.
- No significant dissipation of pore pressures during construction.
- The air in the pores behaves as a perfect gas and is compressed isothermally.
- The vapor pressure in the air and the surface tension of water films are neglected.

For partially saturated soil the basic equation for prediction of pore water pressure, u, in terms of initial volumetric components and vertical strain is;

$$u = P_a \cdot \Delta / (V_a + h \cdot V_w + \Delta)$$

Where: u = pore water pressure
P_a = atmospheric pressure
Δ = vertical strain
V_a = volume of air
V_w = volume of water
h = constant 0.2 is a of product KRT, gas constant, absolute temperature, and proportionality for Henry's law

When the soil just becomes saturated, the pore pressure is

$$u = P_a \cdot V_a / h \cdot V_w$$

Figure 4 illustrates the application of this theory to the predictions of pore pressure increases as a function of compaction moisture content and the height of fill. The predicted curves are shown in roman numerals. As expected soil compacted at two standard deviations wet of optimum water, reaches saturation quickly and the pore pressure increases on a 45 degree line. Drier soils don't generate high pore pressures. Observed pore pressure trends are labeled A though D on the figure. The effects of drainage cause some deviations from the theoretical curves. For example a pieziometer tip located near the edge of the core may have a pore pressure trend shown by curve A. Curve B is in the central part of a core zone, compacted near optimum, and follows the trends perfectly and at the end of construction some dissipation occurs. Curve C represents a soil in a wide central core of the dam. Curve D represents tip conditions near the edge of a core zone, where dissipation toward the shell causes an increase in pressures after construction.

The effects of dissipation were quantified using another series of equations which helped explain the pore pressure behaviors. It was found for most soil conditions the pore pressures inside of dams took rather long to dissipate, in many cases it would take several years to see significant decreases in pressures.

The pore pressure behavior of 26 dams was studied. Pore pressure contour plots were made for the structures. Hydraulic pieziometers were used on most of the dams. An advantage to hydraulic pieziometer is they could measure negative pressures in partially saturated soils. At pressures of minus one atmosphere cavitation could occur.

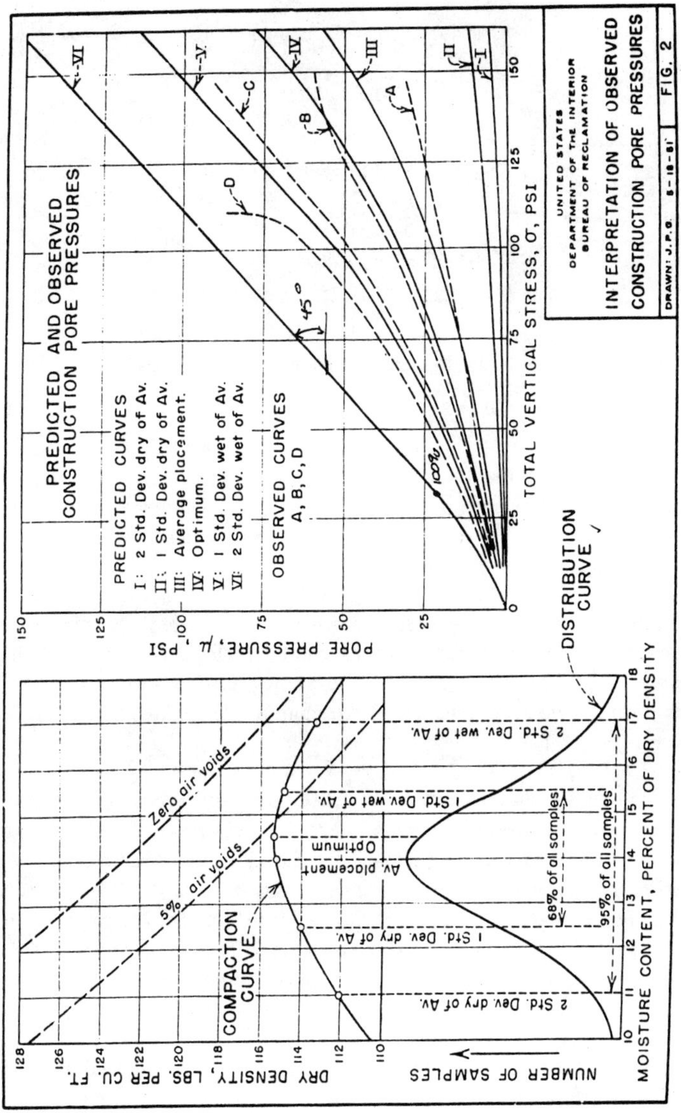

Figure 4 - Interpretation of observed construction pore pressures (1 lb/in^2 = 6.9 Kpa, 1 lb/ft^3 = 16.0 Kg/m^3)

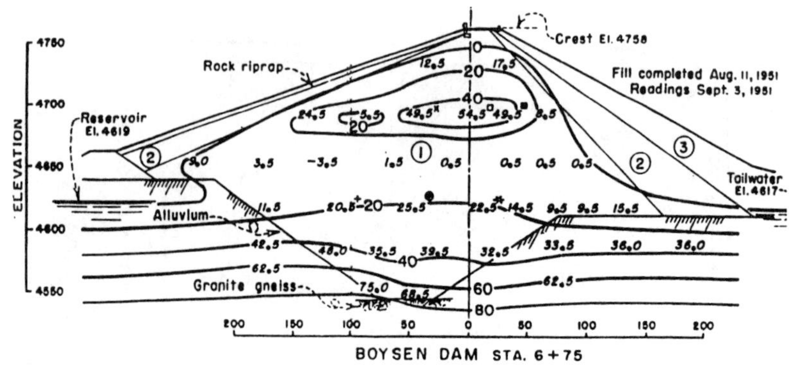

Figure 5 - Example of a wet lift in Boysen Dam - Pressure in feet (1ft = 0.3048 m)

But the cells can be flushed out to remove air. Pore pressures were graphed as a function of fill heights and comparisons were made to the theoretical correlation shown on Figure 4. These studies found some very interesting trends in construction pore pressure development.

An example of the high pore pressure behavior was at Granby Dam during and after construction. The high pore pressures in the cutoff trench were caused by placing soils near or wet of optimum in that location. These pore pressures lowered the static factor of safety during construction to 1.5. These high pore pressures take some time for dissipation.

On some dams, when the borrow became wetter, pore pressure contour plots show high pore pressure zones in the middle of the core - with horizontal distribution. Figure 5 is an example from Boysen Dam which clearly shows a wet lift in the dam. On some dams such as Anderson Ranch, the pore pressures in the core were significant enough to low the factor of safety to values as low as 1.3 to 1.6. These factors of safety are probably the lowest the dam may experience in it's life time. This data indicate that careful control of borrow operations are essential. In most cases Reclamation used irrigation in borrow areas but in some cases water was added during loading.

Reclamation attempted to build dams averaging about 1 % dry of optimum moisture. And when successful, pore pressures were held to low levels. In some cases the embankments registered slightly negative values though out the embankment for a long period of time. The greater the volume of air the slower the dissipation, with pressure lasting several years.

Unfortunately, Reclamation has not made further studies of out data which would show phreatic surface development through the cores of these dams. The data exists in our files but funding for researchers are hard to find. These data can be made available to the public if requested.

CONCLUSIONS

The Conclusions of Gould's study are;

1.) The theoretical prediction of pore pressure and compression behavior were confirmed. Laboratory studies of compression and pore pressure development can be used reliably for design. Deviations in predicted performance could easily be explained by the soil compaction conditions. However, control of the borrow soils are critical.

2.) Drainage conditions could be used to predict the pore pressure patterns. High sand and gravel core materials had lower pore pressure responses. However, Sandy and gravelly soils with plastic fines can develop high pore pressures and compression.

3.) Surface tension has some effect in reducing dissipation of low or negative pressure in dry core materials. Some dams maintained negligible pore pressure in the complete structure.

4.) It is important to perform limiting moisture studies to assure soils aren't placed too dry and result in saturation collapse. Soils placed at an average moisture greater than 0.5% dry of optimum will generate higher pore pressures. For most of the soils tested, compaction averaging 1 % dry of optimum exhibit the best performance.

5.) Wet lifts were readily detected, and they created high pore pressure zones. The effect of differing borrow soils and conditions could be readily seen. Wet soils can be placed in certain areas such as the cutoff trench if the foundation soils have good shear strength.

6) For the prediction of pore pressures, one dimensional consolidation theory for partially saturated soils applies. Compression can be accurately predicted using minus # 4 consolidation tests and applying a simple rock correction. This data can be used to design dam chamber.

7.) If soils are place too wet of optimum, they could result in low factors of safety. It depends on the structure. If strong shells are present, high pore pressures may not be a concern. But for broad cores high pore pressures could be a concern.

8.) The most desireable soils for core construction are GC-SC soils. They have low air content, and if they contain clay fines, excess pore pressures may be alleviated by dilation in shear. Plastic clays can generate high pore pressures and must be carefully controlled and maintained dry of optimum.

APPENDIX I - REFERENCES

ASTM D 698-91 Test Method for Laboratory Compaction Characteristics od Soil Using Standard Effort (12,400 ft-lb/ft^3 (600 kN-m/m^3)), Vol. 04.08, American Society for Testing and Materials, West Conshohocken, PA.

ASTM D 5080-93, "Standard Test Method for Rapid Determination of Percent Compaction," Vol. 0409, American Society for Testing and Materials, West Conshohocken, PA.

ASTM D 2487-93, "Standard Classification of Soils for Engineering Purposes (Unified Soil Classification System," Vol. 04.08, American Society for Testing and Materials, West Conshohocken, PA.

Earth Manual, Part 2, Third Edition, (1990), Bureau of Reclamation, U.S. Government Printing Office, Washington D.C.

Embankment Instrumentation Manual, (1987), U.S. Government Printing Office, United States Department of the Interior, Bureau of Reclamation, Denver CO.

Gould, J.P., (1954), "Compression Characteristics of Rolled Fill Materials in Earth Dams," Technical Memorandum 648, United States Department of the Interior, Bureau of Reclamation, Denver CO, March.

Gould, J.P., (1959), "Construction Pore Pressures Observed in Earth Fill Dams," Technical Memorandum 650, United States Department of the Interior, Bureau of Reclamation, Denver CO, April.

Sherard, J.L., Woodward, R.J., Gizienski, S.F., Clevenger, W.A., (1963), Earth and Earth-Rock Dams, John Wiley and Sons, Inc., New York.

MITIGATION OF SLOPE CREEP BY SUBSURFACE DRAINAGE

By Steven L. Pawlak[1], Member ASCE, Robert Pattillo[2], Member ASCE and Norm Bacheldor[3]

ABSTRACT: The impacts of slope creep on two residences in the Vail ski area were evaluated by floor elevation and wall plumb monitoring, and hillside stability was evaluated by subsurface exploration, multi-depth piezometers and inclinometer casing surveys. Exploratory borings identified a relatively deep, confined water bearing layer with piezometeric levels between about 2 to 4 m (6 to 12 feet) above the ground surface, and a possible basal shear surface between 7½ and 13 m (23 and 42 feet) below the ground surface uphill of the residences. Subsurface drainage by use of low angle (horizontal and declined) drains was installed to depths of 76 to 105 m (250 to 340 feet) in a fan shape below the properties. The piezometer water levels responded quickly after drain construction with a drop of about 14 m (45 feet) uphill of the residences and between about 5 to 10 m (18 to 33 feet) adjacent to the residences. Monitoring of the piezometers, inclinometers and buildings has continued to evaluate the effectiveness of slope creep mitigation. Based on the relatively large drops in the piezometric levels, and the subsequent monitoring results, subsurface drainage appears to have significantly improved the hillside stability and stopped or reduced the slope creep movements.

INTRODUCTION

Two residences located in the Vail ski area have undergone movements for several years which have caused mostly lateral building distortion. The movements have generally been seasonal (cyclic) and small but had progressed to a level of functional and esthetic impairment and potential structural distress. Previous mitigation attempts by building structure modifications and shallow perimeter subsurface drainage had not stopped the building movements based on the continued distress.

The authors were retained in 1997 to review the condition of the residences, conduct subsurface exploration for possible cause of the building distress, and evaluate possible methods of remedial repair. The studies also included level surveying and wall plumb monitoring to evaluate the direction and magnitude of the building movements.

The residences are of relatively modern construction with concrete footings and

[1]President, Hepworth-Pawlak Geotechnical, 5020 Road 154, Glenwood Springs, CO 81601

[2]President, Pattillo Associates Engineers, 715 Grand Avenue, Glenwood Springs, CO 81601

[3]President, Meldor Construction, 1010 Valley Road, Carbondale, CO 81623

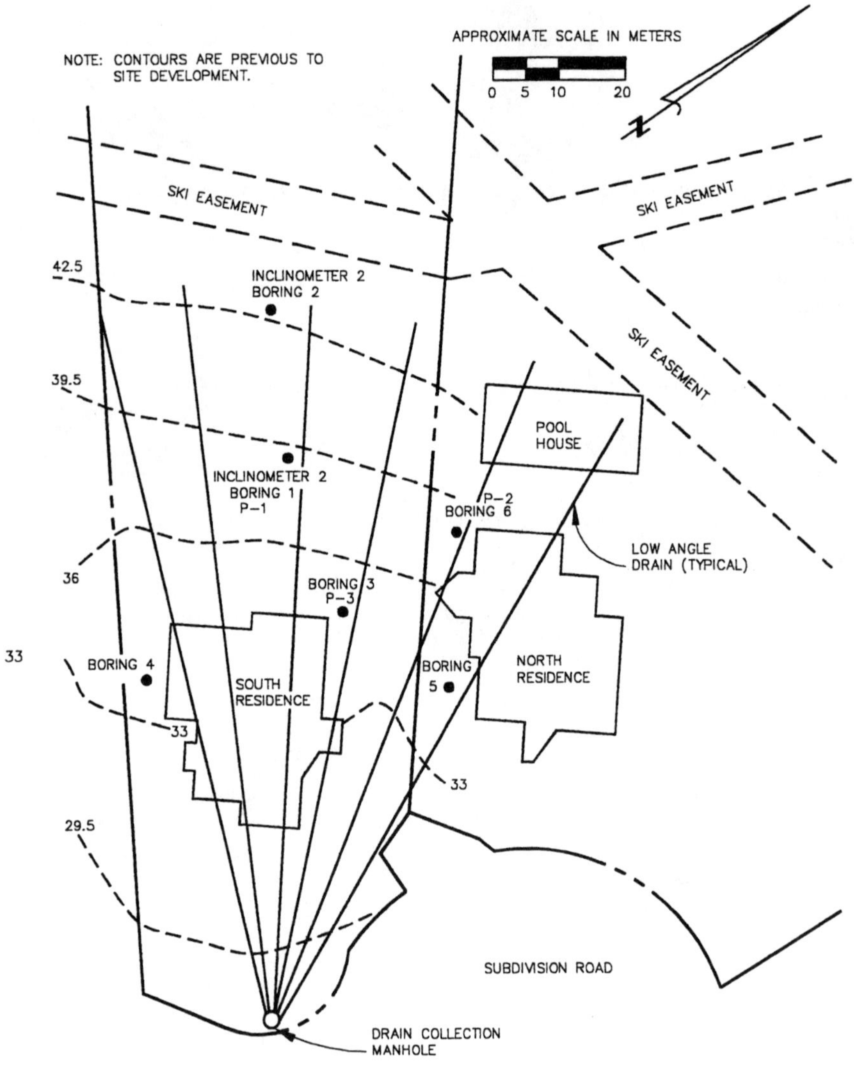

Figure 1 Site Plan

foundation walls but relatively large and complex structures, typical of residences in this area. They are built into a moderately sloping hillside with the lower level a walkout basement. Excavation for the building foundation construction was extensive and required cut depths up to about 3.5 to 4.5 meters (12 to 15 feet) below the natural ground surface. Foundation drainage installed for the original construction has adequately kept the lower levels dry since construction.

The cyclic nature of the building movements, and the lack of success of the previous mitigation attempts, indicated that hillside creep could be a cause of the building distress.

SUBSURFACE EXPLORATION

The subsurface conditions were initially evaluated by drilling 2 exploratory borings (Borings 1 and 2) uphill of the south residence as shown on Fig. 1. A track-mounted drill rig and continuous flight augers were used to advance the borings while taking spoon drive samples on 1.5 to 3 m (5 to 10 feet) intervals. The soils encountered in the borings consisted of valley alluvium derived from glacial outwash and more modern deposits. The soils include very stiff sandy clay and medium dense to dense clayey sand and gravel with cobbles and boulders. Shale of the Pennsylvania-age Minturn Formation was encountered at depths of 13 to 15.5 m (44 to 51 feet). The Minturn consists of gray siltstone and shale with scattered beds of carbonate rock. It has been the author's experience that the subsurface profiles encountered in the borings were generally consistent with those expected in the area. Landslides are not mapped in the project area and deep seated movement that extends into the formation rock was ruled out. If slope creep was occurring, it would be within the overlying soils.

WATER LEVEL MONITORING

Multi-depth, observation well piezometers were installed at each of the boring locations. The lower part of the piezometer casing was slotted, sand packed and sealed with bentonite so that groundwater pressure could be measured at different boring depths. The deeper piezometers were isolated in the zone about 3 to 4.5 m (10 to 15 feet) above formation rock and the shallower piezometers were isolated between about 3 to 6 m (10 to 20 feet) below ground surface. The piezometers showed an increase in piezometric surface elevation with depth. The deeper piezometers had a piezometric surface between about 2 to 4 m (6 to 12 feet) above ground surface. Four additional multi-depth piezometers were subsequently installed as part of the remedial repair (Borings 3-6 located on Fig.1) and found similar water pressure conditions. It has been the author's experience in mountainous areas that water levels can rise during spring runoff and adversely impact slope stability. Therefore, the piezometers were monitored for seasonal water level fluctuations to determine the most severe groundwater conditions.

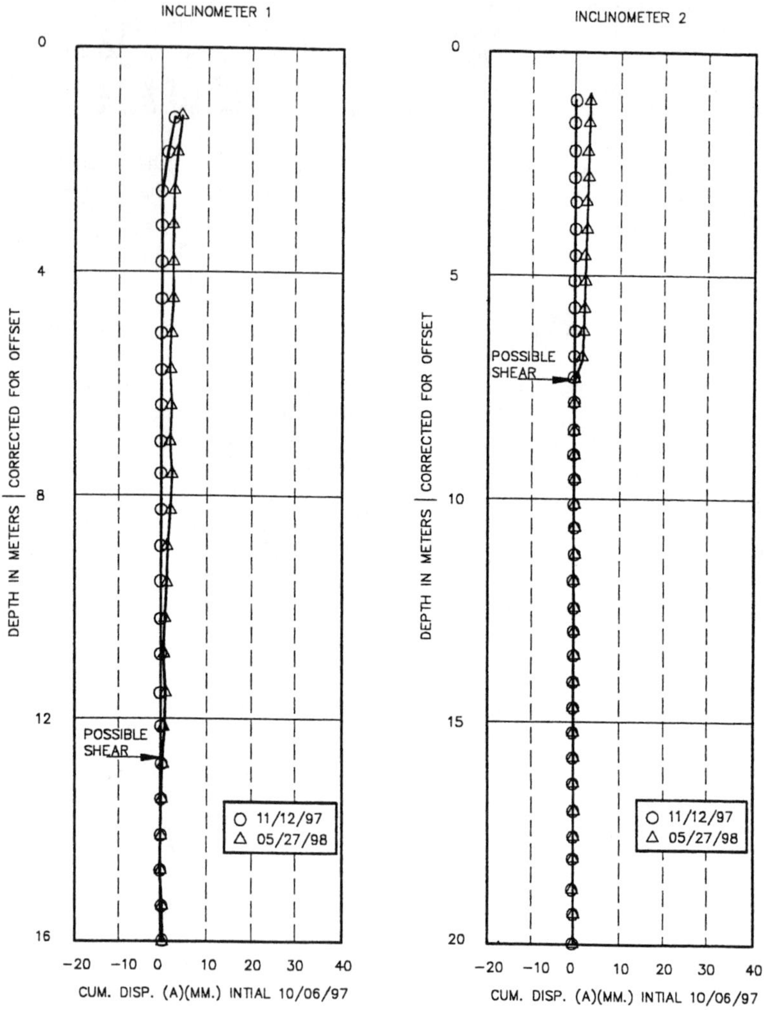

Figure 2 Inclinometers 1 and 2 - Before Drainage

SLOPE MOVEMENT MONITORING

Inclinometer casing was set in 4-inch diameter auger holes at Borings 1 and 2 to measure lateral ground movements. The casing was 2.75-inches in diameter and was set in the formation rock. The annular space between the casing and the boring was backfilled with sand to achieve continuous contact with the ground. The casings were surveyed for lateral movement beginning in October 1997 and continued through the summer of 1998. The inclinometer is lowered into the casing attached to a cable and the vertical plumbness of the casing is measured. By comparing the inclination of the casing to previous readings, the lateral deflection of the casing is measured. The results of the monitoring are shown on Fig. 2 and indicate ground movement and possible shear surfaces at about 7½ and 13 m (23 and 42 feet) below ground surface. The detection of the movement in the casings roughly correlated with piezometric rise measured in the piezometers between about April to May.

BUILDING MOVEMENT MONITORING

Monitoring of building movements can provide information regarding the nature, magnitude, direction and rate of movements. Possible movements of the building were monitored by measuring relative floor and ceiling elevations with a calibrated manometer, vertical plumbness of walls with a plumb bob device and squareness by diagonal room distances. The readings are accurate to within 2.5 mm (1/10 inch) for the manometer, 1mm (1/32 inch) for the plumb bob, and 3 mm (1/8 inch) for the distance measurements. Baseline readings were taken in the fall, 1997. Subsequent readings were taken in the spring and fall, 1998. During this period, the movements were mainly lateral deflection of the uphill foundation walls as indicated by the readings. Building movements were small but more active in the spring and summer during the seasonal high in the piezometeric levels.

REMEDIAL REPAIR OPTION

The inclinometer casing monitoring indicated there may be hillside creep above the residences which could be causing the building distress. The high piezometric levels and artesian conditions are adverse to slope stability and were thought to be causing the hillside creep. By reducing the groundwater pressure, the slope stability would be improved and creep could be mitigated. In the author's experience, slope stability improvements by subsurface drainage in itself can be relatively small, but considering the relatively high water pressures, the stability improvement could be significant with effective subsurface drainage.

Other methods of improving slope stability include reducing the weight of the uphill part of the slope (such as by excavation), and increasing the resistance to sliding at the toe of the slope (such as by additional earth loading or mechanical restraint). These methods would require extensive modification to the landscape and were not considered practical for the current lot developments.

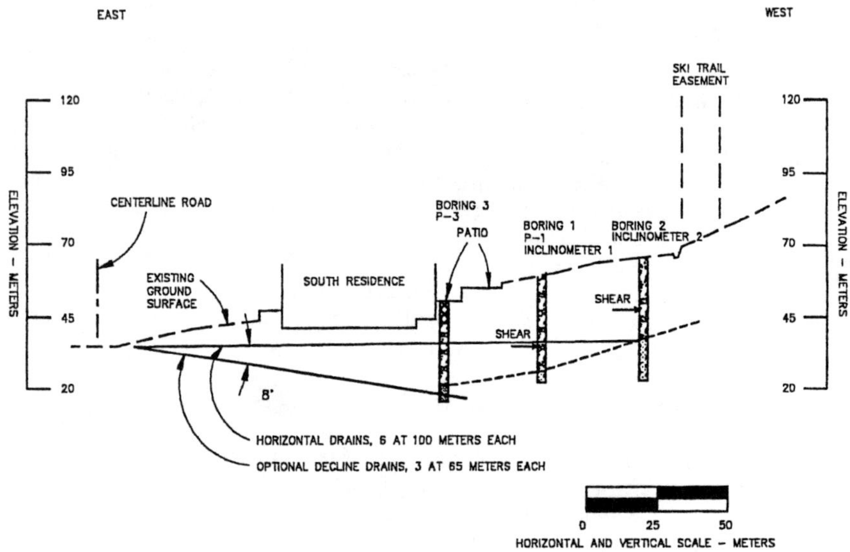

Figure 3 Typical Drain Cross Section

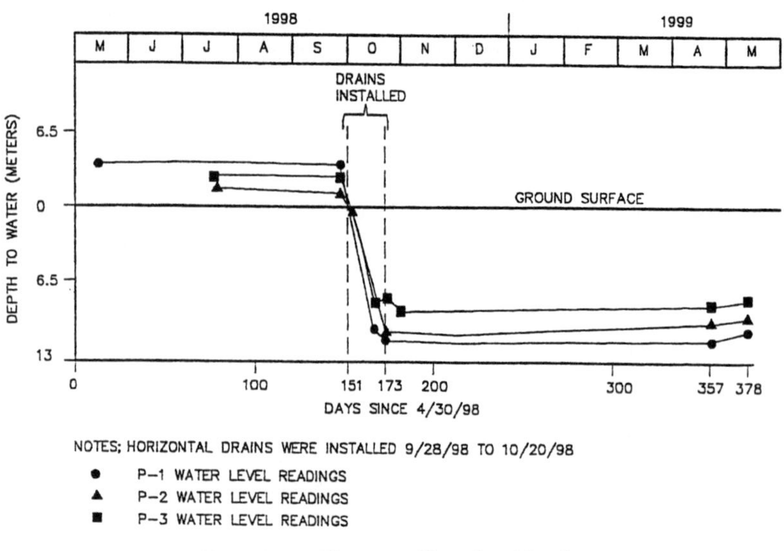

Figure 4 Piezometer Water Level Readings

The authors recommended low angle drains installed in small diameter borings as the subsurface drainage system. The zone of high groundwater pressure was generally within about 3 m (10 feet) of the top of formation rock or a depth on the order of 12 m (40 feet) below the ground surface. Trench drains excavated from the ground surface were not considered practical to reach the high water pressure zone.

Two rows of drains were installed in a fan shape from a downhill location that extended beneath the residences. The first row of drains was extended into formation rock, estimated at a distance of 90 m (300 feet). The second row of drains, declined at about 8° and about 60 m (200 feet) long were proposed to further improve the drainage directly below the residences. A typical cross section of the proposed drain system is shown on Fig. 3.

SUBSURFACE DRAINAGE INSTALLATION AND RESULTS

Nine drains were installed between September 28 and October 20, 1998. The six horizontal drains had lengths of 90 to 103 m (300 to 340 feet) and the three declined drains had lengths of 76 m (250 feet). The drain holes were drilled with 4 inch diameter roller bits using water for circulation. The borings were terminated in the formation rock, 1½ -inch diameter slotted PVC pipe was installed, and drill casing was withdrawn.

Initial flows from each drain pipe were relatively high, on the order of 75 to 150 1pm (20 to 40 gpm) with stabilized rates on the order of 4 to 19 1pm (1 to 5 gpm).

The piezometer levels responded quickly after drain construction with levels dropping up to about 14m (45 feet) uphill of the residences and between about 5 to 10 m (18 to 33 feet) adjacent to the residences. A summary of the water level measurements taken from 3 of the piezometers installed in the deep, high water pressure zone above formation rock, is shown on Fig. 4.

MONITORING RESULTS

The effectiveness of the drain system in reducing the slope creep was evaluated by measuring the drop in piezometric surface, and long term monitoring of the inclinometer casings and building movements. The piezometric level and inclinometer casing measurements were made through the spring. The seasonal rise in the piezometric levels was typically between 0.15 and 1.3 m (0.5 and 4.5 feet), except at Boring 2 (located above the south residence) which had a piezometric level rise of about 4.8 m (16 feet). This piezometer is approximately at the uphill limit of the subsurface drainage system, and is likely influenced by groundwater recharge from uphill. The seasonal water level rise in the piezometers closer to the residences has been relatively minor and considered consistent with expected seasonal fluctuations in this area. The water level readings made through 1999 for the 3 deep piezometers are shown on Fig.4. The water levels in all of the piezometers dropped following the spring peak. These levels were similar to those measured in the fall of the previous year.

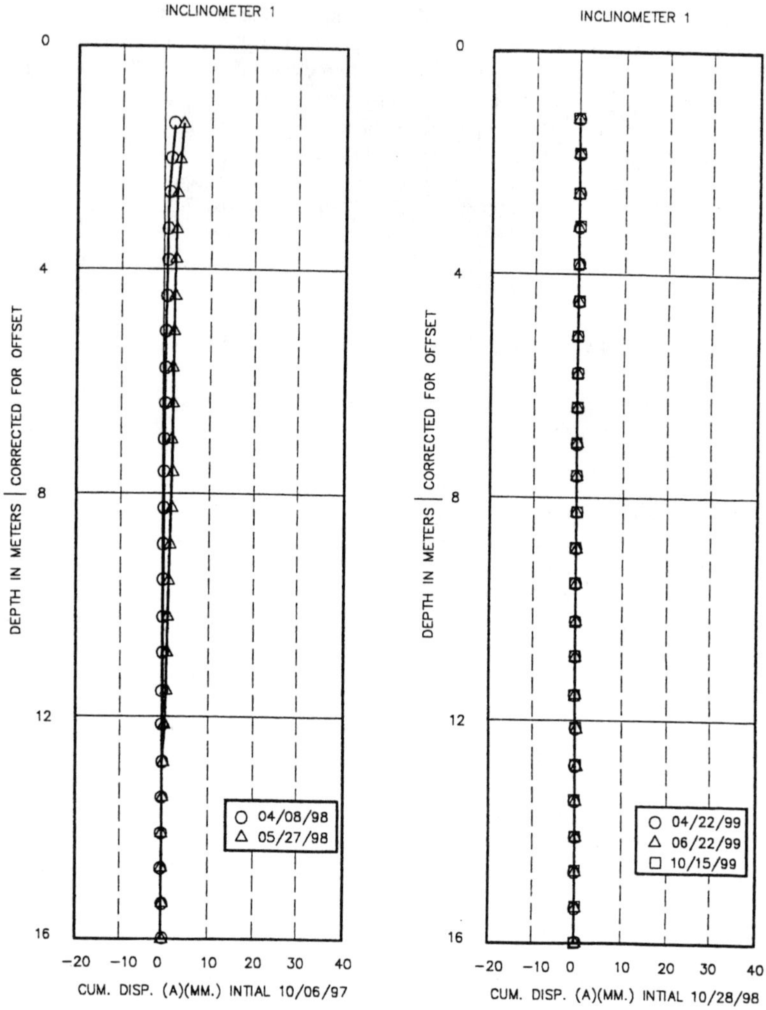

Figure 5 Inclinometer 1 - After Drainage

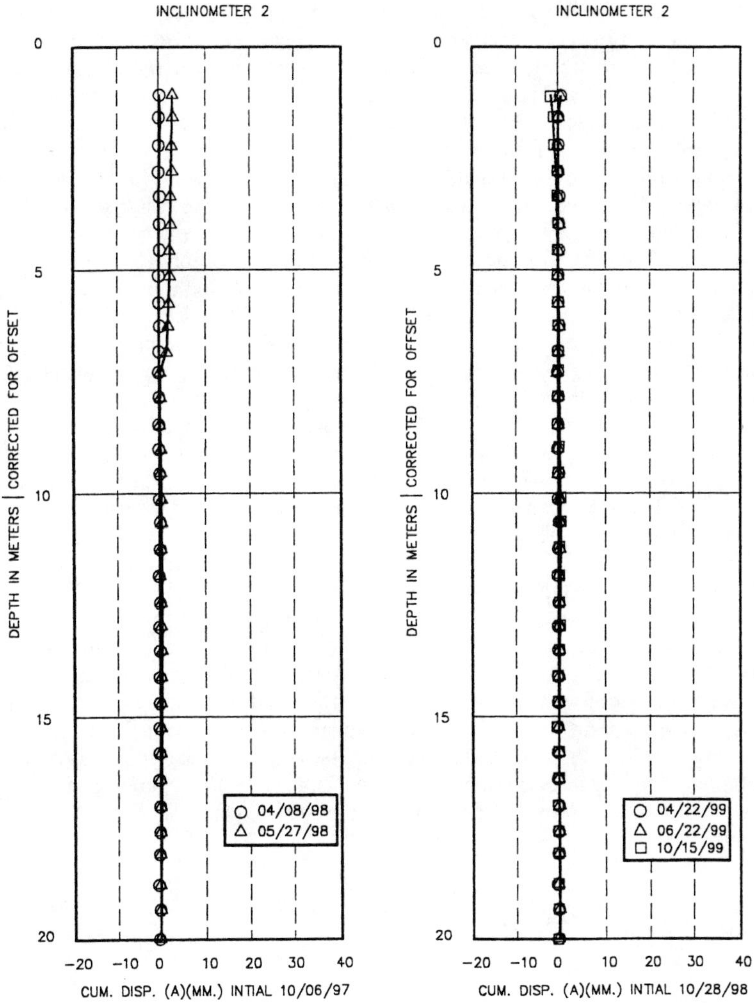

Figure 6 Inclinometer 2 - After Drainage

The individual and total discharge flows from the drains were measured during spring runoff. The rates were typically between 23 and 68 1pm (6 to 18 gpm) with a total drain system discharge of about 275 1pm (72 gpm). The drain flows decreased late in the year and were similar to those measured the fall of the previous year.

The inclinometer readings indicate that there has not been detectable slope movement since the drain system construction in October 1998. The monitoring results are shown on Figs. 5 and 6 and compare the pre-drain and post-drain installation measurements. The pre-drain plots show a lateral offset of the casing relative to the initial reading whereas the post-drain plots generally show no offset relative to the October 1998 reading. The Boring 2 inclinometer may show some lateral movement within 3 to 5 m (10 to 16 feet) of ground surface, which could be related to creep movements uphill of the drained area.

The building movement monitoring conducted since the drain installation, and visual observations of the building conditions, indicate some on-going movement. The movements appear to be relatively minor and may be related to adjustment of the building to the previous over stressed building condition. The south residence, which was about 7.5 cm (3 inches) out of level before the drainage mitigation, showed about 2 cm (0.75 inch) reduction in floor tilt within three months after drainage improvements and some rebound since then. Further monitoring will be needed to evaluate the rates of any on-going movements.

SUMMARY AND CONCLUSIONS

Relatively large and fairly constant drops in the groundwater piezometric levels have been achieved by the subsurface drainage system. The inclinometer measurements generally indicate no additional creep movement of the hillside near the residences since the drain construction. The building movement monitoring is inconclusive at this time and longer-term readings are needed to evaluate the response of the building to the slope stability improvements. Based on these findings, the subsurface drainage appears to have significantly improved the hillside stability and has stopped or reduced the creep improvements.

DEVELOPMENT AND RELAXATION OF COMPACTION INDUCED EARTH PRESSURE OF A CLAY BACKFILL AT TONGUE RIVER DAM, MONTANA

Steve W. Perkins[1], Debora J. Miller[2], Andrew Lindsey[3]

ABSTRACT: Ten earth pressure cells and two pore water pressure transducers have been installed along the face of a 30 foot tall spillway sidewall to measure lateral earth and pore water pressures over time in the compacted clayey backfill. The spillway is a new labyrinth structure constructed as part of the Tongue River Dam Rehabilitation Project in Big Horn County, Montana. The sidewall is a counterfort wall that was assumed to be rigid for purposes of lateral loading estimation and design. Wall rigidity arises from stiffness offered by the counterfort design and the massive roller compacted concrete fill upon which the wall is founded. The portion of the wall instrumented was backfilled with a lean clay constituting the core of the dam. Compaction was achieved with both jumping-jack hand compactors used immediately adjacent to the wall and a vibratory sheep's foot roller. The wall was designed to resist lateral loads induced by the self-weight of the clay, compaction of the backfill clay, water pressure due to soil saturation and swell pressures in the clay. Data collected to-date show the development of earth pressure as compaction proceeds up the wall and the decay or relaxation of this pressure with time. Compaction induced pressures are compared to values calculated using several design solutions. Steady state earth pressures, measured after relaxation is essentially complete, are compared to at-rest values for this material. The time for relaxation to occur is discussed in light of design and operation considerations.

INTRODUCTION

Lateral earth pressure created by the compaction of backfill soil against retaining structures has been measured through numerous studies over the past 65 years, with more recent studies including Sowers et al. (1957), Broms and Ingleson (1971), Coyle and Bartoskewitz (1976), Carder et al. (1977), and Filz and Duncan, (1996). Collectively, these studies have shown that the compaction of soil in relatively thin layers by high-energy compaction equipment creates lateral earth

[1] Associate Professor, Department of Civil Engineering, Montana State University, Bozeman, MT 59717
[2] Associate/Geotechnical Engineer, ESA Consultants, Inc., 2637 Midpoint Drive, Suite F, Fort Collins, CO 80525
[3]Graduate Research Assistant, Department of Civil Engineering, Montana State University, Bozeman, MT 59717

pressures during compaction that may greatly exceed theoretical at-rest values and may approach passive earth pressure magnitudes. Compaction-induced earth pressures have been experienced to depths as great as 15 m when very heavy compaction equipment has been used (Seed and Duncan, 1983). Upon removal of compaction equipment, as much as 90 % of the peak lateral earth pressure induced during compaction may remain as a locked-in residual earth pressure.

The compaction process behind a rigid, non-yielding retaining structure backfilled with cohesionless soil can be modeled by a K_o triaxial test as illustrated in Figure 1 (Clayton et al., 1993). Placement of loose fill results in a comparatively low vertical and horizontal effective stress at a relatively shallow depth along the wall as denoted by point A. The application of compaction pressure increases the vertical and horizontal stress along the K_o line to point B. The removal of the compaction equipment causes the vertical stress to return to its original value at point A. The horizontal stress, however, initially decreases very little and then decreases dramatically as the passive failure state is approached, with the unloading curve eventually moving down the K_r line. Broms (1971) has suggested that K_r=1/K_o. Duncan and Seed (1986) have provided a more elaborate theory based on a hysteretic soil model for loading-unloading behavior of soil elements along the wall as fill placement proceeds to greater depths.

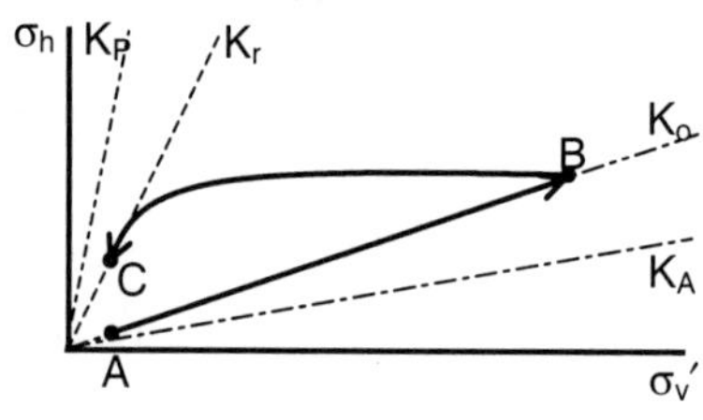

Figure 1: Lateral earth pressure developed in a K_o triaxial test (after Clayton et al., 1993).

The extension of the stress path portrayed in Figure 1 to the full compaction process as fill is placed up to the top of the wall results in the hypothesized residual (i.e., after compaction equipment has been removed) lateral pressure distribution shown in Figure 2 (Broms, 1971). This method assumes that compaction-induced residual pressures do not increase or decrease over time. Measurements of pressures on relatively rigid retaining walls backfilled with clayey soils (Carder et al., 1980; Symons et al., 1989) indicate that these pressures can relax with time, often returning to at-rest values, and that for certain clays, pressures might increase due to pore water pressure equilibration and subsequent swelling.

The purpose of this paper is to present earth pressure measurements from a rigid, non-yielding, full-scale field counterfort wall where a clay has been used as backfill.

The following sections describe the details of the construction project and the wall, backfill materials and placement procedures. Measurements of earth pressure at various points along the height of the wall during and shortly following fill placement are presented.

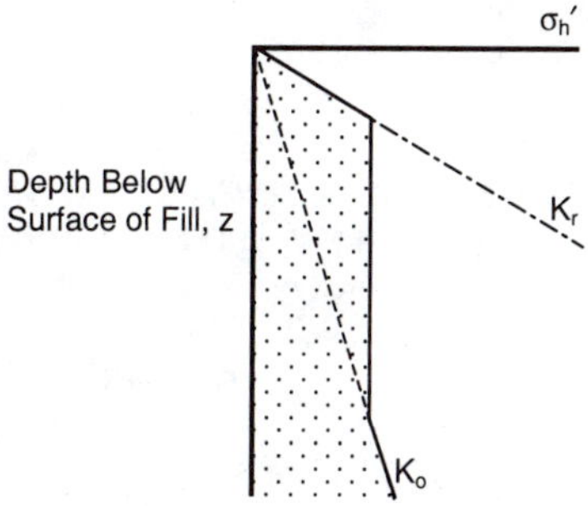

Figure 2: Residual earth pressure distribution due to compaction (after Broms, 1971).

TONGUE RIVER BASIN PROJECT

The retaining wall where earth pressure measurements have been made is part of the Tongue River Basin Project. This major dam rehabilitation project is located in southern Montana, about 10 miles north of the Wyoming/Montana border. The earthfill dam was constructed between 1937 and 1939, and is 28 m high with a crest length of 550 m. The original spillway was a concrete-lined chute in the left (west) abutment, which was nearly destroyed during a 1978 flood event at discharges of only about 10 percent of its theoretical design capacity of 2200 m^3/s. The facility was classified as unsafe by the Army Corps of Engineers in 1980 because of major hydraulic deficiencies in the spillway design, seepage problems in the fractured bedrock on which the spillway was founded, and scour problems around the spillway roller bucket and lower wing walls.

The Tongue River Basin Project, completed in May 1999, alleviates dam safety concerns, preserves existing water rights held in the reservoir, and provides additional water to the Northern Cheyenne Native American Tribe. The completed project provides two new spillways having a combined design discharge capacity of 2830 m^3/s. The primary spillway replaces the original spillway in the left abutment. The emergency spillway is a stair-step Roller Compacted Concrete (RCC) structure constructed on the crest and downstream slope of the dam. A new primary outlet works were constructed and the old outlet was rehabilitated to serve as an auxiliary facility. The maximum reservoir pool has been raised four feet, increasing the total reservoir storage capacity from 83 to 99 million m^3. More information on the overall project, its spillways and other design elements, is available in a number of articles (Kelly and McDonald, 1998; McDonald et al., 1998; McDonald and Yadon, 1998; Miller et al., 1999; Wright, 1998; and Yadon, 1997).

ARTHUR LAKES LIBRARY
COLORADO SCHOOL OF MINES
GOLDEN, CO 80401

The new primary spillway is narrower than the original, requiring the construction of a closure dam on the west side. Fractured bedrock was removed to a depth of about 11 m (35 ft), and a massive backfill of RCC was placed for the foundation of the critical weir section of the new labyrinth spillway. The walls of the spillway consist of a reinforced concrete counterfort retaining structure with the counterfort wings placed on the soil-backfill side of the wall. The counterfort structure along with the foundation of the wall is tied to the roller compacted concrete foundation, provideing for a rigid, non-yielding wall to minimize the potential for seepage through the backfill soils. Figure 3 provides a cross-sectional view of the dam along the primary spillway wall. The section in which earth pressure cell instrumentation has been placed is noted in Figure 3.

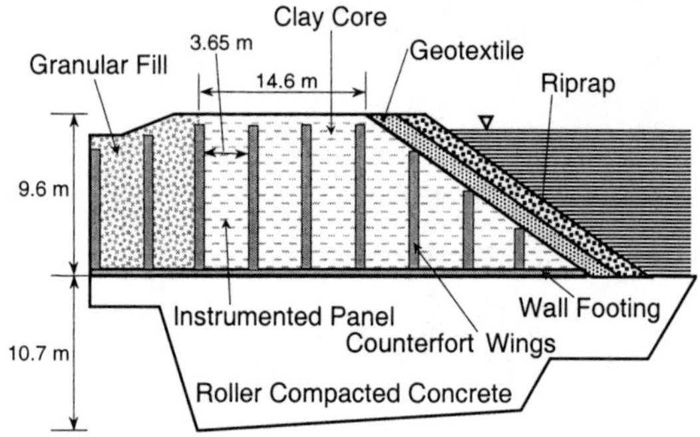

Figure 3: Spillway wall section.

WALL GEOMETRY AND INSTRUMENTATION

Figure 4 provides a schematic diagram of the wall section in which earth pressure instrumentation was placed. Figure 5 shows the locations of the ten earth pressure cells placed along the face of the wall. The cells were placed sufficiently far from the counterfort wings so as to avoid influence of the wing on earth pressure measurements. In addition, two pore water pressure transducers were placed in the locations shown. The earth pressure cells are a Geokon model 4810 vibrating wire earth pressure cell, while the two pore pressure transducers are a Geokon model 4500 vibrating wire piezometer. The instruments were attached to a Campbell Scientific CR10 data logger attached to a telephone modem for remote data recall.

The earth pressure cells were mounted in pairs at varying wall heights to provide redundancy of measurements. The cells were mounted directly to the wall using masonry screws fitted through the mounting lugs provided on the pressure cells. A mortar pad was placed between the pressure cells and the wall to provide for a flexible interface. The pore pressure transducers were placed in the backfill soil approximately 300 mm from the face of the wall.

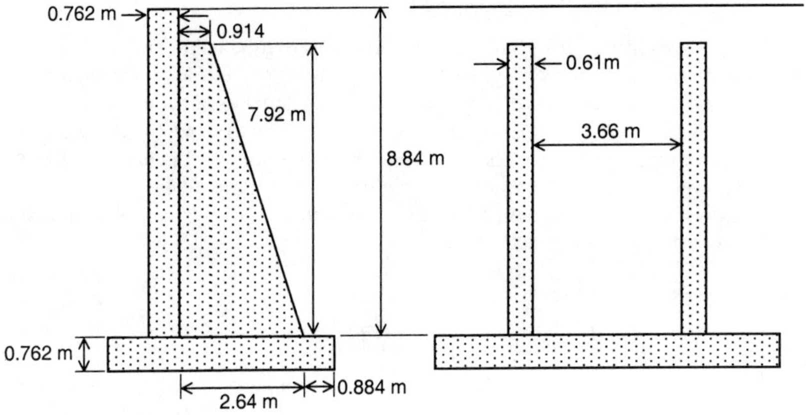

Figure 4: Counterfort wall details.

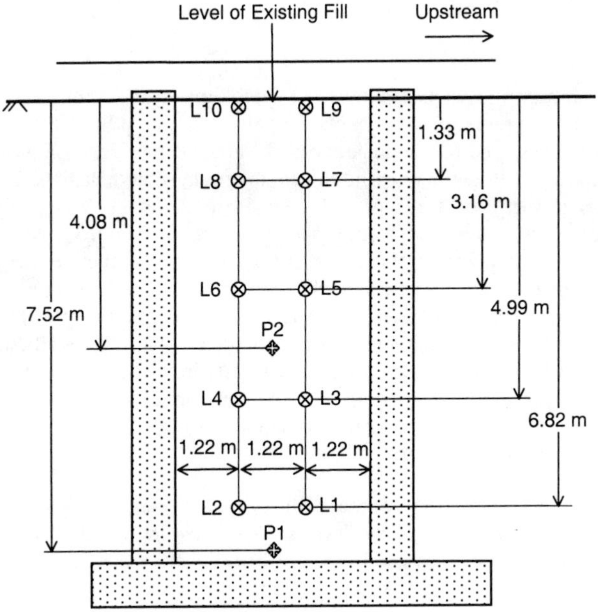

Figure 5: Earth pressure cell placement details.

BACKFILL PLACEMENT AND COMPACTION

Two types of backfill from different sources were used behind the wall and are described in Table 1. The CL material was placed to a height of approximately 5.6 m above the top of the wall footing with the CL-ML material used for the remainder of the wall. During design, drained strength parameters of $\phi'=34.5°$ and $c'=9.6$ kPa were determined for the CL material from consolidated-undrained triaxial compression tests (ESA Consultants, Inc., 1997). A conservative value of $K_o=0.6$ was also used for the wall design based on considerations of overconsolidation ratio and plasticity index of the material.

Table 1: Properties of backfill soil.

USCS Class.	Passes No. 200 (%)	LL (%)	PL (%)	Standard Proctor		In-Place	
				Maximum Dry Density (kN/m^3)	Optimum Water Content (%)	Dry Density (kN/m^3)	Average Water Content (%)
CL	68-92	28-36	10-15	18.7	15.0	18.2-18.7	15.0
CL-ML	60-82	22-25	5-7	17.9	14.6	17.3-17.7	13.8

The wall backfill was placed in lifts of approximately 200 mm. Compaction was achieved with two types of compaction equipment. A Caterpillar CP-563C vibrating padfoot roller was used for the majority of the compaction. The compactor consists of a single drum with two pneumatic tires at the back of the machine. The vibratory drum has a diameter of 1.295 m and a length of 2.134 m. When vibrating, the compactor imparts a maximum force to the ground ranging from 172 to 283 kN. Due to the constraints imposed by the wings of the counterfort structure, the path of the roller was generally perpendicular to the face of the wall. The roller was able to approach the wall such that the axle of the roller was approximately 1.04 m from the face of the wall. The strip of backfill immediately adjacent to the wall that could not be compacted by the roller was compacted in 100 mm lifts by a Wacker BS 65Y rammer plate compactor. The compactor has plate dimensions of 280 by 330 mm and delivers a maximum force to the ground of 15.1 kN, resulting in a maximum plate pressure of 164 kPa.

Compaction at the top of the wall footing began on October 22, 1998. Compaction was carried out to the level shown in Figure 5 and was completed on October 29, 1998. The dates and times at which the fill reached the earth pressure cell instrumentation levels shown in Figure 5 is provided in Table 2.

Table 2. Date and time at which fill level reached earth pressure cell levels.

Earth Pressure Cell Level	Date	Time
L1, L2	10-22-98	16:00
L3, L4	10-24-98	11:30
L5, L6	10-26-98	15:00
L7, L8	10-27-98	17:00
L9, L10	10-29-98	14:30

COMPACTION INDUCED LATERAL EARTH PRESSURE

Information on the compaction equipment provided in the previous section was used to calculate the peak compaction induced lateral earth pressure as a function of depth. The operation of two pieces of compaction equipment required that two solutions be developed and superimposed over one another to arrive at the resultant peak earth pressure distribution. The approach taken for each piece of compaction equipment was to calculate the maximum vertical stress induced by the piece of equipment as a function of depth along the wall, add in the total vertical stress due to overlying soils and then to multiply the sum of these two vertical stresses by a representative value of K_o. Linear elastic theory was used to determine the vertical stress induced by the compaction equipment as a function of depth. Boussinesq's solution for a point load resting on an infinite half-space was integrated over a finite distance to arrive at the solution for the vertical stress induced in the ground due to a finite line load. The solution is for the line load applied parallel to the face of the wall and offset from the wall by a distance given by x. The solution is for points along the face of the wall equidistant from the two ends of the line load and therefore gives the maximum vertical stress induced by the line load. The solution has been doubled according to Mindlin's method of images to account for the effect of the non-yielding wall and is given by:

$$\Delta\sigma_v = \frac{4pLz^3(6x^2+6z^2+L^2)}{\pi(x^2+z^2)^2(4x^2+4z^2+L^2)^{3/2}} \tag{1}$$

where L is the length of the line load, z is the depth below ground surface, x is the perpendicular horizontal distance from the line load to the wall and p is the line load intensity. The solution was calculated for the rated maximum and minimum load delivered to the ground for the roller described in the previous section. Figure 6 provides results from these calculations.

For the rammer plate compactor, the elastic solution for a rectangular load was used and also doubled to reflect the wall rigidity. The vertical stress curve for this situation is also shown in Figure 6. The vertical stress due to the total unit weight of the soil was included in all curves in Figure 6 and was determined from the data presented in Table 1 for the two different soil types. A resultant curve has been sketched in Figure 6 which corresponds to the maximum vertical stress experienced along the wall as compaction takes place at the ground surface.

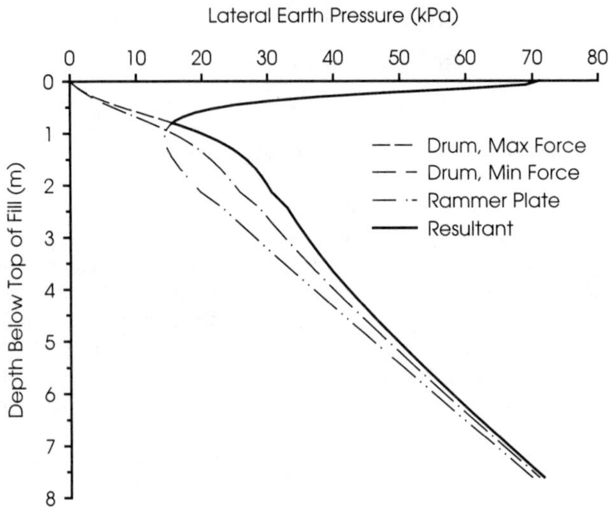

Figure 6: Peak compaction induced earth pressure.

RESULTS

Figure 7 provides records of lateral earth pressure, for each set of earth pressure cells, at the five levels denoted in Figure 5 up until approximately two days after the final fill height shown in Figure 5 was reached. A time of zero in Figure 7 corresponds to 6:00 A.M. on October 22, 1998. In general, compaction took place only during the hours from 6 A.M to 6 P.M. each day. The vertical dashed lines in Figure 7 correspond to times when fill had reached each of the earth pressure cell levels. The data logger was set to read data at intervals of 1.2 minutes up to hour 12. Between the hours of 12 and 51, data were recorded at intervals of 4.8 minutes. Thereafter, data were recorded at 30 minute intervals. Given the relatively slow sampling rate, the data presented in Figure 7 most likely only contains residual stress values (i.e. after the compaction equipment has been removed). The lack of any significant spikes in the data also suggests that these data contain only residual stress values. Given the inappropriateness of vibrating wire gages and data logging equipment for measuring transient effects, no attempt was made to measure and record the peak earth pressure while compaction equipment was operating above or adjacent to a earth pressure cell covered with backfill soil.

Figure 8 illustrates the profile of residual lateral earth pressure versus depth from field measurements on the upstream and downstream set of earth pressure cells shortly following compaction of the last lift of fill. On this figure, an estimate of the at-rest earth pressure has been drawn by assuming a K_o value of 0.43, where this value was calculated from the equation

$$K_o = 1 - \sin\phi' \qquad (2)$$

where ϕ'=34.5°. The use of Equation 2 represents a relatively unconservative estimate with respect to design by not accounting for overconsolidation effects. The line for K_r was obtained by taking K_r=1/K_o. The resultant of the peak compaction induced stress from Figure 6 is also shown in Figure 8. The data in Figure 8 show that earth pressures are lower in the top 0.5 m of the wall than the compaction resultant line, and can be approximated by the K_r line. Below this depth, the field data appear to initially follow the compaction resultant line but then quickly decreases to values less than those predicted by the K_o line. Given that the wall measurements are less than those for the K_o line, and that the equation used to predict this line did not account for overconsolidation effects, the use of an equation for K_o that did account for overconsolidation effects would make this overprediction of lateral earth pressures worse.

Figure 9 further illustrates points raised in Figure 8 by plotting field measurements versus depth at various times throughout the fill placement process. The field data plotted in Figure 9 represent the average of the upstream and downstream set of earth pressure cells. The dates at which these average measures were plotted correspond to the times denoted in Figure 7 when fill had just been placed to the top of the earth pressure cells for the top 4 levels of cells. A depth of zero was taken as the top edge of the cell for each set of field data plotted. Figure 9 shows again that lateral earth pressures are approximated by the K_r line in the upper 0.5 m of the wall and appear to follow the compaction resultant line thereafter. At deeper depths, however, the field data shows earth pressures less than the compaction resultant line and eventually less than the K_o=1-sinϕ line. The field data lines move increasingly to the left with increasing time. This suggests that significant stress relaxation of the clay fill occurs with time, as can also be seen in the individual earth pressure cell data sets in Figure 7.

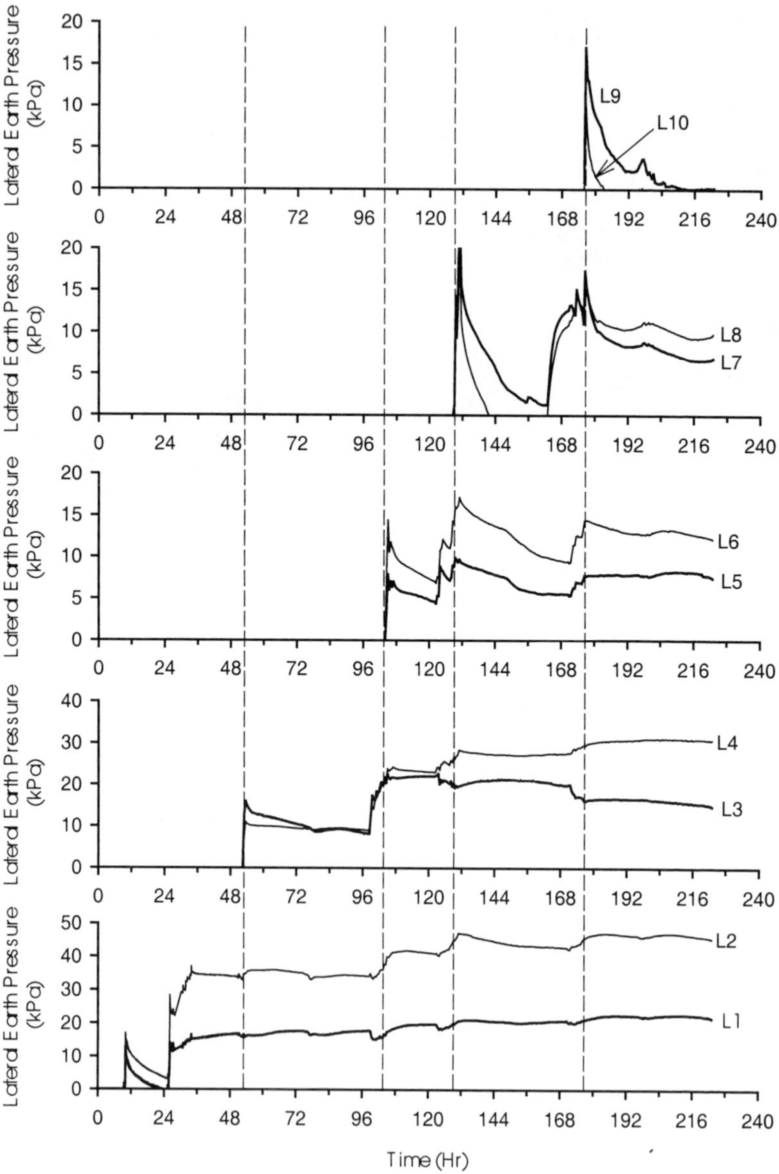

Figure 7: Lateral earth pressure vs. time for each earth pressure cell.

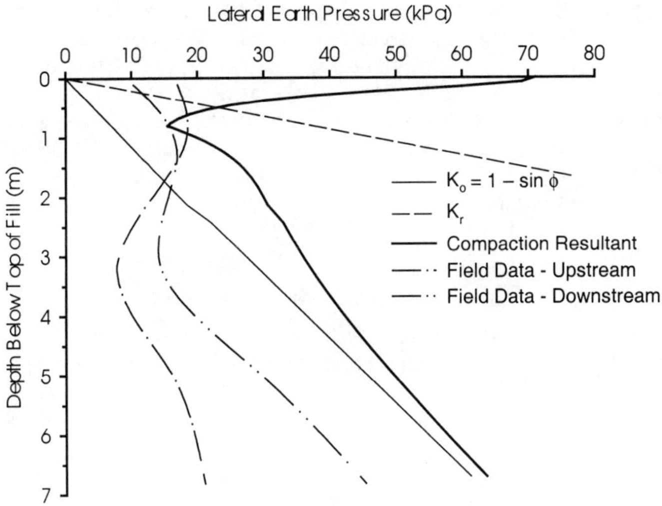

Figure 8: Residual earth pressure at end of fill placement.

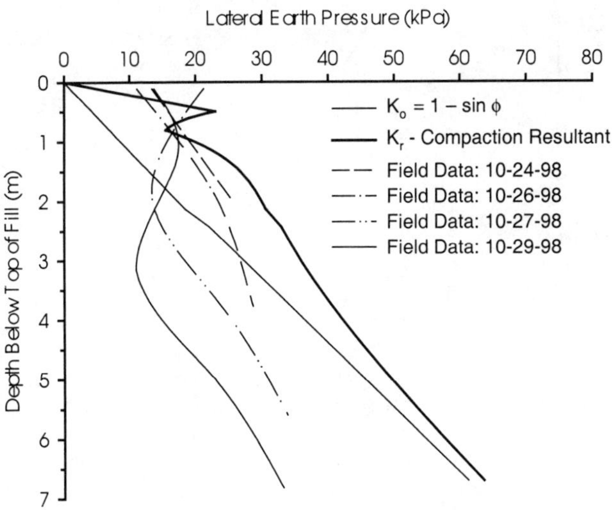

Figure 9: Residual earth pressure at various stages of fill placement.

Figure 10 shows four field data curves corresponding to increasing time following placement of the last lift of fill. The data show that significant stress relaxation occurs above a depth of 3 m and that this relaxation occurs relatively quickly. Additionally, the data show that a relatively low estimate of at-rest earth pressure given by Equation 2 greatly overestimates the residual earth pressure at depth. This may suggest that stress relaxation in cohesive backfill may be better estimated by the use of an active earth pressure coefficient even for conditions where little wall movement occurs.

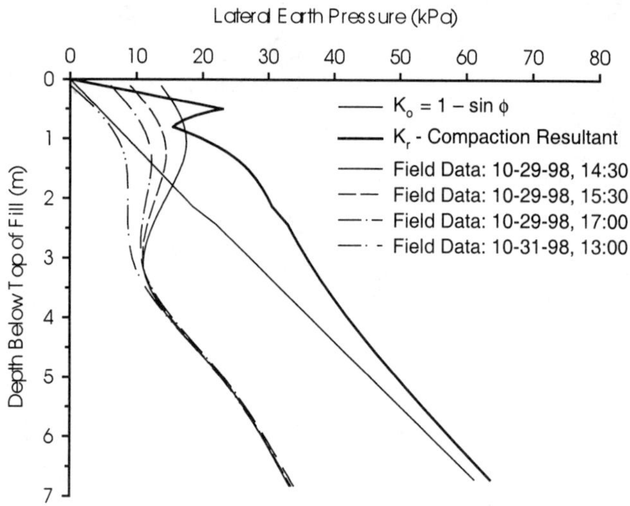

Figure 10: Residual earth pressure following complete fill placement.

CONCLUSION

Earth pressure cell measurements were made against a 8.8 m high rigid, non-yielding counterfort wall which serves as one of the walls for a newly constructed spillway at the Tongue River Dam, Montana. A silty clay soil was used as backfill against the wall. Measurements show that residual compaction induced earth pressure at the top of the fill shortly after compaction is much greater than predicted by at-rest earth pressure coefficients. Stress relaxation, however, occurs relatively rapidly with earth pressures in the upper 1 m of backfill reducing to, or being less than, at-rest values over a 2 day period. For deeper depths in the fill, earth pressures were seen to be considerably less than values predicted by at-rest theory.

ACKNOWLEDGMENTS

The authors gratefully recognize the financial support of the U.S. Bureau of Reclamation in providing funds and equipment for instrumentation. ESA Consultants, Inc. is gratefully acknowledged for their contribution of personnel time and their interest in pursuing this work.

APPENDIX I: REFERENCES

Broms, B. (1971), "Lateral Earth Pressures Due to Compaction of Cohesionless Soils", *Proceedings of the 4th Budapest Conference on Soil Mechanics and Foundation Engineering*, pp. 373-384.

Broms, B. and Ingleson, I. (1971), "Earth Pressures Against Abutment of a Rigid Frame Bridge", *Géotechnique*, Vol. 21, No. 1, pp. 15-28.

Carder, D.R., Pocock, R.G. and Murray, R.T. (1977), *Experimental Retaining Wall Facility – Lateral Stress Measurements With Sand Backfill*, Transport and Road Research Laboratory Report LR766, Crowthorne, United Kingdom.

Carder, D.R. Murray, R.T. and Krawczyk, J.V. (1980), *Earth Pressure Against an Experimental Retaining Wall Backfilled With Silty Clay*, Transport and Road Research Laboratory Report LR946, Crowthorne, United Kingdom.

Clayton, C.R.I., Milititsky, J. and Woods, R.I. (1993), *Earth Pressure and Earth-Retaining Structures*, 2nd ed., Blackie Academic and Professional, London, 398 p.

Coyle, H.M. and Bartoskewitz, R.E. (1976), "Earth Pressure on Precast Panel Retaining Wall", *Journal of the Geotechnical Engineering Division, ASCE*, Vol. 102, No. 5, pp. 441-456.

Duncan, J.M. and Seed, R.B. (1986), "Compaction-Induced Earth Pressures Under K_o-Conditions", *Journal of Geotechnical Engineering*, Vol. 112, No. 1, pp. 1-22.

ESA Consultants, Inc. (1997), *Final Design Report for Primary Spillway and Outlet Works, Phase III Construction, Tongue River Basin Project*, V.1.

Filz, G.M. and Duncan, J.M. (1996), "Earth Pressures Due to Compaction: Comparison of Theory With Laboratory and Field Behavior", *Transportation Research Record 1526*, pp. 28-37.

Kelly, W.R. and McDonald, G. (1998), "Dealing with Complex Constraints in Rehabilitation of the Tongue River Dam Spillway System", *Proceedings of the 1998 Annual Conference of the Association of State Dam Safety Officials*, Las Vegas, NV, October 11-14, pp. 399-410.

McDonald, G., Hughes, H.E., and Yadon, D.M. (1998), "Working Together to Ensure Safety at Tongue River Dam", *Hydro Review*, Sept., V. 17, No. 5.

McDonald, G. and Yadon, D.M. (1998), "Over the Top Concrete", *Civil Engineering*, American Society of Civil Engineers, January, pp. 72-75.

Miller, D.J., Smith, K., and Volpe, R.L. (1999), "Piping and Internal Erosion Control measures at Tongue River Dam, Montana", *Dealing with Aging Dams,* Nineteenth Annual USCOLD Lecture Series, Atlanta, GA, United States Committee on Large Dams, Denver, CO, pp. 443-458.

Seed, R.B. and Duncan, J.M. (1983), *Soil-Structure Interaction Effects of Compaction-Induced Stresses and Deflections*, Geotechnical Engineering Research Report No. UCB/GT/83-06, University of California, Berkeley, CA.

Sowers, G.F., Robb, A.D., Mullis, C.H. and Glenn, A.J. (1957), "The Residual Lateral Pressures Produced by Compacting Soils", *Proceedings of the 4th International Conference on Soil Mechanics and Foundation Engineering*, Vol. 2, pp. 243-247.

Symons, I.F., Clayton, C.R.I. and Darley, P. (1989), *Earth Pressure Against an Experimental Retaining Wall Backfilled With Heavy Clay*, Transport and Road Research Laboratory Report 192, Crowthorne, United Kingdom.

Wright, A.G. (1998), "Montana Dam Proponents Find Common Ground", *Engineering News-Record*, November 9, pp. 60-64.

Yadon, D.M. (1997), "RCC Overtopping Protection in Harsh Conditions," *RCC Newsletter*, Portland Cement Assoc., Fall/Winter V. 13, No. 2.

APPENDIX II: SYMBOLS USED

K_A Active earth pressure coefficient
K_o At-rest earth pressure coefficient
K_P Passive earth pressure coefficient
K_r Residual earth pressure coefficient
L Length of roller line load (m)
p Intensity of roller line load (kN/m)
x Perpendicular distance from roller line load to wall face (m)
z Depth below ground surface (m)
ϕ' Effective stress friction angle
σ_h' Horizontal effective stress (kPa)
σ_v' Vertical effective stress (kPa)
$\Delta\sigma_v'$ Compaction induced vertical stress (kPa)

SETTLEMENTS OF A PRELOADING EMBANKMENT ON PV DRAIN-IMPROVED CHEWELAH CLAY

C.L. Sampaco[1], V.W. Rybel[2], G. Gilman[3], K.Campbell[4], K.D. Sharp[5], K.R. Green[6], and M.J. Warren[7]

ABSTRACT: Embankment preloading, in conjunction with prefabricated vertical (PV) drains, was used to accelerate primary consolidation and to eliminate most of the anticipated secondary settlements of the soft clay foundation due to loading from a proposed wastewater treatment plant (WWTP). Subsurface exploration conducted during preliminary design indicated the presence of up to 60-m thick deposit of soft to very soft clay underlying the site. Estimates of settlement due to filling of the existing lagoon and construction of the WWTP indicated that the soft clay deposits will undergo up to 150 cm of settlement. Approximately 80 cm of this settlement was estimated to occur in the first 10 years of the plant operation and an additional 60 cm was estimated to occur during the following 30 years. UP to 19 settlement plates were installed under the embankment fill to monitor the preload performance. The measured settlements were then used to analyze the behavior of the soft clay foundation, and to develop recommendations regarding the time for preload removal and subsequent construction of the proposed WWTP. This paper discusses the results of settlement analyses and highlights the procedure that was used to predict the magnitude and time-rate of settlements based on field measurements.

INTRODUCTION

The city of Chewelah is located approximately 75 km northwest of Spokane, Washington along U.S. Highway 395. As a part of its ongoing improvement, the City has proposed the construction of a new wastewater treatment plant (WWTP). The new facility will consist of biological secondary treatment system providing nitrification, a disinfection facility, sludge handling and treatment facilities, headworks, a laboratory building, and related electrical and instrumentation control equipments.

The proposed WWTP will be constructed by filling an existing sewage lagoon (called Lagoon #3) by as much as 2 m of soil to bring the grade to the current top elevation (El. 500 m) of the perimeter dikes. The lagoon covers and area of roughly 90 m by 135 m (see Figure 1). From the as-built drawings provided by the City, it appears that Lagoon #3 was constructed by building the perimeter dikes 1.5 to 2.4 m above the original ground surface.

---

[1]Project Geotech. Engr., CH2M Hill, Bellevue, WA; [2]Sr. Geotech. Engr., CH2M Hill Inc., Corvallis, OR; [3]Project Manager, CH2M Hill Inc., Spokane, WA; [4]Principal, Campbell Services Group, Spokane, WA; [5]Sr. Geotech. Engr. and Staff Manager, CH2M Hill Inc., Bellevue, WA; [6]Senior Geotech. Engr., CH2M Hill Inc., Bellevue, WA; [7]President, Drainage and Ground Improvement Inc., Bridgeville, PA

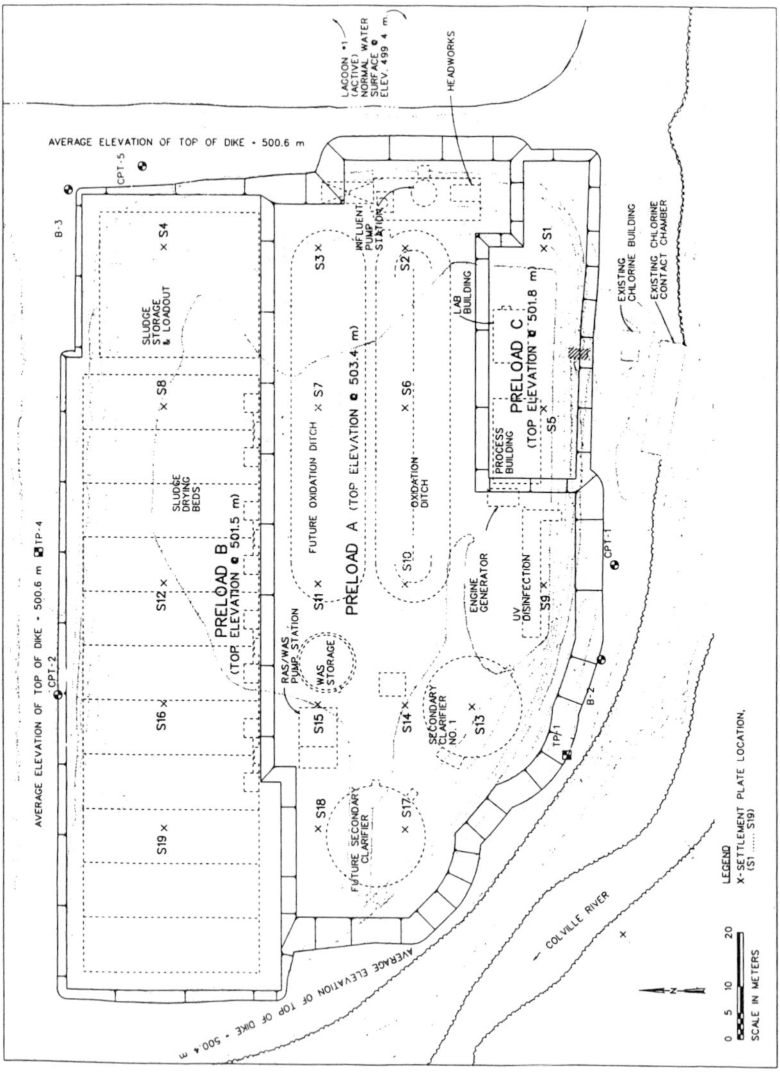

Figure 1. WWTP Site Plan, Preload Layout, and Settlement Plate Locations

Subsurface exploration conducted as a part of the preliminary design and site evaluation of the new WWTP indicated the presence of a thick deposit (up to 60 m) of soft to very soft clay at the proposed site. Preliminary estimates of settlements due to the filling of Lagoon #3 and construction of the WWTP indicated that this highly compressible clay will undergo up to about 150 cm of total settlements. Approximately 50% of these settlements (that is, 80 cm) are estimated to occur in the first 10 years of the plant operation while the next 40% (that is, an additional 60 cm) are estimated to occur during the following 30 years.

Due to the considerable amount of time involved, preloading the site in conjunction with the use of PV drains was recommended. Under this scheme, the preload surcharge simulates the future loading from the WWTP while the PV drains speedup the settlement of the clay.

TOPOGRAPHY AND GEOLOGY

The site is relatively flat with the exception of the embankments of the three existing waste water treatment lagoons. The embankments of the existing lagoons rise above the native ground surface 1.5 to 2.4 m. The side slopes of the embankments are poorly defined in most areas. However, based on visual observations the side slopes of the embankments are about 2.5H:1V to 3H:1V.

During the Pleistocene Epoch, large areas of the present day Stevens County were covered by the Colville lobe of the Cordilleran ice sheet. The glacier advanced southward from Canada over the present day Chewelah area, then retreated, leaving glacial deposits consisting of moraines and kames as well as other associated outwash and ice-contact deposits. The ice-contact deposits are represented by the sand and gravel unit that mantles the basement bedrock at the bottom and sides of the valley. As the glacier retreated northward, it dammed the valley and a large glacial lake was formed. A thick sequence of glaciolacustrine clays, locally referred to as the "blue clay" was deposited in the lake bed. Since the clays were not overridden by the glacier, they are normally consolidated and hence subject to large settlements if loaded. These glaciolacustrine clays are as thick as 76 m about 1.6 km to the southwest of the City of Chewelah and becoming thinner toward the valley margins.

GEOTECHNICAL CONDITIONS

Subsurface conditions at the site were evaluated based on three test borings, nine test pits, and six piezocone (CPT) tests. The test borings were drilled to a maximum depth of about 21.3 m below the ground surface (bgs) while the test pits were excavated to a maximum depth of 2.4 m bgs. The CPT's were advanced to depths of 15 to 58 m bgs. The test borings and CPT's were conducted mainly from the top of the existing perimeter dikes.

The soil encountered during the exploration program primarily consists of the fill for the existing perimeter dikes underlain by a thin layer that mantles the thick deposit of soft to very soft, normally consolidated clay that is a part of the thick sequence of glaciolacustrine clay deposits at the site as discussed in the previous section.

Figure 2 shows the generalized soil profile underlying Lagoon #3. The embankment for the perimeter dikes varies in thickness from 1.5 to 2.4 m and classifies under the Unified Soil Classification System (USCS) as Silty Sand (SM), Sandy Silt (ML), Poorly-Graded Sand (SP), and Silty Gravel (GM). The fill material is medium dense with field SPT blowcounts averaging at 23 blows per 0.3 m.

The native material underlying the embankment generally consists of about 1.5 to 3.7 m of stiff to very stiff clay crust which classifies as CL or CH. The crust is generally underlain by a thick layer of normally consolidated (with OCR close to 1) soft to very soft clay (CH). Figure 3 shows the plots of index properties with depth. The clay has an average liquid limit of 55 and an average plasticity index of 31. The natural water content is generally at or above the liquid limit. The average compression ratio ($C_c/1+e_0$) of the clay is 0.2 while the average coefficient of consolidation (C_v) is 4.7 m^2 per year.

Field SPT blowcounts at the upper 3 to 4.6 m of the clay range from 1 to 4 but generally decrease to as low as zero to a depth of 21.3 m, the maximum depth drilled for the test borings. Both disturbed (SPT) and "undisturbed" (Shelby tube) sampling beyond about 15 to 21 m bgs were extremely difficult due to the softness and sensitivity of the clay, prompting the use of CPT tests at these depths. Results of CPT investigations indicated similar nature of the clay for depths deeper than what had been observed from the test borings. Figure 4 shows a typical variation of slightly increasing cone penetration resistance with depth. Estimates of undrained shear strength (S_u) of the upper 32 m of soft to very soft clay, based on the CPT results and plasticity index, indicate an average S_u of about 31 kPa.

Twenty-four pore pressure dissipation tests were also conducted at various depths in conjunction with the CPT's. Data from dissipation tests were then used to determine the horizontal coefficient of consolidation (C_h) of the clay. Values of C_h determined from the dissipation tests using published correlations (FHWA, 1992) ranged between 4.4 and 15.3 m^2/yr and averaged at about 10 m^2/yr.

Groundwater conditions were assessed from readings of observation wells installed in the test borings. Observation wells installed at depths of 6.1 to 7.6 m indicated groundwater within 1.5 to 3 m bgs. At deeper locations, deep wells installed in the Chewelah area suggest the presence of an artesian aquifer in the dense sand layer and, also possibly, in the upper 1.8-m thick sand layer located at about 35.6 m below the bottom of the preload (see Figure 2). However, the 1.8-m thick sand layer does not appear to be continuous in lateral extent (see Figure 4).

CONSOLIDATION WITH PV DRAINS

Consolidation with PV drains involves the fundamental problem of consolidation of a soil cylinder containing a central drain well under predominantly radial flow. Barron (1948)

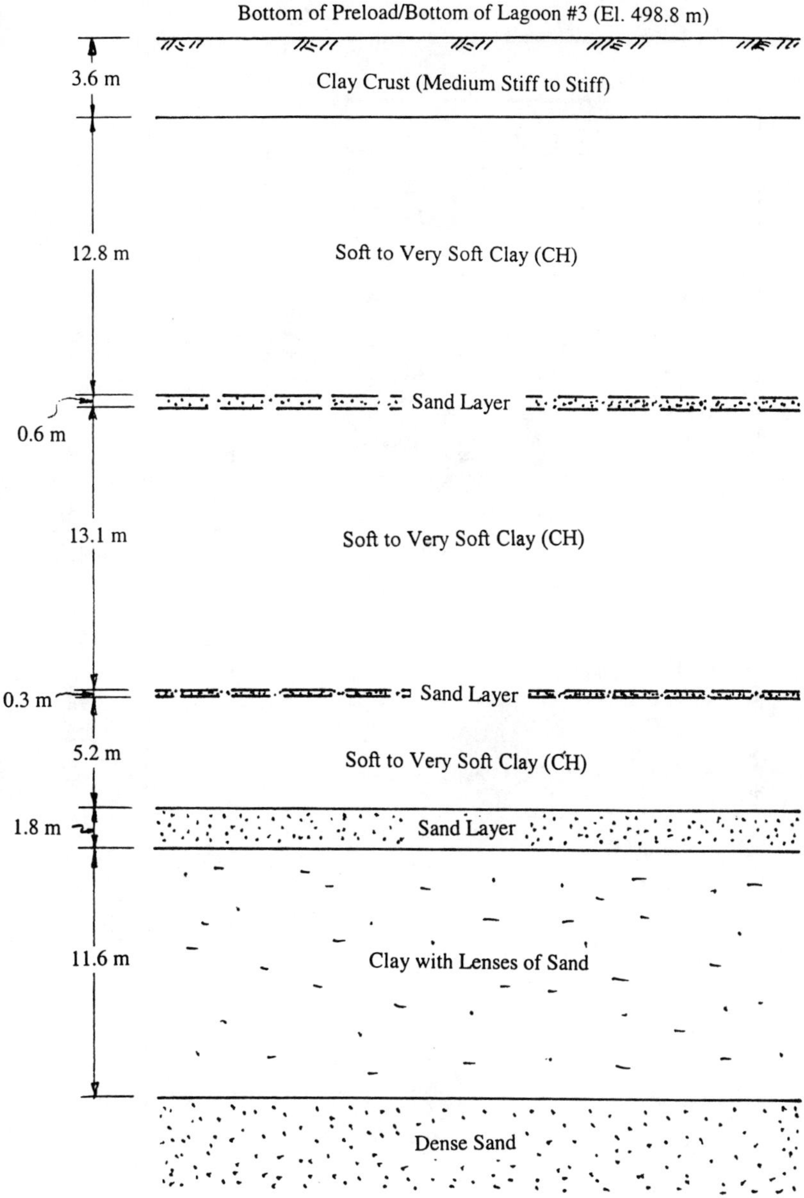

Figure 2. Generalized Subsurface Profile at the Chewelah WWTP Preload Site

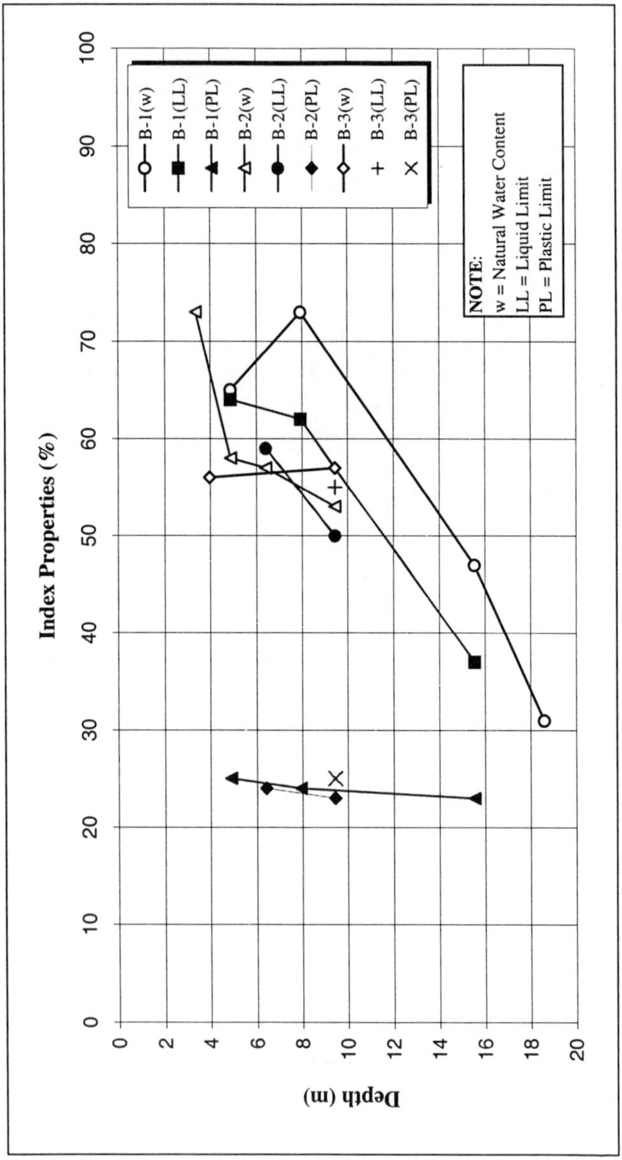

Figure 3. Variation of Index Properties with Depth

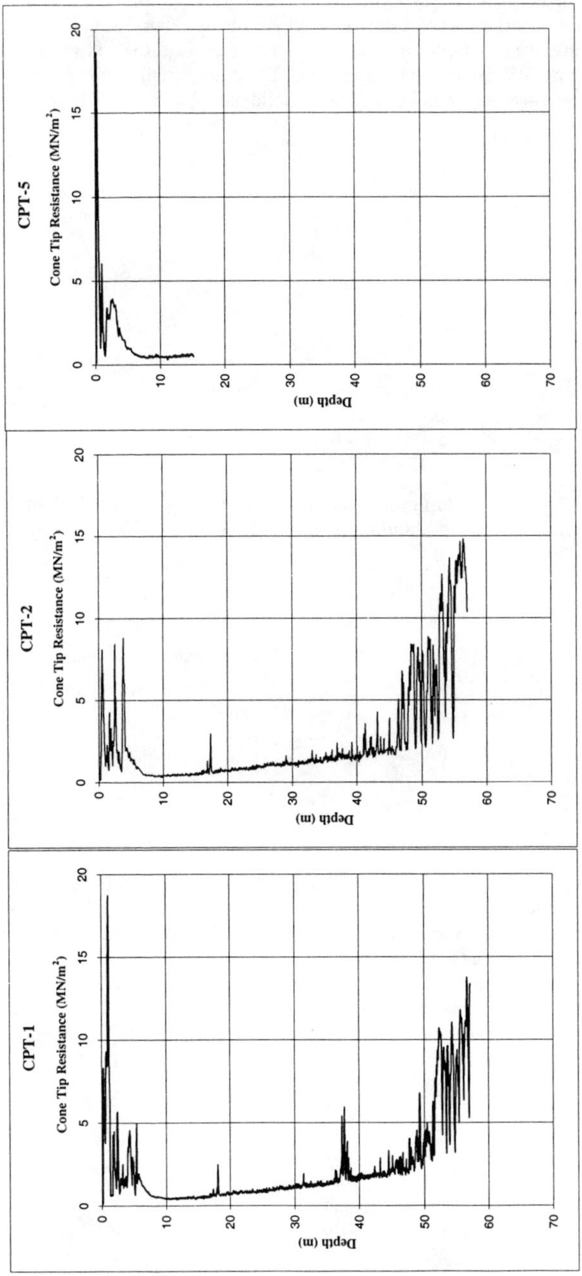

Figure 4. Typical Variation of Cone Tip Resistance (Q_t) with Depth

published the first exhaustive treatment of the radial flow problem. Modifications to the original Barron's solution were later recommended by Hansbo (1979, 1981). These modifications involved simplifying assumptions due to the physical dimensions, characteristics of the PV drains, and the effect of PV drain installation. A commonly used equation for the average degree of horizontal consolidation, U_h is:

$$U_h = 1 - \exp\left(\frac{-8T_h}{F}\right) \tag{1}$$

where:

$$T_h = \frac{C_h}{D^2_e}t \tag{2}$$

C_h is the coefficient of consolidation in the horizontal direction; t is the time after an instantaneous increase of the total vertical stress; and D_e is the diameter of equivalent soil cylinder influenced by a PV drain and is a function of PV drain spacing (S) and pattern. For a square pattern, D_e is equal to 1.13 S while for a triangular pattern, D_e is equal to 1.05 S.

The factor, F, in Eq. 1 expresses the additive effects accounting for the PV drain spacing (F_n), smear or disturbance during installation (F_s), and well-resistance (F_r).

$$F_n = \ln\left(\frac{D_e}{d_w}\right) - 0.75 \tag{3}$$

$$F_s = \left(\frac{K_h}{K_s} - 1\right)\ln\left(\frac{d_s}{d_w}\right) \tag{4}$$

$$F_r = \frac{\pi}{z}(2L - z)\frac{K_h}{q_w} \tag{5}$$

In Eqs. 3 to 5, d_w is the equivalent diameter of PV drain and is equal to (a + b)/2 where a and b are the cross-sectional dimensions of the PV drain (Rixner et al, 1986); K_h is the original soil permeability in the horizontal direction before drain installation; K_s is the reduced soil permeability in the horizontal direction due to smear or disturbance; d_s is the equivalent diameter of the smeared zone and is equal to (2.5 to 3) d_m , where d_m is the diameter of a circle with an area equal to the cross-sectional area of the mandrel; z is the distance from the drainage end of the PV drain; and q_w is the discharge capacity of the PV drain. In this study, PV Mebra drains were used with a = 100 mm and b = 3 mm, such that for a mandrel with d_s

= 25 cm, the value of d_s/d_w is approximately equal to 5. A value of 3 was used for K_h/K_s ratio based on the reported range of 1 to 5 (Rixner et al, 1986).

For an ideal condition, $F_s = 0$ (no smear) and $F_r = 0$ (no well-resistance), such that $F = F_n$. In practice, the effect of well-resistance (F_r) is considered to be very small and is normally neglected (Rixner et al, 1986). In this paper, only the effects of drain spacing (ideal condition) and smear were considered in the analysis.

PRIMARY AND SECONDARY SETTLEMENTS

Estimates of the magnitude of primary consolidation settlement due to the weight of the preload fill and surcharge were made during the WWTP design phase using Terzaghi's one-dimensional consolidation theory using a typical soil profile similar to the one shown in Figure 2. In this method, the magnitude of primary consolidation settlement was calculated using the results of laboratory consolidation tests performed on the recovered Shelby tube samples (that is, $C_\varepsilon = 0.2$). Elastic settlements were neglected in the calculations due to the fact that the bottom of the lagoon will be completely dredged for sludge removal before initial placement of the fill will be made. Also, it was anticipated that no settlement readings will be made until after at least about 0.5 m of gravel pad is in place to properly install the settlement plates. Increase in stresses due to loading from the preload embankment were calculated using Boussinesq stress distribution as implemented in the computer program ZSTRESS (Allen and Duncan, 1987). Stress calculations were performed for each of the 19 settlement plate locations by approximating the irregular preload geometry as a series of rectangular loads with various loading intensities.

The estimated range of primary consolidation settlement for the settlement plate locations under Preload A was between 107 and 178 cm, with an average of 140 cm. For secondary compression settlements, the estimated range was 18 to 30 cm with an average value of 24 cm. Secondary settlement was estimated using C_α/C_c ratio of 0.05 as recommended for soft inorganic clays (Mesri and Castro, 1987). The estimate was based on the assumption that most of the secondary compression will occur within 2 to 5 years after placement of the surcharge. In the estimate of secondary settlement, only the soft to very soft clay (CH) layers were considered based on the assumption that the deeper, 11.6-m thick clay layer (see Figure 2) will undergo very little secondary compression due to the combination of depth, adequate drainage, and relatively high permeability. Secondary settlement was assumed to commence after 95 percent of primary consolidation was achieved.

PRELOADING EMBANKMENT AND PV DRAIN DESIGN

The purpose of the Chewelah WWTP preloading project was to force primary consolidation settlements to occur before the construction of the new WWTP. An additional surcharge load was also included to reduce the long-term effects of secondary compression.

The sequence of preloading included: (a) excavating and sludge removal at the bottom of existing Lagoon #3 to El. 498.8 m; (b) filling Lagoon #3 with gravel drainage blanket to El. 499.4 m; (c) installation of PV drains that extend to an average depth of 36.6 m bgs; and (d) placement of a gravel fill and surcharge to a maximum elevation of 503.4 m. The height of the soil surcharge within the 12,920 m^2 area of Lagoon #3 varied between 1.2 m and 3.4 m above the proposed final grade elevation of about 500 m (see Figure 1), depending on the anticipated weight of the structural facility at that location. Based on the fill height, the preload is divided into 3 zones: Preload A, Preload B, and Preload C (see Figure 1).

PV Mebra drains were installed in triangular pattern at center to center spacing of 1.8 m to accelerate the rate of settlement of the clay. The drains (100 mm wide by 3 mm thick) were installed to an average depth of 36.6 m. It was estimated that the primary consolidation settlement due to the future WWTP could be achieved within a year after placement of the preload surcharge.

Nineteen settlement monuments were installed at various locations of Lagoon #3 to monitor the settlements associated with the weight of the preload and surcharge. The settlement monuments were installed at an approximate elevation of 498.8 m, which corresponds to the bottom elevation of the lagoon. Figure 1 shows the locations of the settlement monuments.

Dewatering of Lagoon #3 began on March 3, 1998 and dredging was completed on March 30, 1998. Placement of initial lift of gravel pad and installation of settlement plates were completed on April 7, 1998. Initial surveys for settlement plate readings were started on April 10, 1998. PV drains were installed between April 22 and May 26, 1998. Filling of the embankment began on June 1, 1998 and was completed on June 18, 1998.

MEASURED SETTLEMENTS AFTER 154 DAYS

Settlements were monitored at various intervals during and after construction by surveying methods. One-hundred-fifty-four days after the start of preload construction (September 11, 1998), the measured settlements were analyzed to develop recommendations regarding the time for preload removal and subsequent construction of the proposed WWTP. The measured settlements for all 19 locations within the preload footprint 154 days after the start of construction are plotted in Figure 5. The results are summarized as follows:

- For Preload A where the fill height is maximum (top-of-fill elevation of 503.4 m), the total settlements ranged between 81 and 130 cm.
- For Preload B (top-of-fill elevation of 501.5 m), the range of measured settlements varied between 40 and 76 cm.
- For Preload C (top-of-fill elevation of 501.8 m), the measured settlements were 68 and 81 cm.

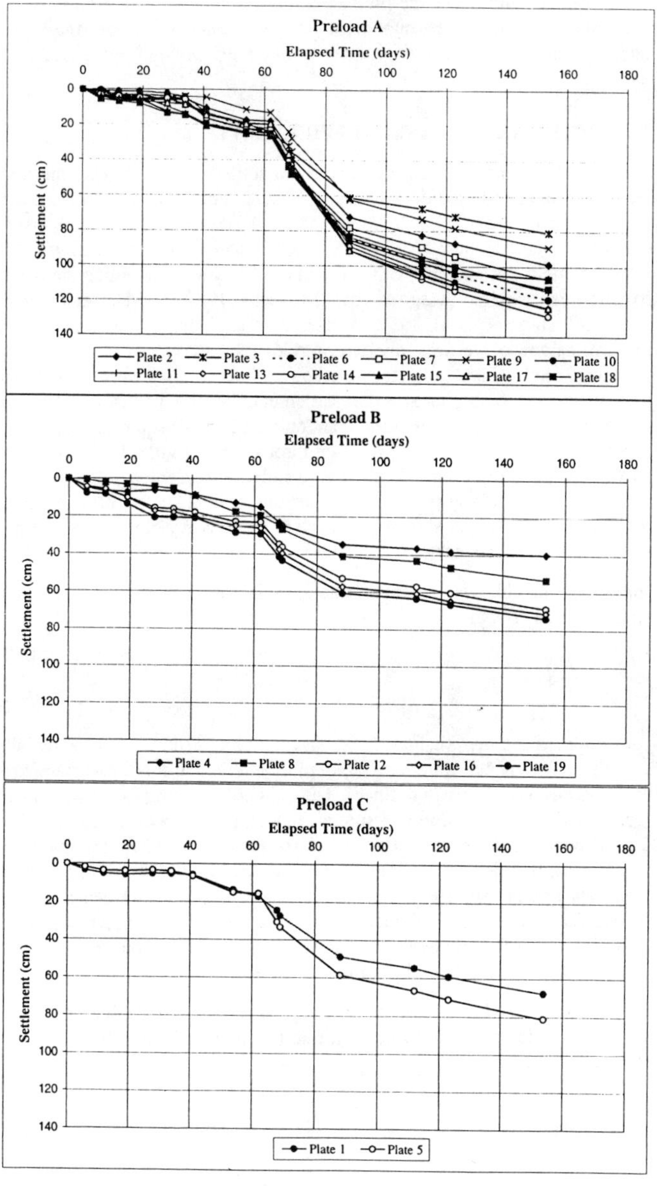

Figure 5. Measured Settlements 154 Days After the Start of Preload Construction

- The settlement profile under the preload embankment follows approximately a dish-shaped configuration with the maximum settlements occurring near the center of the preload. The average measured settlement under the preload embankment is about 100 cm.

PREDICTED SETTLEMENTS BASED ON FIELD DATA

Primary consolidation and secondary compression settlements caused by the preload fill and surcharge were analyzed based on the measured settlement data 154 days after the start of preload construction. Asaoka's Graphical Method (Asaoka, 1978) was used in the settlement prediction. This method is "observational" in nature and has been widely used to predict final embankment settlements (with and without PV drains) if settlement data, at least during construction, is available (Magnan and Mieussens, 1980; Magnan and Deroy, 1980; Bergado et al, 1990; Bergado et al, 1992; Anderson et al, 1994). Using this method, the observed time-settlement curve was divided into 5-day interval and the corresponding settlements (S_i) were read-off and tabulated. These settlement values were then plotted in S_{i-1} versus S_i axis where S_i and S_{i-1} are the settlements in the ith and (i-1)th intervals, respectively. A best-fit straight line was then obtained for the plotted points and was extended to intersect the 1:1 line that was drawn through the origin. The intersection point corresponds to $S_i = S_{i-1}$ and represents the final consolidation settlement (S_f). Figure 6 shows an illustration of the Asaoka's graphical procedure as applied to the maximum measured settlement data for Plate S14. As shown in Figure 6, the Asaoka plot was also extended to interpret the occurrence of secondary consolidation.

The rate of primary consolidation settlement due to radial drainage (C_h) was estimated using the following expression (Bergado et al, 1992):

$$C_h = \left(\frac{-D^2_e F}{8}\right)\left(\frac{\ln \beta}{\Delta t}\right) \tag{6}$$

where β is the slope of the best fit line in the Asaoka plot, Δt is the time interval used to dicretize the settlement-time curve in the Asaoka plot, D_e and F are parameters related to the PV drain properties as previously defined. The value of C_h using Eq. 6 was calculated to be 15.6 m^2 per year if ideal condition is assumed (that is, $F_s = F_r = 0$ in Eqs. 4 and 5) and is equal to 33.4 m^2 per year if smear is included. Therefore, under ideal condition, the back-calculated value of C_h is at least 50 percent faster than the predicted average from the CPT dissipation test data and at least three times faster than that predicted using the laboratory consolidation test data (that is, C_v). If smear is included in the calculations, the rate of radial flow in the field is at least four times faster than the average value from the CPT dissipation test, and about eight times faster than the laboratory C_v value.

In this project consolidation due to vertical flow (U_v) was neglected. Given the Asaoka plot of settlement data, the degree of consolidation due to radial flow only (U_h) was estimated using the following equation:

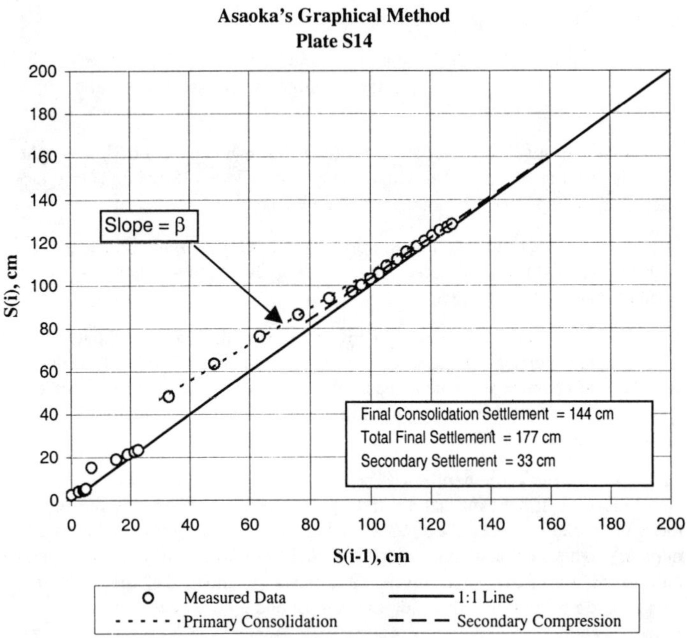

Figure 6. Asaoka's Graphical Method as Applied to Plate S14 Data

$$U_h = 1 - \exp\left(\frac{\ln\beta}{\Delta t}\, t\right) \tag{7}$$

The following is a summary of the results of settlement analyses based on field measurements 154 days after the start of preload construction:

- The predicted total final settlement is 177 cm. This total final settlement is assumed to consist of two parts: (a) primary consolidation settlement and (b) secondary compression settlement.
- The predicted maximum settlement due to primary consolidation of the clay is 144 cm. This is within the predicted range of 107 to 178 cm and very close to the calculated average of 140 cm during preload design.
- The estimated maximum settlement due to secondary compression is 33 cm. This is slightly more than the estimated range of 18 to 30 cm (average of 24 cm) calculated during preload design.
- Ninety-five percent (95%) primary consolidation of the clay is estimated to occur 150 days after the start of construction (that is, September 7, 1998). However, as will be seen later after more settlement readings have been made (Figure 7, for example), it appears that 95% primary consolidation occurred by October 27, 1998 (200 days after the start of construction).
- Judging from the settlement curve and the Asaoka plot of measured settlement data (see Figure 6), it appears that secondary compression of the clay did not occur until the total settlement was about 93 cm (degree of primary consolidation of about 62 percent). This total settlement occurred about 90 days from the start of construction (July 9, 1998). Thus, after July 9, 1998 it was assumed that the settlements were a combination of primary consolidation and secondary compression.
- About ninety-nine percent (99%) of primary consolidation of the clay occurred within 195 days after the start of construction (that is, October 22, 1998). After October 22, 1998, only 1 percent of the primary consolidation is anticipated and the rest of the settlements are due to secondary compression.
- About 95% of the estimated secondary settlements is predicted to occur within 367 days after secondary compression commenced (that is, July 10, 1999).

From the above analyses and prediction of settlements, it was recommended to the City that the preload embankment be left in place until most of the secondary settlements have occurred. It was then suggested that the earliest time in which the preload embankment should be removed was September 1, 1999. This recommendation was prompted by the much higher values of secondary settlements that occurred in the field compared to the estimates used during the WWTP design. It was emphasized that the risk of occurrence of secondary settlements over the design life of the structures can be greatly reduced the longer the preload is left in place.

It was also recommended that settlement monitoring shall be continued until the preload is removed. Due to the fact that the graphical method used to predict the future settlements of the preload embankment is "observational" in nature, the accuracy of the predicted settlements rely heavily on the quantity of observed data available.

Figure 7 shows the measured settlements for Preload A taken on August 3, 1999 (480 days after the start of preload construction). Also shown in Figure 7 is the predicted maximum settlement (PREDICTED-2) using the parameters back-calculated from the Asaoka graphical method (analyzed using data 154 days after the start of construction). This curve was developed assuming that secondary compression commenced 90 days from the start of construction. The predicted curve based on Asaoka's graphical procedure, using data taken 154 days after the start of construction, over-predicts the measured settlements 480 days after the start of construction by as much as 15 percent. This over-prediction is mainly due to the back-calculated rate of settlements (from Asaoka's graphical method) being slightly higher than what is actually occurring in the field.

Also shown in Figure 7 is the predicted settlement curve (PREDICTED-1) if secondary compression is assumed to commence only after 95 percent of primary consolidation has occurred as was originally assumed. Based on this comparison, it appears that closer agreement can be achieved using the latter curve (PREDICTED-1) but may have the drawback that settlements could be underestimated in the later stages.

RECENT DEVELOPMENTS

At the time of this writing, the contract documents for preload removal and site preparation package are going out to bid. This will be followed by bids for the construction of the WWTP. The preload is scheduled for removal sometime in February 2000 before spring thaw load limits are placed on county and state highways in the project area. It is planned to have one more set of settlement readings taken before the preload will be removed. The WWTP construction will likely take place in Summer 2000.

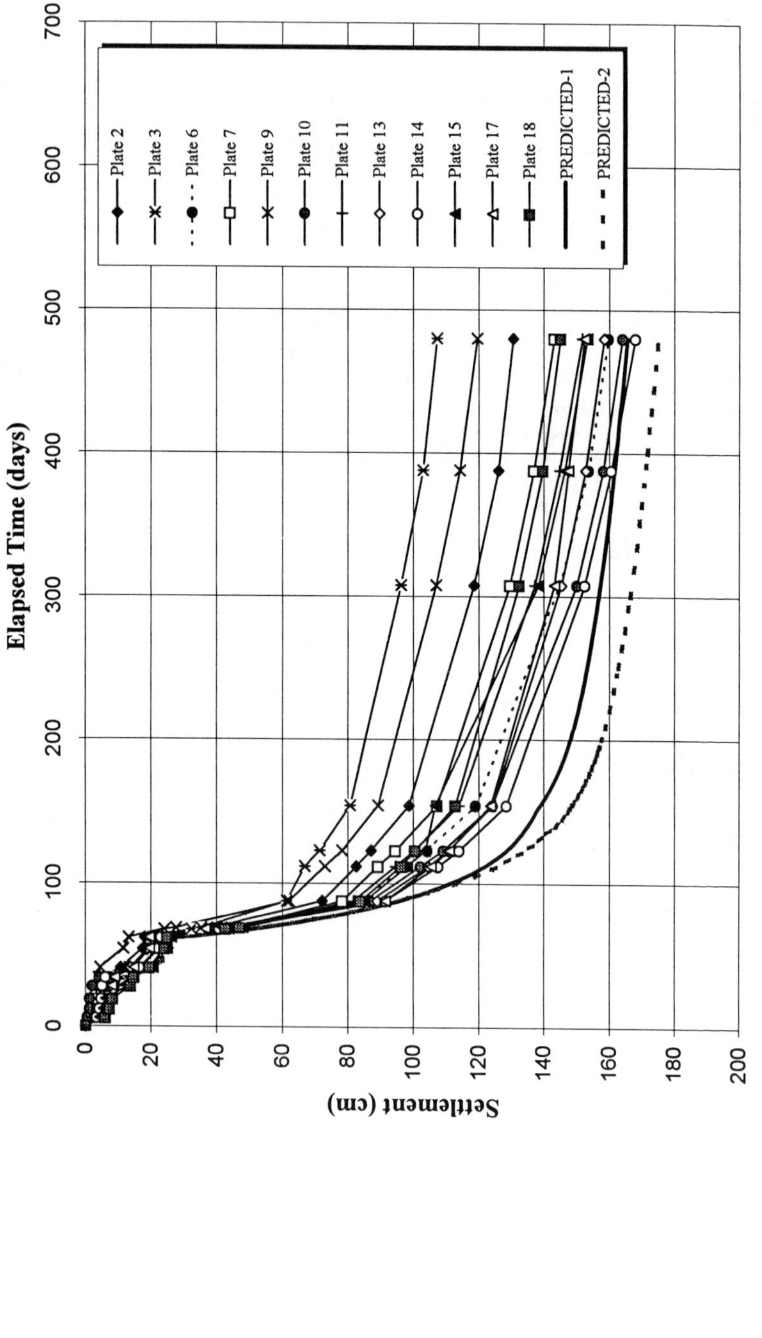

Figure 7. Measured and Predicted Settlements Under Preload A After 480 days

CONCLUSIONS

The following conclusions were derived from the measured settlement data 154 days after the start of preload construction and the prediction of future settlements of the preload embankment:

1. Asaoka's graphical method was a very valuable tool in predicting the magnitude and rate of primary consolidation which are essential in predicting the time for preload removal. Although seldom (if not) used in secondary compression estimates, partly due to its theoretical basis, the method was also applied in this project to provide estimates of the magnitude and occurrence of secondary compression.

2. The predicted final primary consolidation settlement using Asaoka's graphical method is within the range predicted during the preload design. About 95 percent of the predicted primary consolidation settlement was estimated to occur 150 days after the start of preload construction.

3. The estimated amount of secondary compression using the field settlement data is about 40 percent higher than what was estimated during design. Extrapolation of the field data suggests that 95 percent of secondary settlement can be achieved in about a year after it started to occur.

4. Prefabricated vertical (PV) drains, in conjunction with preloading, have been effective in accelerating the rate of primary consolidation settlement of the soft Chewelah clay. Although the positive effects of PV drains on secondary compression remain to be proven, it appears that in this project the combination of surcharging and drainage have enhanced the occurrence of secondary compression.

5. The rate of primary consolidation settlement due to radial flow as observed in the field is at least 50 percent faster than that predicted using the averaged CPT pore pressure dissipation data and at least three times faster than that predicted using laboratory consolidation test data (that is, C_v) if ideal condition (no smear, no well-resistance) is assumed. If smear is included, the rate of radial flow in the field is at least four times faster than the averaged value from the CPT dissipation test, and about eight times faster than the laboratory C_v value. This observation has been widely documented and is partly attributed to the following:

   - The three-dimensional nature of field conditions as opposed to the one- or two-dimensional approximation of theoretical solutions.

   - Localized lenses or layers of relatively pervious materials (such as sand and gravel) interbedded in the clay, which is typical for alluvial deposits such as the soft clay at the site provide drainage boundaries in both the vertical and lateral direction, and thereby helps in accelerating the settlement.

ACKNOWLEDGMENTS

The writers are grateful to the following staff from the City of Chewelah for their support in making this work possible: Ron McCoy (Mayor), Gloria Davidson (former Mayor), Rebecca Blankinship (City Administrator), Bill Provost (former City Administrator), and Robin Hegney (Water and Wastewater Superintendent). The writers would also like to express their appreciation to the following for their services: Nilex Corporation for PV drain installation; and Wilder Construction Co. for preload embankment construction.

APPENDIX. REFERENCES

Allen, R. and Duncan, J.M. (1987). "ZSTRESS: A computer program for calculation of vertical stresses due to surface loads". *Department of Civil Engineering, Virginia Polytechnic Institute and State University*, Blacksburg.

Anderson, L.R., Sampaco, C.L., Gilani, S.H., Keane, E. and Rausher, L. (1994). "Settlements of highway embankments on soft lacustrine deposits". *Vertical and Horizontal Deformations of Foundations and Embankments, Proc. of Settlement '94*, ASCE, College Station, Texas.

Asaoka, A. (1978). "Observational procedure for settlement prediction". *Soils and Foundations*, 18(4), 87-101.

Barron, R.A. (1948). "Consolidation of fine-grained soils by drain wells." *Transactions*, ASCE, 124, 709-739.

Bergado, D.T., Ahmed, S., Sampaco, C.L., and Balasubramaniam, A.S. (1990). "Settlement of Bangna-Bangpakong highway on soft Bangkok clay." *J. Geotechnical Engi*neering, ASCE, 116(1), 136-155.

Bergado, D.T., Enriquez, A.S., Sampaco, C.L., Alfaro, M.C., and Balasubramaniam, A.S. (1992). "Inverse analysis of geotechnical parameters on improved soft Bangkok clay". *J. Geotechnical Engineering*, ASCE, 118(7), 1012-1030.

FHWA (1992). "The cone penetration test". *Report No. FHWA-SA-91-043*, Federal Highway Administration, Washington, D.C.

Hansbo, S. (1979). "Consolidation of clay by band-shaped prefabricated drains". *Ground Engineering*, 12(5), 16-25.

Hansbo, S. (1981). "Consolidation of fine-grained soils by prefabricated drains". *Proc. 10th Intl. Conf. Soil Mech. and Found. Eng'g.*, Stockholm, 3, 12-22.

Magnan, J.P. and Deroy, J.M. (1980). "Analyse graphique des tassements observes sous les ouvrages." *Bulletin Liaison Lab Ponts Chausses*, 109, 45-52.

Magnan, J.P. and Mieussens, C. (1980). "Les remblais d'Essai un outil efficace pour ameliorer les projects d'ouvrages sur sols compresibles." *Bulletin Liaison Lab Ponts Chausses*, 106, 79-96.

Mesri, G. and Castro, A. (1987). "C_α/C_c concept and K_o during secondary compression." *J. Geotechnical Engineering*, ASCE, 113(2), 230-247.

Rixner, J.J., Kraemer, S.R., and Smith, A.D. (1986). "Prefabricated vertical drains, Vol. 1: Engineering guidelines." *Report No. FHWA/RD-86/168*, Nat. Tech. Info. Service, Springfield, Va.

Statistical Analyses

STATISTICAL IMPLICATIONS OF LIMITED FIELD DATA

John T. Christian[1], F. ASCE

ABSTRACT: For reasons of economics and tradition, geotechnical engineers obtain a limited number of data points for most parameters. Well known results from the theory of statistics indicate that uncertainties associated with the small number of points place severe limits on the inferences that can be drawn from the data. This has implications for the design of field instrumentation programs and for the use of field data in verifying computational models. A model of the strength of mine pillars, based on actual data but simplified for present purposes, illustrates the effects.

INTRODUCTION

Making measurements of geotechnical parameters in the field is expensive and time-consuming. Instrumentation also notoriously often fails to function after it has been put in place, so the number of data points is reduced further. Consequently, geotechnical engineers have been accustomed to making do with small numbers of instruments and limited amounts of data. This situation contrasts with that in many other disciplines, where practice demands that enough data be taken to permit meaningful statistical analysis of the results.

Despite the statistical inadequacy of many geotechnical measurement programs, it is quite common to read reports of the verification of computational models with a few data points. Rarely do such reports raise the question of the statistical limitations of the data themselves. As Baecher (2000) observes in his keynote speech to this conference, another set of field or centrifuge measurements could easily yield a different set of values.

This paper describes the statistical implications of limited data, using the margin of safety of mine pillars as an example. The field data are derived from measurements on forty pillars. The results from reliability calculations based on the distributions of the

[1]Consulting Engineer, 23 Fredana Road, Waban, MA 02468.

parameters governing the capacity of the pillars are compared with inferences that would be drawn from different but limited samples of the pillars themselves.

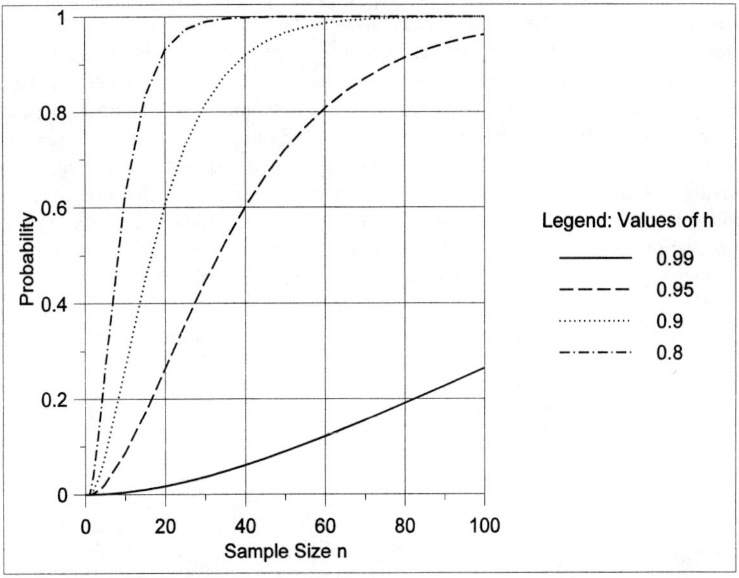

Figure 1. Probability of capturing *h* fraction of the distribution in a sample of *n* statistically independent tests. Results are independent of the distribution of the variable being sampled.

SOME RESULTS FROM THE THEORY OF STATISTICS

Statistical theory (Burington and May 1970) demonstrates that, regardless of the distribution of the variable being observed, the probability that the range of values in a sample of n statistically independent points incorporates a fraction h of the actual variable being sampled is:

$$p = 1 - n \cdot h^{n-1} + (n-1) \cdot h^n. \tag{1}$$

The relation is plotted in Figure 1 for $h = 0.99$, 0.95, 0.90, and 0.80. For example, if a set of 60 independent measurements is made, one can be virtually certain that the range of observed values represents 80% of the population of the variable. On the other hand, there is only a 12% probability that this range includes 99% of population. In geotechnical engineering, it is rare to find 60 measurements of any parameter, so most measurement programs are going to capture limited portions of the actual ranges of the variables.

This has particular implications for engineers' confidence in their knowledge of the range of values that a parameter might have. If one concentrates on the bottom curve of Figure 1, which describes the results when it is desired to capture 99% of population, it follows that 100 tests will yield only 26% probability of capturing the desired range. From equation (1), a suite of 200 tests gives 60% probability, and 388 tests are needed to obtain 90% confidence that one has captured 99% of the distribution. Since it is unlikely that this large a sample is available to support an engineer's judgment, it is not surprising that most estimates of ranges of variables – or of their confidence limits – are overly optimistic. Folayan *et al.* (1970) have given a particularly well known example of this phenomenon – a lack of overlap between experts' estimates of the range of values of the compression ratio of San Francisco Bay Mud and sample observations.

A realistic testing program contains many fewer measurement points than those

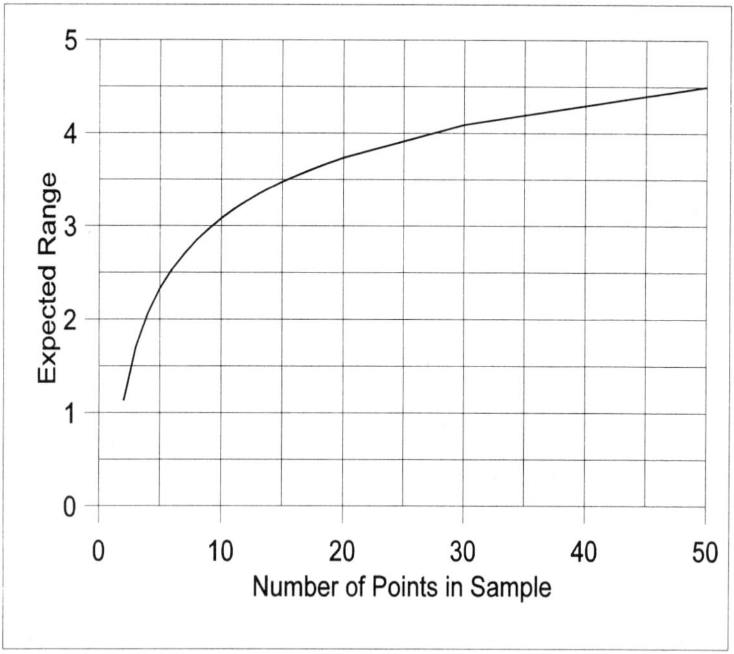

Figure 2. Expected numbers of standard deviations in the range of values in a sample of n points taken from a normal distribution.

called for in the previous paragraph. The usual procedure is for the engineer to obtain a limited number of data points and from these to develop estimates of the statistics of the underlying population, in most cases the mean and the standard deviation or variance. The obvious way to do this is to calculate the mean of the sample and the sample variance or standard deviation. There is also a useful first approximation for estimating standard deviations when some data are available. This comes from the literature on order statistics. Consider that a sample of n observations is made from a normal distribution. One can calculate the expected values of the maximum and minimum of the n values and their difference, which is a function of the standard deviation. Figure 2, drawn from values given by Burington and May (1970) shows the expected range in standard deviation units as a function of the sample size, n. So, if an engineer makes, say, ten tests and observes some range of values, a rough estimate is that the standard deviation is that range divided by about 3.1. The rough estimate should be compared to the sample standard deviation computed from the data themselves.

The statistics of the data not only provide information on the data themselves but also make it possible to estimate the probability of failure. "Failure" in this sense means any unsatisfactory performance, such as excessive settlement, and not only catastrophic collapse. Reliability analysis concerns itself with estimating probabilities of failure from estimates of the statistics of the failure criterion. These are usually calculated from the statistical properties of the variables that enter into the calculation of the margin of safety, but they could also be estimated from the statistics of the field measurements on the geotechnical structures themselves. Knowing the mean and the standard deviation of the margin of safety and assuming a form of the distribution, one could compute the probability that the margin is less than zero.

MINE PILLAR ANALYSIS

One of the formulas for computing the capacity of mine pillars is (Karzulovic 1999)

$$P = k \cdot \frac{W_{eff}^{0.5}}{H^{0.7}}, \tag{2}$$

where k is the design rock mechanics strength (DRMS) in Mpa, H is the height of the pillar in m, and W_{eff} is the effective width of the pillar in m, defined as four times the cross-sectional area divided by the perimeter. The formula is used when $W_{eff} \leq 4.5H$. Since the formula is developed empirically from experimental data, some bias and uncertainty exist, and some uncertainty is associated with the exponents. For the present purpose, these features are ignored, and results are based on the uncertainties in the parameters themselves.

The load in MPa is L. Then the margin of safety, M, is

$$M = P - L. \tag{3}$$

In the following analysis the margin of safety is used instead of the factor of safety because it has more tractable statistical properties and gives more robust results. The data for the individual parameters indicate that the variables can be adequately represented by normal distributions, and the margin of safety's distribution is also close to normal. The factor of safety and the logarithm of the factor of safety are less well behaved, but there is obviously no mechanistic difference between a statement that failure occurs when M = 0 and one that failure occurs when the factor of safety is one.

VALUES OF PARAMETERS AND MARGINS OF SAFETY

Data on forty pillars from several locations at two underground copper mines in the same geologic formation yielded the values of the statistical parameters summarized in Table 1. The load is the estimated tributary laod on each pillar. Four computational procedures were used to compute the reliability: first order second moment (FOSM), Rosenblueth's point-estimate method (Rosenblueth 1975), the linear Hasofer-Lind geometrical method (Hasofer and Lind 1974), and Monte Carlo simulation. Differences reflect the divergence of M and P from normal distributions. The analyses gave consistent results, which are summarized in Table 2. The Monte Carlo results are the averages of ten runs of 1000 tests apiece. These results represent the underlying statistics of the margin of safety for mine pillars in this formation.

Table 1. Statistics for Mine Pillar Variables

Variable	Mean	Std. Dev.	Units	Correlation coefficients	
k	49.13	12.21	MPa	ρ_{kW}	-0.3580
W_{eff}	13.85	2.91	m	ρ_{kL}	0.0363
H	4.00	0.20	m	ρ_{WL}	0.3289
L	33.66	16.44	MPa	All other $\rho = 0$.	

RESULTS FOR DIFFERENT NUMBERS OF POINTS

In the absence of information on the statistical distribution of the parameters, it would be possible to estimate the behavior of mine pillars by examining some or all of the forty pillars. The procedure would be to choose some number of pillars for study, determine the values of the parameters for each pillar, estimate the margin of safety for each pillar, and use these results to estimate the reliability of the general population of pillars. This is similar to estimating landslide risk from a limited number of locations or behavior of retaining walls from data on a few sections. In the present case pillars were selected at random to form groups of five, ten, and twenty pillars, and the entire suite of forty pillars was used as well.

Table 2. Reliability Results for Mine Pillars

Method	Margin of Safety		Reliability Index, β	Probability of Failure
	Mean	Std, Dev.		
FOSM	35.623 MPa	20.887 MPa	1.706	0.044
Point-estimate	34.585 MPa	20.750 MPa	1.667	0.048
Hasofer-Lind			1.664	0.048
Monte Carlo				0.048

The means of the margins of safety were calculated for each suite by averaging

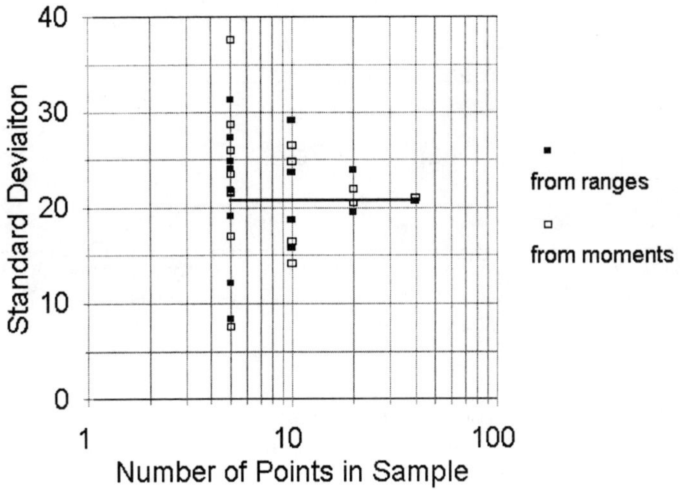

Figure 3. Standard deviation of margin of safety from sample standard deviation and from sample range using Figure 2. Heavy horizontal line is value from point estimate method using distributions of parameters in equations (2) and (3).

the values. The standard deviations were computed in two ways: directly from the samples and estimated from the ranges of values in the samples. Figure 3 shows the estimates of the standard deviation. The estimates from the range of values are comparable to those computed directly as sample standard deviations. As would be expected, the estimates are closer to the correct value when more points are used, and they are virtually identical when all forty points enter the calculation.

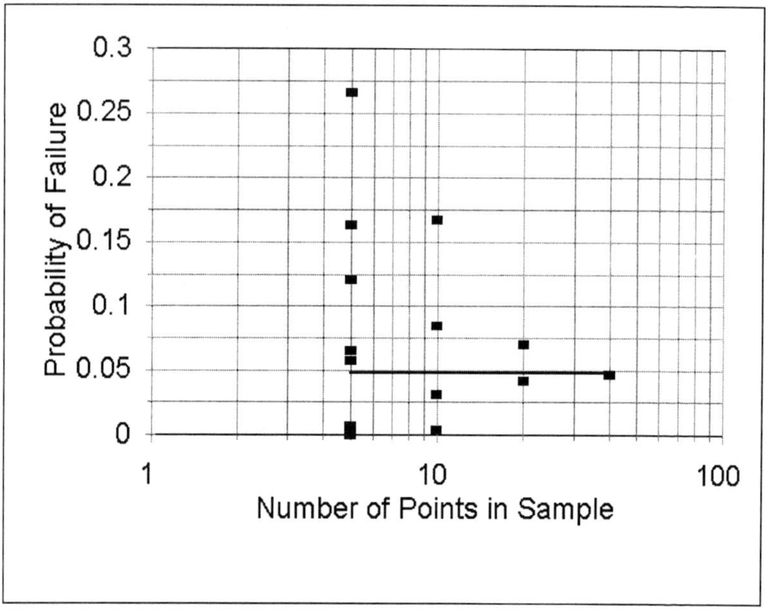

Figure 4. Probability of failure from sample means and ranges of values in the samples. Heavy horizontal line is value from Hasofer-Lind and Monte Carlo methods using underlying parameters.

From the mean and standard deviation of the margin of safety, and with the reasonable assumption that the margin is normally distributed, an engineer can compute the probability that a mine pillar will fail. Figure 4 shows the results when the mean is estimated from the sample means and the standard deviation is estimated from the ranges of the samples. Again, the estimate improves as the number of points in the samples increases.

These results indicate that approximate estimates of the variance in the margin of safety or of the probability of failure require approximately twenty data points. Forty points give very accurate results. It should be remembered that data from all forty pillars were used to develop the statistics of the underlying parameters.

CONCLUSIONS

Some well known results from the theory of statistics show that engineers should not be as assured about the range of a variable as they usually are. This phenomenon has been documented in other fields and with people of widely varying expertise and sophistication in statistical methods (Morgan and Henrion 1990). In particular, many observations are necessary to capture the range of values comprising 99% of the probability distribution.

However, it is possible to make estimates of means and standard deviations using simple techniques. In the present example, the range of observations gives as good an estimate of the standard deviation of a normal population as does the direct calculation. When these estimates are used to compute the reliability of mine pillars, the results are comparable to those using more precise methods. Accuracy improves as the number of points increases; in the present example ten sets of data points seem to be enough to obtain crude estimates of the probability of failure, and twenty sets yield results that are as accurate as most of the data on which the calculations are based.

ACKNOWLEDGMENTS

This paper reflects the results of many hours of discussion with Prof. Gregory B. Baecher. The design procedures and data for the mine pillars are taken from material generously supplied by Dr. Antonio Karzulovic.

REFERENCES

Baecher, G. B. (2000). "Representativeness and Statistics in Field Performance Assessment". *Specialty Conference on Performance Verification of Constructed Geotechnical Facilities, ASCE*, Amherst, Massachusetts, , pp.

Burington, R. S. and D. C. May, Jr. (1970). *Handbook of Probability and Statistics.* 2nd. New York, McGraw-Hill Book Company.

Folayan, J., K. Hoeg and J. R. Benjamin (1970). “Decision Theory Applied to Settlement Predictions”. *Journal of Soil Mechanics and Foundation Engineering, ASCE,* Vol. 96(4): pp. 1127-1141.

Hasofer, A. M. and N. Lind (1974). “An Exact and Invariant First-Order Reliability Format”. *Journal of Engineering Mechanics, ASCE,* Vol. 100(EM1): pp. 111-121.

Karzulovic, A. (1999). Personal communication.

Morgan, M. G. and M. Henrion (1990). *Uncertainty: A Guide to Dealing with Uncertainty in Quantitative Risk and Policy Analysis.* Cambridge, Cambridge University Press.

Rosenblueth, E. (1975). "Point Estimates for Probability Moments". *Proceedings, National Academy of Science,* Vol. 72(10): pp. 3812-3814.

STATISTICAL CONSIDERATIONS IN CALIBRATING PERFORMANCE MODELS WITH FIELD DATA

Robert B. Gilbert[1], Member, ASCE, Mickey C. McBrayer[2], Assoc. Member, Wilson H. Tang[3], Fellow, ASCE

ABSTRACT: This paper highlights and discusses important statistical considerations in calibrating performance models with field data. These considerations include realistically modeling the mean and variance of the data, incorporating multiple sources of data in the calibration, and accounting for uncertainty in the calibrated model when making new predictions with it.

INTRODUCTION

There have been significant strides made recently in predicting the performance of geotechnical systems with analytical and numerical models (e.g., three-dimensional finite element methods), monitoring and measuring performance with field instrumentation (e.g., remote sensing systems), and compiling performance data in databases and information systems (e.g., Geographical Information Systems). However, very little attention has been given to analyzing performance data, and there is a large disconnect in practice between performance modeling and performance measurements. This paper highlights and discusses important statistical considerations in calibrating performance models with field data. These statistical considerations are first introduced. Then, a general methodology for data analysis that is capable of addressing these considerations is summarized. Finally, example results from case studies are presented to illustrate the importance of these statistical considerations.

STATISTICAL CONSIDERATIONS IN MODEL CALIBRATION

The important statistical considerations in calibrating performance models with field data concern realistically modeling the mean and variance of the data,

[1] Assoc. Prof. of Civil Engrg., The University of Texas at Austin, Austin, Texas

[2] Grad. Res. Asst., The University of Texas at Austin, Austin, Texas

[3] Head, Dept. of Civil Engineering, Hong Kong University of Science and Technology, Hong Kong

incorporating data from multiple sources, and accounting for uncertainty in the performance model parameters.

Mean of Data

The mean is the expected result for a measurement. Since a performance model predicts what will be measured, the mean for a data point is obtained from the performance model. There are several important considerations about the mean:

1. The mean value can and generally will vary from measurement to measurement. For example, it is expected that the groundwater concentration of a contaminant will vary with time from the release of contamination (Fig. 1). However, the mean is usually treated as an empirical or a "statistical" model with no physical basis (in many cases, the mean is assumed to be a constant for all measurements). This approach is reinforced by most statistical texts that focus exclusively on parameter estimation for simple analytical models of the mean (e.g., Benjamin and Cornell 1970 and Ang and Tang 1975). Therefore, typical statistical analyses of field data provide little direct understanding of the mechanics, physics or chemistry that govern field behavior, and it is difficult to extrapolate the results to other problems. Furthermore, the value of sophisticated geotechnical modeling is undermined because it is not directly incorporated into statistical analyses of field data.

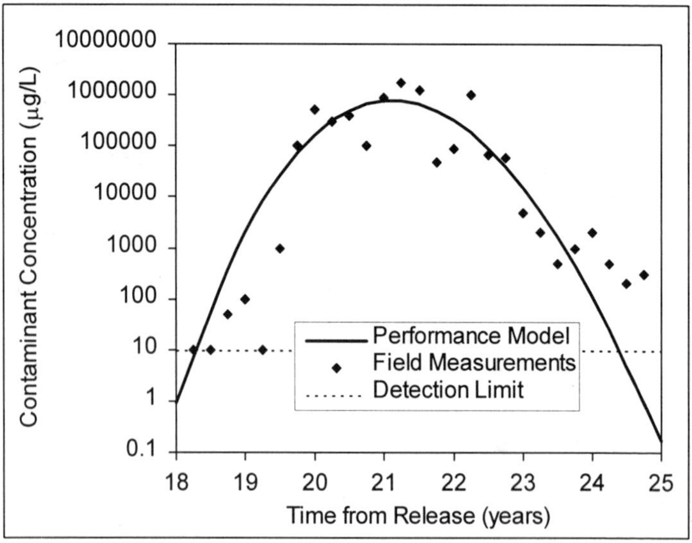

Fig. 1 Example of Performance Model and Field Data for Contaminant Concentration in Groundwater

2. The mean value will depend on the input parameters to the performance model and on the attributes of each measurement (such as the time or location of the measurement). Calibration of the performance model involves identifying the values for the input parameters that give the "best fit" between the expected measurements (the mean values) and the actual measurements. The common definition of "best fit" is to minimize the sum of the squared differences (or errors) between the mean values and individual data points (e.g., Yeh 1986). However, this "least square error" definition of "best fit" implicitly contains assumptions that may not be reasonable, as discussed in the next sub-section.

Variance of Data

The variance represents the variation between each measurement and the mean value (the performance model prediction) for that measurement. There are several important considerations about the variance.

1. The sample standard deviation that is commonly calculated and reported is relevant only if the mean value is constant between measurements. It is variations (or differences) between the measurements and mean values, not simply variations in the measurements, that are relevant. Since the mean value is generally not constant, the sample standard deviation is generally not relevant.

2. The variance for measurements is not necessarily constant between measurements and can (and generally will) depend on the expected or mean value for each measurement. For example, the magnitude of variation in a groundwater concentration as the plume passes a monitoring well when the concentration is about 100,000 μg/l may be on the order of 100,000 μg/l. However, the magnitude of variation long after the plumed has passed when the expected concentration is about 100 μg/l may be on the order of 100 μg/l (Fig. 1). Unfortunately, statistical texts primarily consider only the case where the variance is constant between measurements.

3. The variations between measurements and mean values are not necessarily independent. For example, if the groundwater concentration at a particular well is under-predicted by the performance model during one quarter, it may be more likely to be under-predicted by the performance model at that well in the next quarter (Fig. 1). Hence, there is a correlation with time in the variations between measurements and mean values. There are two important consequences of correlated variations between measurements. First, positively correlated variations in measurements reduce the effectiveness of these measurements in calibrating the performance model. In a sense, fewer effective measurements are available to calibrate the performance model since multiple measurements contain redundant information. Second, the

magnitude of correlation may be important information in decision making. For example, a decision criterion stating that a site requires corrective action if measured groundwater concentrations exceed a specified value in three consecutive quarters may not be appropriate if variations in measurements are correlated over time spans of years. Unfortunately, most statistical texts consider only the case of statistically independent variations (e.g., Benjamin and Cornell 1970, Ang and Tang 1975, and Montgomery and Runger 1999).

4. The distribution of variations about a mean value is not necessarily normal, although it is commonly assumed to be normal. The most significant feature of the normal distribution is that it is symmetrical, indicating that it is as likely for measurements to be below the mean value as above the mean value. Groundwater concentrations are an example of asymmetrical variations about the mean. Since the groundwater concentration cannot be less than zero but can be very large, the distribution of variations tends to be skewed toward larger values. Due largely to mathematical convenience versus technical merit, most statistical analyses rely on an assumption of normality.

5. Because variations are not necessarily constant, not necessarily independent, and not necessarily normally distributed, the use of a "least square error" criterion to calibrate performance models is potentially flawed. The problem is that this criterion puts equal weight on every data point. However, equal weight should not be given if some data points have larger variations, if some data points are correlated, or if data points are more likely to be below versus above the mean value. Groundwater concentrations with variations on the order of 100,000 μg/l should not be treated the same as concentrations with variations on the order of 100 μg/l because they will dominate a "least square error" model calibration. Five groundwater concentration measurements that are clustered together in time should not have the same value in calibrating a model as five measurements that are spread out over time. Concentration measurements below the mean value should not be given the same weight as those above the mean value since it is more likely for measurements to be below the mean value.

Multiple Sources of Data

There may be multiple sources of data available to calibrate a performance model. For example, there may be laboratory and field tests conducted to measure the properties of a groundwater aquifer. In addition, there may be groundwater concentrations measured in monitoring wells to evaluate contaminant transport. Some of these concentration measurements may provide a single value (or data point), while other concentration measurements may only provide an upper bound value because they are below the detection limit (Fig. 1). All of these sources of data should be used, but not necessarily treated equally, in calibrating a groundwater contaminant fate and transport model. Unfortunately, most statistical texts and publications consider only the case of point measurements that are all of the same type (e.g., Benjamin and Cornell 1970, Ang and Tang 1975, Yeh 1986).

Uncertainty in Performance Model Parameters

The parameters that describe the performance model can be divided into two categories. First, there are model parameters that describe the mean value for each individual measurement. For a groundwater contaminant transport model, these parameters would include hydraulic conductivity, gradient, porosity, dispersivity factors, decay factors, etc. Second, there are model parameters that describe the variation between the mean and the measurement for each individual measurement. These parameters include information about the standard deviations for individual measurements, the correlation coefficients between individual measurements, and the shape of the distribution for the variations. The second category of model parameters, which describe variations in the performance, is typically neglected in practice, unfortunately.

Exact values for the model parameters will not be known with certainty, even after the model has been calibrated with field data. In general, the more data available, the less uncertainty there will be in the model parameters. The more complicated the performance model (i.e., the greater number of input parameters), the more uncertainty there will be in the model parameters. This uncertainty in the model parameters will affect predictions using the calibrated performance model; the more uncertainty there is in the model parameters, the more uncertainty there will be in the predictions. In many cases field performance data are scarce in geotechnical engineering, and thus model uncertainty is the dominant source of uncertainty.

GENERAL METHODOLOGY FOR MODEL CALIBRATION

A general methodology exists to capture all of the statistical considerations described above. The approach is to use a first-order approximation to estimate the first two moments (the means, variances and correlation coefficients) for input parameters in a performance model, and it is called the First-Order, Second-Moment Bayesian Method (FSBM). The basic steps in FSBM, which is formulated from maximum likelihood theory (e.g., Rice 1995) and Bayesian theory (e.g., Ang and Tang 1975), are presented below and details are given in Gilbert (1999).

Step 1: Formulate a performance model to predict the data that are measured, including: the expected or mean value for the data, the variance between what is measured and what is expected; the correlation between variations from the mean for individual measurements; and the shape describing the distribution of measurements about the mean. The models for the means, variances and correlation coefficients can be either analytical or numerical, the only constraint is that first and second-derivatives with respect to the input parameters are required. In addition, the shape describing the distribution of measurements about the mean can be any form; non-normal distributions are generally accommodated using transformations with the multivariate normal distribution, such as logarithmic or Hermite polynomial transformations (e.g., Journel and Huijbregts 1978). This model predicting what will be measured is named the likelihood function because it describes the likelihood (or probability) for each possible set of data.

Step 2: Develop preliminary, independent estimates for the input parameters used in the likelihood function (the model parameters). The estimate for each model parameter includes the best guess for that parameter represented by the mean value, the magnitude of uncertainty in that guess represented by the variance, and any correlation between the estimate for that parameter and the estimate for any other model parameter. Laboratory and field measurements of geotechnical properties will generally be incorporated in this step.

Step 3: Calibrate the performance model using the measured field performance data. This calibration is accomplished by maximizing the likelihood function with respect to all of the input or model parameters. Once the maximum likelihood point (or points if there are local maximums of the same magnitude as the global maximum) is found, evaluate the second derivatives of the natural logarithm of the likelihood function with respect to each model parameter at this maximum likelihood point.

Step 4: Combine preliminary (or prior) information from Step 2 with the calibration information from Step 3 to obtain updated values for the first two moments (the mean, variance and correlation coefficients) of each model parameter. This combination process involves calculating matrix sums, products and inverses, and provides a first-order approximation for the updated moments.

EXAMPLE CALIBRATION RESULTS

Example results from calibrating geotechnical performance models with field data are presented in this section. A summary of these case studies is presented in Table 1. These example results are intended to demonstrate the importance of the statistical considerations discussed previously.

Table 1 Summary of Case Studies for Model Calibration

Case Study	Characteristics of Model for Mean Performance	Sources of Data for Calibration	References
Groundwater Contaminant Transport at a Superfund Site	• Numerical model of mean contaminant concentration versus time and horizontal/vertical location • 3-D transport • 1-D steady-state flow	• Laboratory and field measurements of aquifer properties • Measured groundwater concentrations with time at multiple locations around waste lagoon over a ten-year period	McBrayer (1999)
Primary Liner Leakage in a Hazardous Waste Landfill	• Finite-difference model of time-varying mean flow rate into secondary leak detection system • Primary liner leakage related to time-varying flow rate in primary leachate collection system • Consolidation flow from primary clay liner related to time-varying stress application	• Laboratory measurements of lining system properties • Measured monthly flow rates into primary and secondary collection systems with time over a six-year period • Stress application history	Gilbert (1993); Gilbert and Tang (1995)
Design Capacity of Piles in an Offshore Field	• Quadratic models for mean side friction and end bearing with depth • Models related to geology	• Design capacity profiles from boring data at hundreds of boring locations throughout three different geologic conditions	Gambino and Gilbert (1999); Gilbert et al. (1999a and 1999b)

Mean of Data

In all of the case studies in Table 1, the mean value is described with a physically-based model. For the landfill leakage case study, Fig. 2 shows the field-calibrated numerical model describing the mean flow rate into the secondary leak detection system. The field data used to calibrate the performance model are also shown on Fig. 2. In this landfill cell, the secondary flow rate is dominated by consolidation flow from the overlying primary clay liner; the peaks in Fig. 2 correspond to high rates of stress application. As an example of how the field data improve understanding of physical behavior, the field-calibrated coefficient of consolidation for the clay liner is about ten times smaller than the laboratory-measured value. The common statistical assumption of a constant mean value (the sample average in Fig. 2) has little relevance in this case because it leads to an under-predicted flow rate at early times (during waste filling) and an over-predicted flow rate at later times (after closure).

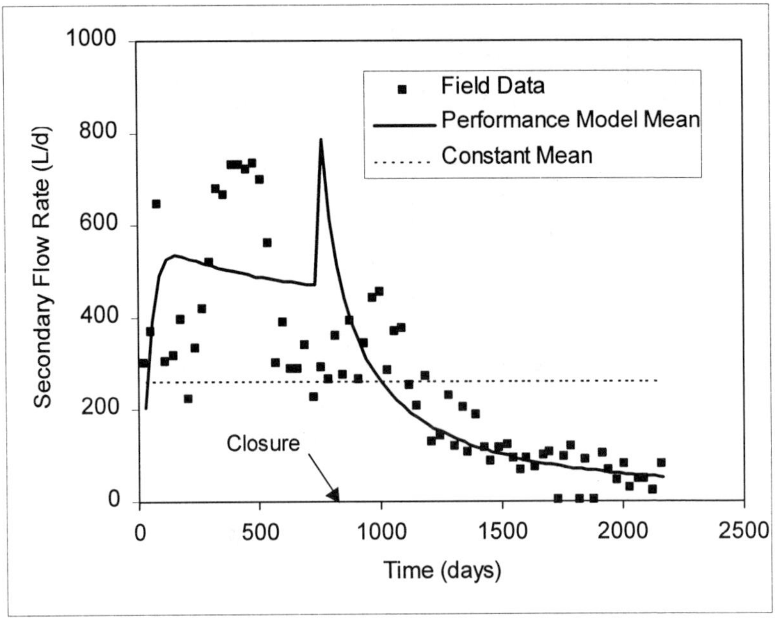

Fig. 2 Calibrated Mean Performance Model for Landfill Leakage

Variance of Data

The pile capacity case study provides an example where the assumption of a constant variance is not reasonable. The standard deviation in axial pile capacity is shown on Fig. 3 as a function of pile penetration. The standard deviation, like the mean capacity, increases with increasing pile penetration. Also, the standard deviation at any depth depends on the geology, with a significantly larger standard deviation in the heterogeneous alluvial geology versus the more homogenous marine geology (Fig. 3).

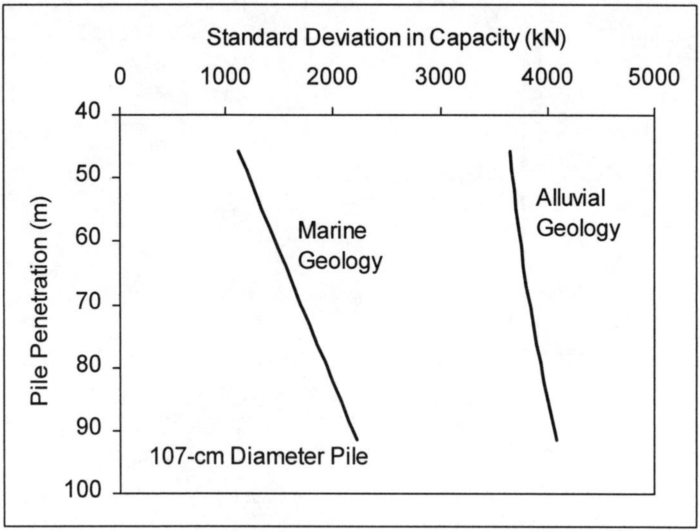

Fig. 3 Standard Deviation in Axial Pile Capacity versus Pile Penetration

An example of correlation between data points is shown on Fig. 4 for the pile capacity case study. The correlation in design pile capacity between two sites decreases as the horizontal distance between the sites increases. The "correlation distance" is the horizontal distance within which two sites are correlated. Specifically, the correlation coefficient, which has a magnitude of 1.0 when the two sites coincide and is 0.0 when the two sites are very far apart, has a magnitude of approximately 1/3 when the sites are separated by the correlation distance. Figure 4 shows that the correlation distance can be significant. For sites in the marine geology, pile capacities at sites within several kilometers from one another will be correlated. This result is important because it indicates how far information from a boring can be extrapolated if the location of a platform does not coincide with the location of the boring (Gilbert et al. 1999a). This result is also important because nearby borings provide redundant information in calibrating the pile capacity model and should not be treated equally with borings that are far apart. In addition, Fig. 4 shows how the horizontal correlation distance in pile capacity varies with pile penetration and with geology.

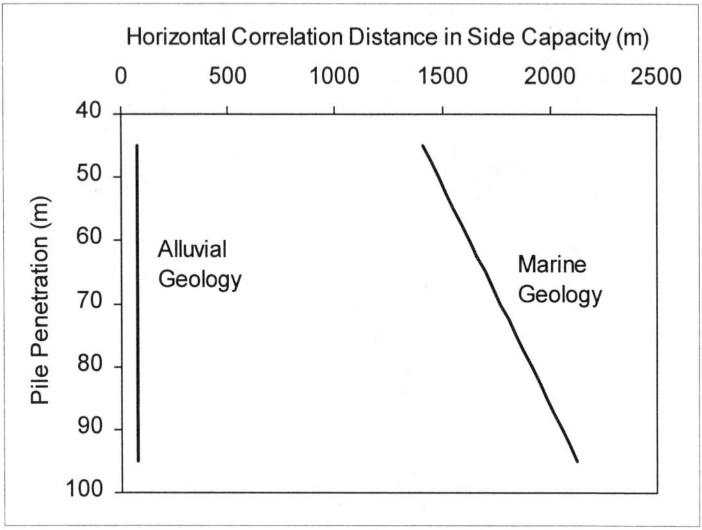

Fig. 4 Horizontal Correlation Distance in Side Capacity versus Pile Penetration

Other examples of correlation between measurements are presented in Table 2. In all cases, the correlation distance has the same meaning whether the units of distance are time or space. Note that the variations between measurements and model predictions (the mean value) are correlated over substantial distances and times for the groundwater transport model. This result is reasonable because the transport model, while fairly complicated with nearly twenty input parameters, still does not realistically capture geologic features (such as layers and channels) and time-varying environmental conditions (such as the hydraulic gradient). The correlation with time in the landfill leakage model is seen graphically on Fig. 2; the field measurements are clustered above and below the model predictions.

Table 2 Example Results for Calibrated Correlation Distances

Case Study	Correlation Distances
Contaminant Transport Model of Groundwater Concentrations	Horizontal (longitudinal with flow): 19 m Horizontal (transverse to flow): 21 m Vertical: 1.5 m Time: 1200 days
Model of Landfill Leakage with Time	Time: 30 days

Information about the variance of the data is valuable in quantifying how well a performance model fits the field data. As an example, field-calibration results for two different groundwater transport models with the same data set are summarized in Table 3. The variations between the measurements and the model were modeled with a standard deviation that was proportional to the mean value. Specifically, the coefficient of variation, which is the standard deviation divided by the mean, was modeled as a constant. The coefficient of variation and the correlation distances are substantially greater for the simple model, indicating that it does not explain variations in the field data as well as the more complicated model. Note that even with the more complicated model, the measured concentrations still vary by more than an order of magnitude from the model predictions (that is, the calibrated coefficient of variation is greater than ten).

Table 3 Comparison of Calibration Results for Two Groundwater Models

Parameter	Simple (Analytical) Model	Complicated (Numerical) Model
Coefficient of Variation	2400	26
Horizontal Correlation Distance (longitudinal)	73 m	19 m
Horizontal Correlation Distance (transverse)	66 m	21 m
Vertical Correlation Distance	1.5 m	1.5 m
Temporal Correlation Distance	4400 days	1200 days

Multiple Sources of Data

The pile capacity case study provides an example of combining multiple sources of data. In this case, borings before about 1980 were drilled using wire-line percussion methods to take samples (driven sampling) in contrast to the more common sampling techniques using hydraulically pushed, thin-walled tubes (pushed sampling). The objective with the performance model was to predict the design capacity that would be obtained if a modern boring (with pushed sampling) were drilled at the site.

In order to incorporate the older borings into calibrating this model, the design capacity obtained from pushed-sampling borings was related to that obtained from driven-sampling borings as follows. First, the pushed-sampling mean was modeled to be a factor of the driven-sampling mean. The calibrated factor for the mean was 1.10, meaning that the design profile obtained from driven sampling tended to be lower (more conservative) than that obtained from pushed sampling. This result is consistent with the effects of sample disturbance due to driving a sampling tube, which would reduce the shear strength of the soil. However, the reduction in average strength was on the order of 30 percent for driven samples (versus a 10 percent reduction in the design capacity), indicating that the geotechnical engineers implicitly accounted for the two different sampling methods to some extent in developing design profiles for pile capacity. Specifically, they tended to develop a design shear strength profile based on the average strength with driven samples and on a lower-bound strength with pushed samples. Second, the pushed-sampling standard deviation was modeled to be a factor of the driven-sampling standard deviation. The calibrated factor for the standard deviation was 1.16, meaning that the design profiles obtained from driven sampling were less variable about the mean than those from pushed sampling. This result is also consistent with the effects of sample disturbance, which would tend to mask variations in soil structure. Finally, a correlation coefficient was introduced to describe how variations in the field data are related for pushed and driven sampling at the same location. This calibrated correlation coefficient was 0.6. Design profiles from the two sampling methods are positively correlated: if the design profile obtained from a driven-sampling boring at a particular site was higher than the mean, then the design profile obtained from a pushed-sampling boring would also tend to be higher than the mean. However, the correlation is not perfect: it is not possible to predict with certainty the design profile that would be obtained from a modern boring based on the design profile from an older boring.

There are several benefits to including both sources of information in calibrating the model for design pile capacity. First, a significant portion of the field data in these offshore fields was from older borings. Therefore, while the data are not of as high of quality, it would be unreasonable to neglect these data. In the model calibration, the older data were included but given less weight than the newer data since the correlation between the two design profiles was not perfect (a correlation coefficient of 0.6). Second, decisions still need to be made on the basis of the older data so understanding how they relate to the newer data is important. For example, offshore platforms that were designed with older borings may require re-qualification today to extend their design life or to increase their deck capacity.

The groundwater contamination case study provides another example with multiple sources of data. Laboratory and field measurements were available to estimate the aquifer properties before calibrating the contaminant transport model with field-measured groundwater concentrations. Furthermore, more than twenty percent of the measured concentrations were below the detection limit and only provided information about the upper bound for the concentration. The formulation used to include these detection-limit measurements in the model calibration is described in McBrayer (1999).

Table 4 shows the prior estimate of the mean, the value that maximized the likelihood function with the field data, and the updated mean for several aquifer properties. For hydraulic conductivity and longitudinal dispersivity, the updated mean (which combines all of the information together) was between the prior mean and the maximum likelihood value for the field data. In both cases, the field data did not provide very convincing evidence relative to the prior information, and the updated value is closer to the prior mean. The source concentration provides an interesting example of what can happen when multiple data sources are combined to estimate multiple model parameters. The updated mean is smaller than both the prior mean and the maximum likelihood value. This result occurs because the source concentration is correlated with other input parameters and its updated mean is affected by the updated means for those parameters. Specifically, the source concentration is negatively correlated with the hydraulic conductivity since the groundwater concentration is directly related to the source concentration and inversely related to the hydraulic conductivity (the faster the water passes across the source, the less mass is dissolved into the groundwater). Therefore, if the hydraulic conductivity were underestimated, then the source concentration would be overestimated. Since the updated mean for the hydraulic conductivity is pulled up from its maximum likelihood estimate by the prior information, the updated mean for the source concentration is pulled down from its maximum likelihood estimate. This effect is particularly significant on the source concentration because the prior value is very uncertain (see Table 5).

Table 4 Combining Multiple Data Sources for Groundwater Model

Model Parameter	Prior Mean Value	Maximum Likelihood Value from Field Performance Data	Updated Mean Value
Hydraulic Conductivity (m/s)	$1x10^{-5}$	$5x10^{-8}$	$5x10^{-6}$
Longitudinal Dispersivity (m)	10	1400	21
Source Concentration (mg/L)	1100	230	16

Uncertainty in Model Parameters

The groundwater contamination case study illustrates how uncertainty in the calibrated (or updated) model parameters is an important consideration. Results for several model parameters are summarized in Table 5. Note that a significant amount was learned about the hydraulic conductivity and the source concentration, while relatively little was learned about the longitudinal dispersivity. The effect of uncertainty in the updated model parameters is demonstrated if the calibrated model is used to predict a new concentration (at a new location and a future time). In one example, the coefficient of variation for this new prediction is about twenty percent smaller if uncertainty in the updated model parameters is neglected, even though more than 200 concentration measurements were used to calibrate the model. Therefore, the total uncertainty in new predictions with the performance model is under-valued if uncertainty in the model parameters is neglected.

Table 5 Uncertainty in Model Parameters for Groundwater Model

Model Parameter	Prior Coefficient of Variation	Updated Coefficient of Variation
Hydraulic Conductivity (m/s)	7.0	0.3
Longitudinal Dispersivity (m)	0.5	0.4
Source Concentration (mg/L)	270000	0.4

SUMMARY

In summary, the following important points should be considered in analyzing field data to calibrate performance models:

1. The field data should be analyzed within the context of the performance model, not some simplified analytical model used for the sake of statistical convenience, so that the field data are useful in understanding and predicting performance.

2. Variations (or differences) between predicted and actual measurements should be modeled realistically, including variations that depend on the magnitude of the predicted measurement, variations that are not necessarily independent between measurements, and variations that are not necessarily normally distributed about the predictions.

3. Least-square error calibration (or regression) techniques should be used with caution because they may unreasonably give each data point the same weight in calibrating the model.

4. All sources of information should be used in calibrating the performance model, including multiple types of laboratory and field data, in order to utilize as much information as possible.

5. Uncertainty in the parameters of the performance model should be evaluated, reported and included in predictions made using the calibrated performance model.

All of these points should also be considered in designing field measurement programs in order to maximize their value.

ACKNOWLEDGMENTS

The authors wish to acknowledge the National Science Foundation, the Texas Advanced Research and Technology Programs, the Gulf Coast Hazardous Substance Research Center, and the Illinois Department of Energy and Natural Resources for supporting the work presented in this paper.

APPENDIX I - REFERENCES

Ang, A. A-S. and Tang, W. H. (1975), Probability Concepts in Engineering Planning and Design, Volume I – Basic Principles, John Wiley & Sons, Inc., New York.

Benjamin, J. R. and Cornell, C. A. (1970), Probability, Statistics, and Decision for Civil Engineers, McGraw-Hill, Inc., New York.

Gambino, S. J. and Gilbert, R.B. (1999), "Modeling Spatial Variability in Pile Capacity for Reliability-Based Design," Analysis, Design, Construction and Testing of Deep Foundations, Roesset Ed., ASCE Geotechnical Special Publication No. 88, 135-149.

Gilbert, R. B. (1993), "Performance Reliability of Landfill Lining Systems," Ph.D. Dissertation, University of Illinois, Urbana, Illinois, 299 pp.

Gilbert, R. B. (1999), "First-Order, Second-Moment Bayesian Method for Data Analysis in Decision Making," Geotechnical Engineering Center Report, Dept. of Civil Engineering, The University of Texas at Austin, Austin, Texas, 50 pp.

Gilbert, R. B., Gambino, S. J. and Dupin, R. M. (1999a), "Reliability-Based Approach for Foundation Design without Site-Specific Soil Borings," *Proceedings*, Offshore Technology Conference, Houston, Texas, 631-640.

Gilbert, R. B., Stong, T. J., Lang, J. T., Albrecht, R. S. and Dupin, R. M. (1999b), "Optimizing Investigation Programs for Offshore Foundations – Effect of Geology on Axial Pile Capacity," *Proceedings,* 2nd Intl. Conf. on Seabed Geotechnics, London.

Gilbert, R. B. and Tang, W. H. (1995), "Reliability-Based Design of Waste Containment Systems," *Proc.,* Geoenvironment 2000, ASCE, New Orleans, Louisiana, 499-513.

Journel, A. G. and Huijbregts, Ch. J. (1978), Mining Geostatistics, Academic Press, Harcourt Brace & Company, San Diego.

McBrayer, M. C. (1999), "Calibration of Groundwater Flow Models for Modeling and Teaching," Ph.D. Dissertation, The University of Texas at Austin, 403 pp.

Montgomery, D. C. and Runger, G. C. (1999), Applied Statistics and Probability for Engineers, Second Edition, John Wiley & Sons, Inc., New York.

Rice, J. A. (1995), Mathematical Statistics and Data Analysis, Second Edition, Duxbury Press, Belmont, California.

Yeh, W. W-G. (1986), "Review of Parameter Identification Procedures in Groundwater Hydrology: The Inverse Problem," *Water Resources Res.*, Vol. 22, No. 2, 95-108.

RELIABILITY OF MEASURED GEOTECHNICAL PROPERTIES

N. O. Nawari[1] and R. Liang[2]

ABSTRACT

Geotechnical problems are unique and hard to pin down using exact mathematical models. In such environment there are numerous sources of ambiguity and imprecisely defined data. Conventionally, in the construction of mathematical models, the uncertainties are treated as random variables. Probability theory is perceived as the unique methodology to handle uncertainty in these parameters. Nevertheless, our preliminary investigations with geological and engineering geological maps, bore holes, field tests, sampling, laboratory work up to the analysis and design, are all combined with vague, indeterminist and to a great extent non statistical data. The application of decision theory and stochastic models in such an environment is linked up with many drawbacks and perplexing features. In this paper, the reliability evaluation of measured properties will be investigated applying the fuzzy-reliability techniques. Within this conception setting, the measured parameters are rendered into fuzzy variables and the structural safety is characterized by fuzziness. The procedure is proposed as an alternative or complementary postulation of reliability in geo-engineering. This approach will be illustrated with computational algorithms and a numerical example.

[1] Assistant Professor, Department of Civil Eng., University of Akron, Akron, OH 44325-309, nawari@uakron.edu
[2] Professor and Chair, Department of Civil Eng., University of Akron, Akron, OH 44325-309, rliang@uakron.edu

INTRODUCTION

Unlike in other disciplines of civil engineering where the construction materials, such as steel, timber, concrete ...etc can be largely fabricated to exact specifications, goetechnical materials in situ need to be determined using a variety of field and laboratory procedures. Geo-materials are normally measured and assessed using geological and engineering geological maps, field tests, sampling, and laboratory tests. Hence, the levels of information available at a given site vary substantially. These uncertainties are greatly associated with vague, indeterminist and non-statistical nature of the data.

The reliability analysis of geotechnical structures is directly affected by the soil or rock properties. In both the ASD and the LRFD design concepts, plays the determination of the nominal (characteristic) material property a crucial role (Figure 1). Therefore it is imperative that the uncertainties in the design parameters be adequately quantified and their effect carefully evaluated.

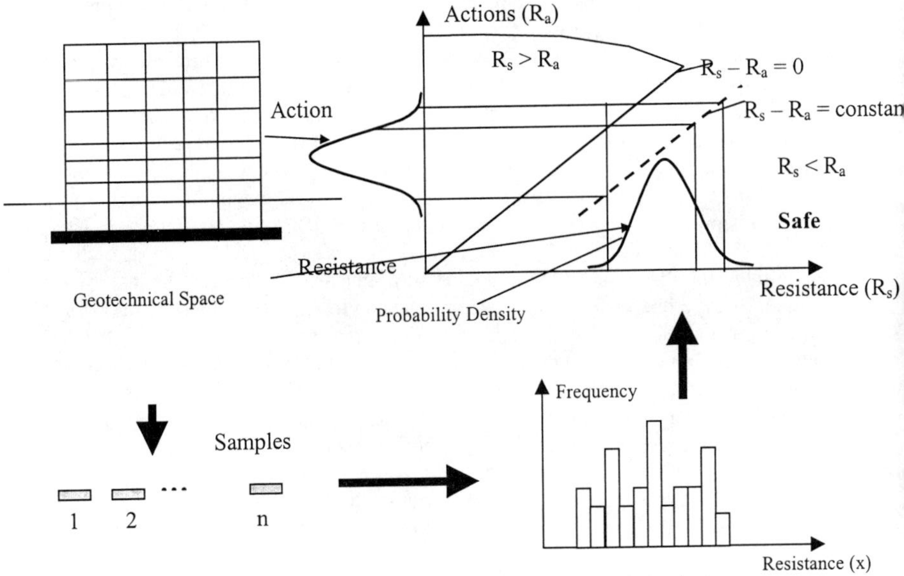

Figure 1: The relation between the measured Property and the design concept

The general approach for determining the nominal material properties is based upon the statistical results of a large number of samples, for which the mapping of the sample mean $\bar{x}$ and standard deviation s to the population values μ and σ is obtained through the confidence intervals and the likelihood functions. Figure 1illustrate the range of the nominal material property assuming normal distribution.

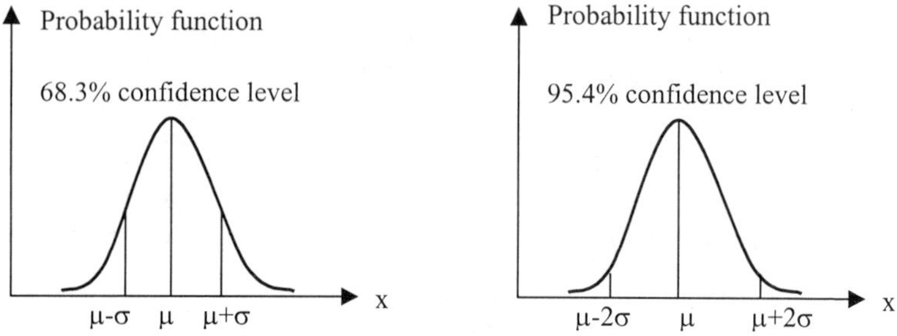

Figure 2: Typical ranges for the nominal value of a random parameter (x)

Three main components of the uncertainties on geotechnical properties have been generally recognized (Tang, 1993): (i) intrinsic variability from spatial variation and (ii) statistical uncertainty from testing error and insufficient test samples for estimating the material property (iii) uncertainty contributed mainly by the inability of a test to reproduce the in situ property due to factors such as sample disturbance, limited specimen size, etc. Present modeling methods can capture only the second component of the material uncertainties. Namely, by modeling the material property as random variables, described by the mean value and the standard deviation along with a probability distribution function (Figure 1). Based upon this concept many investigators suggested different stochastic models dealing with the reliability of measured soil parameters (e.g. Krumbein and Greyhill, 1965; Ang and Tang, 1984; Tang, 1984; Vanmarke, 1977, 1984; DeGroot and Baecher, 1993). Most of these models fell into disfavor due to a number of criticisms. One of these is that a large number of experimental results are required, which in some occasions are impossible to perform or connected with high cost. Von Mises (1919) pointed out that this approach is applicable whenever we have sufficient reasons to believe that the relative frequency of the observed attribute would tend to a fixed limit if the observations were indefinitely continued. Evidently the probabilistic approach can not model appropriately the first and the second components of uncertainty.

Tang (1984) suggested a simple formula for combining the components of uncertainties in determining the mean and the coefficient of variation of average soil properties X_i:

$$\bar{X}_i = N\,X_i \tag{1}$$

where

X = the mean soil property estimated from laboratory or field tests

N = systematic bias.

$$\Omega_i = \sqrt{\Delta^2{}_o + \Delta^2{}_n + \Gamma^2(D)\delta^2} \tag{2}$$

where the first two coefficient of variations on the right side of equation (2) denote respectively the statistical and systematic uncertainties; the third term is a product of δ, the coefficient of variations from tests and Γ, a reduction factor depending on the average domain D and the spatial correlation characteristics of the soil properties. The mean soil parameter determined from field or laboratory experiments is also modified by the systematic bias N of the respective test as indicated in equation (1)

Vanmarke, (1977) and DeGroot and Baecher, (1993) developed probabilistic models for the spatial variability by introducing the autocorrelation. The variation of a soil property in space is given by a function of trend T(x) and residuals, $\varepsilon(x)$ (Figure 3). The trend function is obtained by regression analysis and residuals are correlated. The spatial variation can be modeled as the sum of a trend and a residual:

$$Y(x) = T(x) + \varepsilon(x) \tag{3}$$

where Y(x) is the measurement at location X. In this technique the soil property is controlled by a random process.

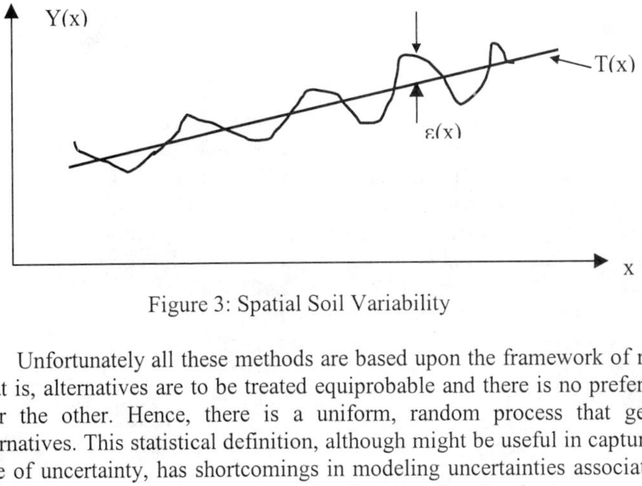

Figure 3: Spatial Soil Variability

Unfortunately all these methods are based upon the framework of randomness. That is, alternatives are to be treated equiprobable and there is no preference of one over the other. Hence, there is a uniform, random process that generates the alternatives. This statistical definition, although might be useful in capturing specific type of uncertainty, has shortcomings in modeling uncertainties associated with the measured geotechnical properties of soils or rocks. In short, some observations obtained from the measured material properties are precise, while some are measurable only in statistical sense, and others can not be measured at all. So the data procured in this manner are hybrid; that is their components are not homogenous but a blend of precise, statistical and fuzzy information.

The fuzzy-reliability techniques offer different models that can capture type of uncertainties associated with the geotechnical material properties, which are not random in nature. These models are armed with engineering judgement and experience along with objective data. This approach is described next.

STATEMENT OF THE PROBLEM

Consider the geotechnical property X of soils or rocks, which its nominal value needs to be estimated. X is treated as the universe of discourse of the numerical values x_i (friction angle, cohesion, compressibility coefficient, etc.) of that property. Information available for determining the nominal (characteristic) value of this property has the following components:

(i) Observations of test results on a limited number of sample (usually a few numbers).

(ii) Experience with similar material or some data from other sources.

It is required to determine the numerical nominal value C_x such that the probability of having lower values than or at most equals to C_x is very small.

COMPUTATION ALGORITHM

First, given are the experimental results $x_1, x_2, \ldots x_n$ of sample size n; $x_i \in X$. Then the computation procedure is:

(A) Every object x_i will be treated as a fuzzy parameter (fuzzy variable). For more details about fuzzy variables and fuzzy techniques refer to, for instance Zadeh (1965), Nahmias (1978), Dubois and Prade (1980), Kosko (1992), Nawari and Liang (1999), and Nawari and Hartmann (1998).

(B) The existing extra information (pre-information) from experience, engineering judgement or other sources will be implemented in the definition of the mathematical function for x_i. This is generally linked to the concept of confidence interval and the level of presumption.

(C) The weighted mean M_x is determined using x_i and the extension principle (Zadeh, 1965, 1973).

(D) Using M_x, the nominal value C_x will be estimated using one of the following techniques:

  (i) The centroid method

  (ii) Fuzzy-deviation method

In this computation algorithm, the following parametric reference functions are introduced for the definition of the fuzzy variable X:

Type (1): Linear LR-Function

The LR(Left-Right) functions were first defined by Dubais and Prade (1980) for describing fuzzy sets. Given X as a fuzzy variable, with $L(\xi)$ and $R(\xi)$ as two linear reference functions, then X is a linear LR-function, when,

$$
\lambda_x = \begin{cases} L(\xi) = L((a_1 - x)/u) & x \le a_1, u > 0 \\ 1 & a_1 \le x \ge a_2 \\ R(\xi) = L((x - a_2)/v) & a_1 \le x_1 \le a_2 \\ 0 & x < a_1 - u,\ \mathbf{x} > a_2 - v \end{cases} \tag{4}
$$

The coefficients a_1 and a_2 are known as the supports at the height (hgt) of the fuzzy variable X, with hgt(X) = $\lambda_X(a_1) = \lambda_X(a_2) = 1.0$, u and v are called the left and right interval of X respectively (see Figure 4). If $a_1 = a_2 = a$, then the function is a unimodal one having a unique global maximum point. The following short notation is adopted in the rest of the paper:

$$
X = \langle a_1; a_2; u; v \rangle \tag{5}
$$

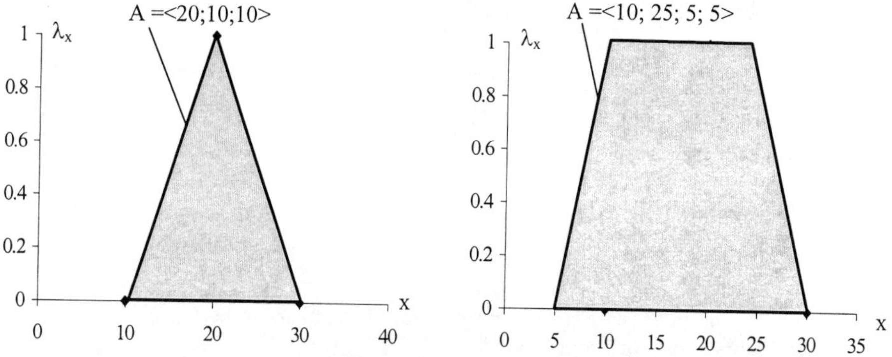

Figure 4: Linear LR-function

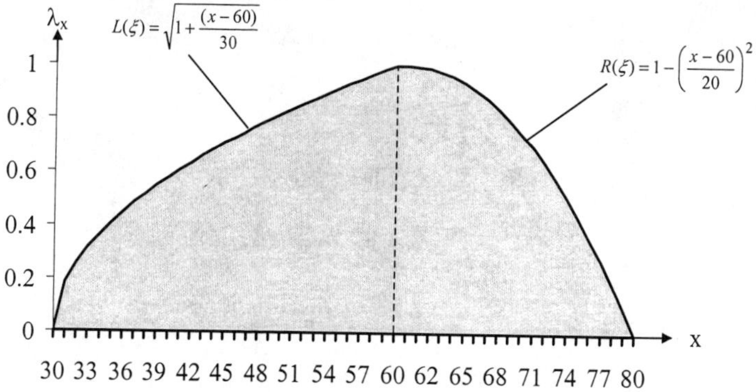

Figure 5: NonLinear LR-function

Type (2): NonLinear LR-Function

The definition of these functions resemble that of equation (4) with the exception that L(ξ) and R(ξ) are any arbitrary nonlinear analytical functions (see Figure 5). Again, here the following short notation is used in the subsequent sections:

$$X = \langle a_1; a_2; u; v \rangle_{NL} \tag{5}$$

The Centroid Method

This procedure serves as a tool for determining the required nominal value C_x from the weighted mean M_x. This process is in general terms a decision-making methodology in a fuzzy environment. In this case, a set of alternatives M_x and a fuzzy subset λ_{Mx} over M_x indicating the degree to which each alternative satisfies the desired decision criteria and goals. The centroid method (center of area method) is defined by:

$$C_x = \frac{\int_{Mx} \lambda_{Mx} \cdot M_x dM_x}{\int_{Mx} \lambda_{Mx} dM_x} \tag{6}$$

The Fuzzy-Deviation Method

To illustrate this method, consider the LR-function for the fuzzy variable A (see Figure 6). Then the expression

$$\delta_L(A) = \int_0^1 [a - X_L(\alpha)] d\alpha \tag{7}$$

is termed the left mean fuzzy-deviation and graphically represents the area to the left of a. The quantity

$$\delta_R(A) = \int_0^1 [X_R(\alpha) - a] d\alpha \tag{8}$$

is called the right mean fuzzy-deviation. And

$$\delta(A) = \delta_L(A) + \delta_R(A) \tag{9}$$

is known as the mean fuzzy-deviation of the fuzzy variable A.

Now, the nominal value can be determined in terms of the fuzzy-deviation:

$$C_x = a - \delta_L(M_x) \tag{10}$$

For the unimodal type (1), the mean fuzzy-deviation can be determined as follows:

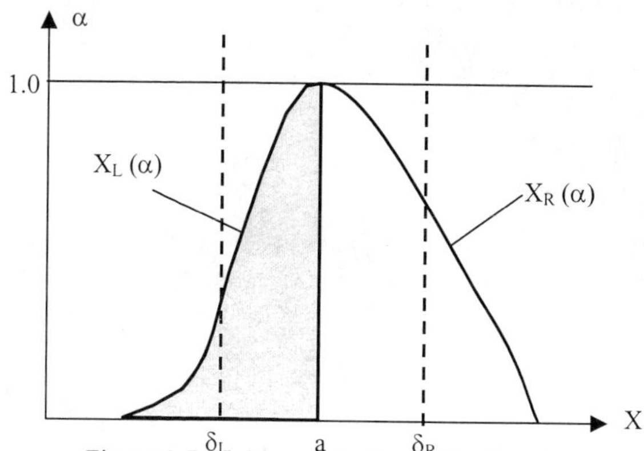

Figure 6: Definition of the Fuzzy-Deviation

$$\delta_L(A) = u\int_0^1 (1-\alpha)d\alpha = u/2 \tag{11}$$

In this case, the nominal value of the soil property is obtained from the support value M_{xi} for $\alpha = 0.5$.

FUZZY MEASURE

Fuzzy measure is customarily evaluated using the fuzzy index or fuzzy entropy. These actually measure the level of imprecision of a fuzzy variable. It answers the question: How fuzzy is a fuzzy variable? And it is a matter of degree. The fuzzy entropy of **A**, Ent(**A**) varies from 0 to 1. The mathematical description of fuzzy entropy can be obtained by different approaches. The classical definition uses the analog to SHAENON's random entropy:

$$Ent(A) = -\int\left[A\ln(A) - (1-A)\ln(1-A)\right] = -\int\left[A\ln(A) - (A^C)\ln(A^C)\right] \tag{12}$$

Yager(1979/80) introduced a fuzzy measure based upon the fact that for proper fuzzy sets the intersection C = $\mathbf{A} \cap \mathbf{A^c}$ is different from the empty set $\varnothing$, whereas for crisp sets one know that $\mathbf{A} \cap \mathbf{A^c} = \varnothing$. Hence the cardinality of C (Card(C)) is a suitable fuzziness measure for **A**. With

$$\lambda_C = \min\{\lambda_A(x), 1-\lambda_A(x)\}$$
$$= \frac{1}{2}\left[1 - \left|\lambda_A(x) - (1-\lambda_A(x))\right|\right] \tag{13}$$

For the universe X, Card (C) is defined by

$$card(C) = \frac{1}{2} \int_X \left[1 - \left| \lambda_A(x) - (1 - \lambda_A(x)) \right| \right] dx \tag{14}$$

Then, the fuzziness index is given as

$$\mathbf{F_1} = \frac{\int_X \left[\left| \lambda_A(x) - (1 - \lambda_A(x)) \right| \right] dx}{card(C)} \tag{15}$$

The nominator of this equation can be interpreted as a distance of the two sets **A** and $\mathbf{A}^c$, called the Hamming distance $\rho_1(\mathbf{A}, \mathbf{A}^c)$, with which equation (15) can be rewritten in the form:

$$\mathbf{F_1} = 1 - \rho_1(A, A^c) / card(C) \tag{16}$$

Alternatively one can use, e.g. the corresponding Euclidean distance

$$\rho_2(A, A^c) = \left[\int_X \left| \lambda_A(x) - (1 - \lambda_A(x)) \right|^2 dx \right]^{\frac{1}{2}} \tag{17}$$

to define the fuzziness index $\mathbf{F_2}$.

Based upon the above definitions, the fuzzy indices for type (1) and (2) fuzzy variables are:

$$\mathbf{F_1} = \frac{1}{2} \left(\frac{u + v}{a_m} \right) \tag{18}$$

$$\mathbf{F_2} = 1 - \sqrt{1 - \frac{2}{3} \left(\frac{u + v}{a_m} \right)} \tag{19}$$

Where $a_m = (a_1 + a_2)$; with $u < a_m/2$ and $v < a_m/2$

PRACTICAL EXAMPLE

It is required to determine the nominal (characteristic) value of the cohesion soil property in undrained condition (C_u). The results from the laboratory tests are given in table (1) below.

The fuzzy model will be implemented using the computation algorithm outlined above. First, Type (1) with a unimodal ($a_1 = a_2 = a$) function is defined:

$$C_{ui} = \langle a_i;\ u_i;\ v_i \rangle \tag{20}$$

Table 1: Sample results

(i)	C_u (kPa)	Frequency
1	180	3
2	140	2
3	170	2
4	230	1
5	260	1
6	270	1

To describe the intervals (u_i and v_i), one needs the additional information from experience and engineering judgement. For instance, assuming values between 0.25x(sample Standard Deviation) and 1x(sample Standard Deviation) represents a reasonable start.

$C_{u1} = <180; 15; 5>$
$C_{u2} = <140; 5; 5>$
$C_{u3} = <170; 10; 10>$
$C_{u4} = <230; 20; 10>$
$C_{u5} = <260; 30; 10>$
$C_{u6} = <270; 30; 5>$

The weighted mean M_x is determined from equation (21):

$$M_x = \frac{\left(\sum_{i=1}^{n} k_i (\langle a_i ; u_i ; v_i \rangle) \right)}{\sum_{i=1}^{n} k_i} \tag{21}$$

where k_i = weight, i.e. the frequency of C_{ui}. Applying equation (21) the fuzzy mean value of C_u is

$M_{Cu} = <192; 15.7; 7>$

The fuzzy indices for this mean value of C_u according to equation (18) and (19) are 0.06 and 0.04 respectively.

The graphical representation of this computational result is depicted in Figure (7). The next step is to estimate C_{Cu} from M_{Cu}. After the Centroid-Method, the nominal value of the undrained cohesion is C_{Cu} = 189 kPa and the Fuzzy Deviation Method gives C_{Cu} = 184 kPa.

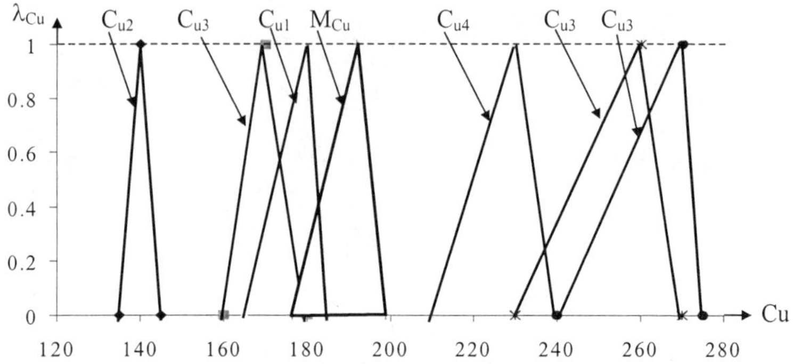

Figure 7: Type (1) Model for the nominal cohesion

Second, consider that $a_1 \neq a_2$, then type (1) of the fuzzy model is described by

$$C_{ui} = \langle a_{1i};\ a_{2i};\ u_i;\ v_i \rangle \tag{22}$$

$$M_x = \frac{\left(\sum_{i=1}^{n} k_i \left(\langle a_{1i}; a_{2i}; u_i; v_i \rangle \right) \right)}{\sum_{i=1}^{n} k_i} \tag{23}$$

$C_{u1} = \langle 170;\ 190;\ 15;\ 5 \rangle$
$C_{u2} = \langle 135;\ 145;\ 5;\ 5 \rangle$
$C_{u3} = \langle 160;\ 180;\ 10;\ 10 \rangle$
$C_{u4} = \langle 210;\ 235;\ 20;\ 10 \rangle$
$C_{u5} = \langle 230;\ 260;\ 30;\ 10 \rangle$
$C_{u6} = \langle 240;\ 270;\ 30;\ 5 \rangle$

Applying equation (23), $M_{cu} = \langle 178;\ 198.5;\ 15.5;\ 7 \rangle$. This is displayed graphically in Figure 8. Under this setting, the nominal value of the cohesion is

$C_{Cu} = 182.7$ kN/m^2 after the Centroid-Method
$C_{Cu} = 170.3$ kN/m^2 after the Fuzzy-Deviation Method

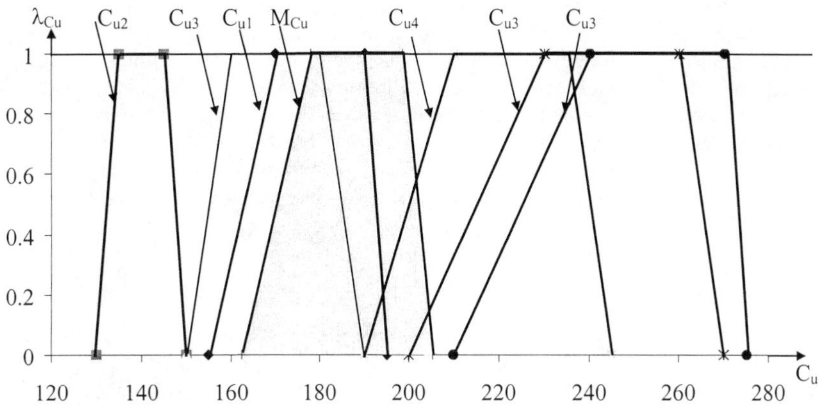

Figure 8: Type (1) Model (second alternative) for the nominal cohesion

The third alternative approach for estimating the nominal value of C_u is to apply Type (2) with the assumptions (i) $a_1=a_2=a$ and (ii) $a_1 \neq a_2$. In (i) the reference functions are given by (see Figure 5)

$$L(\xi) = \sqrt{1+(x-a)/u} \tag{24}$$

$$R(\xi) = 1-\left(\frac{x-a}{v}\right)^2 \tag{25}$$

The weighted mean M_{Cu} is computed using equation (21)

M_{Cu} = <192; 15.5; 7>$_{NL}$

The fuzzy index in this case based upon the Hamming distance and the Euclidean distance is given by $\mathbf{F_1} = 0.07$ and $\mathbf{F_2} = 0.05$ respectively.

The nominal value of C_u is then determined employing the Centroid-Method and the Fuzzy-Deviation method: 185 kN/m^2 and 171 kN/m^2 respectively.

The last case (ii) resembles the previous one except that each function has two global maximum points such that $\lambda_{Cu}(a_1) = \lambda_{Cu}(a_2) = 1.0$. The weighted mean M_{Cu} is determined by applying equation (22):

M_{Cu} = <178; 198.5, 15.5; 4>$_{NL}$

The desired nominal value is 176 kN/m^2 after the Centroid-Method and 168 kN/m^2 according to the Fuzzy-Deviation method.

Comparing these results with mean value (192 kN/m^2), it is seen that taking the mean as the nominal value of the cohesion is misleading and clearly overestimate the cohesion strength of the soil. On the other hand, the fuzzy-reliability approach suggests values ranging between 184 and 168 kN/m^2, in which the Centroid-Method tends to produce higher values than the Fuzzy-Deviation method. The linear models give values greater than its nonlinear counterpart. In this conjunction, it is worth mentioning that the fuzzy indices of linear models are less than the nonlinear models. Thus, a nonlinear model with the Fuzzy-Deviation method suggests being the most plausible tool.

CONCLUSIONS

Ambiguity in measured geotechnical properties is normally an acknowledged source of uncertainty that has been modeled using probability theory. According to this representation, assigned density functions describe the likelihood that each possible value or a variable currently is or will become the true value of that variable in the state of nature. This approach requires a sufficient amount of data and presumes a random nature of the geotechnical properties. These requirements are difficult to satisfy within the geo-materials context.

To characterize the reliability of the measured geotechnical properties the engineer needs to combine, in addition to the actual data, information on the quality of the data, knowledge about the geology, ground water, and other environmental aspects. Engineering judgement plays here heavily the crucial role.

The fuzzy modeling techniques suggested for the determination of the geotechnical nominal values are patronized upon the inclusion and handling of the uncertainty in the measured geotechnical properties through a combination of objective and mostly explicitly expressed data and from information that is eminently subjective. The results indicate that the proposed concepts and algorithms provide a feasible procedure for estimating the characteristic values of geo-materials.

APPENDIX-I. REFERENCES

Ang, H-S., and Tang, W.H. (1984). Probability Concepts in Engineering Planning and Design. V.2, John Wiley and Sons, New York.

De Groot, D. J. and Baecher, G.(1993). Estimating autcovariance of in-situ soil properties. ASCE, Journal of Geotechnical Engineering, Vol.119, No.1, pp.147-167.

Dubois, D. and Prade, H. (1980). Fuzzy Sets and Systems. Academic Press, London.

Kosko B. (1992). Neural networks and fuzzy systems. Prentice Hall-Englewood Cliffs.

Krumbein, W.C. and Greyhill, F.A. (1965). An introduction to statistical models in geology, MacGraw-Hill, New York.

Nahmias, S. (1978). Fuzzy variables. Fuzzy sets and systems I, pp.97-110.

Nawari, O. and Hartman, R. (1998). Fuzzy Logic Concept in the Limit States of Geotechnical Structures. The XI Danube-European Conference, 25-29 May, , Porec, Croatia, pp.855-862.

Nawari, O. and Liang, R. (1999). Characterization of Uncertainty in Underground Structures. Accepted for publication in the Third National Conference of the Geo-Institute of ASCE, June 1999.

Tang, W. H. (1984). Principal of probabilistic characterization of soil properties. Proc. of the Symp. On Probabilistic characterization of soil properties, ASCE, National Convention, Atlanta, GA, pp.74-89.

Tang, W. H. (1993). Recent development in geotechnical reliability. Proc. of the Conf. On Probability Methods in Geotechnical Engineering, Canbera, Australia, pp.3-27.

Vanmarcke, E.(1977). Probabilistic modeling of soil profiles. ASCE Journal of Geotechncial Engineering Division, Vol.103, NO. GT11, pp.1227-1246.

Yager, R. (1979): On the measure of fuzziness and Negation. Int. J. Gen. Sys.5, pp.221-229.

Zadeh, L. A.(1965). Fuzzy Sets. Inform. And Control, Vol. 8, pp. 338-353.

Zadeh, L. A. (1973). Outline of a New Approach to the Analysis of Complex Systems and Decision Processes, IEEE Trans. on Systems, Man and Cybernetics, Vol. SMC-3, pp. 28-44.

APPENDIX-II. SYMBOLS

a_1, a_2 = Supports at the height (hgt) of the fuzzy variable X.
C_x = Nominal (characteristic) value of a geotechnical material property
$\mathbf{F}_1$ = Fuzzy index based upon Hamming distance
$\mathbf{F}_2$ = Fuzzy index based upon Euclidean distance
M_x = Weighted mean value
X = Sample mean
S = Sample standard deviation
u, v = Left and right interval of the fuzzy variable X.
δ_L = The left mean fuzzy deviation
δ_R = The right mean fuzzy deviation
λ_X = Membership function of the fuzzy variable X.
μ = Population mean
σ = Population standard deviation

Performance and Evaluation of Roadways and Railways

'THE 'BOW-WAVE' EFFECT IN SOFT SUBGRADE BENEATH HIGH SPEED RAIL LINES'

Michael E. Heelis[1], Andrew C. Collop[2], Andrew R. Dawson[3], David N. Chapman[4], Victor V. Krylov[5]

ABSTRACT: This paper discusses the critical speed (the 'bow-wave' effect) and how it is influenced by soil stiffness. In order to reach the centers of major conurbation, modern high-speed rail lines are increasingly utilizing marginal land avoiding previously built housing areas. This land often has low soil stiffness. The critical speed of waves across such soils can approach the speed of modern high-speed trains. The large deflections caused can increase environmental noise and the track maintenance required and decrease ride quality. This paper discusses how the effect of train speeds can be predicted using computer modeling and an instrumented impact device. The modeling of possible remedies to the problem are discussed.
A convolution technique is used to model axles travelling at different speeds and the effects of multiple axles. The resulting displacements are highly dependent on the amount of damping introduced into the system modeled. In-situ testing, laboratory testing and finite element modeling are compared and are shown to provide consistent results. The conclusion is that a powerful analytical tool for the prediction of the 'bow-wave' effect has been devised.

INTRODUCTION

One method to reduce the visibility and noise pollution caused by high-speed rail lines is to limit the height of the embankments. However, these smaller embankments tend to be flexible and, hence, on soft ground, track-soil bending waves caused by the loading from the train may result in significant transient soil deflections. Ride quality and maintenance costs have a direct relationship to the magnitude of these deflections and, hence, there is a requirement for their accurate prediction.

Several analytical models can be used to predict the propagation of bending waves in the track/embankment system. A model of a Euler beam on a Winkler foundation was studied and used to demonstrate how the magnitude of the

[1]Dr Michael E. Heelis, Research Fellow, University of Nottingham, UK, NG7 2RD.
[2]DrAndrew C. Collop, Lecturer, University of Nottingham, UK, NG7 2RD.
[3]Andrew R. Dawson, Lecturer, University of Nottingham, UK, NG7 2RD.
[4]Dr David N. Chapman, Lecturer, University of Birmingham, UK, B15 2TT.
[5]Dr Victor V. Krylov, Professor, Nottingham Trent University, UK, NG2 7BU.

resulting displacements are dependent on train speed and track damping. The effect of the number of axles per bogie and the axle distances were also studied. In this paper it was shown that the model predicts realistic results when compared to data measured from in-situ tests. The methods by which the model parameters may be calculated are discussed and their relative advantages and disadvantages are considered. Alternative methods of calculation from vibration and impact tests are also investigated. From this study it is possible to form practical suggestions on methods by which a maximum train speed can be recommended to limit the maximum transient deflections as a train passes.

EULER BEAM ON A WINKLER FOUNDATION MODEL

The simplest approach to the problem outlined above is a model of a point load moving on an infinite beam supported on an elastic foundation comprising a series of discrete spings. This model is linear elastic and, therefore, superposition can be used to provide the solution for a multi-axle train. Because the springs are discrete there is no shear coupling along the axis of the beam. This is not the case for real (or even idealised) soils. However, it does provide a model that can be solved relatively simply in closed form (Fryba, 1972). The defining differential equation of this problem is,

$$EI\frac{d^4v(x,t)}{dx^4}+\mu\frac{d^2v(x,t)}{dx^2}+kv(x,t)=P\delta(x-ct) \tag{1}$$

where EI is the flexural rigidity of the beam resting on the foundation, v is the vertical displacement, x is the distance along the beam, t is time, k is the stiffness coefficient of the Winkler foundation, μ is the mass per unit length of the beam, P is the load and $\delta(x - ct)$ is the Dirac function of a point load moving with velocity c. The variables should be defined so that they are dimensionally consistent It has to be decided whether EI represents the bending stiffness of the rail only, of the rail and ballast, or even, of the rail, ballast and embankment. In order to produce finite oscillations at critical conditions a damping factor is introduced which is proportional to the vertical velocity of the deflection at any point and, hence, the equation used becomes (Fryba, 1972),

$$EI\frac{d^4v(x,t)}{dx^4}+\mu\frac{d^2v(x,t)}{dx^2}+C\frac{dv(x,t)}{dx}+kv(x,t)=P\delta(x-ct) \tag{2}$$

where C is the damping in the system.

The solution to the above equations is highly dependent on the velocity, c and the damping, C, of the system. In the solution, these are characterised by two non-dimensional parameters, α and β respectively, defined by:

$$\alpha = \frac{c}{c_{cr}} = \frac{c}{2\lambda}\left(\frac{\mu}{EI}\right)^{\frac{1}{2}} \text{ and } \beta = \frac{C}{2(\mu k)^{\frac{1}{2}}} \quad \text{(3 and 4)}$$

where λ is the non-dimensional stiffness of the soil and is given by:

$$\lambda = \left(\frac{k}{4EI}\right)^{\frac{1}{4}} \quad (5)$$

In this solution α is a non-dimensional measure of the train velocity taking the value of unity at the critical velocity c_{cr} (the speed of travel of bending waves), and β is a non-dimensional measure of damping taking a value of unity when the beam is critically damped.

Determination of Winkler model properties.

The problem then arises as to how to relate *k* and C to the properties of soils on which the railway will operate. Relationships between *k* and other elastic soil parameters have been made by Vesic (1963), Biot (1937) and Vlaslov (1957). It is also possible to use a Finite Element Model in which the soil is modelled more realistically than a 'bed' of springs, and compare deflections under a known applied load. In Heelis et al (1998) it was shown that the above approaches give answers which are of the same order of magnitude. For a particular embankment on the East Coast Mainline in the United Kingdom basic track and subgrade data has allowed the coefficient of subgrade reaction to be estimated. The elastic parameters of the subgrade are, E_s = 10 MN/m^2, and ν_s = 0.4 and the embankment is taken to be a beam with EI = 152.5 MNm^2 and width, B = 3 m (Hunt (1994), Figure 1). Hunt used a layered finite element model, to calculate a coefficient of subgrade reaction of 16.5 MN/m^2.

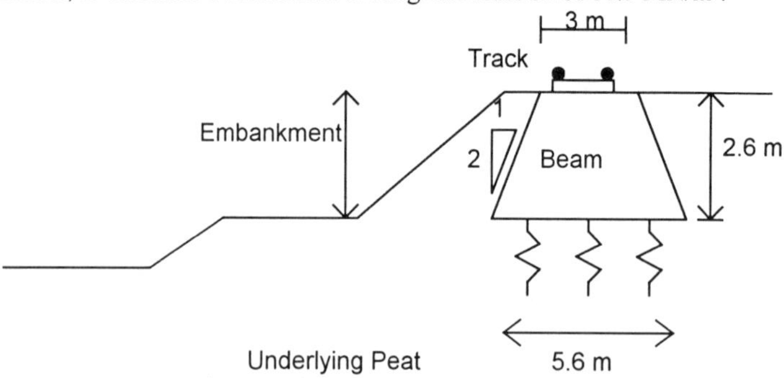

Figure 1. Cross-section of the embankment

Initially, if Fryba's model and solution are accepted as sufficiently valid, then it will be necessary to find the maximum displacement, slope, shear force, and hence, bending moment as functions of α and β in order that worst cases can be predicted. The use of a multi-axle model must also be considered, as it could be that certain axle spacing might effect the solution (for instance if the spacing coincides with the wavelength of the bending waves in the structure.

The relationship of damping models to real soil parameters is much more difficult. There appears little agreement on what model to use or what the parameters signify. Damping, it has been argued, is both dependent and independent of frequency (Crandall, 1970). Hunt (1994) concluded that material damping is associated with shear deformations and that damping losses associated with change in volume are negligible. Therefore, the mechanism by which the vertical deformations of the rail track produce shear deformations in the soil requires investigation. Heelis et al (1999) performed resonant column tests on soft ground from the site of the proposed Channel Tunnel Rail Link, UK, in order to measure the damping as well as the stiffness of the underlying peat material. An alternative method of measurement of overall system damping is to measure the attenuation of vibration with distance or time from an impact load. This measurement will be taken from actual rail/embankment systems and is, therefore, likely to lead to a more accurate estimation of damping of bending waves travelling along the rail track. This will be discussed in a later section of this paper.

Results from Winkler model.

The numerical solution of deflection directly beneath the travelling load can be plotted in terms of the non-dimensional parameters, α and β, as in Figure 1. This allows the prediction of the dynamic amplification factor due to increasing train speed. Note that within 15% of the critical speed the displacements can be approximately doubled for low levels of damping.
Closer inspection of the deflection patterns under a single axle, (considered in more depth later) shows that the maximum deflection in the system is not always directly underneath the applied load.

In Figure 3 the maximum displacement at any point in the system is plotted for a single travelling load. Note that the reduction in displacements for super critical trains is not as pronounced as when the displacement is measured directly underneath the load (Figure 2). This leads to the conclusion that for sub-critical speed trains the maximum deflection is underneath the applied load but as critical speed is approached, and for speeds in excess of the critical speed, the maximum deflection does not occur under the applied load.

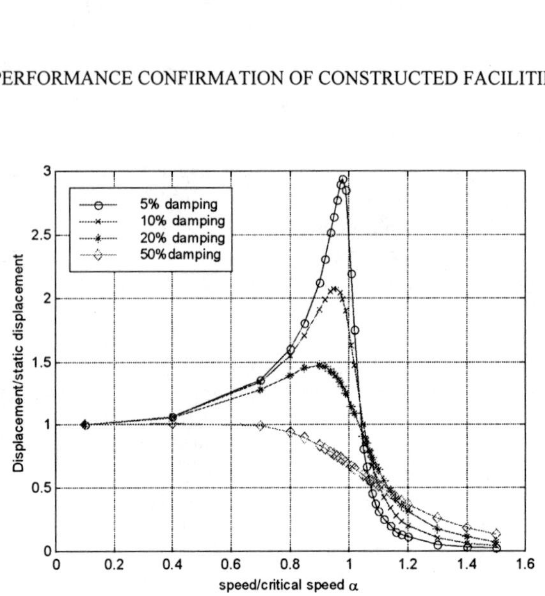

Figure 2. Non-dimensional plot of vertical displacement directly under a single axle.

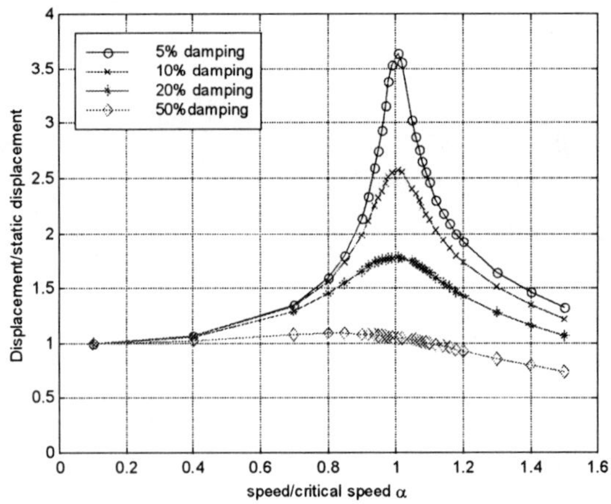

Figure 3. Non-dimensional plot of maximum vertical displacement.

However, it is possible using a convolution technique, outlined in Hunt (1988), to model several axles of different loads crossing a particular foundation. Figure 4 shows a plot of the predicted deflection characteristics for a load travelling at 90% of critical speed for the ECML site discussed by Hunt

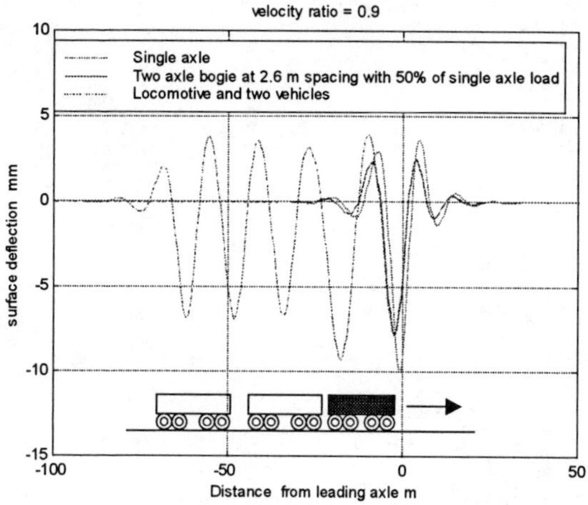

Figure 4. Response from travelling loads.

(1994) (see above). The dashed line represents the deflection due a single axle (of 40 tonnes) travelling at this speed and the solid line represents the response due a two axle bogie sharing the same load as before. In both cases the leading axle is at 0 m. The dash-dot line is the line for a locomotive with two 40 tonne bogies and two following double bogie vehicles of 64 tonnes or 16 tonnes per axle. (in each case axle spacing in the bogies is 2.6 m and, for the train bogies are spaced at 0,14,18,32,46 and 60 m. As expected, distributing the load over two axles decreases the deflections in the subgrade, however, increasing the number of bogies increases the deflection, especially under the trailing bogie of the locomotive. Between the bogies, there is also an increase in the upward deflection experienced by the railway line. It is hypothesised that large upward deflection will tend towards a tension condition in the top-most ballast due to bending and, therefore, increased maintenance problems.

Similar analyses, although at different speeds to those performed to produce Figure 4, were carried out in order that the effect of train speed and axle configuration could be analysed. In Figure 5 these are plotted with displacement being normalised by the deflection under a single static axle. The track structure is the same as that considered for Figure 5. In this case, the maximum displacement under the simulated train is a maximum at a train speed in excess of the critical speed of the foundation. This is due to the cumulative interference of the responses due to the different axles at this particular speed and on this particular foundation.

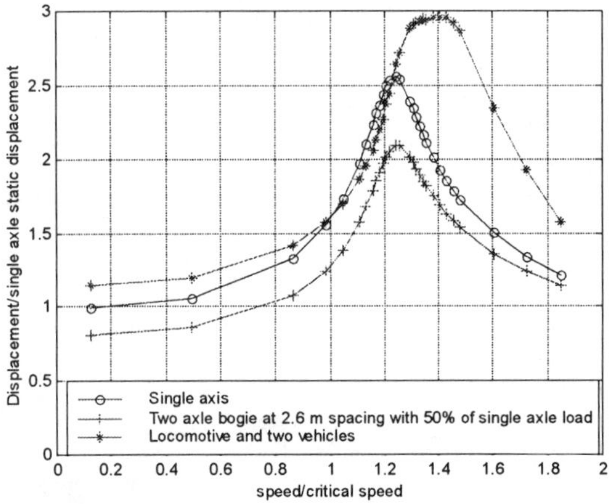

Figure 5. Maximum displacements against train speed for different axle arrangements.

IN-SITU MEASUREMENTS

Field trials have been conducted both in Ireland and the U.K. to measure the critical speed of railway lines as well as the displacements caused by trains as they approach the critical speed of a particular section. The acceleration of the track was measured by a series of accelerometers attached to the ties on the track. The output of the accelerometers was double integrated in order to give displacements. In Figure 6 the displacement from a locomotive with 7 passenger cars is plotted. The similarity with the displacement calculated for Figure 3 should be noted. The train was crossing a section of track under which there was over 5 metres of soft peat material and thus the train was travelling at approximately 50-70% of the critical velocity. This can be estimated by comparing the predicted deflection patterns at a wide variety of speeds with the measured pattern. It should be noted that, as expected, the maximum displacement is under the heavy locomotive and the upward displacement between the bogies of the locomotive is almost equal in magnitude to the downward displacement under the leading bogie. The measured accelerations peaked at approximately 5 g, however this was for a relatively high frequency vibration. Investigations are under way to examine if peak deflections or peak accelerations contribute more to misalignment and therefore increased maintenance costs. Using such measurements from a variety of European sites Woldringh and New (1999) have plotted the ratios of maximum/static displacement against train speed and found that they line on a single curve. This supports the use of the Winkler model for the present problem.

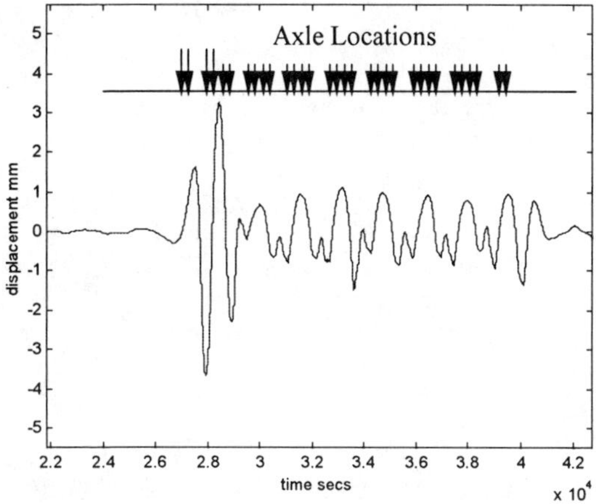

Figure 6. Measured Displacements under a 70 mph train.

The beam on a Winkler foundation model is used in this analysis to predict the impulse response of the rail/track system. Obviously, there are other means by which this impulse response may be found. Finite element modelling is an effective tool in analysing such a problem. With such a model the effect of different materials and layered systems can be easily analysed. Such work is ongoing at the University of Nottingham, particularly with a view to verifying the predictions based on the Winkler model previously discussed.

It is also possible to measure the impulse response of the track system directly. This can be done either by a vibration system or by the use of an impact hammer. The University of Nottingham has developed an impact system using an array of accelerometers approximately 5 – 8 metres away from the impact device. The impact tests performed have used an array of 4-8 accelerometers to pick up the response of the foundation from a 75 kN impact of approximately 10 ms duration. Figure 7 presents the traces obtained at a distance of 0.7 and 5 m from the impact for another site in England. From the time delay between the maximum accelerations, a wave speed can be calculated as (1.4-0.7)/(0.0223-0.0198) = 180 m/s = 400 mph. It has been suggested previously that line speeds should have a 30% safety margin to this wave speed (Woldringh and New, 1999). This means that the maximum line speed for this particular section of track would be 280 mph.

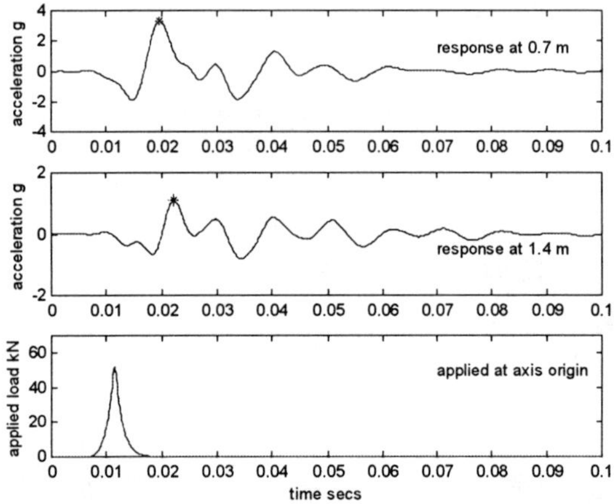

Figure 7 Response to a hammer loading of 75 kN for 10 ms.

If the magnitude of the acceleration Transfer function normalised by the applied force for an impact test is plotted (Figure 8) then it is possible to make an estimate of the damping involved for any particular peak (Inman, 1994). For example, the maximum peak of Figure 8 is at 101 Hz, the frequencies whose amplitudes are 3 dB down from the maximum of the peak are at 95 and 106 Hz. This gives a damping ratio for this particular peak as (106-95)/(2x101) or 0.054. The peak at approximately 50 Hz can be shown to have 0.12 damping ratio. This is equivalent of taking the damping ratios of $\beta = 0.12$ for the Winkler model. This leads us to be confident about the proportion of damping used in the simulations of the multiple axle model. If displacement measurements are required then a double integration is required. This means the lower frequency components of the acceleration will become more significant therefore it can be argued that the damping coefficient should be taken from the peak at 50 Hz.

POSSIBLE SOLUTIONS

The routine solution adopted to the problem of excessive deflections caused by high speed trains crossing soft ground is to limit the train speed, With increasing demand by operators for high speed travel construction rather than operational solutions must be found.

The non-dimensional nature of the solution suggested by Fryba (1973) means that suggestions can be provided about how to increase this critical speed of bending wave propagation in the rail/track system so as to limit the magnitude of transient vertical displacements at the train operating speed. The critical speed is given by the following formula, (Equations 3 and 5):

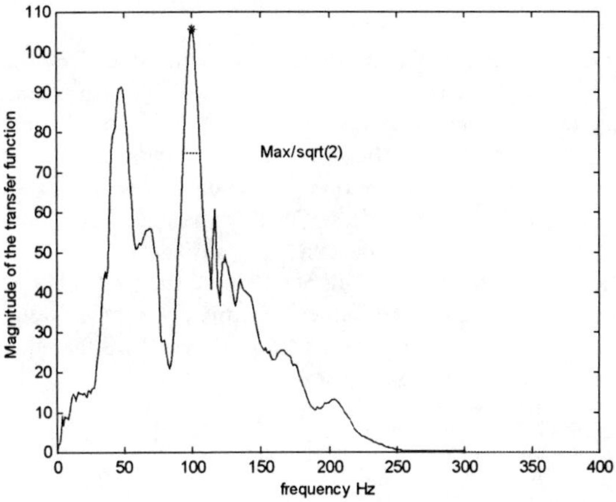

Figure 8. Typical acceleration Transfer function for an impact test.

$$c_{cr} = \frac{1}{\sqrt{2}}\left(\frac{kEI}{\mu^2}\right)^{1/4} \tag{7}$$

Hence, in order to increase the critical speed (and thus reduce the deformations to levels at the left of Figure 1) either, the stiffness of the subgrade can be increased, or the flexural rigidity of the beam can be increased, or the mass per unit length of the beam can be reduced. Increasing the thickness of the embankment (for example, by partial subgrade replacement) could increase the flexural stiffness. However, this will also increase the mass per unit length of the beam. An optimal design method could be to increase the thickness of the effective beam by the insertion of a lightweight material of sufficient durability, thus increasing the second moment of area of the embankment without adding additional mass. Alternatively, the damping in the system could be increased, this might be achieved by adding a viscous damped layer. It has been suggested that a bitumen bound ballast or sub-ballast layer might achieve this effect. Another practical means of achieving such a solution is to use piles through soft soils whereby loads are transferred downwards to a stiff 'subgrade' level. This effectively moves the vibrations to a layer in which the velocity of bending waves would be very high so that the critical speed is raised beyond the range of concern.

CONCLUSIONS

An analytical solution to the problem of dynamic amplification of the vertical displacements of a railway track has been outlined. The displacement of the track has been shown to increase with increasing train speed until a 'critical' velocity is exceeded. The critical velocity depends on the soil stiffness, the track stiffness and the mass of the foundation system. Using parameters derived from a soft site on the East Coast Mail Line in the United Kingdom, the displacements due to a single axle, a bogie and a hypothetical train have been predicted and compared to surface displacements from a soft site in Ireland. The use of impact and vibration tests for the direct measurement of the critical velocity for particular sections of track has been investigated and is on-going work at the University of Nottingham.

The different approaches suggest that Fryba (1973) model can produce responses to dynamic and travelling loads of the same type of those experienced in-situ. There are difficulties in finding the right values of the parameters to put into the model but in-situ testing and back analysed response to train loading gives confidence that the values obtained by the means described are applicable.

ACKNOWLEDGEMENTS

The authors would like to acknowledge the support of the Engineering and Physical Science Research Council for funding the project, and to Iarnrod Eireann (Irish Rail) and Railtrack for site availability.

APPENDIX I. REFERENCES:

Biot A.M. (1937), 'Bending of an infinite beams on an elastic foundation', J. Appl. Mech., 4, 1.

Crandall S.H. (1970), 'The role of damping in vibration theory', J. Sound Vib, 11 (1), 3-18.

Fryba L. (1973), Vibration of solids and structures under moving loads, Noordhoff Pub.

Heelis M.E., Dawson A.R. Chapman D.N. Krylov V.V., & Collop A.C., Predictions of vertical track displacements caused by the passage of high speed trains over soft soils, Proceed. 10th International Scientific Conference, Communications on the edge of the Millennium, Zilina, Slovak Republic.

Heelis M.E., Dawson A.R. Collop A.C., Chapman D.N. & Krylov V.V., Resilient Modulus of soft soil beneath high speed rail lines', TRB 1999.

Hunt H.E.M. (1988), 'Measurement and modelling of traffic induced ground vibration', Ph.D. Thesis, Cambridge University, England.

Hunt G.A. (1994), 'Analysis of requirements for railway construction on soft ground', British Railway Report No. LR TM 031, Derby, England.

Inman D.J. (1994), Engineering Vibration, Prentice Hall Int.

Vesic, A.B. (1963), 'Beams on elastic subgrade and Winkler hypothesis', Proc. 5th Intnl Conf. Soil Mech. Found. Engng, Paris, Vol. 1, 845-850

Vlaslov V.L. and Leontiev N.H. (1956), Beams, planes and shells on elastic foundations'.

Woldringh R.F. and New B.M, (1999), 'Embankment design for high speed trains on soft soils' Geotechnical Engineering for Transportation Infrastructure, Balkema Rotterdam

APPENDIX-II. SYMBOLS USED

μ = the mass per unit length of the beam
λ = the non dimensional foundation stiffness parameter
β = the non-dimensional damping parameter
α = the non-dimensional speed parameter
$\delta(x - ct)$ = a point load moving with velocity c
c = velocity (m/s).
C is the damping coefficient in (Ns/m/m)
c_{cr} = the critical speed of the bending waves in the foundation (m/s)
EI = the flexural rigidity of the beam resting on the foundation (Nm^2)
k = the stiffness coefficient of the Winkler foundation (N/m/m)
P = the load (N)
t = time (s)
v = the vertical displacement (m)
x = the distance along the beam (m)

GEOSYNTHETIC CAPILLARY BARRIERS IN PAVEMENTS

Karen S. Henry, Ph.D., P.E., Member[1], John C. Stormont, Ph.D., P.E., Member[2] and Robert D. Holtz, Ph.D., P.E., Fellow[3]

ABSTRACT: "Capillary barrier" is the term for a layer of large-pored material (soil and/or geosynthetic) placed in a finer-pored soil to 1) reduce unsaturated flow of soil water upwards and/or 2) reduce or prevent water from infiltrating from the overlying fine-pored unsaturated soil into the soil below the capillary barrier. In the latter case, the capillary barrier is sloped so that the infiltrating water flows downwards along the interface between the capillary barrier and the overlying soil. The same physical principles govern the behavior of capillary barriers in both cases. This paper reviews capillary theory and what is known about capillary barriers (made of soil and/or geosynthetics) in a variety of applications. This information is then applied to the potential use of geosynthetic capillary barriers in pavements to reduce frost heave and to divert infiltrating water. The use of a geocomposite capillary barrier drain (GCBD), consisting of a drainage net sandwiched between geotextiles that possess certain hydraulic properties, appears promising. In addition to reducing frost heave, a GCBD may also help drain the soil above it before it becomes fully saturated. It may also be designed to drain saturated pavements at critical times of the year, such as spring thaw.

INTRODUCTION

"Capillary barrier" is the term for a layer of large-pored material (soil and/or geosynthetic) placed in a finer-pored soil to 1) reduce unsaturated flow of soil water upwards and/or 2) reduce or prevent water from infiltrating from the overlying fine-pored unsaturated soil into the soil below the capillary barrier. In the latter case, the capillary barrier is sloped so that the infiltrating water flows downwards along the interface with the capillary barrier (e.g., Ross, 1990) (Fig.1). Because both cases involve unsaturated flow, the same physical principles govern the capillary barrier behavior. We are interested in the use of geosynthetic capillary barriers in pavements to perform both functions: to reduce frost heave by preventing upward water flow

[1]U.S. Army Cold Regions Research and Engineering Laboratory, Hanover, NH.
[2]Department of Civil Engineering, University of New Mexico, Albuquerque, NM.
[3]Department of Civil Engineering, University Washington, Seattle, WA.

and to divert infiltrating water, thus protecting underlying soil layers from becoming saturated. In this article we summarize the knowledge gained from theoretical and experimental studies with capillary barriers, then apply it to the use of capillary barriers in pavements.

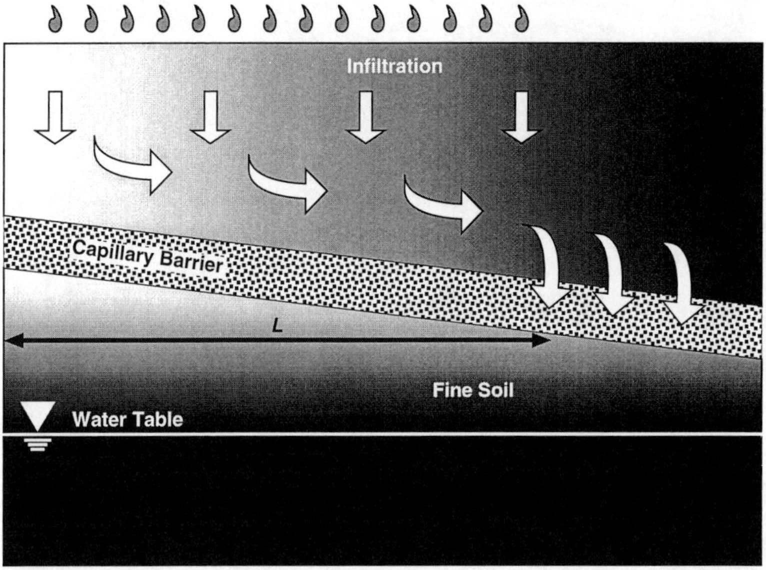

Figure 1. Capillary barrier to cut off upward flowing water and divert infiltrating water.

THEORY

Capillary Barriers

Capillary barriers function because there is a contrast in hydraulic conductivity at the interface between the finer and coarser-pored materials (usually soils) when they are unsaturated (Fig. 2). Suction head is the tension (negative pressure) of water in the pores of an unsaturated soil, expressed as a head of water. Suction head decreases with increasing water content and becomes zero when a soil is fully saturated. As shown in Figure 2, at relatively high soil suction heads, the fine soil has a measurable hydraulic conductivity, whereas the hydraulic conductivity of the coarse layer is immeasurably small. With increasing water content and decreasing suction head, the hydraulic conductivity of the fine layer increases gradually while that of the coarse layer remains negligible. Under these conditions, water flowing downward will not move from the fine layer into the coarse layer but will instead

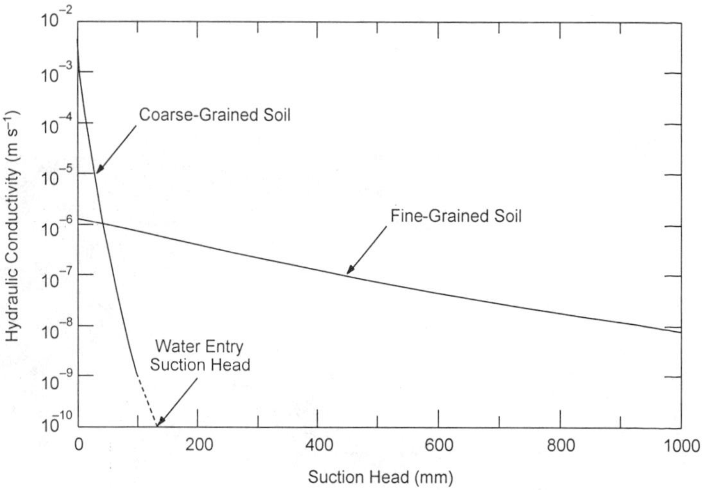

Figure 2. Typical hydraulic conductivities of fine- and coarse-grained soils used in capillary barrier. (After Stormont and Anderson, 1999.)

increase the water content of the fine layer. If the interface is tilted, water flows downslope in the finer soil above the interface in response to the gravitational potential gradient (e.g., Miyazaki, 1988).

Breakthrough of water into the coarse layer occurs when the suction head at the contact equals the water entry suction head or breakthrough head of the coarse layer (Hillel and Baker, 1988). The water entry suction head is the suction head associated with the movement of water into the smallest pores that form a continuous network through an initially dry soil. The larger the pores, the lower the water entry suction head. For clean, uniform gravel the water entry suction head may be on the order of a few centimeters. When the suction head decreases below the water entry suction of the coarse soil, the hydraulic conductivity of the coarse layer will increase rapidly and eventually exceed that of the fine layer. Under these conditions, the capillary barrier layer no longer impedes water movement.

Preventing Upward Rise of Water

Henry and Holtz (1997) presented theory pertinent to the use of porous capillary barriers to stop the upward, unsaturated flow of soil water. Briefly, the tension (or suction) head, h_t, of soil water is defined by:

$$T = (-P_w) = \lambda_L \, h_t \qquad (1)$$

where T is the soil water tension or the negative pore water pressure (Pa), and γ_L is the unit weight of water (N m^{-3}).

Based on the assumption that water above the water table is due only to capillary rise (i.e., no infiltration has recently occurred), h_t is simply the distance above the water table for any location within the capillary fringe. Lambe (1950) noted that there is a maximum height at which soil water is still in hydraulic contact with the water table (i.e., a maximum height to which Eq. 1 can be applied). Assuming that soil pores can be modeled as cylinders, this maximum height, $h_{t_{max}}$, will be controlled by the smallest pores in the soil, according to Eq. 2 (Castellan, 1983):

$$h_{t_{max}} = \frac{2\psi_{aw}(\cos\alpha)}{r\gamma_L} \qquad (2)$$

where ψ_{aw} is the surface tension of the liquid (in this case, water) (N m^{-1}), α is the contact angle of the liquid with the pore wall and r is the pore radius (m). Therefore, the maximum soil moisture tension at which water can be held in the pores, T_{max}, is defined by substituting Eq. 2 into Eq. 1:

$$T_{max} = \gamma_L h_{t_{max}} = \frac{2\psi_{aw}(\cos\alpha)}{r} \,. \qquad (3)$$

Thus, if a capillary barrier is placed above the water table in a soil, water will not enter it unless the water has a moisture tension less than T_{max} as defined by Eq. 3 for the capillary barrier. In other words, if the base of the capillary barrier layer is placed above the water table at a distance greater than $h_{t_{max}}$, the water from below cannot enter the smallest pores of that material because the pores are too large to support the water at that tension. The capillary barrier thus "breaks" the capillary rise of water.

Based on the above analysis, the maximum capillary rise of water into a porous material should be equal to its water entry suction head. However, this is an idealized concept—soil pores are not cylindrical and there is considerable wetting–drying hysteresis in porous materials. Baker and Hillel (1990) found that the water entry suction at which water entered a coarse-pored soil in measurable amounts was somewhat less than the water entry suction estimated by observing the capillary rise of water in it. Therefore, water entry suction head should not be equated to observed capillary rise until more is known about this physical process. However, these equations serve as an initial conceptual model.

Diverting Downward Infiltrating Water

Lateral diversion of downward infiltrating water provides a means to remove or drain soil water from the fine layer above a capillary barrier. Lateral diversion is gravity-driven unsaturated drainage within the fine layer. Because the water content in the fine layer is usually greatest near the interface with the coarse layer, and the hydraulic conductivity of an unsaturated soil increases with water content, lateral diversion is concentrated near the fine/coarse interface. Laterally diverted water will increase the water content downslope. The diversion length is the distance along the interface that water is diverted before there is breakthrough into the coarse layer (Fig. 1).

The diversion length for conventional fine-over-coarse soil arrangements have generally been limited to small values (e.g., much less than 10 m, Nyhan et al., 1993) when soil such as silt is used as the fine layer. The relatively low hydraulic conductivity of these soil layers limits the amount of water that can be transported under unsaturated conditions, and thus limits lateral drainage capacities.

Lateral drainage can be increased substantially by placing an intermediate layer between the overlying soil and the underlying capillary barrier layer (Morris and Stormont, 1997). The intermediate layer, referred to as a transport layer, should be conductive enough to laterally divert or drain downward moving water, yet remain unsaturated so as to preserve the capillary break with the underlying coarse material. Fine sand (K_S of 2.1×10^{-4}) has successfully been used as the transport layer in experiments (Stormont, 1995, 1996) and in numerical simulations (Morris and Stormont, 1997) of landfill capping applications. The success of lateral drainage using this configuration for landfill capping has led to consideration of other applications, including the use of capillary barriers with transport layers within pavements (e.g., Stormont and Stockton, in review).

GEOSYNTHETIC CAPILLARY BARRIERS

In order to design with geosynthetics as capillary barriers in unsaturated environments, their unsaturated hydraulic properties should be measured. The first consideration is whether a particular geosynthetic material is "wetting" or "non-wetting." The wetting of a solid by a liquid is quantified by the contact angle that the liquid makes with the solid. Properties of the outermost layers of the solid and anything that affects the surface tension of the liquid influences the contact angle (e.g., Aveyard and Haydon, 1973). Chemical and physical heterogeneity and the presence of films on the solid's surface will influence the contact angle, as will the presence of solutes in the liquid and temperature. Soil particles (e.g., silica) typically wet easily with contact angles of 0°.

Henry and Patton (1998) measured the dynamic (i.e., moving) contact angle of water on fibers from two geotextiles as received from the manufacturer and after they had been cleaned (dry cleaning followed by rinsing in water). They found that both fiber type and surface finish influence contact angle; and, that there was hysteresis between advancing and receding (wetting and drying) contact angles. Henry and Patton (1998) also measured water capillary rise in strips of the geotextiles for which the fiber contact angles had been measured and found that both contact angle and pore geometry measurably influence the capillary rise of water in the geotextiles. Henry and Holtz (1997) placed soil fines in geotextiles and found that this decreased the contact angle of water on the fibers (as opposed to changing pore size significantly), making them more wettable. The soil fines apparently cling to the surfaces of the fibers (e.g., Henry et al., 1990).

For materials that wet, the value of the water entry suction head is a critical parameter. Stormont et al. (1997) determined soil moisture retention functions of four needle-punched polypropylene geotextiles. From the moisture retention functions (saturation vs. suction head relationship) measured on geotextiles received from manufacturers, the cross-plane water entry suction heads were estimated to range from 0 to 30 mm. However, water entry suction heads estimated for the same geotextiles placed in soil columns ranged from 60 to 70 mm (Henry and Holtz, 1997). This difference may be due to differences in the measurement techniques or possibly to some soil fines that entered the geotextiles in the tests by Henry and Holtz, resulting in increasing the water suction entry head. Nonetheless, the water entry suction heads measured by both methods indicate that the soil moisture suction head required to induce water flow into the geotextiles tested, when they are initially dry, lies between 0 and 80 mm.

Stormont et al. (1997) also measured significant hysteresis in water retention functions of all four of the geotextiles. For two of the geotextiles, suction heads had to reach 0–20 mm before water would enter dry geotextiles; when the geotextiles were wet, suction heads had to reach at least 300 mm to initiate significant desaturation.

In summary, wetting–drying hysteresis, contact angles (as influenced by fiber type, surface properties or finish, the presence of soil fines, and whether or not it is moist), and pore geometry all affect the capillary behavior of water in porous materials. French et al. (1982) described the capillary behavior of a heat-bonded polypropylene geotextile that illustrates the influence of all of these factors. The geotextile that they tested initially retained water until a positive pressure head of 150 mm was reached; however, once the geotextile was wet, a positive pressure head of 80 mm could easily induce flow through it (wetting–drying hysteresis). Any localized pressure (decreased pore size) or wetting agent (decreased contact angle) reduced its ability to support a positive head of water, and once the fabric was wetted and dried, it would not sustain the same head (contact angle as affected by a surface finish).

Because geotextiles will be subjected to moistening and the emplacement of soil fines when used in pavements, both of which significantly reduce water entry suction in geotextiles, they may not be good capillary barriers. We discuss several experimental observations below that also support this observation. Therefore, a geocomposite, consisting of a drainage net with geotextile filters, is likely to be a better capillary barrier than geotextiles alone. In this case, if one or both of the geotextiles on the geocomposite become moist and/or soil fines become emplaced in them, the drainage net is the capillary barrier. A geocomposite may also help drain water at times that positive pore water pressures induce flow. If positive drainage is desired, the system should be designed so that the drainage net layer dries in order to function again as a capillary barrier once infiltration stops (or is greatly reduced) and the adjacent soil becomes unsaturated. Stormont and Anderson (1999) demonstrated that after water breaks through gravel capillary barriers, they are restored when the soil-capillary barrier system is dried.

CAPILLARY BARRIER APPLICATIONS

Reducing Frost Heave in Pavements

Taber (1929) used a layer of Ottawa sand (25 mm thick, but size of sand not given) located above the water table in soil-freezing experiments to prevent ice segregation in a frost-susceptible clay. Indeed, coarse sand and gravel capillary barriers have successfully reduced frost heave of roads and minimized the cost of construction (Casagrande, 1938; Beskow, 1946; Rengmark, 1963; Taivenen, 1963). In this application, the capillary barrier is placed on in-situ frost-susceptible soil, then backfilled with the same frost-susceptible soil. This is an alternative to replacing the frost-susceptible soil with non-frost-susceptible granular material for a greater thickness and at significantly greater expense.

When soil-freezing tests were first conducted in the laboratory using geotextiles as capillary barriers, results indicated that certain geotextiles significantly reduced frost heave (e.g., Hoover et al., 1981; Allen et al., 1983; Henry, 1991). However, Henry (1998) later found that moist geotextiles containing soil fines did not prevent frost heave, as discussed below.

To study the reduction of frost heave caused by geosynthetic capillary barriers, Henry (1998) froze cylindrical, frost-susceptible soil specimens from the top down with water freely available at the base and then measured frost heave. The tests were conducted at thermal conditions representative of a wide range of climates in which pavements seasonally freeze. Disks of either geocomposites or geotextiles were placed in a horizontal position to function as capillary barriers. The soil specimens were 150 mm high. The geosynthetic, if present, was about 30 mm above the bottom, and the water level was about 25 mm above the bottom. Four geotextiles and two geosynthetic nets were used to form geocomposite capillary barriers for use in freezing tests. The geotextiles were needle-punched polypropylene with mass per unit area values of 340 and 540 g m^{-2}. The nets had a minimum thickness of 6 mm and minimum pore dimensions of 6 mm. Geocomposites were made by placing geotextiles on each side of a net.

Henry (1998) found that geotextiles containing soil fines and moistened to a saturation of 30–40%, by being initially saturated then dried, failed as capillary barriers—i.e., they did not reduce frost heave. Furthermore, this result was not product dependent. However, even when geocomposites were subjected to the same moistening and soil fine emplacement as the geotextiles, they significantly reduced frost heave when the fine-grained soil above it was 75% saturated or less (Fig. 3). For the test soil, 75% saturation corresponds to a soil moisture suction head of about 2000 mm. However, the geocomposite capillary barriers did not reduce frost heave when the percentage of saturation of soil above the geocomposite was initially greater than or equal to 81%. At this degree of saturation, the soil moisture suction head is approximately 1000 mm. Again, the results were not dependent on specific geotextiles or geosynthetic nets used in the geocomposites tested.

Unsaturated Drainage with a Geocomposite Capillary Barrier Drain

Previous studies on lateral diversion of infiltrating water in unsaturated soils above a capillary barrier have led to the development of the Geocomposite Capillary Barrier Drain (GCBD) (Henry and Stormont, in prep.). The GCBD provides a means to drain water while soil pore pressures remain negative and may have potential to drain pavement layers. The GCBD comprises three layers that are, from top to bottom, a *transport* layer, a *capillary barrier* layer, and a *separator* layer. Certain nonwoven geotextiles can be used as transport layers, while a geonet with relatively large, open pores is the capillary barrier. The separator layer, such as a nonwoven geotextile, simply prevents underlying soil from intruding into the pores of the capillary barrier. A nonwoven geotextile may serve this function. (This configuration may also be able to laterally drain upward moving water. In this case, the lower layer would serve as the transport layer.)

The GCBD and the geocomposite to limit frost heave described above are similar in that they both consist of a geonet (serving as a capillary barrier) sandwiched between two geotextiles. The transport layer geotextile of the GCBD, however, is intended to become wet and consequently hydraulically transmissive at as large a suction head as possible. Stormont and Stockton (in review) report measurements of transmissivity under suction of wetting geotextiles.

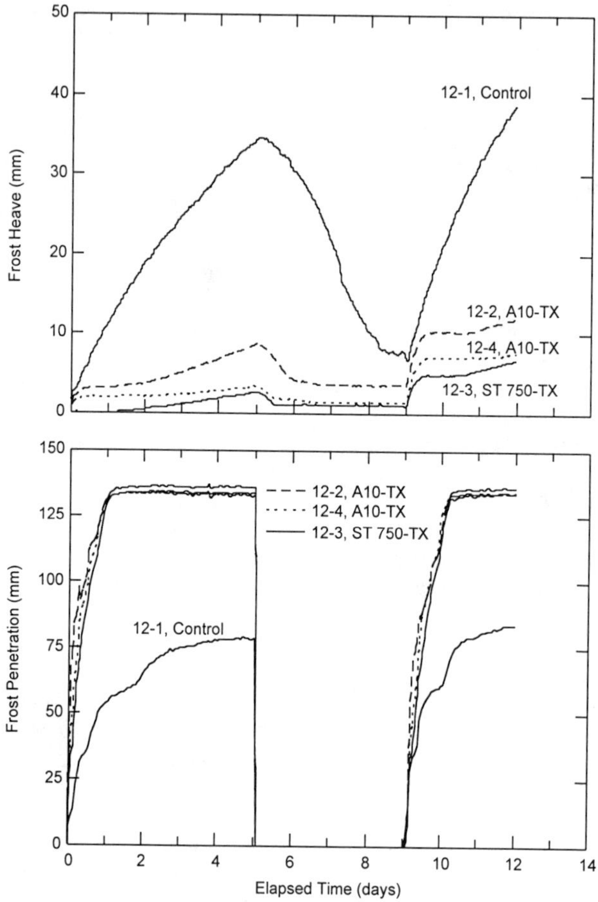

Figure 3. Frost heave and penetration vs. time for frost-susceptible soil initially compacted to 75% saturation. Specimen 12-1 was a control specimen and 12-2, 12-3 and 12-4 contained geosynthetic capillary barriers consisting of nonwoven geotextiles sandwiching a drainage net. Henry et al. (1998) provide specific information about the construction of the geocomposite capillary barriers as well as test details.

Although this geocomposite outwardly resembles a conventional geocomposite drain, a GCBD is designed to drain water in the geotextile (not the geonet) under negative water pressures (not positive water pressures). Further, it does not require the underlying impermeable layer that a conventional drain requires. In the GCBD configuration, it is the unsaturated hydraulic properties of the transport layer and capillary barrier that are of principal importance.

A potential application of such a system in pavements would be between a base course and subgrade. In this configuration the transport layer would accept and laterally drain downward moving water within the base course before positive pressures are reached. Further, such a system would prevent or limit the amount of water that reaches the subgrade.

Stormont and Stockton (in review) tested the drainage capacity of two GCBD systems in a 3-m-long soil box that was tilted to achieve various slopes from 2.5% to 9%. The profile tested was 100 mm of subgrade soil, a prototype GCBD, and 50 mm of pavement base material. Water was added to the overlying base material at constant rates as high as 5×10^{-5} cm/s (4.3 cm/day), which represents the 24-hr storm intensity with a 2-yr return period in much of the southwestern U.S. (Anderson and Stormont, 1997). Measurements were made of water drained out of the GCBD, water laterally drained in the overlying soil, and water produced out of the bottom of the subgrade soil. Measurements were also made of soil suction above and below the GCBD. Water content measurements were made in the subgrade soil after the tests.

For most test conditions, the GCBD system drained water under negative pressures at a rate sufficient to prevent any positive water pressures from developing in the base course. The system also prevented the infiltrating water from moving beneath it into the subgrade soil. The GCBD drained water under suction heads of about 50 mm. A test conducted with just the transport layer (without the underlying capillary barrier layer) revealed that lateral drainage does not occur without the composite system. Ongoing testing is evaluating different GCBD systems as well as the effect of the adjacent soil type, slope, and unsteady infiltration.

DISCUSSION

If applied to pavements, the GCBD should probably be tied into a drainage system or a ditch as long as the capillary barrier level remains above the water level in the ditch (Fig. 4). This would help drain the water in the transport layer. In addition, when the capillary barrier layer does experience breakthrough, this type of configuration will also help rapidly drain water and promote “healing” of the capillary barrier. Recall that Stormont and Anderson (1999) demonstrated that capillary barriers can be restored after failure (i.e., water breakthrough) with a subsequent drying cycle. This is important because relatively high rates of infiltration occur in pavements constructed in temperate regions. This infiltration may cause temporary failure of capillary barriers placed in pavements for the purpose of reducing frost heave.

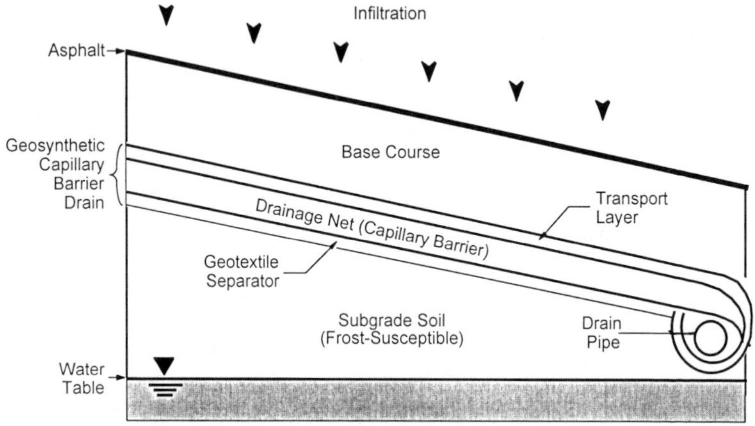

Figure 4. Schematic of a geosynthetic capillary barrier drain in a pavement.

Based on freezing test results, the transport layer should be hydraulically conductive at soil moisture suction heads of 1000–2000 mm in order to prevent failure of the geocomposite during freezing. Therefore, the water entry suction head of the transport layer should be somewhat greater that 2000 mm. The geotextiles tested to date have water entry suction heads significantly less than 1000 mm when initially dry, such as the product used in the tests by Stormont and Stockton (in review), which drained water at 50 mm of suction head. This indicates a need for more product development.

Note that although the freezing test results indicate that capillary barriers successfully reduce frost heave and laboratory experiments with GCBDs show their ability to divert infiltrating water, the combination of these two functions in a GCBD has not yet been validated in the laboratory or the field.

CONCLUSION

There is some evidence to support the idea that GCBDs can be used to improve pavement performance. Freezing test results show that drainage geocomposites significantly reduce frost heave under a wide range of conditions. Laboratory experiments with geosynthetic GCBDs indicate that they divert a significant amount of infiltrating water that would have otherwise reached the underlying soil. However, the combination of lateral water diversion and reduction of frost heave by a GCBD has not yet been validated in the laboratory or the field. In addition, more development of the transport layer is needed to develop a product that is hydraulically conductive between 1000 and 2000 mm suction head of water.

APPENDIX-I. REFERENCES

Allen, T., Bell, J. R., and Vinson, T. S. (1983). "Properties of geotextiles in cold regions applications." *Trans. Res. Rep. 83-6,* Transportation Research Institute, Oregon State University, Corvallis, Oreg.

Anderson, C. E., and Stormont, J. C. (1997). "Prediction of long-term erosion from landfill covers in the Southwest." *Proc., Int. Containment Tech. Conf.*, St. Petersburg, Fla., 389–395.

Aveyard, R., and Haydon, D. A. (1973). *An introduction to the principles of surface chemistry.* Cambridge University Press, New York.

Baker, R. S., and Hillel, D. (1990). "Laboratory tests of a theory of fingering during infiltration into layered soils." *Soil Sci. Soc. Am. J.,* 54, 20–30.

Beskow, G. (1946). "Supplement: Some results of Scandinavian soil frost research 1935–1946." In translation of *Soil freezing and frost heaving with special applications to roads and railroads.* Swedish Geol. Soc., C, No. 375, Year Book no. 3 (Translated by J. O. Osterberg). Technological Institute, Northwestern University, in Spec. Rep. 91-23, U.S. Army Corps of Eng. Cold Reg. Res. and Eng. Lab., Hanover, N.H., 161–169.

Casagrande, L. (1938). "Examination of the sub-soil of roads." *Proc., Eighth Int. Road Cong.,* The Hague.

Castellan, G. W. (1983). *Physical chemistry.* 3rd ed., Benjamin/Cummings Pub. Co., Menlo Park, Ca.

French, W. J., Poole, A. B., Ravenscroft, P., and Khiabani, M. (1982). "Results of preliminary experiments on the influence of fabrics on the migration of groundwater and water-soluble minerals in the capillary fringe." *Quart. J. of Eng. Geol., London,* 15, 187–199.

Henry, K. S. (1991). "Effect of geotextiles on water migration in freezing soils and the influence of freezing on performance." *Proc., Geosynthetics, '91,* Atlanta, Ga., 469–484.

Henry, K. S. (1998) "The use of geosynthetics to mitigate frost heave in soils." PhD thesis, Civil Engineering Department, University of Washington, Seattle, Wa.

Henry, K. S., and Holtz, R. D. (1997). "Capillary rise of water in geotextiles." *Proc., Int. Symp. on Ground Freezing and Frost Action in Soils,* Luleå, Sweden, 227–233.

Henry, K. S., Holtz, R. D., and Ellis, E. (1998). "Reducing frost heave with capillary barriers: Interim results." *Proc., Ninth Inter. Conf. on Cold Reg. Eng., ASCE.*

Henry, K. S., and Patton, S. (1998). "Measurement of the contact angle of water on geotextile fibers." *ASTM Geotech. Test. J.,* 21(1), 11–17.

Henry, K. S. and Stormont, J. C. (in prep.) "Geocomposite capillary barrier drain." Patent application, COE-470 (U.S. Army Corps of Engineers).

Henry, K. S., Taylor, S. and Ingersoll, J. (1990) "Effects of freezing on the microstructure of two geotextiles." *Geosynthetics: Microstructure and Performance,* ASTM STP 1076, 147–164.

Hillel, D., and Baker, R. S. (1988). "A descriptive theory of fingering during infiltration into layered soils." *Soil Sci.,* 146(1), 51–56.

Hoover, J. M., Pitt, J. M., Handfelt, L. D., and Stanley, R. L. (1981). "Performance of soil–aggregate systems in frost-susceptible roads, Linn County, Iowa." *Trans. Res. Board,* 827, 6–14.

Lambe, T. W. (1950). "Capillary phenomena in cohesionless soils." *Proc., ASCE,* 116, 1–32.

Miyazaki, T. (1988). "Water flow in unsaturated soil in layered slopes." *J. of Hydr.,* 102, 201–214.

Morris, C. E., and Stormont, J. C. (1997). "Capillary barriers and Subtitle D Covers: Estimating equivalency." *J. of Env. Eng.,* 123(1), 3–10.

Nyhan, J. W., Langhorst, G. J., Martin, C. E., Martinex, J. L., and Schofield, T. G. (1993). "Hydrologic studies of multilayers landfill covers for closure of waste landfills at Los Alamos, New Mexico." *Rep. No. LAUR 931914,* Los Alamos National Laboratory, Albuquerque, N.M.

Rengmark, F. (1963). "Highway pavement design in frost areas in Sweden." *High. Res. Rec.*, No. 33, 137–157.

Ross, B. (1990). "The diversion capacity of capillary barriers." *Water Res. Res.,* 26(10), 2625–2629.

Stormont, J. C. (1995). "The performance of two capillary barriers during constant infiltration." *Landfill closures,* R. J. Dunn and U. P. Singh, ed., *ASCE Geotech. Spec. Pub. No. 53,* 77–92.

Stormont, J. C. (1996). "The effectiveness of two capillary barriers on a 10% slope." *Geotech. and Geolog. Eng.,* 14, 243–267.

Stormont, J. C., and Anderson, C.E. (1999). "Capillary barrier effect of fine-over-coarse soil layers." *ASCE J. of Geotech. and Geoenv. Eng.,* 125(8), 641–648.

Stormont, J. C., and Stockton, T. (in review). "Preventing positive pore water pressures with a geocomposite capillary barrier drain." *Testing and Performance of Geosynthetics in Subsurface Drainage*, J. Goddard, D. Suits and J. Baldwin, ed., ASTM STP 1390.

Stormont, J. C., Henry, K. S., and Evans, T. M. (1997). "Water retention functions of four nonwoven polypropylene geotextiles." *Geosyn. Int.,* 4(6), Industrial Fabrics Association International, St. Paul, Minn., 661–672.

Taber, S. (1929) "Frost heaving." *J. of Geol.,* 38. In "Historical perspectives in frost heave research." CRREL Special Report 91-23, U.S. Army Cold Regions Research and Engineering Laboratory, Hanover, N.H., 29–35.

Taivenen, O. A. (1963). "Preventive measure to reduce frost action on highways in Finland." *High. Res. Rec.,* 33, 202–216.

APPENDIX-II. SYMBOLS USED

h_a	Air entry suction of the fine soil overlying the capillary barrier, expressed as a (negative) head of water (m)
$h_w{}^*$	Water entry suction of the coarse soil, expressed as a (negative) head of water (m)
h_t	Soil water tension, expressed as a (negative) head of water (m)
i	Flux of water entering the soil surface (m s^{-1})
P_w	Positive soil water pressure (Pa)
r	Radius of the smallest pores (m)
T	Soil water tension (Pa)
α	Contact angle of the liquid with the pore wall
γ_L	Unit weight of liquid (N m^{-3})
ψ_{aw}	Surface tension of the water (N m^{-1})

φ Slope of the interface between fine and coarse soil in a capillary barrier

K_s Saturated hydraulic conductivity (m s^{-1})

IN-PLACE STIFFNESS AND MODULUS MEASUREMENTS

Scott A. Fiedler[1], Melvin Main[2], and Albert F. DiMillio[3]

ABSTRACT: There is a broad interest and certain expectations in knowing the in-place engineering and performance properties of the compacted or stabilized soil. This has been the case since soil was thought of as an engineered structure. The processes of removing small numbers of specimens and then conditioning the specimens to represent the actual site are estimates at best. A great deal of skill is required to accurately prepare the specimens to approximate field conditions and provide meaningful and believable laboratory test results. Density based measurements have been surrogates for material modulus and structural stiffness. Available tests to measure the in-place properties are cumbersome, destructive and time consuming. Recently, a new device has been developed that does what the geotechnical community has desired, measure the in-place engineering and performance properties of the soil structure. Conceived and developed partially by funding from FHWA, it is now commercially available. Fast and immediate measurements of stiffness and modulus allow the compaction process to be controlled by the contractor without delay or outside feedback and should provide a more uniform structure. Control of the compaction process via stiffness and modulus permits construction to performance specifications derived directly from design. The use of one test method to link design, specification, construction, and maintenance appears to complete a loop that ultimately enables contractor warranties. This paper will examine the use of the device and the practical implementation of various functions or processes that the device offers.

---

[1]Humboldt Mfg. Co., 7300 West Agatite Avenue, Norridge, IL 60706, fiedler_hmc@email.msn.com

[2] Main Associates, 16 Vegas Drive, Hanover, PA 17331, main@cyberia.com

[3] FHWA, USDOT, Turner-Fairbanks Highway Research Center, McLean, VA 22101, al.dimillio@fhwa.dot.gov

INTRODUCTION

Measuring the stiffness or modulus of each pavement component: subgrade, subbase, and base independently are crucial to establish the engineering and performance properties of the pavement system as a whole. The costly performance margins inherent in density or empirically based construction can no longer be tolerated. Until recently, in-place modulus measurements were not considered due to the impracticality of available methods. Now, there is a hand-portable device that not only measures stiffness and modulus, but also does it in a fashion that does not delay the construction process and can be thought of as the means to control the compaction process.

ADVANTAGES

The advantages of measuring the in-place stiffness and modulus are readily apparent.

1. Pavement designs are based on the moduli of the available soils and crushed aggregates. It is therefore most meaningful and precise that the soil structure, or the compaction of soil, be constructed and evaluated based on modulus. In this way, the relationship between design and construction is as tight as construction materials and conditions will allow.
2. The implementation of optimizing compaction with minimum effort is placed in the hands of the contractor, enabling effective and long-term warranties.
3. Design over-specification is avoided and over-compaction is minimized or eliminated. Designs are often over-specified due to the uncertainties in soil performance and the relative lack of correlation between design parameters and measured in-place quantities. Typically, contractors prefer to compact with greater effort to avoid rework or penalties.
4. More structural uniformity is possible by evaluating in-place modulus. Modulus is a more sensitive measure of the uniformity in soil compaction than is density, even when moisture content is accounted for. It is well understood by design and construction professionals that the uniformity of compaction is directly related to structural failures and life cycle cost.

The benefits to earthwork reliability, maintainability and life-cycle costs are obvious. It is estimated that as much as 30 percent of compactive effort and its associated costs can be eliminated with compaction process control. This can be enabled by using modulus or stiffness to directly link design, specification, construction, inspection, maintenance, and warranty.

THE HUMBOLDT STIFFNESS GAUGE (GEOGAUGE)

The Humboldt Stiffness Gauge (GeoGauge) directly and simply measures stiffness. That is, it measures the resistance of a layer of material to deformation. It imparts a measured force to the surface of the layer and measures the resulting deflection. The ratio of force to deflection is stiffness.

If the area over which the force is applied is known and Poisson's Ratio is assumed, the modulus of the material can be directly derived from equation (1) (Poulos et al., 1974).

$$E = \frac{K(1 - \upsilon^2)}{1.77\,R} \tag{1}$$

Where:

E = Young's Modulus
R = Outside radius of GeoGauge's foot (57.15 mm)
υ = Poisson's Ratio
K = Stiffness

The GeoGauge is a 254 mm round, 279 mm tall, 10 kg. cylindrical instrument with a ring shaped foot on the bottom (Figure 1). The heart of the GeoGauge is a hollow metal cylinder rigidly attached to the inside of the foot. To the bottom of the cylinder is attached a displacement sensor. To the top of the cylinder is attached a force sensor and a shaker. The technology in this hardware, which is borrowed from the US military, allows for the measurement of extremely small forces and displacements (e.g., 0.00127 mm).

Press the "on" button, the GeoGauge displays battery voltage, date & time, runs an electronic self test and displays the last measurement made.

Next is the most important part of a GeoGauge measurement, seating the foot to the surface to be measured. At the beginning of the day, when first arriving at a new construction site or when site conditions significantly change, a minute or two is needed to practice seating the gauge. Place the gauge on the surface, lift it up and inspect the footprint to determine if ~ 60% of the ringed surface was in contact. If not, place the gauge down again and turn it slightly back and forth. If ~ 60% of the surface is still not in contact, place the gauge down again and turn it back and forth while pushing. If ~ 60% of the surface is still not in contact, chances are the surface is irregular (e.g., aggregate) or it is hard and smooth (e.g., highly compacted clay, asphalt or concrete). In this case, use < 6 mm of moist sand or local fines to visibly achieve the needed contact. Use this practice to guide seating of the gauge for the rest of site.

Figure 1: The GeoGauge

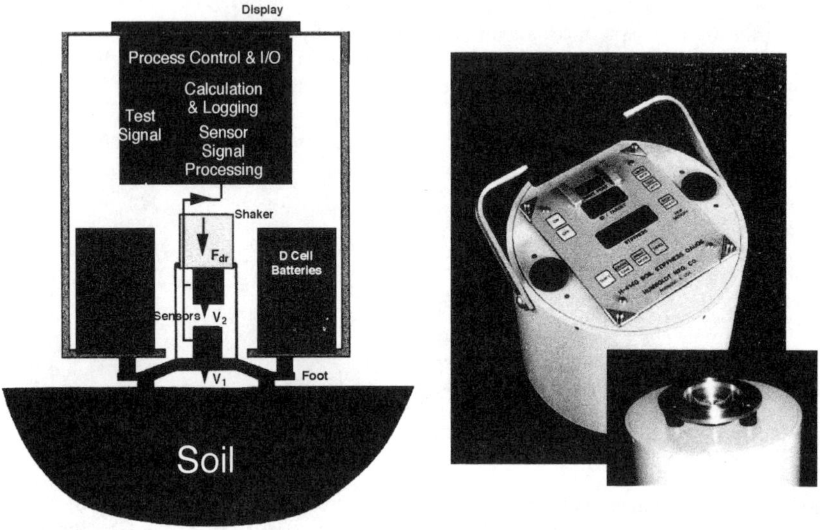

Once the gauge is seated, press the "MEAS" button. For 11 seconds, the gauge will display "nois" and measure the output of both sensors at 25 frequencies between 100 and 196 Hz without the shaker turned on. This data is stored for use later in judging the quality of the measurement. The gauge next displays "data". With the shaker on, the GeoGauge dwells at each of the same frequencies for 2 sec. Force and deflection are measured at each frequency. If at 20 or more frequencies, the signal to noise ratio is > 10 dB (3 times) for both force and deflection and no signal has overloaded the gauge, the measurement is judged good and the average stiffness over frequency is displayed in MN/m (or klbf/in). The entire measurement is completed in 1 minute, 15 seconds.

The algorithms used in processing the data are based on the assumption that construction materials have both the properties of a permanently deformable and resilient material. Force and deflection are measured as complex quantities and the calculation of stiffness is matrix algebra.

In order to place the meaning of GeoGauge measurements in the simplest terms, a field, quasi-static plate load test, where the plate is identical to the foot of the GeoGauge, was developed by CNA Consulting Engineers of Minneapolis, MN (Nelson et al, 1999). The loading and measurements of deflection were under precise computer control. Figure 2 shows a quasi-static load (the shallow sloped line) with working loads (steep sloped loading and unloading) periodically imposed. A GeoGauge measurement was made at the same location prior to the plate load test.

Figure 2: Quasi-Static Field Plate Load Test Results

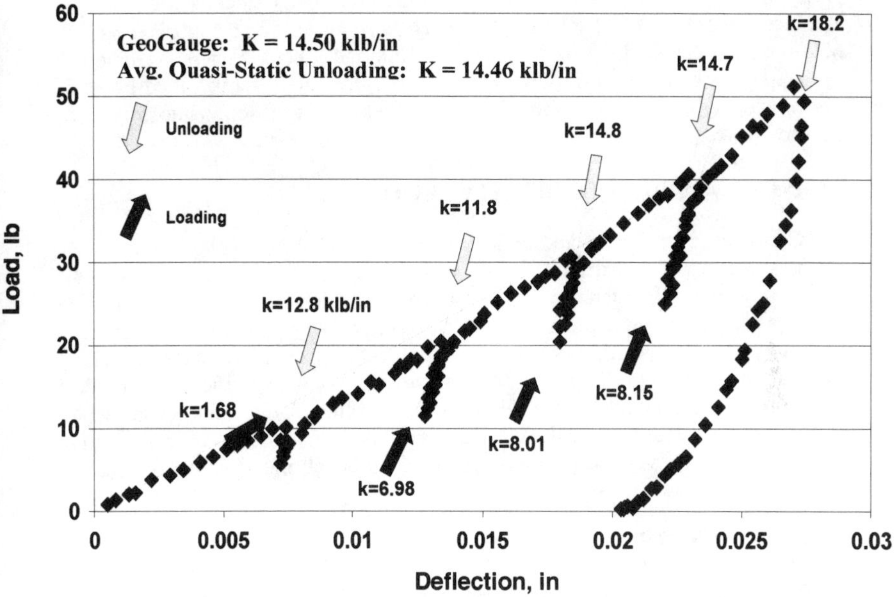

(1 lb = 4.448 Newtons, 1 in = 25.40 mm)

The GeoGauge measurement of stiffness corresponds to the average reaction of the soil when the working load is removed. The soil measured was located on a construction site at the junction of Highway 55 and Minnehaha Parkway in Minneapolis, MN. The soil was fill material composed mostly of sand, with gravel of various size up to 2 inches. CNA is gathering data on a wide variety of soils to confirm this relationship.

The University of New Mexico recently conducted a study to determine if the GeoGauge truly measures stiffness (Lenke, et al, 1999). A highly uniform and compacted granular media was constructed from silica sand of known specific gravity and mean particle size. The media was pluviated into a 762 mm X 762 mm foam lined steel bin to a height of 457.2 mm. The surface was carefully leveled and the weight of the media tracked during construction. From estimates of the mean effective stress imparted by the GeoGauge, void ratio, Poisson's ratio and the outside radius of the GeoGauge's foot, the stiffness of the top 228.6 mm (9") of the media was estimated to be 5.92 MN/m. The mean effective stress was estimated from the dimensions of the GeoGauge's foot, the GeoGauge's weight and estimates of how deep the stress would penetrate the media. The void ratio was determined during the placement of the sand. Poisson's ratio was estimated from the sand's angle of repose. The mean stiffness measured by the GeoGauge, when placed on the media, was 6.19 MN/m with a standard deviation of 0.04 MN/m and a coefficient of variation of 0.7%. The estimated and GeoGauge measured values of stiffness varied by <5%. The University's continuing evaluation will include similar determinations on clay, silt and layered media.

CURRENT MODULUS TO DENSITY PROTOCOL

Currently, many states and countries use material design properties, such as stiffness and modulus that are inferred from the available properties such as CBR and density. It is this inference that causes concern. Converting CBR or density to modulus can include an uncertainty of 100% or more. This encourages conservative values in density specifications to account for the uncertainty and the natural variation in soil quality. Contractors take these conservative specifications and then over compact to account for the relatively uncontrolled nature of the compaction process, to avoid rework and penalties. It is the combination of conservative specifications and over compaction that increases direct and indirect construction costs.

MODULUS TO MODULUS PROTOCOL

Current activities, such as the development of the 2002 AASHTO Pavement Design Guide, strongly suggest that in-place modulus will be used in the future to evaluate compaction. It is not acceptable to measure the moduli of materials in the lab and from the pavement surface after construction has completed, when it is too late to make corrections. Modulus will need to be measured more frequently during the construction, lift by lift, material by material. Construction specifications will follow directly from design with no need for conservative allowances. The variability within each structural component will be reduced, the proper load transfer of pavement systems will be enhanced, and over compaction will be minimized. All of which leads to a better and longer life pavement system with substantially lower life-cycle cost.

GEOGAUGE USAGE

Since early 1998, the GeoGauge has seen evaluation or use with more than 30 separate organizations. Table 1 summarizes some of the uses that the GeoGauge is being evaluated for. In general, all of the organizations evaluating the GeoGauge have found merits towards in-place stiffness and modulus measurements. More importantly is the potential in the GeoGauge for production testing, where measurements can be made without interference or interpretation from third parties or interfering and delaying construction and at the same time in sufficient volume to enable optimum compaction with minimum effort.

Table 1. GeoGauge Uses Being Evaluated

- Improved characterization of highway soil compaction
- Non-nuclear evaluation of compaction
- Specification & installation of Controlled Low Strength Materials
- Lime Stone Bearing Ratio replacement
- Optimizing Compaction while minimizing effort
- Performance specification development
- Evaluation of fills stabilized with cement, fly ash, lime stone & polymers
- Control of cold mix asphalt installation
- Compaction evaluation via in-place modulus
- Improved installation of buried plastic, concrete & corrugated steel pipe
- Validation of an Unbound Aggregate Base model
- Replacement of Standard Proctor

GEOGAUGE CORRELATION TO OTHER METHODS

Many measurements have been made in companion with known methods of determining modulus. These methods included Falling Weight Deflectometer (FWD), Dynamic Cone Penetrometer (DCP), Point FWD (PFWD, Finnish Loadman), Seismic Pavement Analyzer (SPA), and the Spectral Analysis of Surface Waves (SASW). In general, the values of modulus derived from the GeoGauge agree with the other methods about as well as the values from the other methods agreed with themselves. Figures 3 (Siekmeier et al., 1999) and 4 (Chen et al., 1999) illustrate typical results. The simple manner and speed of GeoGauge measurements suggests it is the most suitable of these methods for production testing.

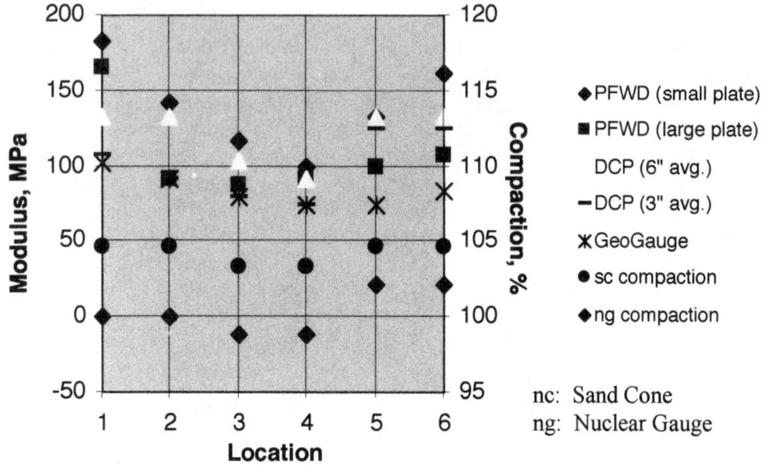

Figure 3: In-Place Modulus, Mn/ROAD, Section 17
Granular Base, Sand/Gravel Mix With < 10% Fines
Typical of13 Pavement Sections at 5 MnDOT Sites

DEMONSTRATED BENEFITS

Following are examples of the possible benefits of in-place stiffness and modulus measurements.

Compaction Testing

It is not practical to know the maximum Proctor density for every mixture of soil at a given construction site. This is especially true when the compacted material is non-uniform. Under these conditions, in-place measurements of density are not a true measure of the soil's performance. The in-place modulus of the soil appears to better track the observed soil performance. For this reason and the adoption of mechanistic design practices, such organizations as MnDOT will in the future include in-place measurements of modulus in its quality assurance testing (Siekmeier et al., 1999).

Figure 4: In-Place Modulus
Various Subgrade & Base Materials
In 6 TXDOT Districts

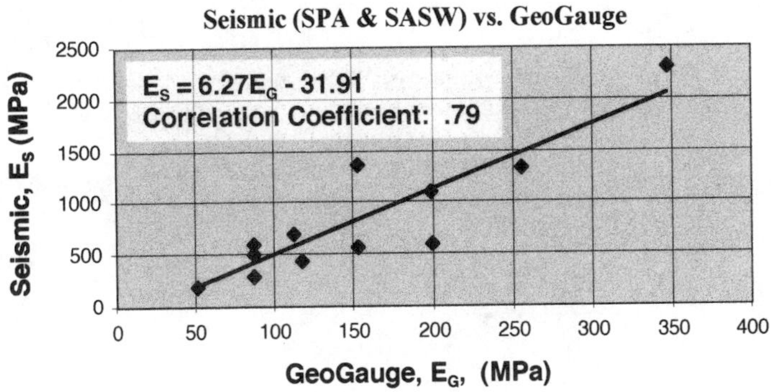

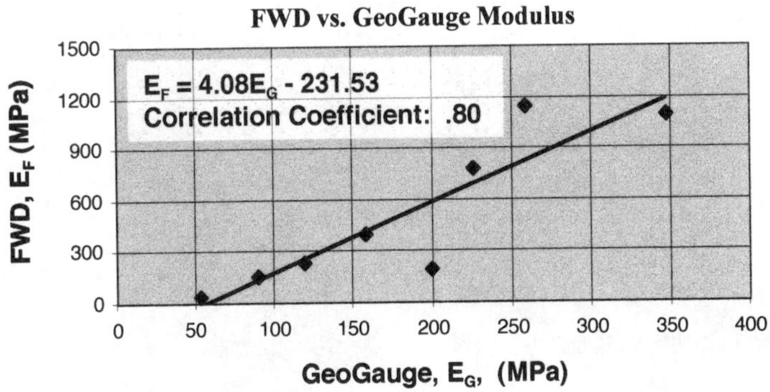

Concrete Pipe Design & Installation

In recent years, the concrete pipe industry has gone to great lengths to help designers better understand the behavior and interaction of buried concrete pipe and the supporting soil. The results are such new methods for design and installation as Standard Installation Direct Design (SIDD). These methods use finite-element soil structural modeling to reduce flexural moments in the pipe. This kind of modeling requires the stiffness of the pipe and the surrounding soil as inputs. This information to date has been indirectly determined from soil classification, placement, and density. The MnDOT has recently completed a project aimed at combining the benefits of the SIDD method with new methods of directly obtaining in-place soil stiffness (Hill et al., 1999). The project concluded that directly knowing the in-place soil stiffness would greatly improve the performance of concrete pipe. The project also concluded that if its findings regarding soil stiffness and design were supported by subsequent studies, the design and installation of underground pipeline structures would change dramatically.

Figure 5: Base Stiffness vs. Time
Rehabilitated Limestone/Clay Base
Stabilized with Organic Chemicals
Lafourche Parish, LA, July, 1999
Fluid Sciences, LLC, Lafayette, LA

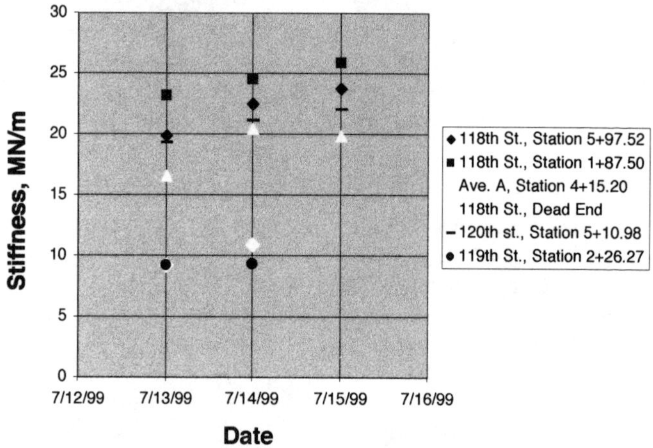

Pavement Quality Control

The TXDOT is one of a growing number of organizations that are actively looking for the missing link between pavement design and construction. TXDOT is evaluating methods of measuring in-place the same modulus they use to design subgrades, bases and pavements (Chen, et al, 1999). TXDOT believes that directly linking design and construction via material modulus will greatly improve pavement quality. It has gone so far as to quantitatively define pavement quality in terms of modulus.

Non-Standard Stabilized Fills

Fluid Sciences, LLC of Lafayette, LA is one of a number of companies producing an organic polymer system to compete with other soil stabilization methods (e.g., cement & lime stone). What they lack is a method to answer such questions as, "Is the fill hard enough to release it to traffic or the application of pavement?" or "Has the last few days of rain inhibited stabilization or rendered it non-uniform?" They intend to use in-place stiffness as their primary tool for monitoring the installation of their product. The use of in-place stiffness is an important part of a current CERF/HITECH evaluation of their product. As shown in Figure 5, the GeoGauge can be used to monitor the hardening of a stabilized fill, helping to answer the questions above.

Cold Mix Asphalt

The use of cold mix (emulsion) asphalt as a paving material has been limited due to a lack of a reliable method of measuring its in-place properties. This is potentially changing with the advent of methods for measuring in-place stiffness. The cold mix chemical reaction can be monitored via changing stiffness in real-time to quantitatively determine when compaction can begin. Like wise, stiffness can be monitored as the asphalt is compacted to achieve optimum compaction with minimum effort, Figure 6. Koch Materials is one company that is evaluating this method.

Figure 6: Cold Mix Asphalt Installation
Koch Materials, Conley, GA
October, 1999

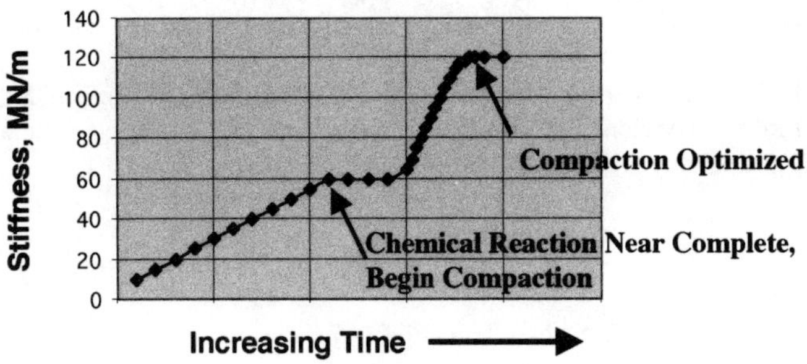

CONCLUSIONS

There is a fundamental change underway in how the construction industry does business. Organizations should heed the direction towards in-place stiffness and modulus measurements. New instruments such as the GeoGauge hold potential for enabling the continuity of design and construction and for true process control for the industry. The real benefits of this will be increased product quality at reduced cost.

APPENDIX – I. REFERENCES

Poulos, H.G., and Davis, E.H., 1974, Elastic Solutions For Soil & Rock Mechanics, pages 167-168.

Nelson, C.R., and Sondag, M., 1999, Comparison of the Humboldt GeoGauge with In-Place Quasi-Static Plate Load Tests, CNA Consulting Engineers, Minneapolis, MN, page 4.

Lenke, L.R., Grush, M. and McKeen, R.G., 1999, Evaluation of the Humboldt GeoGauge on Dry Cohesionless Silica Sand in a Cubical Test Bin, ATR Institute, University of New Mexico, pages 2-8.

Siekmeier, J.A., Young, D. and BeBerg, D., 1999, Comparison of the Dynamic Cone Penetrometer With Other Tests During Subgrade and Granular Base Characterization in Minnesota, Nondestructive Testing of Pavements and Backcalculation of Moduli: Third Volume, ASTM STP 1375, S.D. Tayabji and E.O. Luanen, Eds., American Society for Testing and Materials, West Conshohocken, PA, page 9.

Chen, D.H., Wu, W.,Rong, H.and Arrelano, M., 1999, Evaluation of In-Situ Resilient Modulus Testing Techniques, Recent Advances in the Characterization of Transportation Geo-Materials, ASCE 0-7844-0437-2, 62 pp., E. Tutumluer and A.Y. Papagiannakis, Eds., American Society of Civil Engineers, Geoinstitute, Reston, VA.

Hill, J.J., Kurdziel, J.M., Nelson, C.R., Nystrom, J.A. and Sondag, M., 1999, MnDOT Overload Field Tests of Standard and SIDD RCP Installations, Transportation Research Board Annual Meeting, pages 8-9.

APPENDIX – II. SYMBOLS USED

E = Young's modulus
K = stiffness
υ = Poisson's Ratio
R = outside radius of the ring-foot

Geotextile Separators for Hike and Bike Trail

Elisabeth Freeman[1], Student Member, J. Erik Loehr[2], Associate Member and John J. Bowders[2], Member

Abstract: Intrusion of surface aggregate into the underlying soil of hike and bike trails creates undesirable conditions for trail users and exacerbates the problem of trail rutting. A study is underway to evaluate the effectiveness of geotextile separators for rails-to-trails applications. Non-woven needle-punched and spunbonded geotextiles were placed between the subsoil and the aggregate surface in three test sections along a trail. The test sections are being monitored for visible intrusion of aggregate into the subsoil, drainage of surface water off the trail, and enhanced performance and durability of the aggregate surface compared to control sections. Destructive samples of the geotextile separator and aggregate were taken and tested periodically to evaluate field performance of the geotextiles. The characteristics of the surface aggregate and geotextile separators, installation of the three test sections, and results of performance monitoring and laboratory testing activities are described herein. The findings indicate the geotextile separators to be performing well and the test sections have maintained a well-drained, smooth surface after one and a half years. The performance of these sections vastly exceeded that of the control (no separator) sections.

INTRODUCTION

The objective of this effort is to demonstrate the advantages of using a geotextile separator on hike and bike trails. This requires documentation of a higher quality, longer lasting surface course along with decreased frequency for maintenance of the trail surface. The ultimate goal is to demonstrate that a geotextile separator can be beneficial to both trail users and owners. Geotextiles are normally used in pavements to provide separation and filtration (Koerner, 1998). The separator function in pavements has been well documented by Black and Holtz

---

[1] Undergraduate Research Assistant, Civil & Environmental Engineering, University of Missouri, Columbia, MO, 65211-2200

[2] Assistant and Associate Professor, respectively, Civil & Environmental Engineering, University of Missouri, Columbia, MO, 65211-2200, Ph 573-882-6380, Fx 573-882-4784, eloehr@missouri.edu

(1999) in their article assessing the performance of geotextile separators five years after installation. In the application presented herein, the geotextile is installed between the subsoil and aggregate to prevent the subsoil and aggregate from mixing (Figure 1). This design reduces the amount of rutting and muddy spots.

The City of Columbia Missouri Parks and Recreation Department maintains 7.5 km of a hike and bike trail, constructed on the former Missouri-Kansas-Texas (MKT) railroad line (Figure 2). The MKT trail connects to the KATY trail, also a rails-to-trails project, which stretches approximately 240 miles across Missouri from Sedalia to St. Louis.

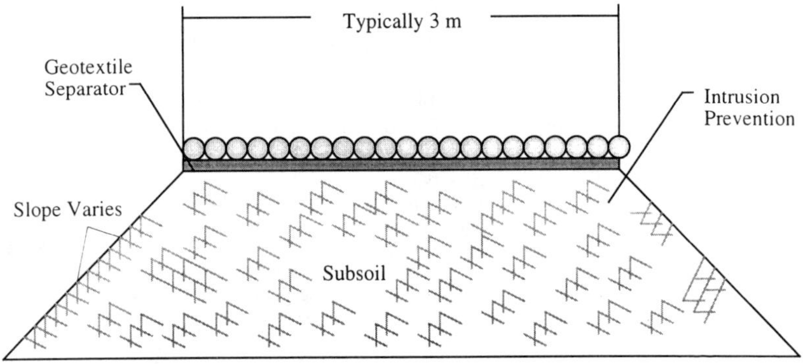

Figure 1: Typical Cross-Section of Geotextile Stabilized Area of the MKT Hike & Bike Trail.

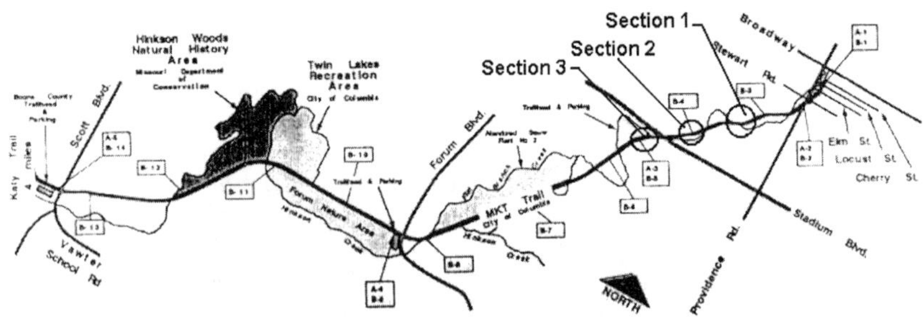

Figure 2: The 7.5 km MKT Trail Hike and Bike Trail Maintained by Columbia (Missouri) Parks and Recreations Department.

Railroad ballast, outcrop rock, and clayey soils make up the subsurface of the trail. The trail surface (10 mm crushed limestone) is generally five to ten centimeters thick placed directly on the subsurface to provide a firm surface for trail users to run, walk, or bicycle.

Several problem areas have been encountered on the trail. Excessive rutting occurs where users repeatedly travel along the same paths. Intrusion of aggregate into the underlying soil causes water to pond and muddy spots on the trail surface. Aggregate also tends to wash off the sides of the trail leaving only fines on the trail surface.

Columbia Parks and Recreations Department annually places fresh aggregate on the entire trail and quarterly performs spot maintenance along portions of the trail. Approximately $17,000 (U.S., 1999) is spent per year on labor and aggregate to upkeep the 7.5 km MKT trail (Griggs, 1998). The overall objective of this project is to demonstrate the reduction in long-term maintenance cost and improvement in the trail's surface performance by introducing the use of a geotextile separator.

PRE-INSTALLATION TESTS

Laboratory tests were performed on samples of the MKT trail aggregate to determine its classification and to investigate the performance of two geotextiles for filtration purposes. Aggregate used for the trail is a 10-mm crushed limestone from Boone Quarry in Columbia, Missouri. The two geotextiles used include a non-woven needle-punched geotextile and a non-woven spunbonded geotextile.

Sieve analyses (ASTM D422) and Atterberg limits tests (ASTM D424) were performed on samples of the MKT aggregate to determine its grain size distribution and soil classification according to the Unified Soil Classification System (USCS) (ASTM D2487). The fine-grained portion of the aggregate was determined to be non-plastic. The grain-size distribution for the surface aggregate is shown in Figure 3. The aggregate is dually classified as a poorly graded sand and silty sand (SP-SM) with eleven percent of the aggregate passing the #200 sieve (0.075mm).

A laboratory test was performed to evaluate whether the finer particles from the aggregate would migrate into or through the candidate geotextiles. A loose sample of aggregate (dry unit weights varied from 3.72 kN/m^3 to 5.29 kN/m^3) was placed in a rigid-wall permeameter over a geotextile. Aggregate was subsequently removed and divided into thirds. Sieve analyses were performed on each third to establish baseline data ("Pre-Test") taking into account segregation due to sample preparation (Figure 4). A new sample of aggregate was then placed in the mold and a constant-head permeability test was performed. The average hydraulic conductivity was measured to be 0.014 cm/s. After the permeability test, the aggregate was retrieved from the permeameter and divided into thirds. Sieve analyses were again performed on each third to determine the degree of particle migration through the sample and into the geotextiles (Post-Test). The fine particles tended to migrate in the direction of flow. Some fines passed through the geotextiles but the majority stayed in or on the geotextiles. As shown in Table 1, the non-woven needle-punched geotextile collected more fines within the geotextile compared to non-woven spunbonded. The fines within the geotextile may result in a slight reduction in the permittivity of the geotextile; however, a slight reduction is not expected to diminish the performance of the geotextile as a separator.

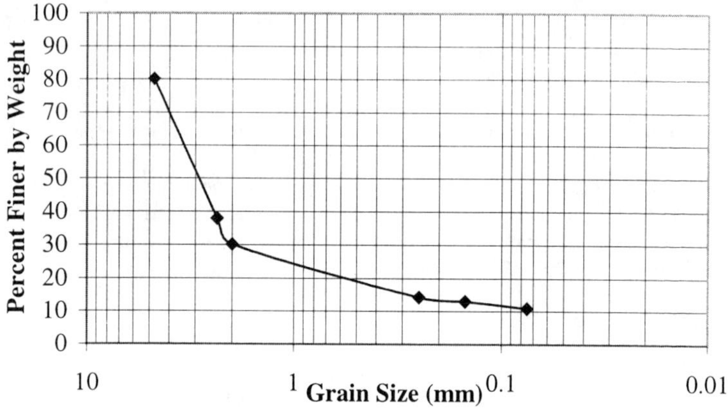

Figure 3: Grain Size Distribution Curve of a Sample of Trail Surface Aggregate.

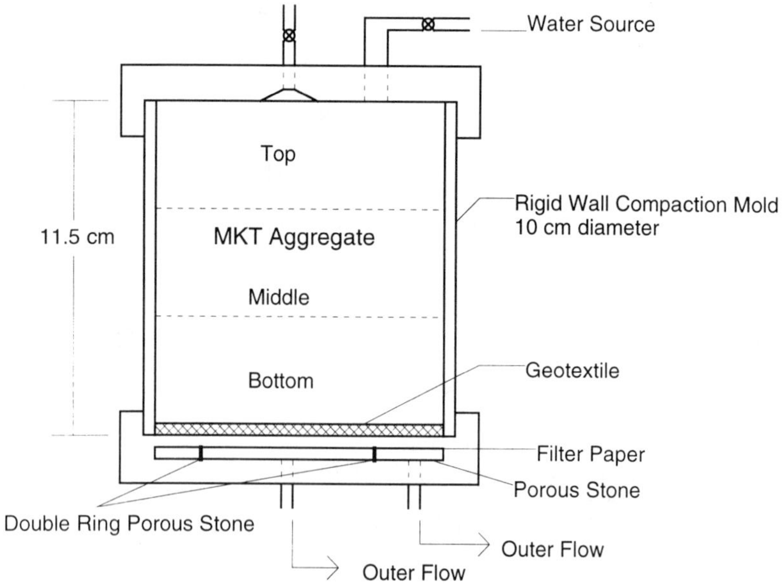

Figure 4: Test Setup to Evaluate Migration of Aggregate Fines Into and Through the Underlying Geotextile Separator.

Table 1: Dry Mass of Geotextiles Pre- and Post-Test.

Geotextile	Dry Mass of Geotextile		
	Pre-Test (g)	Post-Test (g)	Post-Test Fines Shaken From GT (g)
Non-woven Needle-punched	1.3	11.3	4.7
Non-woven Spunbonded	1.1	13	1.3

FIELD INSTALLATION OF GEOTEXTILE SEPARATORS

In June 1998 three test sites were selected and installed along the MKT trail for evaluation of geotextile separator performance. Test sections were selected to meet one or more of the following criteria:

- the aggregate had a history of intruding into the subsoil,
- water was ponding on the surface, or
- excessive rutting was occurring.

Each test section is approximately 30-m long by 3-m wide (Figure 2). The control sections are located at the end of each test section and are approximately 3-m long. Columbia Parks and Recreation staff and researchers from the University of Missouri-Columbia performed the installation. The installation procedure consisted of removing the top 5-10 cm of aggregate and subsoil using a box scraper. Each section was checked for debris that could puncture the geotextiles. Next, the geotextile was cut to size (≈ 30-m by 3-m) and placed directly on the subgrade of each test section. A non-woven needle punched geotextile (172 g/m^2) was installed in test sections 1 and 2. A non-woven spunbonded geotextile (132 g/m^2) was installed in section 3 (Figure 2). Wrinkles were removed from the geotextile. The surface aggregate was placed using a 10-ton dump truck. The truck traveled backwards while placing the aggregate so that it did not travel directly on the exposed geotextile. Aggregate was then manually distributed over the sections to ensure a uniform cross-section and compacted using a 1-ton walk-behind vibratory roller.

After installation, depth measurements and representative samples of aggregate were taken at four or five locations along each test section. Visual observations during installation indicated that the aggregate was uniformly mixed of large and fine particles. Sieve analyses were performed on the samples to quantify any variation of grain-size distribution along the test sections (Figure 5). The percentage of finer particles did not vary as greatly as that of the coarser particles. Table 2 shows the in-place aggregate course thickness that ranged from 2.5-cm to 7.6-cm depending on the test section.

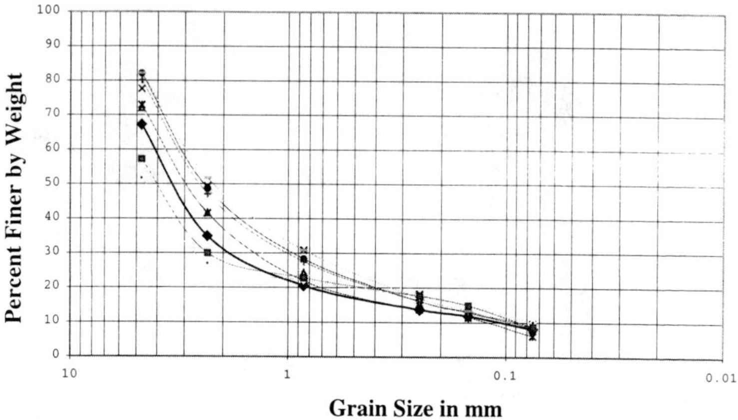

Figure 5: Grain Size Distribution Curves for the Surface Aggregate Sampled Along Various Locations During Installation of the Test Sections.

Table 2: Depth Measurements for each Test Section

Site	Depth (cm)		
	Average	Low	High
1	5.0	2.5	7.6
2	5.5	5.0	6.4
3	4.0	2.5	5.0

FIELD PERFORMANCE

Monthly visual observations were taken of the three test and control sections for one and a half years (July 1998 through November 1999). The sections are performing very well compared to the control sections (sections without geotextiles). There is no sign of intrusion of aggregate into the underlying soil. Rutting and water ponding has decreased in the geotextile-separated sections. In the control sections, the aggregate is covered with soil and the amount of loose lying aggregate has significantly reduced compared to the time of installation indicating that the unstabilized section aggregate has intruded into the subsoil which has lead to frequent ponded water and rutted sections.

Destructive samples were taken to investigate the possible migration of fines through the geotextiles - both from the subsoil upward and aggregate fines downward. The permittivity of the sampled geotextiles was measured to assess any clogging that might have developed due to migration of fines into or onto the geotextile. The samples were taken from the three test sections approximately one year after installation. The depth of surface aggregate at the time of installation and the depth after one year are shown in Table 3. The surface course thickness at Site 2 decreased while the thickness increased at Sites 1 and 3. It is most likely that the increase in thickness was due to scattering of surface aggregate due to trail user activity; although, the Parks Department did add aggregate to some non-test sections during this time.

Table 3: Surface Aggregate Course Thickness Taken at the Time of Installation and One Year After Installation.

Site	Surface Aggregate Thickness		
	Installation (cm)	At 1 year (cm)	Change in Thickness (cm)
1B	5	5.4	0.4
2A	6.4	4.13	-2.27
2B	5	3.02	-1.98
2C	5	3.02	-1.98
3A	2.5	4.45	1.95
3B	3.8	5.08	1.28
3C	5	5.72	0.72

The masses of the exhumed geotextiles were measured and the results are shown in Table 4. The non-woven needle-punched geotextile had a minimal amount of mass gain over time while the spunbonded geotextile, in some cases, quadrupled it's mass. The increased mass of the geotextile is a result of fines collecting in and on the geotextiles. Table 5 shows that the percentage of voids filled by migrating soil is less for the needle-punched geotextile than the spunbonded geotextile. This indicates that the needle-punched geotextile would clog at a slower rate and would be a better geotextile to use with this type of aggregate in this application.

Table 4: Dry Mass of Geotextiles at Time of Installation and Destructive Sampling.

Site	[1]Geotextile Type	Area (cm^2)	Mass at Installation (g/m^2)	Mass at Destructive Sampling (g/m^2)	Mass of Soil Gained (g/m^2)	% Mass Gain
1B	NP	169.1	172	183.4	11.4	6.6
2A	NP	173.6	172	155.5	-16.5	-9.6
2B	NP	321.5	172	174.0	2.0	1.2
2C	NP	338.7	172	239.1	67.1	39
3A	SB	81.1	132	518.1	386.1	293
3B	SB	251.6	132	230.5	98.5	75
3C	SB	169.1	132	201.1	69.1	52

[1]NP = Non-woven Needle-punched, SB = Non-woven Spunbonded

Table 5: Percentage of Geotextile Voids Filled with Soil.

Site	Geotextile Type	Area (cm^2)	Volume of Voids in Geotextile (cm^3)	Mass of Soil (g)	Volume of Soil (cm^3)	% Voids Filled
1B	NP	169.1	22.0	0.189	0.071	0.3
2A	NP	173.6	22.6	0.194	0.073	0.3
2B	NP	321.8	41.9	0.360	0.136	0.3
2C	NP	338.7	44.1	0.379	0.143	0.3
3A	SB	81.1	2.0	0.091	0.034	1.7
3B	SB	251.6	6.2	0.282	0.106	1.7
3C	SB	169.1	4.2	0.189	0.071	1.7

For each destructive sampling location, the surface aggregate, geotextile, and underlying soil were exhumed and in situ balloon density tests (ASTM D2167) were performed (Table 6). Sieve analyses were also performed on the surface aggregate and underlying soil to evaluate the migrations of fines after one year of field exposure. The amount of fines on the geotextiles increased over time. The mass of the geotextiles increased with time as shown in Table 4; indicating that the fines migrated into the geotextile. Clogging of the geotextile may occur if the fines continue to increase over time. The subsoil fines may also have migrated upwards toward the geotextile. Alternatively, part of the increase in fines may be a result of slight variation in sampling locations, degradation of aggregate particles, or collection of fines from other sources.

Table 6: Unit Weights from Destructive Sampling.

Site	Moist Unit Weight		Dry Unit Weight	
	Above Geotextile (g/mm^3)	Below Geotextile (g/mm^3)	Above Geotextile (g/mm^3)	Below Geotextile (g/mm^3)
2B	2.07	1.81	1.98	1.69
2C	1.82	1.79	1.76	1.66
3B	--	2.22	--	2.12

Typically, the underlying soil had a greater amount of fines than the surface aggregate. Data in Table 4 indicate that fines are being collected on the geotextile. A thin residue of fine particles covering the subsoil was observed after removing the destructive samples indicating that fines also traveled through the geotextile. The water contents at sites 3A, 3B, and 3C of the surface aggregate and underlying soil are all practically the same (Table 7). The water contents taken from sites 1B, 2A, 2B, and 2C vary significantly between the underlying soil and the surface aggregate (Table 7). Samples of sites 1B, 2A, 2B, and 2C were taken a few days after a significant rainfall occurred. Surface aggregate contained about half as much water as the underlying soil indicating that the geotextile is not clogging but rather allows the water from the surface aggregate to quickly drain from the surface thereby reducing the effects of ponding and improving the surface of the trail.

Table 7: Water Contents of Surface Aggregate and Subsoil

Site	Water Content	
	Surface Aggregate (%)	Subsoil (%)
1B	5	9.1
2A	3.3	6.5
2B	4.3	7.4
2C	3.6	7.6
3A	4	5.8
3B	5	4.7
3C	3.8	6.2

Permittivity tests were also performed on the exhumed geotextiles in accordance with ASTM D4491 (Table 8). Fines were shaken off the surface of the geotextile before permittivity tests were performed. The values in Table 8 therefore represent internal permittivity and not in situ conditions. For the non-woven needle-punched geotextile, there was a consistent decrease in the permittivity for the exhumed geotextiles compared to new material. The decrease ranged from 20 to 30 percent of the original permittivity. Fines migrating into the geotextile are believed to have caused the reduction in permittivity; however, when the exhumed geotextile was held up to the light it appeared much thinner than the new geotextile.

The permittivity tests on the exhumed non-woven spunbonded geotextile resulted in one decreased permittivity and one increased permittivity (Table 8). No conclusions can be extended until additional samples of the spunbonded material are exhumed and tested.

Table 8: Permittivity Results

Material	Site	[1]Manufacturer's Permittivity (sec^{-1})	New Permittivity (sec^{-1})	Exhumed Permittivity (sec^{-1})
Non-woven Needle-punched	1B	1.5	1.1	0.81
	2A	1.5	1.1	1.0
	2B	1.5	1.1	0.83
	2C	1.5	1.1	0.83
Non-woven Spunbonded	3A	0.70	0.36	--
	3B	0.70	0.36	0.29
	3C	0.70	0.36	0.52

[1]Geotechnical Fabrics Report, 1998 Specifier's Guide, Vol. 15, No. 9.

SUMMARY

The MKT hike and bike trail in Columbia Missouri is a 7.5-km rails-to-trails system owned and operated by Columbia Parks and Recreation Department. The surface of the trail along some sections is muddy, ponded with water, or rutted due to intrusion of the surface aggregate into the subsoil. These sections must be repaired multiple times each year by the Parks Department. Repair consists of adding additional aggregate to the trail surface.

Three test sections were installed and are being monitored to demonstrate the effects of geotextile separators on the performance of the trail. Each section is approximately 30-m long and 3-m wide. A non-woven needle-punched geotextile and a non-woven spunbonded geotextile were employed as the separators. An average thickness of five centimeters of aggregate (10-mm diameter) was placed on top of the geotextiles as a surface course. The installations were completed in June 1998.

Observations of the test and control sections were made monthly and destructive sampling was performed in April 1999. The findings indicate that the geotextiles are undergoing some clogging likely due to fines from the overlying aggregate migrating into the geotextiles. There has been excellent drainage of surface water from all three of the test sections. In spite of the slight reduction in permittivity the geotextile sections are outperforming the control sections. The Parks Department has not had to apply any additional aggregate to the test sections for more that one and a half years. Other sections of the trail have had to receive additional aggregate during this time.

Field performance monitoring is ongoing. Provided the geotextile stabilized trail sections continue to outperform the unstabilized sections, the City of Columbia can consider use of geotextile separators on the remainder of their trail system.

ACKNOWLEDGEMENTS

The cooperation and collaboration of Mike Griggs, Superintendent of Parks Management and Operations for Columbia Parks and Recreation Department is gratefully appreciated. This project was made possible through the generous contributions of William Hawkins of Reemay Inc. and Stephen Velaro of Synthetic Industries Inc. We extend our deepest gratitude to these individuals.

APPENDIX-I. REFERENCES

American Society for Testing and Materials (1998), Annual Book of Standards, Section D18.

Black, P.J., and R.D. Holtz (1999) "Performance of Geotextile Separators Five Years After Installation," *Journal of Geotechnical and Geoenvironmental Engineering*, Vol. 125, No. 5, pp. 404-412.

Griggs, M. (1998), *Personal Communications*, Parks and Recreation Department, City of Columbia, Missouri.

Koerner, R.M. (1998), *Designing with Geosynthetics*, 4th Edition, Prentice-Hall, Upper Saddle River, New Jersey.

Comparative Evaluation of Stabilized Subbase Courses

Gerald A. Miller[1], Member ASCE

ABSTRACT: On a rural road near Ada, Oklahoma, four test sections were constructed with treated subbases. Treated subbases in three sections were constructed using Portland cement kiln dust (CKD) from three different sources and one was constructed using granular quick lime. The purpose of the study was to compare the effectiveness of the three CKD sources and granular quick lime for modifying and stabilizing moderate to highly plastic soils. To evaluate the effectiveness of the various additives, treated soil was collected immediately before compaction to prepare unconfined compression test (UCT) samples that were used to assess strength gain after 7 and 28 days of curing. In addition, dynamic cone penetration (DCP) tests were conducted at different times after compaction, and falling weight deflectometer (FWD) tests were conducted on the finished flexible pavement. A comparison of results indicates there is agreement between the trend of cone index (CI) values from DCP tests and unconfined compression strength data along the roadway alignment. Furthermore, the DCP results give an indication of the variation in strength with depth, an advantage that the other forms of testing lack. Unconfined compression test results and DCP data do not correlate as well with the FWD data, apparently because the FWD results are less sensitive to differences in the treated subbase moduli for each section. This study shows that a reliable comparison of stabilized subbase courses for different test sections requires field testing to account for variations in field compaction and soil type, and laboratory testing to assess the relative effect of the soil additives on strength.

INTRODUCTION

Recent interest in using cement kiln dust (CKD) for soil modification and stabilization prompted the Oklahoma Department of Transportation (ODOT) to support a field and laboratory research project to evaluate CKD. On a rural road near Ada, Oklahoma, four test sections were constructed, each with a different soil stabilizer. Treated subbases in three sections were constructed using CKD from three different sources and one was constructed using granular quick lime. The purpose of the study was to assess and compare the effectiveness of the three CKD sources and granular quick lime for modifying and stabilizing moderate to highly plastic soils. To evaluate the effectiveness of the various additives, a multi-faceted

[1]Assistant Professor, School of Civil Engineering and Environmental Science, University of Oklahoma, 202 W. Boyd St., Rm. 334, Norman, OK 73019

approach was taken. First, treated soil was collected immediately before compaction to prepare unconfined compression test (UCT) samples that were used to assess strength gain after 7 and 28 days of curing. Second, dynamic cone penetration (DCP) tests were conducted at selected locations at various elapsed times after compaction. Finally, falling weight deflectometer (FWD) tests were conducted on the finished flexible pavement. This paper compares and contrasts the results of the field testing and the lessons learned.

BACKGROUND

During manufacturing of Portland cement, a small percentage of the material in the form of dust (CKD) is collected and removed as an industrial waste. More than 3,500,000 Mg (3,860,000 T) of CKD is disposed of annually in the United States (Todres et al. 1992).

Previously, CKD, which has cementitious properties, has been found to improve the unconfined compressive strength of both cohesive and cohesionless soils (Baghdadi 1990, Fatani and Khan 1990, Zaman et al. 1992, Baghdadi 1995, Azad 1998). In addition, CKD has been found to reduce the plasticity index of cohesive soils (Baghdadi 1990, Zaman et al. 1992, Azad 1998). In another study, Miller et al. (1997) found CKD effective at reducing the collapse potential and compressibility of compacted shales. In Oklahoma, CKD has been used on county roads and recently interest has turned to using CKD for soil stabilization on state highway projects. The latter has prompted the current study.

FIELD TEST SECTIONS

To evaluate the effectiveness of CKD relative to lime, four test sections were constructed between January and March of 1998. One of the test sections was stabilized with 4% (by weight) granular quick lime and the three remaining test sections were stabilized with 15% CKD. Granular quick lime was used in this study and is referred to as quick lime or lime in this paper. Each of the three CKD test sections was designated for a different CKD supplier, labeled CKD 1, CKD 2, and CKD 3. Due to scheduling difficulties and actual material quantities delivered, the test sections vary slightly in length, and CKD 3 was used in two sections. The contractor decided to use CKD 3 to complete the west-end of the CKD 1 test section since the CKD 3 source is located much closer to the site. Also, note that the lime test section is separated from the beginning of the CKD test sections between stations 151 to 155+80 (1 Sta. = 100 ft. = 30.5 m).

Subgrade samples were collected by the ODOT Materials Division at regular intervals along the test sections prior to subgrade stabilization, and subjected to index property testing including Atterberg Limit tests and sieve analyses. Soil samples were obtained from the surface to a depth of 610 mm (24 in.) below the top of subgrade. Results of these tests are summarized in Fig. 1.

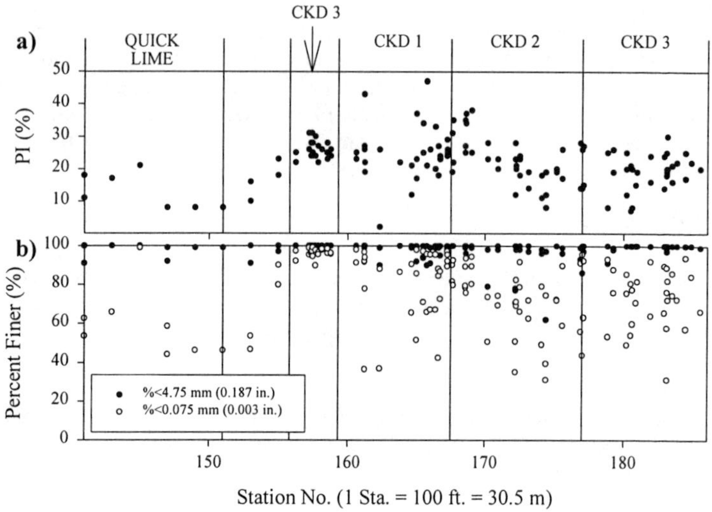

FIGURE 1 Index Property Test Results: a) Plasticity Index, b) Sieve Analysis Data

Based on the plasticity index (PI) and sieve analysis data shown in Fig. 1, most of the soils along the alignment classified as lean clay, lean clay with sand, or sandy lean clay with occasional fat clay. The subgrade soils were derived partly from weathered sandstone and shale from the cut sections and partly from natural in-place soils.

Within a test section, the subgrade soil was somewhat variable as shown in Fig. 1, but overall the soils appear similar between sections, particularly for the test sections beyond Sta. 160. The soil in the short test section for CKD 3, before Sta. 160, is noticeably more uniform than other sections. Overall, this short test section appears slightly more plastic and contains more fines than in other sections. On the other hand, the lime test section appears to contain fewer fines on average, compared to other test sections. There is less data for the lime test section because construction scheduling caused the original lime section to be abandoned in lieu of the one shown.

UNCONFINED COMPRESSION TESTS

Immediately prior to compacting the soil in the test sections, samples were collected and unconfined compression test (UCT) specimens were prepared in the field using a Harvard miniature apparatus. The Harvard miniature compaction procedure was calibrated with soil collected from the roadway alignment in accordance with the procedure in ASTM Standard D 4609, "Standard Guide for Evaluating Effectiveness of Chemicals for Soil Stabilization." Calibration was performed so that Harvard miniature samples compacted at the standard Proctor optimum moisture content (OMC) would achieve the standard Proctor maximum dry unit weight. The same compaction effort was used on all specimens prepared with the Harvard miniature device. For each sampling location, a minimum of three compacted samples were prepared and in some cases up to 12 samples were prepared depending on the time available and the number of locations sampled on a given day. Samples were subjected to UCTs after curing periods of 7 and 28 days. Curing involved wrapping samples in cellophane and placing them in a humidity chamber. Unconfined compression tests were performed in triplicate to quantify statistical variation in the UCS results.

Values of UCS for treated soil and the range of UCS for representative samples of the raw soil are shown in Fig. 2. Several important observations are made as follows:

1) Unconfined compression strength values for samples treated with CKD 3 are considerably higher than for other CKD sources and lime, for both 7-day and 28-day curing periods.
2) The performance of granular Quick Lime and CKD 1 appears to be similar, although the increase in strength from 7 to 28 days is generally more distinct for the lime-treated soil. The CKD 2 samples showed the worst performance relative to other additives.
3) Additives other than CKD 3 produced little or no improvement in UCS relative to the range of UCS for untreated soil. The small or negligible increases in UCS observed for CKD 1, CKD 2, and lime after 28 days of curing indicate generally unsatisfactory performance with regard to strength (ASTM D 4609).
4) Generally, the percent increase in UCS between 7- and 28-day curing periods is greater for lime than CKD. This is consistent with observations that CKD-treated soil gains most of its strength within the first 7 to 14 days while lime-treated soils typically gain strength steadily over longer periods.

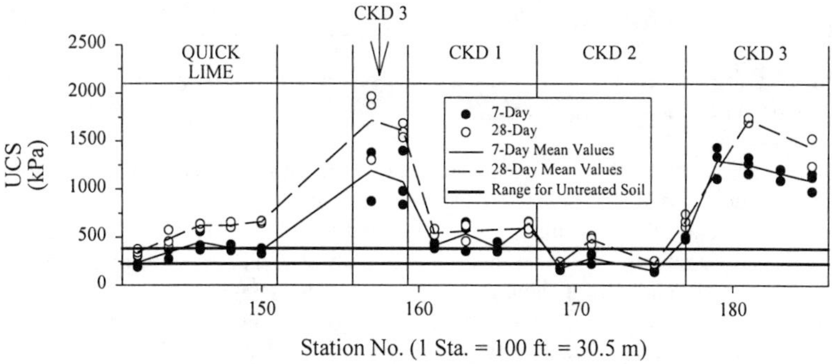

FIGURE 2 Results of Unconfined Compression Tests

To summarize, results of UCTs on field samples indicate that the performance of CKD 3 was superior to other CKDs and lime. The CKD 1 performed similar to lime, while CKD 2 performed the worst. The observed differences in performance can be attributed to three major factors. First, there are notable differences in the chemical composition of the different CKD sources as shown in Table 1, that tend to corroborate the superior performance of CKD 3. In particular, the calcium oxide (CaO) content of CKD 3 is higher than other CKD sources, CKD 3 exhibited lower loss on ignition (LOI), and the pH of treated soil is higher for CKD 3. These differences point to a higher free lime content and a potential for greater cementation and pozzolanic activity in soils treated with CKD 3.

TABLE 1 Chemical Composition of CKD

Provided by CKD Suppliers[a]			
Additive Type	CKD 1	CKD 2	CKD 3
Approximate Date Tested	08/08/1997	08/12/1997	02/20/1998
Silicon Dioxide, SiO_2 (%)	11.9	13.8	16.0
Aluminum Oxide, Al_2O_3 (%)	4.7	4.1	3.6
Iron Oxide, Fe_2O_3 (%)	1.8	1.5	2.3
Total Calcium Oxide, CaO (%)	42.5	44.1	52.8
Magnesium Oxide, MgO (%)	1.4	1.4	2.2
Sulfur Trioxide, SO_3 (%)	7.3	3.0	6.0
Sodium Oxide, Na_2O (%)	---	0.4	0.3
Potassium Oxide, K_2O (%)	---	1.6	3.5
Loss on Ignition (%)	25.8	29.1	---
Determined by OU[b] and Independent Laboratory[c]			
Loss on ignition (%)	24.0	27.8	17.5
pH (Pure CKD + H_20)	12.65	12.48	12.52
pH (15% CKD + 85% Shale + H_20)	11.76	11.72	12.23
pH (15% CKD + 85% Sand + H_20)	12.33	12.27	12.48

Notes:
[a]Oxides determined by x-ray fluorescence. [b]pH by OU personnel. [c]Loss on Ignition by QuanTEM Laboratories, LLC.

A second factor to be considered is that the impacts of some variables during construction are difficult to assess. In particular, environmental factors, the uniformity and depth of mixing, moisture conditions, and CKD losses due to wind are generally unknown. Unconfined compression test samples were prepared in the field and then cured in a laboratory under constant temperature and humidity so environmental factors are less important relative to others. Observations during construction indicated that mixing of additives and soil was consistent from section to section and the weather was similar on days when test sections were constructed. While some CKD was blown off-site on windy days, the losses appeared to be minor.

A third factor to consider in comparing the UCT results is that the subgrade soils were somewhat variable. However, the response of soil in the two different test sections with CKD 3 was similar though the soil in one section had a larger fines fraction. It is also important to consider that lime lacks intrinsic cementitious properties and therefore will not be as effective for stabilizing low PI soils. Therefore, if lime was used in another test section with higher PI and more fines it may have exhibited better performance.

Mixing, sample preparation, and unconfined compression testing in a laboratory with the same soil can eliminate many of the uncertainties associated with variability in field conditions (mixing, temperature, moisture, soil type). While not discussed in this paper, additional testing was conducted under controlled laboratory conditions. In general, similar results were obtained with controlled laboratory testing, however, CKD 1 and 2 performed better relative to the strength of untreated soil and lime-treated soil.

DYNAMIC CONE PENETRATION TESTS

The DCP apparatus is simple, lightweight, and can be readily deployed by two persons - one for operating the equipment, and the other for recording data. The basic elements of the device are a conical tip attached to a rod, which is driven into the ground using a donut shaped hammer and anvil attached to the top of the cone rod (e.g. Burnham and Johnson 1993). A test is conducted by driving the cone vertically into the ground and recording the depth of penetration that is accomplished with each hammer strike. The cone rod is permanently marked in 5.1-mm (0.2-inch) increments to facilitate the depth measurement. The driving energy is controlled by maintaining a consistent drop height. Test data are presented in the form of a plot of Cone Index (CI) versus depth, where the CI is simply the depth of penetration per blow. The CI is typically expressed in millimeters or inches per blow. Lower CI values indicate stronger material.

Practically speaking, the DCP test is suitable for investigating the upper 1 m (40 inches) of a pavement or soil profile. The DCP test has been used successfully in a number of states within the U.S. and in other countries to establish reliable correlations between California Bearing Ratio (CBR) values and DCP Cone Index (CI) (e.g., Burnham and Johnson 1993, Livneh 1988, Livneh 1989). One would expect that such a correlation would generally be very good considering that both the CBR and DCP test involve measurements that reflect the energy necessary to force a penetrating body into a soil mass. Also, the penetration resistance depends on the shear strength of the soil. Sowers and Hedges (1966) were among the first to recognize the relationship between DCP resistance and soil shear strength.

Dynamic cone penetration (DCP) tests were conducted in the test sections after 28 and 56 days following compaction of the treated subbase. Results of 28- and

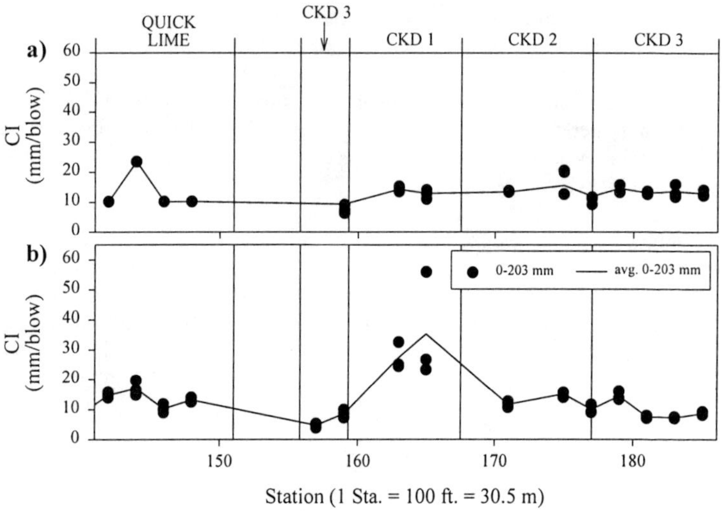

FIGURE 3 Results of Dynamic Cone Penetration Tests: a) 28-Day, b) 56-Day

56-day DCP tests are shown in Fig. 3. Each sounding was conducted to a depth of approximately 610 mm (24 in.) below the ground surface. The data reported in Fig. 3 covers the top 203 mm (8 in.) since this represents the design thickness for the stabilized subbase. Reported CI values for each test were calculated by averaging individual CI values (per each blow) for depths of 0-203 mm (0-8 in.). In addition, shown in Fig. 3 are CI data obtained by averaging values from each profile at a given location. Generally, three soundings were performed for each test location. Important observations follow:

1) At 28-days, average CI values for the full depth of stabilized base (0-203 mm, 0-8 in.) indicate that the subbase strength in the lime and first CKD 3 (near Sta. 159+00) test sections are similar, except for values from Station 144+00 in the lime section where a weak spot was encountered. On average, the three CKD sections from about Station 160+00 to 187+00 performed similarly based on 28-day CI values.
2) By 56 days, average CI values for the stabilized subbase (0-203 mm, 0-8 in.) are lower in the CKD 3 test sections. Lime and CKD 2 appeared to give similar performance and changed little compared to 28 days. In the CKD 1 test section, CI values increased significantly from 28 to 56 days for unknown reasons. This difference may have occurred partly because cool weather conditions slowed the curing process in the field relative to the laboratory environment.

Complete DCP profiles for selected stations are shown in Fig. 4. Several interesting observations are made from these plots:

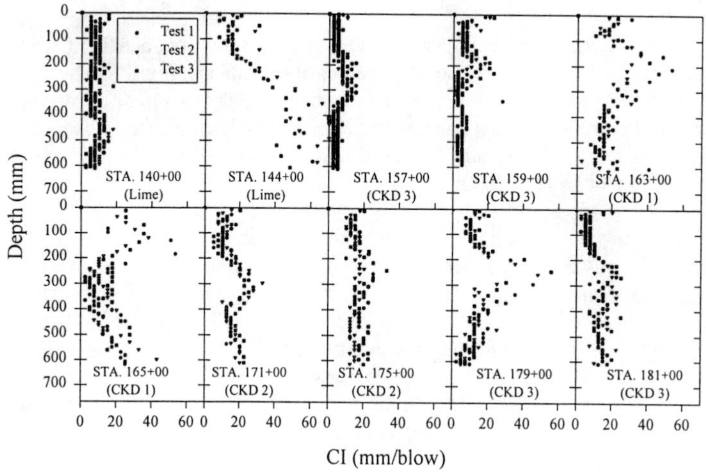

FIGURE 4 56-Day Dynamic Cone Penetration Test Profiles

1) In many of the profiles, a distinct change is seen near a depth of 200 mm (8 in.). See for example Sta. 144+00, 159+00 171+00, 179+00. This indicates that the actual stabilized depth was close to the target of 203 mm (8 in.).
2) Where the subgrade below the stabilized subbase is competent, as indicated by low CI values, the results of stabilization generally appear better. For example, at Sta. 140+00 in the lime section, CI values below 203 mm (8 in.) indicate a competent subgrade whereas at Sta. 144+00 a weaker subgrade is indicated by higher CI values. Corresponding CI values in the top 203 mm (8 in.) indicate a stronger and weaker subbase, respectively. Similar observations were made in other test sections, for example, compare Sta. 179+00 with 181+00 in the CKD 3 section.
3) The superior performance of the CKD 3 section is further reflected by the DCP profiles. While the subgrade appears to be relatively softer at some locations in the CKD 3 section (beyond Sta. 178+00), CI values in the stabilized base are among the lowest.

To summarize, the results of 56–day DCP tests generally corroborate the conclusions drawn from UCS testing on field samples. Performance of the CKD 3 test sections is superior while the lime section tends to be on par with CKD 1 and 2 test sections. One difference in the outcome of UCS and DCP testing is that the DCP results suggest that CKD 2 generally out performed CKD 1.

FALLING WEIGHT DEFLECTOMETER TESTS

Falling weight deflectometer (FWD) tests were conducted to provide additional information about the structural integrity of the stabilized subbase. Tests were conducted by ODOT personnel on September 9, 1998, shortly after completion of pavement construction. The completed pavement was designed for 152 mm (6 in.) of asphalt concrete over 101 mm (4 in.) of aggregate base over 203 mm (8 in.) of stabilized soil (subbase).

Falling weight deflectometer tests were conducted on 30.5-m (100-ft.) intervals in the eastbound and westbound lanes throughout the test sections using Dynatest equipment. During FWD testing, after a seating load of approximately 29 kN (6,500 lb.) was applied, a nominal impulse load of 31 kN (7,000 lb.) was applied four times to a 300-mm (11.8-in.) plate atop the pavement. For each impulse load, pavement deflections were measured with sensors located 0, 203, 305, 610, 914, 1,219 and 1,524 mm (0, 8, 12, 24, 36, 48, and 60 in.) away from the center of the loading plate. These sensors were designated D1 through D7, respectively.

For evaluating the composite character of the subbase and subgrade, the sensors located at 610, 914, and 1,219 mm (24, 36, and 48 in.) were selected for determining a composite subbase/subgrade resilient modulus. Resilient modulus was calculated using the AASHTO (1993) procedure, which is based on layered elastic analysis. The analysis assumes that selected sensors are far enough away from the loading so that the corresponding deflection is primarily a function of subgrade deformation. However, it must be close enough to the loading so that the deflections are measured accurately. Using AASHTO (1993) recommendations and comparing values of modulus calculated for the different sensors, it was found that the D4, D5, and D6 sensors gave similar results, which are assumed to be representative of the composite response of the stabilized subbase and subgrade.

Results of FWD analysis are shown in Fig. 5 where sensor deflections and corresponding back-calculated modulus values are presented for eastbound and westbound lanes. Each point represents an average of values obtained from the four impulse loads at a given location. Inspection of Fig. 5 reveals some noteworthy observations:

1) Eastbound and westbound FWD results are very similar at a given station, which indicates that the pavement is uniform across lanes.
2) Modului along the alignment were quite high, in a range of 100 to 800 MPa (15 to 116 ksi) with a majority falling between 140 to 350 MPa (20 and 50 ksi). These values are indicative of a competent subbase/subgrade.
3) Modulus values are quite high in the CKD 3 test section between Stations 155+80 and 159+60. This is consistent with the results of DCP and UCS tests and reflects both the integrity of stabilized soil and superior strength of the underlying subgrade at these locations.

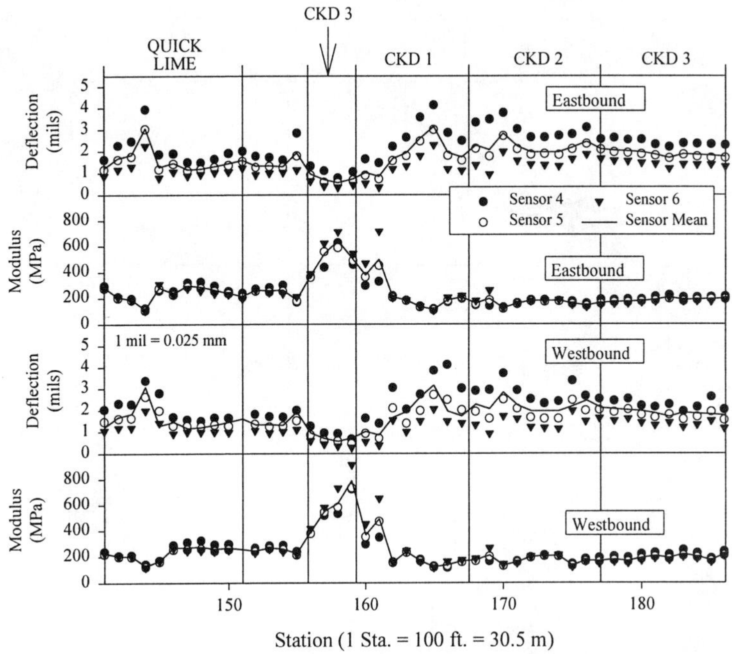

FIGURE 5 Falling Weight Deflectometer Deflection Measurements and Back-Calculated Modului

4) Following the CKD 3 test section mentioned in 3), the next highest modulus values were found in the lime section. The remaining CKD test sections beginning at about Station 163 gave similar FWD results.

To summarize, the FWD results do not seem to be as sensitive to differences between the test sections. In particular, the FWD moduli for the three CKD sections beyond station 160 are similar, whereas UCT and DCP data showed that the stabilized subbase in the CKD 3 section (near Sta. 180) was distinctly different than CKD 1 or CKD 2 sections.

DISCUSSION

A comparison of results from the UCT, DCP and FWD tests conducted along the alignment is presented in Fig. 6. Inverse CI (1/CI) values are shown in Fig. 6 to allow for an better comparison with UCT and FWD data, i.e. higher values of inverse CI indicate higher strength. The results in Fig. 6 indicate that generally the UCT and 56–day DCP results show similar trends except with respect to the CKD 1 and 2 test

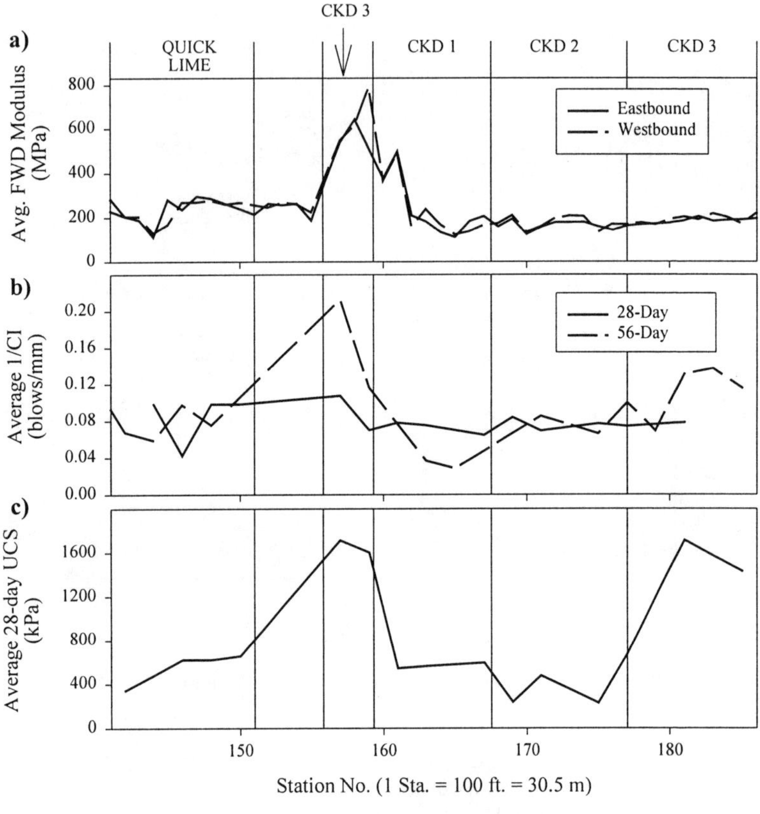

FIGURE 6 Comparison of Field Test Results: a) Back-Calculated Resilient Modulus from Falling Weight Deflectometer, b) Inverse of Cone Index from Dynamic Cone Penetration Test, c) 28-Day Unconfined Compression Test Data

sections. In particular, the CKD 3 is clearly the best performer based on UCT and DCP results. The performance of CKD 3 was corroborated by additional laboratory testing under controlled conditions, which also revealed that CKD 1 and CKD 2 performed similarly. The UCT and DCP test results are expected to show similar trends because both of these tests depend on shear strength, although the DCP test also reflects variations in construction.

On the other hand, the FWD test is non-destructive and results in an estimate of elastic modulus. Since the FWD results depend on small-strain measurements and the DCP and UCT impose large-strain shearing, it is not surprising that similar trends are not seen between FWD and the other test results. Furthermore, the FWD measurements and analysis are influenced by the whole pavement system and therefore the procedure is not as sensitive to differences in the subbase layer. It is likely that FWD tests directly on the stabilized subbase would better reflect the subbase character.

CONCLUSIONS AND RECOMMENDATIONS

The following conclusions and recommendations are presented based on the field study described in this paper.

Conclusions

1) The dynamic cone penetration test (DCP) can be used to assess the relative degree of improvement in a stabilized subbase. Comparisons with unconfined compression test results indicate that the DCP cone index (CI) is a good indicator of the relative shear strength of the subbase.
2) The DCP can be used to assess variations in strength with depth. In many cases, the depth of the stabilized subbase was detected using the DCP.
3) Falling weight deflectometer measurements on the finished pavement were less sensitive than the DCP measurements to differences in subbase characteristics.
4) Differences between 28- and 56-day DCP results indicate that the DCP is sensitive to the curing process and changes that occur under field conditions.

Recommendations

1) The DCP is recommended as an excellent tool for doing relative assessments of treated subbases in the field.
2) The FWD on the finished pavement should not be used for doing relative assessment of treated subbases without doing additional testing (e.g. DCP or UCT on stabilized subbase).
3) A multifaceted approach to assessing stabilized subbases is recommended, including field and laboratory testing.

ACKNOWLEDGEMENTS

The author is grateful to graduate students Jameel Rahman, Norman Tan, and Kin-Phan Ng, for their contributions to this project. Thanks also to ODOT personnel including Wilson Brewer, William Kenneda, and Chuck Donavon for their assistance in the field and to Jim Nevels who provided additional field and laboratory data. Additional thanks to Lawrence Senkowski and David Ooten of ODOT for their support and guidance. Thanks also to Steve Shawn of Silver Star Construction who was responsible for construction of the test sections.

APPENDIX-I. REFERENCES

American Association of State Highway and Transportation Officials (1993), *AASHTO Guide for Design of Pavement Structures*, Washington D.C., 640 p.

American Society for Testing and Materials (1998), *Annual Book of Standards*, Section 4, Construction, Vol. 4.08, West Conshohocken, PA, 962 p.

Azad, S. (1998), *Soil Stabilization of Three Different Soils Using Cement Kiln Dust*, Master's Thesis, University of Oklahoma, Norman, OK.

Baghdadi, Z.A., Fatani, M.N., and Sabban, N.A. (1995), "Soil Modification by Cement Kiln Dust," *Journal of Materials in Civil Engineering*, Vol. 7, No. 4, pp. 218-222.

Baghdadi, Z.A. (1990), "Utilization of kiln Dust in Clay Stabilization," *Journal of King Abdulaziz University.: Engineering and Science*, Jeddah, Saudi Arabia, Vol. 2, pp.153-163.

Burnham, T. and Johnson, D. (1993), *In Situ Foundation Characterization Using the Dynamic Cone Penetrometer*, Report No. MN/RD-93/05, Minnesota Department of Transportation, St. Paul, MN.

Fatani, M.N., and Khan, A.M. (1990), "Improvement of Dune Sand Asphalt Mixes for Pavement Bases," *Journal of King Abdulaziz University: Engineering and Science*, Jeddah, Saudi Arabia, Vol. 2, pp. 39-47.

Livneh, M. (1988), "The Relationship Between In Situ CBR Test and Various Penetration Tests," *Penetration Testing 1988*, Proceedings of the First International Symposium on Penetration Testing in Orlando, Florida, A.A. Balkema, Rotterdam, Netherlands, Vol. 1, pp. 445-452.

Livneh, M. (1989), "Validation of Correlations Between a Number of Penetration Tests and In Situ California Bearing Ratio Tests," *Transportation Research Record 1219*, Transportation Research Board, National Research Council, Washington, D.C., pp. 56-67.

Miller, G. A., Azad, S., and Dhar, B. (1997), "The Effect of Cement Kiln Dust on the Collapse Potential of Compacted Shale," *Testing Soil Mixed with Waste or Recycled Materials*, ASTM STP 1275, American Society for Testing and Materials, West Conshohocken, PA, pp. 232-245.

Sowers, G.F. and Hedges, C.S. (1966), "Dynamic Cone for Shallow In Situ Penetration Testing," *Vane Shear and Cone Penetration Resistance Testing of In Situ Soils*, ASTM STP 399, American Society for Testing and Materials, pp. 29-37.

Todres, H.A., Mishulovich, A. and Ahmad, J. (1992), *Cement Kiln Dust Management: Permeability*, Research and Development Bulletin RD 103T, Portland Cement Association, Skokie, IL, pp. 1-6.

Zaman, M., Laguros, J.G., and Sayah, A. (1992), "Soil Stabilization Using Cement Kiln Dust," *Proceedings of the 7th International Conference on Expansive Soils*, Dallas, Texas

RESILIENT MODULUS BACKCALCULATION TECHNIQUES FOR TRACK

Theodore R. Sussmann[1] and Ernest T. Selig[2]

Abstract

The layer moduli of track substructure components are important factors in the evaluation of track condition and estimation of maintenance costs. Layer moduli can be estimated from in-situ tests such as the cone penetration test (CPT); however, for layers of open graded aggregate, such as railway ballast with large individual particles, the results of the CPT are significantly influenced by individual particles and may not reflect the behavior of the mass. A method of evaluating layer properties using downhole measurement of layer deformation is presented in this paper.

Tests were conducted on track using downhole measurements of layer deformation under both quasi-static test loads and passing traffic. The applied loads were measured by instrumenting the track superstructure. The top of rail load and substructure deformation data were used with a geotechnical model of track to backcalculate the moduli of individual layers. The process is similar to the procedure for processing data from the falling weight deflectometer; however, data from actual traffic can be evaluated. The results are compared to the CPT results from the same site and also to expected results for similar materials.

Introduction

The moduli of the various geotechnical components of railway track (ballast, subballast, and subgrade) need to be evaluated to ensure that corrective rehabilitation of the track provides a suitable structure to support the imposed traffic loads. In addition, design of strategic sections of track, such as bridge-track transitions, requires knowledge of the strength and stiffness of the substructure layers so that the proper materials can be specified to provide suitable track support.

----

[1]Civil Engineer, US DOT/Volpe National Transportation Systems Center, Kendall Square, Cambridge, MA 02142 (previously Senior Engineer, Ernest T. Selig, Inc.) and [2]Principal, Ernest T. Selig, Inc. 105 Middle Street, Hadley, MA 01035

Methods are available to measure the resilient modulus in the laboratory, but samples of many of the substructure components are difficult or impossible to obtain without a high degree of disturbance. Therefore, correlations for resilient modulus based on other tests have been developed to provide a reasonable indication of the resilient modulus. As an alternative, an innovative measurement system was used on existing railway track to measure layer deflections under load. Then a model was used to evaluate the layer resilient modulus in a method analogous to the falling weight deflectometer analysis procedure for pavements.

This technique to evaluate the layer stiffness properties provides a method to determine the properties of the materials in-situ. The actual value that is provided by the model will be referred to as the apparent resilient modulus, since it represents the slope of the stress-strain curve after many load cycles.

Measurements

Elastic and plastic vertical deformations were measured by a multi-depth deflectometer (MDD) installed to a maximum depth of 3 m below top of tie. The MDD was developed in South Africa by the CSIR and adapted by Spoornet (South African National Railroad) for railway use. The MDD consists of a series of LVDT transducers attached to the interior of a flexibly lined borehole. A schematic of the MDD is presented in Figure 1 and a general track cross section showing the typical placement of the measurement modules is shown in Figure 2. The numbers 1-5 along the MDD borehole are the locations of the sensor attachment points in the substructure layers. The letter A designates the bottom reference anchor. The sensors measure the change in vertical deflection between two adjacent attachment points (i.e., 1-2, 2-3, 3-4, 4-5, 5-A). The MDD installation is accompanied by strain gauges on the rails to measure the wheel load and rail seat loads causing the deflections.

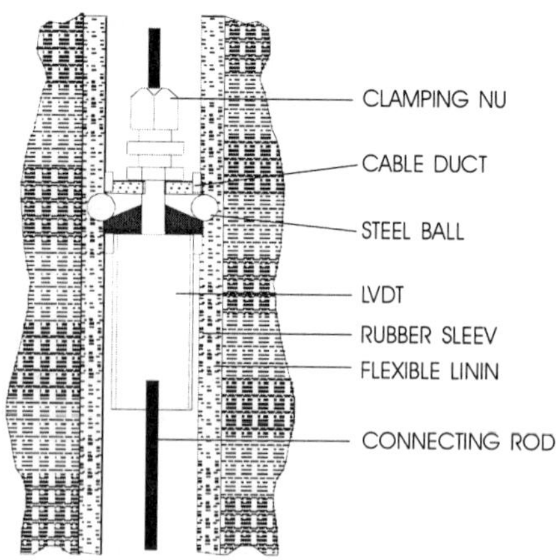

Figure 1 The Multi-Depth Deflectometer Sensor

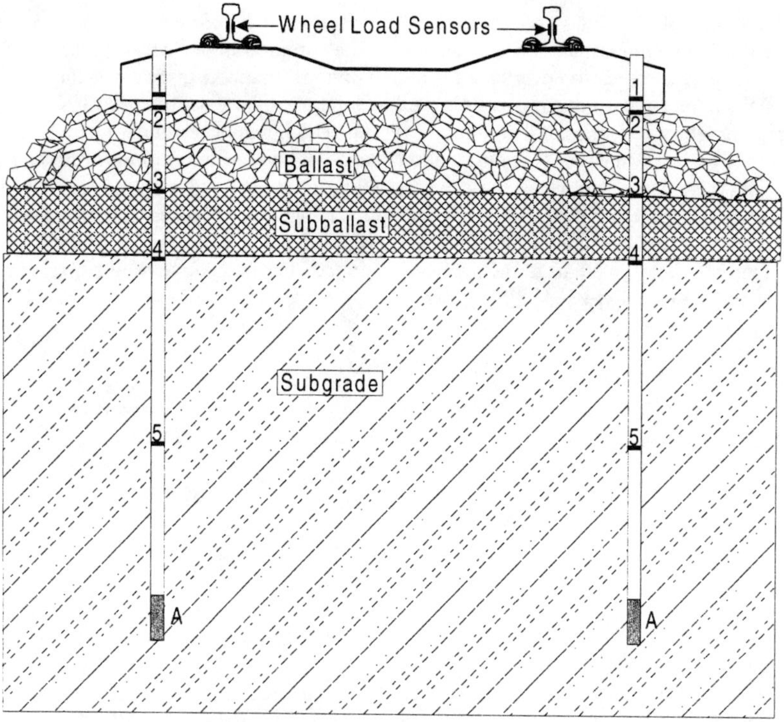

Figure 2 MDD Measurement Locations

The MDD is installed by drilling two borehole of approximately 40 mm diameter through the ends of the concrete tie, ballast, and subballast into the subgrade. Once the borehole is drilled, the anchor and borehole lining are installed. A liquid rubber adhesive is used to bond the flexible lining to the borehole wall. Then each of the LVDT sensors are positioned at layer interfaces, which are recorded during the drilling operation.

Following the installation, static testing was conducted to calibrate the MDD wheel load sensors. During the static testing, the top of rail track deflection was monitored using a digital surveying level and the static offsets of each LVDT within the MDD's were monitored to evaluate individual layer compression. The known load and top of rail deflection data were used to evaluate the track stiffness. Track stiffness is a structural measure that indicates the load-deflection characteristics of the track.

A wide variety of other measurements were taken at the site, but they do not relate to the backcalculation of the layer moduli and will not be discussed in this paper.

The MDD utilizes a high-peed data collection system and therefore the output is practically continuous, as shown in Figure 3. The output from the MDD (Figure 3) is a plot of layer compression over time for each individual borehole that depicts the deflection due to each passing axle above that borehole. The data in Figure 3 is approximately 3 seconds of data from the west tie end borehole, a similar plot exists for the east tie end borehole. This information coupled with knowledge of the imposed load based on the wheel load measurements from the instrumented rail provides the basic data that was used to evaluate the layer resilient moduli. Moduli estimates were made using maximum wheel load and layer compression values for each axle. A total of 4 track sites (8 total boreholes) were instrumented for this project, the cumulative results from one site (2 boreholes) are discussed in this paper.

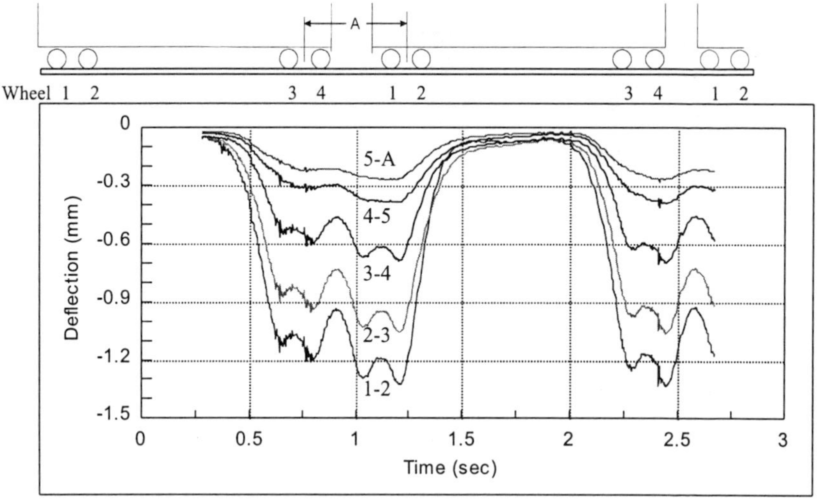

Figure 3 Typical MDD Results

Modeling

To analyze the layer deformation and wheel loads measured by the instrumentation at the test site, the track structure was modeled using Geotrack. Geotrack is a three-dimensional, multi-layer model for determining the elastic response of the track structure and is more fully described by Chang et al. (1980).

Geotrack does not consider moving or changing loads, so the data from the instrumentation was analyzed to determine the maximum deflections and associated wheel loads to use as input to Geotrack. This load and deflection data were then used with initial estimates of the layer resilient moduli to determine the initial layer deflections and track stiffness. The track stiffness was compared to the data obtained from the static calibration and the deflections were compared to the dynamic layer deflection MDD data to evaluate the accuracy of the initial resilient modulus estimates for each of the layers. Then, a trial and error iterative process was followed whereby each layer resilient modulus was varied in order to obtain model results that matched as closely as possible with the track stiffness and dynamic layer deflection measurements.

Results

The results of the static calibration tests conducted to calibrate the instrumented rail section for wheel load were used to determine the deflection variation with depth for each MDD, as shown in Figure 4 and to evaluate the track stiffness based on the load-deflection response of the track, shown in Figure 5. These parameters were matched as closely as possible using Geotrack with estimated values of the resilient modulus for each layer. Then, the maximum dynamic deflection for each layer and corresponding wheel load from the passing traffic were determined based on the continuous output of the MDD and used to refine the results to ensure dynamic response of the track was reflected in the results for the predicted apparent resilient modulus.

Table 1 presents the apparenet resilient modulus for each of the layers (shown in Figure 6) along with comparison to the results obtained from correlation to the cone penetration test (CPT) results and published or expected values. The CPT data used for the correlation to resilient modulus are presented in Figure 7. Correlations between CPT tip resistance and resilient modulus have been developed by Selig and Ebersohn (1996) and El-Sharkawi (1991) specifically for railway track materials. Based on this research a correlation was found between resilient modulus and CPT tip resistance even for large diameter ballast particles.

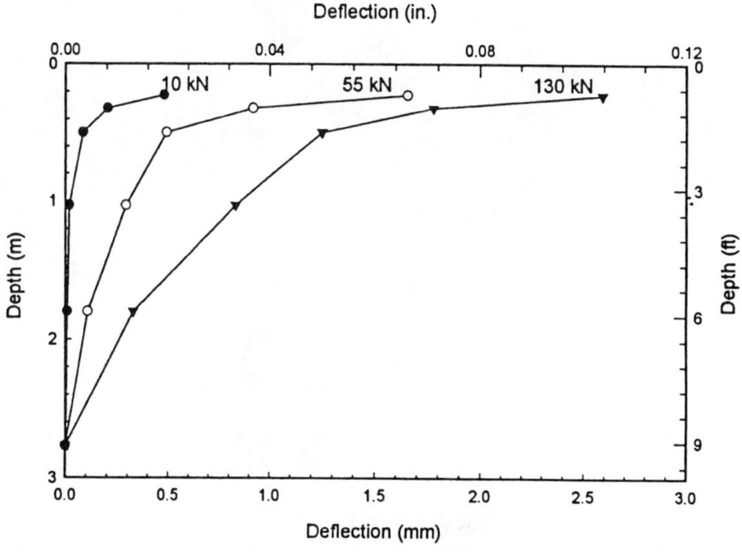

Figure 4 Measurements of Deflection with Depth

The clean granite ballast material is a standard AREMA (American Railway Engineering and Maintenance of Way Association) gradation material with maximum particle size of approximately 37 mm. The modulus back-calculated from the MDD data was on the order of 55 MPa, which does not compare well with published results (Selig and Waters, 1994) of 310 MPa or the results from correlation with the CPT data of 325 MPa for similar rock type, gradation, and void ratio.

The low modulus from the MDD measurements may be attributed to separation of the flexible borehole lining from the borehole wall which was observed during testing. This separation does not usually occur, but since the lining passes through the tie, any longitudinal or lateral tie movement can cause separation of the lining and the borehole wall. The track at the test site was surfaced (tamped) prior to the collection of the MDD data and the tie may have been inadvertently moved at that time. Once the lining and the borehole wall are separated, the first and second sensors in the MDD may measure increased deflections due to tie seating. The high deflection measurement results in low layer stiffness.

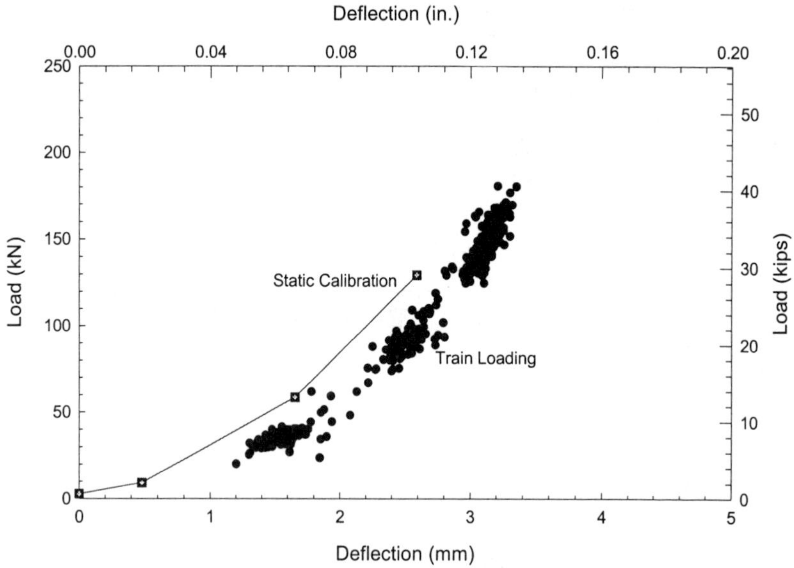

Figure 5 Track Load-Deflection Characteristic

The wet fouled ballast layer was a ballast material that has degraded over time and traffic resulting in a well-graded material containing ballast particles up to 37 mm and fines that may consist of both silt and clay. The wet fouled ballast from the test site contained fine material with high plasticity indicating a fairly high clay content resulting in a clay-gravel mixture. The modulus back-calculated from the MDD data was on the order of 34 MPa, which compares well with the results from correlation with the CPT data of 30 MPa, but not with published results of 140 MPa (Selig and Waters, 1994). The expected results are for poor quality fouled ballast of similar rock type and gradation. The field identification of the sample indicated that the material was a poor quality fouled ballast, but the high clay content at the test site resulted in a lower modulus than published for typical fouled ballast material.

Table 1 Track Substructure Modulus Results

Description (Layer Location/ Field Description)	MDD Modulus MPa (psi)	CPT Modulus MPa (psi)	Published Value MPa (psi)
Top Ballast/ Clean Granite Ballast	55 (8000)	325 (47000)	310 (45000) (A)
Bottom Ballast/ Wet Fouled Ballast	34 (5000)	30 (4350)	140 (20000) (A)
Subballast/ Moist Sand and Gravel	58 (8400)	65 (9400)	70 (10000) (A)
Fill or Upper Subgrade/ Moist Stiff Clay	76 (11000)	55 (8000)	80 (11600) (B)
Subgrade/ Moist Soft Clay	16 (2300)	15 (2200)	17 (2500) (B)

Notes:

A = Selig and Waters, 1994.

B = Li, Sussmann, and Selig, 1996.

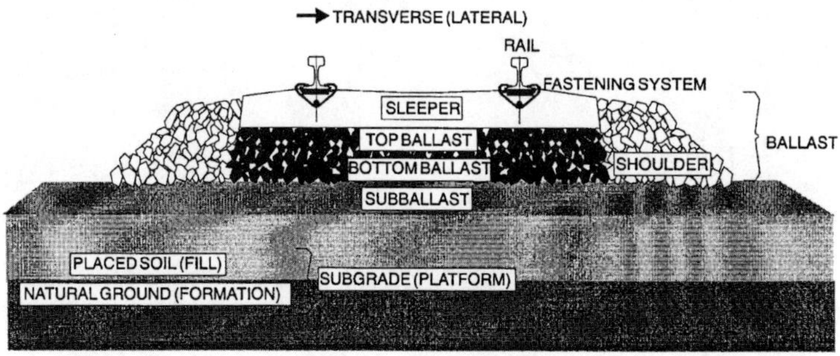

Figure 6 Track Substructure Layers (from Selig and Waters, 1994)

The sand and gravel layer was below the wet fouled ballast layer. The clay content in the wet fouled ballast resulted in a low hydraulic conductivity that limited penetration of moisture into this lower sand and gravel layer, allowing the sand and gravel to remain in a dry to moist condition rather than wet or saturated. The layer modulus from the MDD backcalculation was approximately 58 MPa, which is lower than the value from the correlation with the CPT of 65 MPa and the published value of 70 MPa, but on the same order of magnitude. The MDD value is approximately 11% lower than the modulus from the CPT data, which is likely to be within the error associated with the CPT correlation. The published values are considered to compare favorably with the MDD results since the difference between the two is on the order of 21% and only the field description was available to establish expected results.

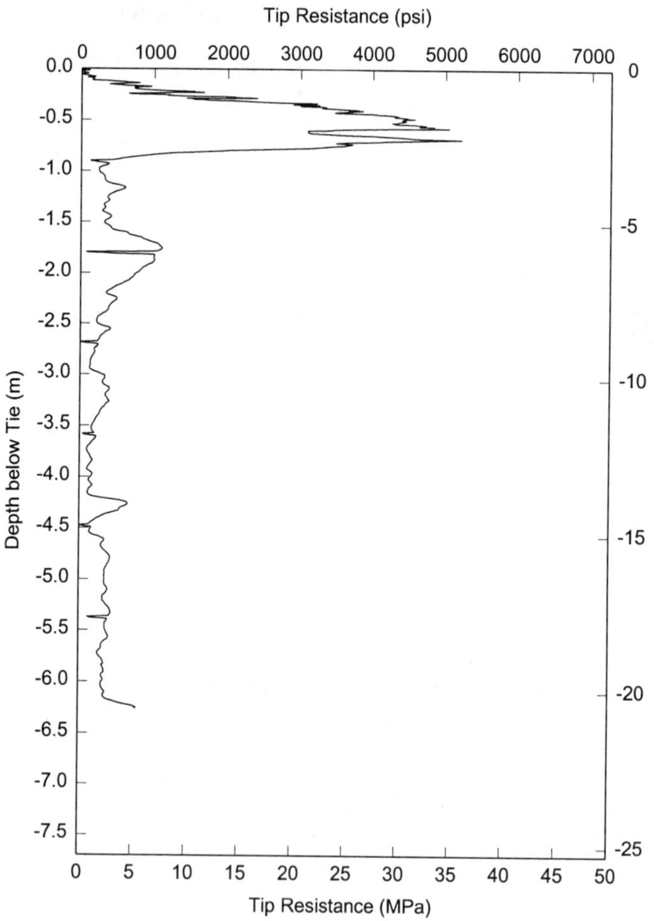

Figure 7 Cone Penetration Test Results

The moist stiff clay located in the upper subgrade or fill section, based on Figure 6 was below the sand and gravel. The stiff nature of the clay is most likely the result of compaction from traffic loading and possibly track construction. The apparenet resilient modulus results obtained from the MDD data and backcalculation of 76 MPA compares well with the published value of 80 MPa (Li et al, 1996) and is higher than the results of the correlation with the CPT tip resistance of 55 MPa (El-Sharkawi, 1991). The difference between the MDD modulus and the CPT correlation value is on the order of 27% and the difference may be the result of local variability of the clay layer along the track and possible inaccuracy of the correlation.

The moist soft clay layer was the deepest layer tested and appeared to be the only layer not placed or disturbed during construction. The measured apparent resilient modulus based on the MDD data and backcalculation was on the order of 16 MPa, which compared well with both the CPT correlation (El-Sharkawi, 1991) and published data (Li et al, 1996).

Summary and Conclusions

An innovative method to measure dynamic track substructure response and applied load was utilized with a geotechnical model of railway track to back-calculate the in-situ resilient moduli of the various substructure layers. In addition, a correlation to resilient modulus based on cone penetration test (CPT) tip resistance was used to evaluate the results of the backcalculation procedure. The backcalculation technique produced results that compared favorably with both the values obtained from correlation with CPT and expected published results for similar material with one exception.

The results from the uppermost layer appeared to be in error most likely due to separation of the flexible borehole liner from the soil at the borehole wall. This can occur when the tie is moved laterally or longitudinally during maintenance. The separation will result in transmission of tie deflection to lower sensors due to the stiffness of the liner even though a flexible liner was used to ensure compliance of the measurement with soil movement. The results from all other layers appeared reasonable and differences between the results from the CPT and MDD were attributed to local soil variability and errors inherent to the analysis or correlation techniques.

The downhole measurement of layer deflection is very useful for evaluating elastic response, and the data can be used to develop more advanced models for track response to load. The backcalculation of resilient modulus is only one application of the data from the MDD.

Acknowledgements

The authors wish to acknowledge Amtrak and the former Conrail, since both have supported the use of geotechnical investigations for evaluation of railway track. The analysis methods presented in this paper were developed at the University of Massachusetts, Amherst and the instrumentation was developed and installed by Spoornet. The cone penetration test (CPT) data presented in this paper was obtained using the Association of American Railroads CPT vehicle.

References

Chang, C. S., Adegoke, C. W., and Selig, E. T. (1980). "The GEOTRACK Model for Railroad Track Performance," Journal of the Geotechnical Engineering Division, ASCE, Vol. 106, No. GT11, November, pp. 1201-1218.

Ebersohn, W. E., Selig, E. T. (1996). "Evaluation of Substructure Using Field Tests," Association of American Railroads Transportation Technology Center Report No. LA-009, Pueblo, CO, December.

El-Sharkawi, A. (1991), "Correlation of Railroad Subgrade Resilient Modulus with Cone Penetration Test Data," University of Massachusetts, Amherst, Geotechnical Project Report No. AAR91-389P, September.

Li, D., Sussmann, T. R., and Selig, E. T. (1996). "Procedure for Railway Track Granular Layer Thickness Determination," Association of American Railroads Transportation Technology Center Report No. R-898, Pueblo, CO, October.

Selig, E. T., Waters, J. M. (1994). Track Geotechnology and Substructure Management, Thomas Telford Services Ltd, London.

Grouting in Geotechnical Engineering

DESIGN VERIFICATION INTO THE GROUTING PROGRAM

Richard M. Berry, M ASCE[a]

Abstract

Verification of the success or failure of grouting is rarely possible as an afterthought when a project is complete. Also, verification often is difficult and expensive. If potentially effective methods of verification are considered right along with the project and its objectives during the planning stage, there is a good chance of coming up with a reasonably effective verification scheme.

The paper outlines example projects where performance was its own verification. It also covers areas where direct measurement and monitoring can prove effectiveness, where use of tracer materials will track performance, and where locating the grout seems to be the only way to ascertain performance and verify quality. Two projects are described where complex and combined methods were used, methods which showed some of the possibilities which exist if the proper fact gathering is designed into the system from the start.

Introduction

Oftentimes, a grouting project can constitute its own verification when the effectiveness of the entire grouting process can be measured. One example is when the grouting is done to stop water inflow or to re-direct underground water migration. This requires that sampling points be established for measuring the effectiveness of the grouting.

Another verification method is the self checking situation which results when the grouting is done to stabilize settled structures and possibly to compensate for settlement by lifting. Both the immediate lift and the future settlement can be monitored.

However, in many other cases, grouting is undertaken as a protective measure when there is no direct way of measuring the effectiveness of the overall process. A secondary method is to verify if the grout is in the locations where it is supposed to be.

[a] Chairman, Rembco Engineering Corporation, PO Box 23009, Knoxville, TN 37933-1009, 865/690-6917, email berry@rembco.com.

This can involve drilling into the zone and measuring properties before and after the grouting. Cross hole tomography can be used in some cases. Gas and water measurement using tracers is sometimes a useful procedure.

Innovative thinking and planning of the verification strategy all the way from the initial planning can result in cost effective methods being used. This paper summarizes some projects where innovative and unusual approaches were effective. A valuable report on the subject of verification methods was published by a task force of the ASCE Grouting Committee in 1995.[1]

Performance of the Grouted System Using Tracers

●● **Leaking Storm Sewer Outfall.** Leakage at storm sewer Outfall 302 at the Oak Ridge National Laboratory (ORNL) showed a buildup of radioactive materials that was coming from a nearby liquid waste storage pond.[2] The sewer was in good condition structurally but had leaking joints. Part of the sewer was relined by a cured-in-place pipe (CIPP) liner which cut off immediate leakage. However, the pipe trench gradually became a new conduit allowing pond leakage to continue to the same small stream. It soon reached a contamination level about equal to that which had been measured before the new liner was installed. Chemical grouting using acrylamide gel grout was chosen to seal the trench.

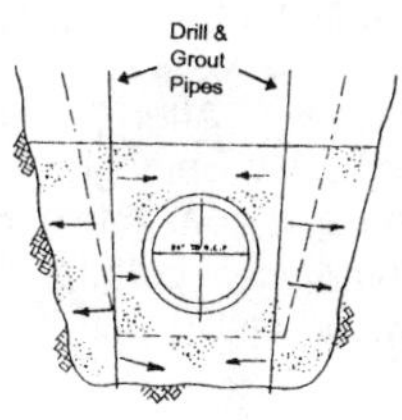

Fig. no. 1

The presence of radioactive materials in a stream can provide an instant and automatic method of monitoring. In addition to measuring radioactivity, it was possible to observe and measure water flow at the end of the sewer pipe trench. Thus, both flow and radiation level were directly measured. Figure 1 shows the grouting scenario. Figure 2 is a photo taken during the work. Based on measured grout takes as the grouting operation moved along the trench, the grout not only filled the trench but also was forced back into leaking areas of the pond berm, renewing its seal. After the grouting was completed, all radioactive material drainage at the outfall had stopped.

Drilling in the Radiation Area

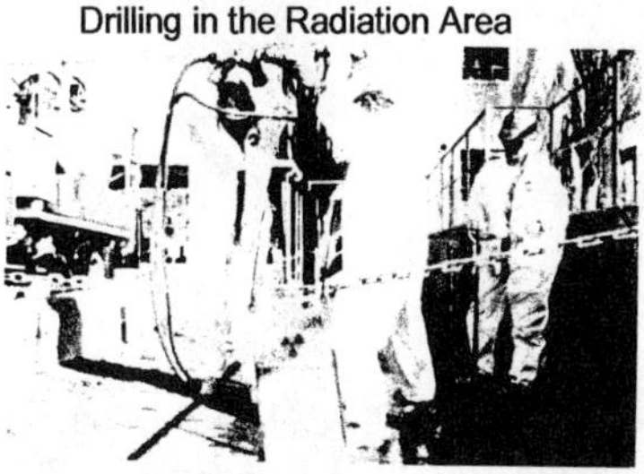
Fig. no. 2

●● **Coal Impoundment Dam.** United Coal Company runs a large coal washing plant in southern Virginia. They have the usual dam and impoundment area typical of the Appalachian mountain coal area. In extending the capacity of the impoundment, often times large volumes of rock are excavated to increase the volume of the fines settling pond. This leaves a high

Cross Section at the High Wall

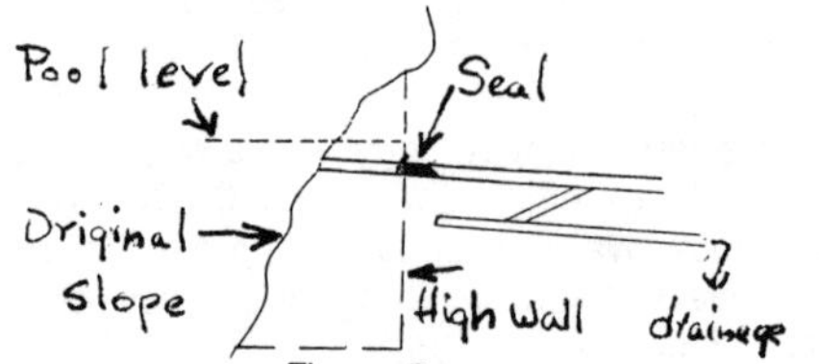

Fig. no. 3

wall for at least part of the settling pond as the level of sediment rises. In this case, the excavation exposed four formerly underground drifts from an old and abandoned mine. The ends of the drifts were filled with rock and shale and then had clay seals placed over the outside surface. Those clay seals would become the impervious inside of the impoundment.

When the coal fines slurry from the washing operation reached a vertical level of about twenty feet above the top of the drifts, a significant flow of black coal fines appeared in the stream at the base of the mountain. These fines were coming through the old mine from the impoundment. A grouting remediation program was considered and accepted. Figure 3 shows both the situation and the grouting scenario.

Operating from a bench at the edge of the impoundment, compaction grout piles were installed across the face of the drifts in the rock fill to prevent a major and possibly catastrophic breakthrough. The installation of the grout pipes through the rock and rubble field which had been placed across the front of the drifts, as shown in Fig. 4, was not an insignificant task as it had to be done through a boulder field. After the compaction piles were in place, the clay seals were restored using Bentogrout, a time release form of Bentonite. Bentogrout can be pumped at a concentration of about 45% by weight (compared to a maximum concentration of perhaps 12 or 15% for ordinary Bentonite). Bentogrout does not expand until it wets out and breaks down the coating, which takes 30 or 40 minutes.

Compaction Grouting

Fig. no. 4

To monitor the work, a weir was set up where the old mine drainage water exited to the river. This allowed recording of the flow and sampling of the drainage. Coal wash fines are basic as a result of the flocculents and washing chemicals used; they have a pH of about 8.5. The drainage from the old abandoned mine was an acid waste having a pH of about 5.5. Samples for pH testing and flow readings were taken each shift from the weir during the work.

During the compaction grout pile installation, flow dropped somewhat and by the finish of that work, the liquid drainage had turned clear. However, the pH was still basic. As the Bentogrout mixture was pumped in, the pH of the drainage water gradually went acidic taking on its old and obnoxious sulfur odor and amber color, leaving a polluted stream but a secure reservoir. Another case where the verification was in the measurement of the entire process.

Where Grouting Acts as its own Proof and Verification

●● **A Leaking Seawall.** At a BP Oil refinery near Philadelphia, a leaking seawall had begun to allow oil and petroleum chemicals to create a sheen on the river. The amount of oil gradually increased, and containment booms were laid out to absorb the oily seepage. Oil pickup frequency from behind the booms had increased to where service was required every two to three days. The seawall was a concrete gravity wall constructed on top of an old wharf modified to expand the refinery size during World War II. Through aging and tidal action, soil had washed out from under the base allowing voids to form behind the wall. The base of the wall was near the tidal line itself. When the tide came in, water washed under and into the voids and when the tide went back out, it carried oil with it.

Wharf Cross Section

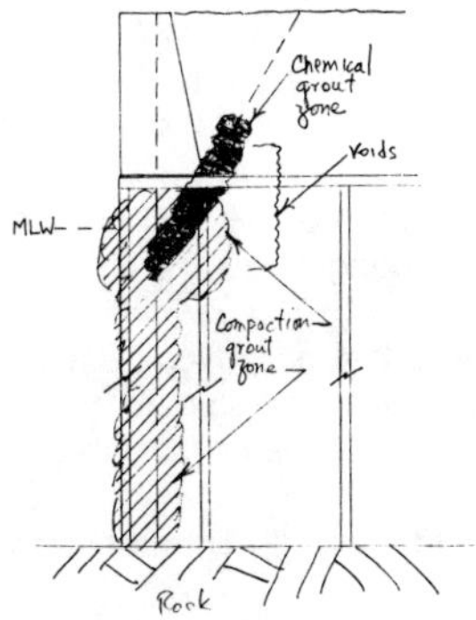

Fig. no. 5

Grouting to create a barrier and lock the oil behind the seawall was a suggested remedy. The plan was for a three step operation. Expansion joints, holes where anchors came through, and some leaking cracks in the concrete were sealed with a polyurethane grout on a spot basis. Large voided zones were found behind the wall and these were grouted using a thick low mobility cement based grout. The grout filling these voided zones was set on grouted columns formed in the soil from this same grout which extended down at intervals to the bedrock to prevent settlement of that heavy grout of its own weight. Then the area along the toe of the wall was permeation grouted with water thin acrylamide gel grout to provide a low permeability (10^{-8} cm/sec) barrier. See Fig. 5 for a cross section view.

The leakage had occurred along about 100 feet (30 m) of the wharf, but the grouting was carried out along the entire 500 foot (150 m) length as a preventative measure against possible recurrence. Containment booms were left in place for about a week to be sure that all seepage had stopped, at which time they were removed.

●● **Chemical Plant Column Settlement.** At a caustic-chlorine plant operated by LCP Chemicals in Brunswick, GA continued column settlement caused operating problems as well as a structural problem with the building. Caustic drainage from the plant had eaten away the organics in the silty sand and may have been responsible for some or all of the settlement.

The four story building was supported on spread footings. The upper stories contained the electrolytic cells. Cell alignment in such a plant is important in that the cells depend on a slow gravity flow across the electrodes for proper operation. Misalignment meant inefficiency and early electrode burnout. Many of the cells had been shimmed to keep them in level operation, but the plant was reaching its limits for additional shimming.

A hardpan layer of soil existed below the muddy ocean bottom about 60 feet (18 m) down. Compaction grouting was chosen as the remedial process. Four grout piles were created in the soil under each footing and extended to the hardpan. The proof of effectiveness was to be small controlled lift of the footing as measured in the columns, but that lift was limited to less than 1/8th inch (3 mm); greater lift would create new operating problems. The work was done while the plant was in operation.

The process selected was one of stage grouting, where the first stage is placed directly under the footing, the next stage below that, and so on. Stages were about five feet in depth. At about 20 feet (6 m) below the footings, the lift took place. Then to avoid the large amount of drilling associated with stage down grouting all the way to 60 feet (18 m), a two inch (50 mm) casing was set down to the hardpan bottom and the remainder of the grouting took place as the casing was withdrawn. During this latter grouting, a 5/8 inch (16 mm) rebar was grouted in place as the casing was withdrawn, so that the rebar extended into the footing.

The proof of process effectiveness was in the lift. Elevations of all columns were benchmarked to provide a verifiable set of readings in case of future settlement. None has occurred after 15 years.

Verification by Direct Methods

●● **A Hotel Ballroom Addition.** The Radisson Hotel in downtown Knoxville, TN was undergoing extensive modifications to add a new ballroom.[3] The ballroom was to occupy the open top deck of a three story parking garage built into the hotel. The original parking garage was on spread footings. With the new addition, each affected footing had to carry an additional 100 kips (45,000 kg). The construction plan was to keep as much of the parking structure open as possible during the work and not to lose any parking spaces when it was completed.

A three stage program was designed. First, four compaction grout piles were installed through each footing and carried down to competent rock which turned out to be located from 15 to 65 feet (5 to 20 m) below. This gave a 400 kip (185,000 kg) increase in load bearing capability (calculated as ultimate loading). Then the interior columns on the first and second floors were wrapped with a four inch (8 mm) reinforced concrete section to increase their load bearing capacity. This was done with bolt together, wrap around forms which were filled with coarse aggregate as they were assembled and then grouted with a slightly expanding cement based grout using a pre-placed aggregate process. Finally,

60 foot (18 m) long beams, which were adequate for the final loads but would deflect seriously when the new room was built, had external reinforcement applied using a steel plate bonding process.

The control and monitoring system was one of precise measurement at selected points during the entire construction period. Columns on both the first and second floors were benchmarked. On the lower floor, this was to monitor for footing movement, and on the second floor it was to observe load transfer from the new enhanced wrap around sections to the columns directly above.

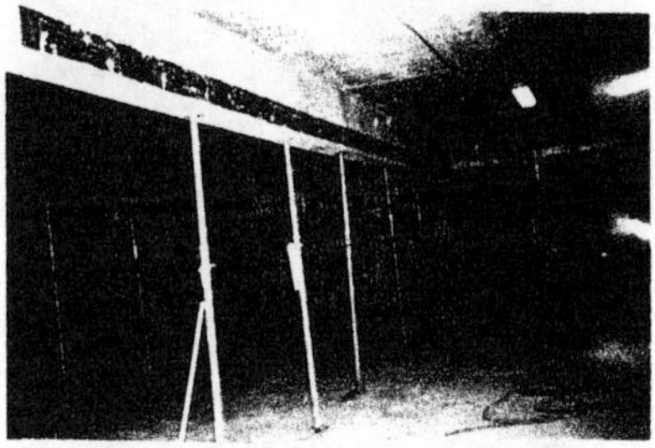

Fig. no 6

To install the steel plate reinforcement, epoxy bonding was used. The 60 foot (18 mm) long beams were jacked upward giving them a positive camber of 3/4 inches (19 mm), which was the calculated deflection which would occur when the new room loads were added. Figure 6 shows the steel reinforcement with the jacks used to camber the beams upward. The new construction loads ultimately returned the beams to within 1/16th inch (2 mm) of their original position.

Fig. no. 7

Steel plate bonding was also used to strengthen one 20 foot (6 m) wide floor section which had a new wall planned on the floor directly in the center of its unsupported span. The steel there was applied in a passive manner. Still, the measured deflection was only about 1/16th inch (2 mm). The floor reinforcement plates are shown in Fig. 7. Verification was in the measurements and the overall performance of the system.

●● **Tank Settlement at a Process Plant.** Long term history can become a verification method. At an A. E. Staley corn syrup and alcohol plant in Loudon, TN, five product storage tanks had a continuous record of settlement.[4] They were built of stainless steel, 12 feet (3.5 m) in diameter and 33 feet (10 m) tall. They were constructed on a "mud" mat located just inside an outer wall of a manufacturing building. The floor on which they sat was on fill soil, placed to give a dock height floor about 4 1/2 feet (1.4 m) above the natural ground elevation. The tank shells extended through holes in the concrete second floor of the plant about 20 feet (6 m) above. The tanks began to settle shortly after the plant went into operation and were monitored weekly by plant forces. As the tanks settled, they leaned outward toward the outer wall of the plant so that the shells contacted the floor above.

Working on the theory that the settlement was shallow and that the problem was in the fill section between ground level and dock height, slurry cement grouting was used under the empty tanks to raise them and correct their tilt. However, the settlement was deep seated and continued after that grouting, in fact, at an increased rate due to the added weight of concrete. Bedrock was some 40 feet (12 m) down and the soil was marginal all that way.

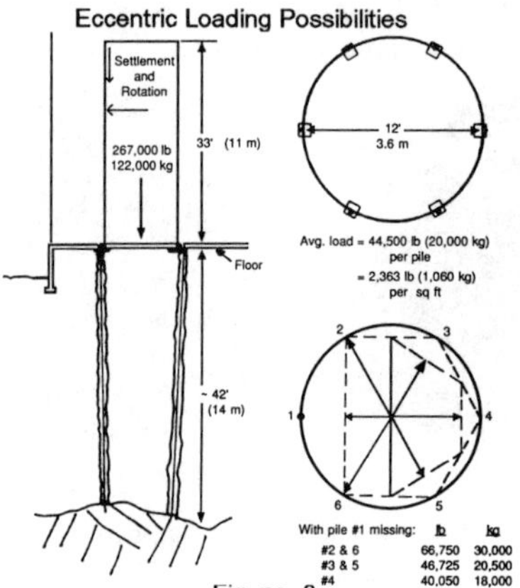

Fig. no. 8

In what we believe is an intriguing fix, a six pile compaction grout pile array calculated to be adequate even if one of the six were to fail. Figure 8 shows the arrangement and calculations. In a fit of conservatism, the plant engineers decided on eight piles. During the compaction piling, the tanks were re-trimmed to vertical. The work was done while the tanks were in use and filled to various and random levels but never less than 1/4 full. During the drilling for the compaction grout casings, lenses of the previous cement grout were randomly encountered. Verification was in the deep seated lift and was confirmed by continued monitoring. Recording of tank elevations was continued for at least a year, and no additional settlement was measured.

Is the Grout is where it is Supposed To Be?

●● **Mine Shaft Soil Grouting.** The ten foot (3 m) diameter Service shaft located at the Weeks Island, LA site of the Strategic Petroleum Reserve (SPR) had small continuous leakage which created a potentially catastrophic failure situation. The 96 year old shaft had been installed using the pneumatic method; that is, using compressed air to hold back the sand and water, and tiering the excavation and liner down from the top. About forty feet (12 m) below the sand/salt interface, there had been a partial collapse during construction which dissolved and/or eroded out a substantial amount of the surrounding salt. This resulting void was funnel shaped around the shaft and was filled with ocean bottom silty sand.

After repairing this area of cave-in, the shaft had been excavated another ten feet (3 m) down into the salt by blasting. During construction, asphalt was installed as the seal between the salt and the concrete liner. During many years, small but continual leakage dissolved out salt and enlarged the funnel shaped void to where it came to within one to four feet (0.3 to 1.2 m) of the bottom of the shaft. The incentive for a protective fix was the $5 billion worth of crude oil stored below which would not be able to be pumped out if the shaft collapsed and the drifts and pump room flooded. Figure 9 shows the before-the-project design. The fix was to solidify the sands so that a catastrophic failure could not take place. The grouting was to be complete enough so that there was time to mobilize to seal off any small leakage should it begin to develop.

The running sands outside were of such consistency that a 1/2 inch (13 mm) diameter hole drilled through the liner would weep water, but if the sands were disturbed, they would run in. The groundwater pressure in this zone varied from 7 to 20 psi (48 to 140 kPa). Grain size tests were made and the quantity of fines in the soil indicated a water thin gel grout was needed. Permeation grouting was called for. The soil was estimated to have 35% voids using a Corps of Engineers based formula. Groutability tests were run at three depth locations using saturated brine as the medium, and the results indicated a maximum pumping rate of about 3.5 gpm (16 l/min) [5]. The grout was a stabilized 15% acrylamide gel. Grout gel times and quantities for each injection were set using the Karol "short gel time theory." [6]

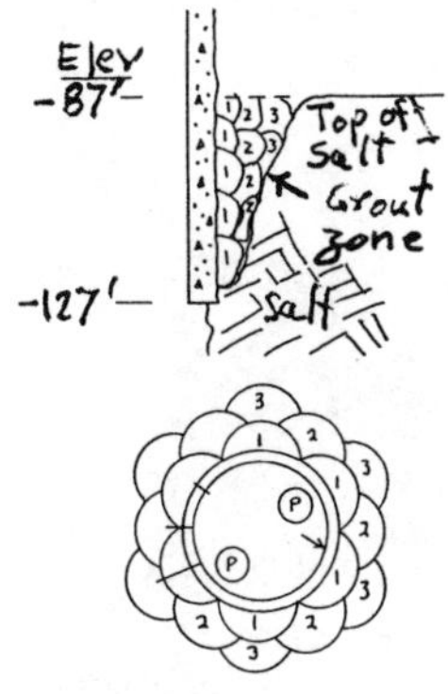

Fig. no. 9

Grout Takes, % of total		Sector	
Bottom to top		Horizontal	
A	0.6%	NE	46.7%
B	9.7	SE	11.7
C	44.0	SW	14.8
D	30.7	NW	11.8
E*	15.0		15.0
	100.0%		100.0%

* Amount above top of salt

Fig. no. 10

The shaft grouting results are shown in Figure no. 10. The pattern of drilling was irregular because of a congested mass of oil and ventilation pipes, utilities, and a Halon fire protection system installed in the shaft.

To verify the completeness of the grout coverage, the only reasonable scheme seemed to be to drill back into the pattern at those locations which were in-between where the grout bulbs were centered, and that was done. There were four pockets where ungrouted soil was found, but they were small pockets and surrounded by grouted sand in all cases. In a recalculation of the actual void volume filled from the grouting logs, there was an apparent over grouting of about 15%. This excess material probably forms a grouted mat above the "technical" top line of the void. The project was deemed a success by the client and there never has been any measurable leakage in the 13 years since.

●● **An Underwater Pipeline Repair.** A six foot (1.8 m) diameter fiber glass pipeline at a Kimberly Clark paper plant carries effluent to a waste treatment plant under the Mobile River in Alabama.[7] It had recurring leakage at joints along a 100 foot (30 m) section. The pipeline is located under about thirty feet

(9 m) of water and thirty more feet (9 m) of river mud. Borings adjacent to the pipeline suggested that a section of the pipe trench had collapsed when the sand fill was being placed preparatory to installation.

Drilling/Grouting Mast

Fig. no. 11

A fix was designed to stop the movement. The program included concrete saddles over the top to weight the pipe down and prevent it from lifting when it was less than full. To prevent lateral movement, grouting close alongside was chosen. Grout was placed in a zone running from about two feet (0.6 m) below the bottom of the pipeline to about two feet (0.6 m) above the top. Figure 11 shows a view of the drilling mast from the barges. The grouting process called for permeation grouting of the sand-mud. The grout selected was a water thin polyacrylate gel material to obtain maximum penetration.

Work Setup on Barges

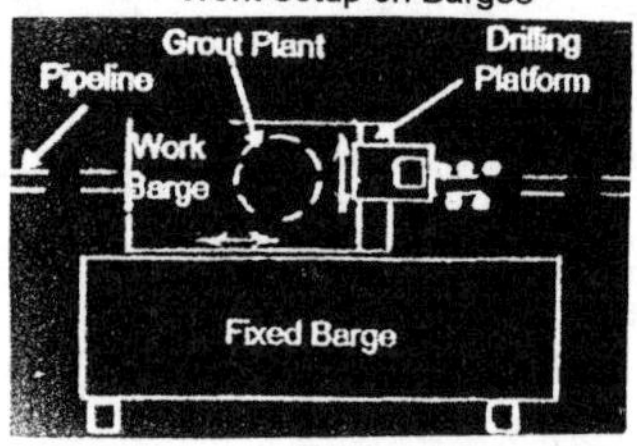

Fig. no. 12

The work was done from a two barge setup where the drilling platform could be moved along the pipeline, see Fig. 12. Finding the pipeline was a tricky procedure, re-done about every ten feet, since the river is tidal and the anchor barge was subject to considerable shift with the tidal current even when spudded in. Confirming the precise location of the edge of the pipe was done using the grout pipe itself. That pipe was tipped with an open center bit. Jetting in grout pipes cannot be done when fine silty or clayey sands are present as movement of fines during the jetting will plug off the sand's ability to accept the grout. When the center of the pipe was located by gentle probing, the injection point was moved to a location about a foot (0.3 m) off the edge of the pipe to the proper position for grouting. The grouting was done from the top of the zone downward to create a cured layer of grout on top thus preventing flotation of the new liquid introduced.

There was no good method of sampling that suggested itself, but a consensus of the project group team thought that some determination of the grout location should be attempted; that is, see if the grout was where it was supposed to be and that it had cured. So, before the grouting began, SPT tests were run at a number of locations reasonably close to the pipeline to obtain "before" data. Then as the grouting proceeded, after the fact SPT tests were run about every ten feet (3 m) along the pipe. The grouting created a mass with 300 to 800 psi (2000 to 5500 kPa) unconfined compressive strength and brought the very low blow count of the river bottom mud up to a clearly discernible number. There have been no additional leakage problems with the pipe.

Two Projects with Complex Verification Schemes

●● **Sinkhole at Weeks Island.** In 1994, a large sinkhole developed just above the edge of the SPR upper level oil storage cavern, which was located about 650 feet (200 m) below ground level.[8] It was large as settlement holes located on the top of salt domes go, about 40 feet (12 m) in diameter and 35 feet (11 m) deep, circular in shape, and had a volume of about 1050 cubic yards (850 m^3) after it fell in. That fall-in was of short duration, narrowed down to a two hour period. At that location on the dome, it is about 175 feet (50 m) to the top of the salt and groundwater level is normally about twenty feet (6 m) above that.

The fill pipes to the storage cavern have a crude level monitoring capability. All during 1994, there had been a general rise in the leakage level, but with a cavern capacity of about 95 million barrels (532 million m^3), taking measurements using a single standpipe is not very accurate. However, on a long term basis, the leakage pattern was positively up. The problem sorted itself out as very complex. Was the apparent sinking related to flow going into the cavern at all? If so, fresh water leaks into salt mines can grow very rapidly. If it was going into the cavern, where was the sinkhole throat? Or was there one discreet sinkhole throat? Was grouting a possibility to seal the sinkhole? Was major shifting at the edge of the cavern a continuing possibility? To answer these problems, the DOE set up a task force with members from Sandia National Laboratory, mining consultant PB-KBB, grouting consultant/contractor Rembco, and DynMcDermott, who operate the SPR for the DOE.

A large number of solutions were raised and assessed quickly, and the following actions were initiated:

- The sinkhole was filled with sand to make working on the surface easier.
- Four vertical wells were established equally spaced around the apparent center of the sinkhole. These were installed for logging and cross hole tomography.
- Then two angular holes were drilled, based on initial cross hole data, to intersect the presumed sinkhole area and determine its configuration. With meticulous care in measurement and control of drilling (or blind luck), both slant holes went through portions of the throat of the sinkhole and came out the other side.
- Dye was injected, and sampling was set up at the fill hole pipes to physically detect that dye. Rhodamine WT was selected as the dye based on its visual detectability down to about 0.1 ppm and instrument detectability down to 1 ppb.
- A directional flowmeter developed by Sandia National Laboratory was installed at the bottom of one slant hole which placed that meter directly in the throat of the sinkhole itself. It was to monitor both direction and quantity of flow.

Cross Section of Sinkhole Area

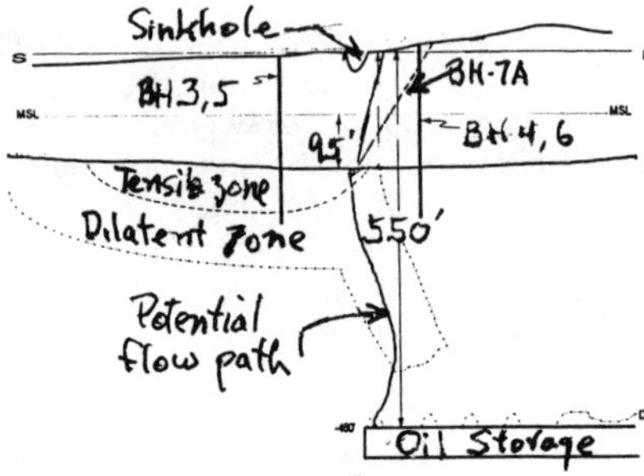

Fig. no. 13

The situation after the exploratory drilling was complete is shown in Fig. 13. The slant holes, BH-7A and 9, were drilled from two directions and intercepted the sinkhole at two different elevations. The injected dye was never detected. The well logging and cross hole tomography shown in Fig. 14 was done across boreholes BH-5 and BH-6, and showed a highly eroded or dissolved surface at the salt/sand interface. The directional flow meter indicated the flow was vertical and flowing at a rate of about 2 gpm (8 l/min), with that amount running into the cavern and not off the top of the salt as was thought probable. The flow meter could be read continuously at the well head. Data was stored in memory and could be retrieved and processed remotely. A plot of the leak rate vs. time showed it increasing and could possibly reach an out of control rate within a few months.

P wave Velocity Tomogram

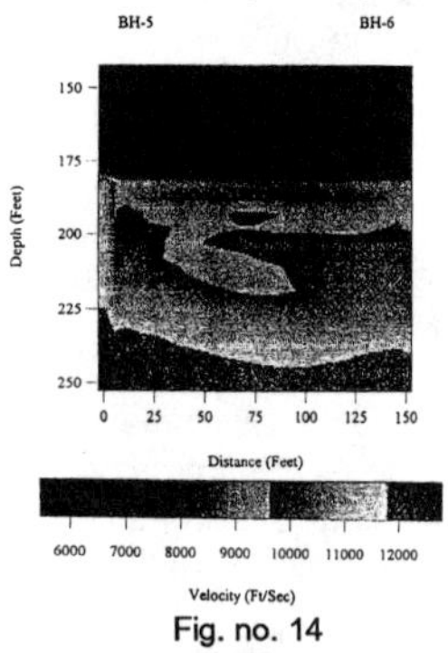

Fig. no. 14

Coincident with obtaining this information, the DOE shifted its already announced schedule to get out of the oil reserve business and placed the closing of Weeks Island first instead of a planned third among its sites. This shifted the program to one which would prevent flooding for a few years instead of the requirement of a more or less permanent fix. By introducing saturated brine into the pipe casing where the flow meter was located, the dissolution of salt could be stopped. This was done by introducing saturated brine down slant borehole no. BH-7A at a rate of 4 gpm (15 l/min), about twice the leakage rate into the cavern. The brine leakage continued but the increase in flow rate stopped.

At this writing, all the oil has been removed from the cavern and the service drifts are being prepared for brine flooding to the ground water level, which will allow eventual abandonment. The introduction of saturated brine continues. Detection and verification using a variety of complex measurements was the solution, an analysis easy only in retrospect.

●● **Maintenance of Above Ground Storage Tanks.** Repair of the bottoms of above ground storage tanks is an enormously important and expensive thing

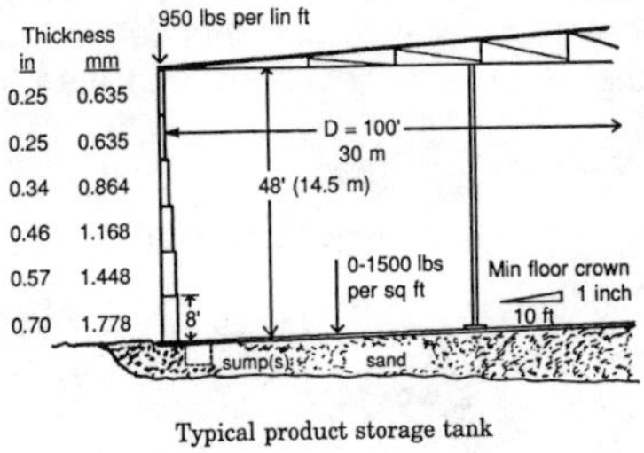

Fig. no. 15

particularly for the oil and chemical industry who own so many tanks.[9] The bottoms of such tanks usually are made of thin gauge steel 1/4 inch (6 mm) thick. The heavy shell, often over two inches (50 mm) thick, sits on the edge of that thin bottom which is either on a ringwall foundation or placed directly on a sand mat. The tank contents weigh evenly on the bottom and that weight varies from completely empty to completely full. Figure no. 15 shows how a tank is constructed and its structural situation.

Water sinks to the bottom. The tank bottoms are installed to a grade which channels water to the perimeter where it is drawn off. Tank sizes are typically 100 to 300 feet (30 to 70 m) in diameter and 50 feet (15 m) tall. The tank bottom flexes during filling or emptying. Flexure is worst at the edges where the heavy shell is welded to the light bottom. As the tank ages and flexure continues, stress concentrates at the outer edge and deformation can become extreme at the weld area in one of two fashions as shown in Fig. 16. Stress corrosion and ordinary rusting combine to create leaks there.

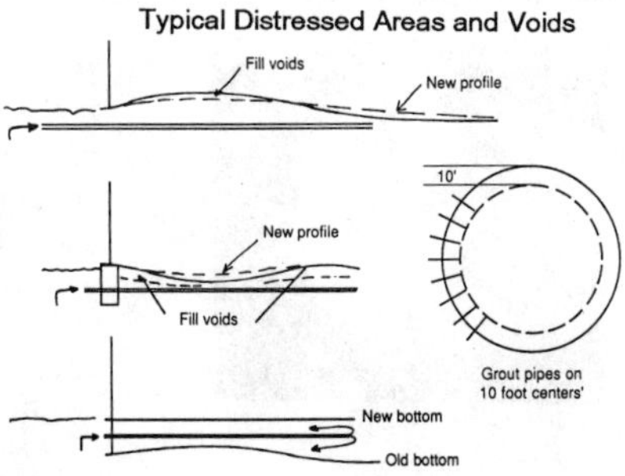

Fig. no. 16

According to industry sources, the cost of a new tank is in the $2-4 million range today. One alternative to a new tank is to put in a new bottom. But this will cost $200-400 thousand. Just opening a tank and preparing it for inspection can cost $15,000 today with all the environmental and safety problems which exist. If bottom movement and sharp flexing can be stopped before the tank has rusted through, then tank life can be prolonged for a long time. One procedure to do that is to grout a zone around the perimeter void as shown in the same Fig. 16. This needs to be done with the bottom held down on the sand mat, requiring the tank to be half full or better. It takes a careful approach but we have done it many times.

Since this is a maintenance feature which cannot be carried out if the bottom has corroded too much, the real problem is to identify which tanks need such work before they are too far gone. Working with a medical group from the Rice Medical Center, acoustic rebound information has been processed through a CAT scan tomography program. This is to locate voids and isolate deflection zones in a full or partly full tank. Procedurally, (where a ringwall is present) a free fall hammer raps on the concrete ringwall and geophones are used to pick up the rebound signals. The geophones are spaced about eight feet (2.5 m) apart around the perimeter. The acoustic free fall hammer weighs twenty pounds (9 kg) and is housed in a six foot (2 m) long sleeve. It is dropped at four foot (1.2 m) intervals around the tank, that is, near a geophone and then in between.

Acoustic Rebound Data

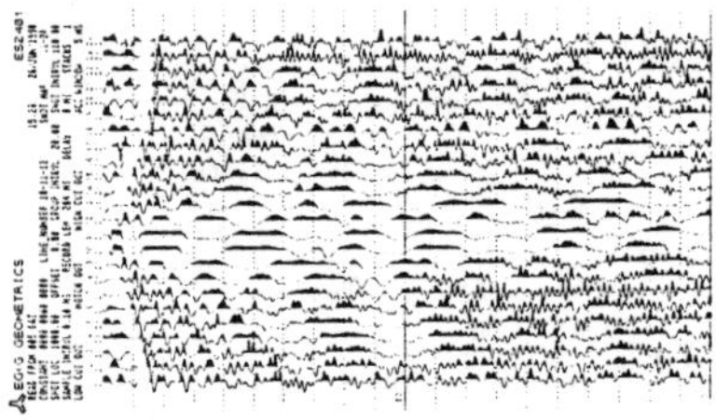

Fig. no. 17

Rebound information from a single drop generates a tracing like that shown in Fig. 17. The information near the drop point is useless, but in a 100 foot (30 m) diameter tank, there are about 45 sets of drop data. After processing through the tomographic program, the result is a pattern and image such as that shown in Fig. 18. This particular tank had a center roof support, hence the depression near the center, and two other areas had developed high spots and there were voids at the perimeter.

Processed Tomograph

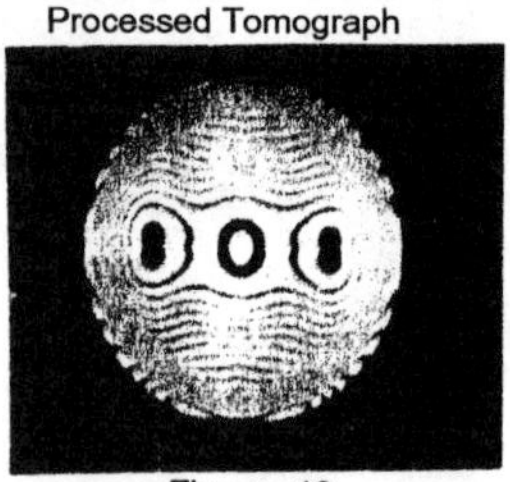

Fig. no. 18

The tank bottom was grouted while the tank was reasonably full but varying in depth as it remained in service. Grout pipes were wash bored into place through holes in the ringwall and slid in under the tank bottom. Maximum grouting pressure was kept below that of the static pressure of the liquid; that is, about 1500 psf (70 kPa) at the 40 foot (12 m) liquid height.

Grout quantity distribution logs in specific sectors of the tank confirmed the configuration of the void pattern in general. Our assumption was that grout filled all the voids under the bottom while the tank bottom was pressed down, thus stopping movement. The configuration of the tank bottom is shown at the top of Fig. 19. Verification of the work was by rerunning the acoustic logs. This indicated that there were no voids left or acoustic features present. It would have been desirable to confirm the results by coring some example locations, but the client did not want to incur that expense until there was some other reason to empty the tank.

Physically Located Voids

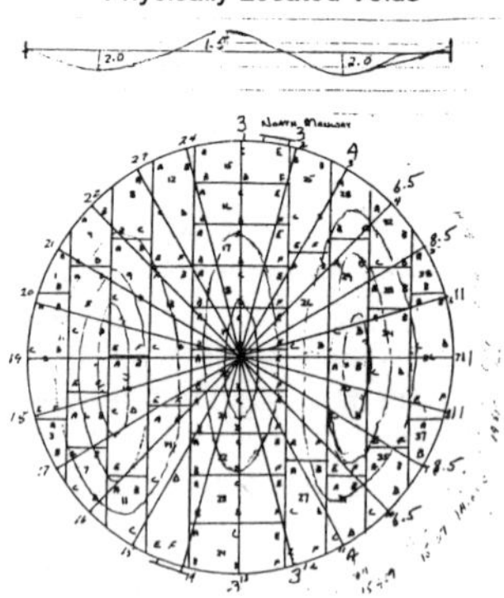

Fig. no. 19

Conclusions and Observations

This brief review of methods for verifying the effectiveness of grouting operations shows that many methods are available. In the short time range, planned verification can avoid the very consequences of what project delay can mean. And in the long run, it can provide some measure of slow moving effects.

Of the methods available for grouting verification, performance measurement or proof testing of the entire process is probably the most effective if that is possible at reasonable cost and without undue destruction of the facility.

For projects with immediately measurable effects such as grouting undertaken for controlling groundwater movement or gas leakage, tracers may be an excellent verification technique. If the tracers are built into the system, as may be the case with contaminated sources, so much the better. Otherwise, there are many sophisticated materials available today which can be used to make quantitative measurements as well as qualitative, and which will allow one to differentiate among specific sources of flows.

Then there are the projects wherein the only solution seems to be to identify that the grout has reached the areas where it is intended to be, and that it has properly cured. Considerable ingenuity often is required even to do this.

Investigating the source of a problem to properly identify its real source sometimes can become a task in itself. As drilling into a structure (or the medium) begins in preparation for grouting, reassessment of the situation often is necessary based on new findings. Reassessment of the verification program should go right along with program changes. One should have the best innovative and talented people assigned to a project for that to happen easily.

If verification of grouting is to be effective, it cannot be an after thought. It must be designed into the project. Without careful consideration of everything from soil or rock properties all the way to knowledge of particular instrumentation or testing schemes, it is unlikely that hastily conceived or after-the-fact programs will be conclusive or satisfying. Such considerations inevitably are most effective when all parties to the situation have an early and continuing input. This means a partnering of interests between the client, engineer, and contractor at an early date.

References

[1] Byle, Michael J. and Borden, Roy H., Editors (1995) *Verification of Geotechnical Grouting* ASCE Geotechnical Special Publication No. 57.

[2] Berry, Richard M. and Farrar, Robert R. (1997) *Grouting Around Trenchless Repairs* Trenchless Pipeline Projects ASCE Conference Proceedings, p347-348.

[3] Rembco publication (1994) *Steel plate bonding provides solution to complex hotel addition.*

[4] Berry, Richard M. and Buhrow, Robert P. (1992) *Settlement, Structural Failure, and In-place Repair of Above Ground Storage Tanks* Grouting, Soil Improvement and Geosynthetics ASCE Conference Proceedings, p247-249.

[5] Rembco publication (1987) *Porosity Determination and Groutability Tests.*

[6] Karol, Reuben H. (1990), *Chemical Grouting* Second Edition, Marcel Dekker, Inc. Chapter 5, Section 5.4 *Short Gel Times*, p158-181.

[7] ibid. Berry and Farrar. *Grouting Around Trenchless Repairs*, p346-347.

[8] Linn, James K. and Culbert, Jon (1999) *Experience in Underground Storage of Crude Oil in Salt* Geo-Engineering for Underground Facilities, Geotechnical Special Publication No. 90, ASCE Conference Proceedings, p810-819.

[9] ibid. Berry and Buhrow. *Settlement, Structural Failure, and In-place repair of Above Ground Storage Tanks*, p240-251.

GEOTECHNICAL INVESTIGATIONS FOR GROUTING IN SOIL

Michael J. Byle, F. ASCE[1]

Abstract

One of the most common complaints about grouting is that quantities cannot be estimated in advance, to a reasonable level of accuracy. Some feel that cost of grouting always overruns the estimate and by a lot. A second common complaint is that after a significant sum of money had been spent it became obvious that the type of grouting used just was not going to work. Both of these complaints originate with the planning of the grouting program. Planning begins with the subsurface geotechnical investigation. A good investigation is essential to the success of a grouting project. While the contractor's execution of the contract is a significant factor, a reputable and knowledgeable contractor will be more likely to be successful with a well written specification and appropriate and sufficient subsurface information. For a design engineer to write a good specification, knowledge of the appropriate grouting methods and thorough subsurface information is essential. This paper will focus primarily on grouting in soil but may have application in rock grouting.

Introduction

Many efforts have been made to identify various aspects of geotechnical investigations for grouting. A popular continuing education course in grouting begins with geotechnical engineering principles as its first session. Henn (1996) in his Practical Guide to Grouting includes a description of a geotechnical design summary report (GDSR) and discussion of the key elements necessary for a good characterization of subsurface conditions. Baker (1983) outlines the steps involved in the planning of a grouting program as follows:

1. Establish specific objectives for the grouting program.
2. Define the geotechnical conditions requiring treatment.

----

[1]Regional Geotechnical Manager, Gannett Fleming, Inc. 1010 Adams Avenue, Audubon, PA 19403-2402 e-mail: mbyle@gfnet.com

3. Develop an appropriate program design, performance prediction and specification.
4. Develop and execute a detailed construction work plan with QA/QC.
5. Evaluate the results of the grouting program.

The above steps were outlined for chemical grouting but are equally applicable to other types of grouting. A concise definition of the subsurface investigation requirements is given by Xanthakos et al (1994) as follows:

> "The overall objective of the site investigation is to obtain a detailed geotechnical classification of the different ground types, together with their locations, extents, and thicknesses"

Bell (1993) adds some additional insight into the importance for evaluating the history of a site for evaluating the presence of fills and past uses that may affect the subsurface stratigraphy and local variation in the subsurface conditions. ASCE (1997) has printed an excellent guidelines and practices publication for Geotechnical Baseline Reports for Underground Construction. The standard U.S. Army Corps of Engineers manual EM 1110-1-1804 (1984) and numerous other texts provide guidance on evaluating geologic and subsurface conditions and methods of exploration.

What most of the literature does not acknowledge is that often, the need for grouting is not identified until after the geotechnical field investigation is complete. In some instances the need for grouting is not determined until during the final foundation design phase of the geotechnical investigation. The case where the purpose of the initial geotechnical investigation is for design of a grouting program is generally limited to dam foundations and a few specialized cases where conditions are well enough understood to include grouting as a feasible and preferred option.

In most cases the geotechnical investigation is intended solely to prepare recommendations for foundations for a building, a bridge, a tunnel or some other structure, not for fundamental site characterization. This is the result of poor business practices on the part of geotechnical engineers. Geotechnical engineering consultants have become so competitive that they have had to reduce the scope of their investigations to a very minimal almost cursory level to be able to get the price down low enough to win competitive contracts. Of necessity, these limited scope investigations are based on a preconception of what the final recommendation is going to be. This makes the field and lab work more confirmatory rather than exploratory. Engineers have become production oriented and many have eliminated engineering observation and logging of test borings and choose to rely on drillers logs to reduce costs. Many geotechnical engineers routinely do not provide any laboratory testing as part of their geotechnical investigations. For many routine sites with straight forward conditions that are reasonably well known in advance, this cursory approach to subsurface exploration may be adequate, but not without a degree of risk. This cursory type of investigation is not sufficient to characterize a site. The product of such an investigation might be represented by Figure 1.

The competitive nature of geotechnical consulting and the advent of loss prevention language in reports has developed an environment where a cursory type investigation is the norm. The dilemma comes for an engineer, who knows about a problem site and how it should be investigated, when he has to decide whether to limit his scope and price the work low to get the job or risk losing the contract because his investigation is more costly than his competition. Clients have also come to expect low cost investigations and balk at paying the cost for a good characterization. The disclaimers and similarity between professional proposals, lead clients to select the lowest cost proposal for what seems like the same thing, a "soils report".

For all of the forgoing reasons, it is difficult to get the necessary geotechnical information for a grouting program from the standard geotechnical investigation. Grouting is often only considered after the initial investigation identifies a condition that can be remedied or improved by grouting more effectively than other methods. In these cases the design geotechnical engineer is faced with one of two options (a) convince the owner/client to pay for additional investigation, or (b) do a better job of getting information in the initial investigation.

Option (a) is the preferred alternative since the investigation may be directed specifically to evaluate conditions pertaining to grouting. The owner/client may not be amenable to this option, especially when it requires drilling borings at or near the locations of borings of the initial investigation, or drilling to what may seem to be unreasonable depths or ridiculously close spacings. It can take a substantial amount of effort to convince a client to pay for double the number of borings in the original investigation. The engineer may feel embarrassed at giving the client the impression that something was missed in the first investigation. However, the client should be made aware that the variability of the site can not

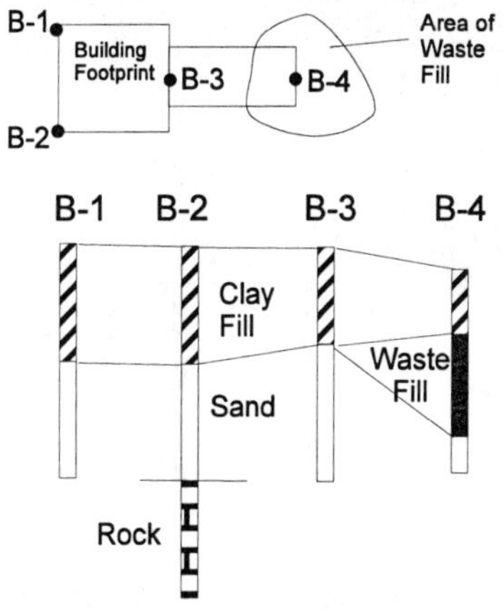

Figure 1 A representation of results from a typical geotechnical investigation for a building.

be completely anticipated. The geotechnical engineer is equally entitled to extra work for unforeseen conditions just as a construction contractor would be. Many geotechnical engineers include clauses to this effect in their service contracts. Clients should not expect the same results and cost regardless of the conditions. After all the site is theirs, not the engineer's.

It may even be difficult to define the scope of a second investigation, since what seems to be a simple problem can be revealed as a more complex one with additional investigation. Using the example of Figure 1, adding one boring can confuse the picture. Adding two can reveal a much different picture (Figure 2). It can be difficult also to determine how much is enough information.

Option (b) is always available but requires more diligence and a higher level of expertise in conducting the initial investigation. The secret to a successful investigation is in adapting the investigation to conform to the conditions encountered. This can sometimes be done without adding cost, but frequently requires timely coordination with the client to authorize additional expenditures as the need arises. This approach can be more cost effective, since a second mobilization can be eliminated, but requires the engineer to respond quickly and have a good relationship with the client.

Geotechnics of Grouting

Grouting takes on a great many forms each having its own effect on the subsurface and each having different geotechnical parameters for design. However, the geotechnical parameters of concern can be summarized in a relatively short list and include:

Permeability
Grain Size Distribution
Compressibility
Void Ratio
Specific Gravity
Relative Density
Unit Weight
Shear Strength

Frequently index properties are used to estimate the above parameters and to assess variability/uniformity of the soil. Index properties such as Atterberg limits in

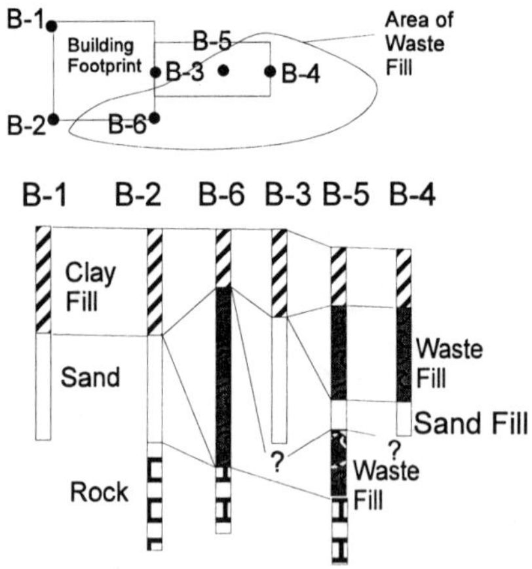

Figure 2 Adding borings B-5 and B-6 completely change the picture revealing more complicated stratigraphy and a larger problem.

conjunction with gradation analyses (sieve and hydrometer) together with the Standard Penetration Test (SPT) are often adequate to estimate permeability, compressibility and shear strength. In situ tests such as Standard Penetration Test (SPT), Cone Penetromenter Test (CPT), Pressure meter (PMT) are very useful in estimating strength, compressibility and void ratio, as well as, variability of the soils. Special problem soils such as expansive or collapsing soils may require specialized tests to evaluate their behavior, such as swell/consolidation tests. Many geotechnical texts can give good guidance on measurement and estimation of these parameters.

Heterogeneity and variability of the subsoils directly effect the effectiveness of all grouting methods. The most significant factor in a geotechnical investigation is determining the extent and variability of the subsurface horizons. A sure road to disappointment in a grouting program is to inadequately delineate the extent of a problem soil layer or to over estimate the extent of a horizon that is amenable to grouting.

(a) Permeation Grouting

Permeation grouting is the injection of a fluid grout into the pore spaces of a soil without disturbing or displacing the soil particles. Permeation grouting is probably the kind of grouting the average person thinks of when thinking of grouting and refers to as pressure grouting. The effectiveness of permeation grouting is controlled by the permeability of the soil. The grout viscosity must be matched to the soil permeability in order to control the injection. A very fluid grout will travel great distances in highly permeable soils. This can lead to a loss of control, large quantity overruns and failure to fill the soil pores with grout in the area desired. Where conditions are adequately characterized by the geotechnical investigation, grout travel can be controlled by using a more viscous or rapid setting grout.

Permeation is controlled by primary and secondary permeability of the soil. Primary permeability is the permeability of intact soil or rock, usually measured in a laboratory test. The primary permeability of a soil is the resistance to fluid flow offered by the soil structure and is related to the grain size distribution, porosity, void ratio, plasticity of the clay fraction, grain shape and the nature of the soil fabric (e.g. flocculated or dispersed structure).

Secondary permeability may be considered the structural permeability or the permeability of the fissures, fractures, joints, bedding planes or other structural features of the deposit. Secondary permeability is a significant factor for residual soils, rock and thinly bedded soils. The fluid grouts used for permeation grouting will follow the path of least resistance (i.e. highest permeability) and permeation may not be uniform where there are variations in soil types or significant secondary permeability.

(b) Limited Mobility Displacement Grouting

The controlled injection of a very stiff grout with high internal shear strength to displace the soil, without permeating the soil is limited mobility displacement grouting (LMD). This type of grouting is designed to form controlled masses of grout that either densify the soil or form structural shapes and can be used to fill voids. The effectiveness of this type of grouting is controlled by compressibility and permeability (if wet). The ability of the grout to achieve controlled displacement depends on how easily the soil can be displaced and the ability of the soil to dissipate pore pressures generated during displacement. The failure of the ground to dissipate excess pore pressure results in a loss of control that can result in ground heaving, lensing of the grout and grout traveling long distances from the point of injection. This mechanism is responsible for numerous stories of relatively stiff grout traveling and filling up utility lines, buried tanks and basements. Void ratio and relative density are necessary to determine the amount of grouting needed or achievable.

Adequate prediction of the success of LMD grouting depends upon the purpose of the grouting and the variability of the soil stiffness, permeability and degree of saturation. The presence of obstructions or large particles, such as boulders or rubble, increase the difficulty of advancing casings and interfere with the formation of bulbs and columns. Where densification is the goal, the extent and compressibility of the zone to be densified must be known in three dimensions.

(c) Jet Grouting

Jet grouting is a partial replacement method that washes away a portion of the soil with a rotating high pressure jet of grout or water, and mixes the remaining soil with grout to form soilcrete columns. The success of this method depends on the ability of the jet to cut away and mix grout with the soils. This cutting ability is related to the cohesive strength and gradation of the soils and the presence of foreign materials that can obstruct the jet and prevent the formation of uniform columns. More granular soils typically make stronger soilcrete and can generally be jetted to form larger columns. The method has been used effectively in some fine grained soils, but appropriate soil parameters such as cohesive strength, grain size distribution, moisture content and Atterberg limits must be obtained to properly design the grouting.

Variations in the subsoil can lead to less than satisfactory performance of jet grouting. Thin layers of clay in sands can create a discontinuity in the jet grout mass and cause a reduction of diameter of the jet grouted column in the clay zone. A reduction in column size will affect the grout effectiveness both in achieving seepage cutoff and structural capacity. Boulders, rubble, pipes or other structures can produce a shadow effect where they block the jet, to create discontinuities. Other subsurface conditions including the presence of peat, organic soils, cobbles, weathered rock, voids may affect the material and injection procedures as well. Highly permeable zones of rubble or boulders, may cause loss of circulation and greatly increase the amount of grout required. Loss of circulation could lead to large displacements and damage to structures. Slow or very hard to cut soils will reduce the achievable column size and increase the volume of grout and waste.

(d) Special Grouting

There are many types of special grouting methods that focus on special applications and may be hybrids of other types of grouting. Chemical grouts and cement grouts can be mixed to provide a grout that will both permeate and displace soils. Other specialty grouting methods may include the injection of hot bitumen, fluid grouts to form controlled fractures, sand suspensions and other materials for various purposes.

Investigation Planning

The ASCE Guidelines and Practices for Geotechnical Baseline Reports for Underground construction provides useful guidance in the preparation and planning for geotechnical investigations. As discussed above, one difficulty in planning an investigation for grouting is that the geotechnical investigation has often been completed before grouting is determined to be an option for the site. Two alternate approaches are described below. The first, outlines some minimum requirements that can easily be incorporated into any geotechnical investigation. The second approach outlines more specific requirements that should be incorporated into a geotechnical investigation specifically for grouting design.

General Investigations

<u>1) Data Review:</u> For all geotechnical investigations, some preliminary evaluation of site conditions is warranted. A review of past use of the site, through a client interview or by reviewing an environmental site assessment (ESA) report can give valuable information relating to possible locations and extent of prior facilities and land uses that could produce conditions where grouting may be required. ESA's are commonly required for real estate transactions and usually contain detailed information on the past use of a property and may include aerial photographs and fire insurance maps that give a good picture of prior site conditions. Geotechnical engineers should routinely ask for this information, since it may also give an indication of the presence of contaminants that might be encountered in borings or other field investigations. Geology maps inform of the site geology. United States Geological Survey (USGS) topographic maps can give an idea of areas of fill or regrading (depending on the date of the topography) and can be useful in evaluating the geomorphological environment. United States Department of the Interior, Soils Conservation Service (SCS) Maps provide information on the surface soils and often

include vintage aerial photographs that can be useful in evaluating change in the site over time. The client should always be asked for any previous studies and subsurface data available.

2) Sampling: All investigations should include sampling of all strata and should obtain and retain samples from each soil type encountered. Every investigation should have provision for obtaining undisturbed Shelby tube samples, in case they are required. Undisturbed samples should be taken in any strata where they are feasible. The cost is nominal, and, it is easier to convince a client to pay for additional laboratory tests than to fund a remobilization to get samples that could have been obtained the first time out. Even if no testing is planned, samples should be taken and retained so that they can be tested later, if necessary. Representative samples should be taken from test pits as well as borings. Photographs of test pits should also be taken, since they provide a specialty contractor a much better view of the subsurface conditions than do logs alone.

3) Investigation Limits: Establish limits and criteria for the investigation as follows:

a. If not using continuous sampling, sample at any change in strata. This is evidenced by any change in the rate or resistance to drilling or a change in the cuttings returned.
b. Never terminate a boring in fill or soft soil.
c. Verify refusal. Do not log auger or spoon refusal as bedrock unless this is verified by coring, or other means.
d. Where significant differences in subsurface conditions are found between adjacent borings, probes, pits, etc., consider additional exploration to define the limits and location of the transitions. For larger investigations, this can sometimes be accomplished simply by adjusting the exploration plan without adding more test locations.
e. Make sure that the field engineer/inspector knows what the expected conditions are and to check with the engineer in charge if anything different is encountered.
f. Fully describe rock where encountered. Include hardness, joint type and any infilling, strike if possible, dip of bedding and fractures, rock type, formation if known, degree of weathering, solution features if any, etc.
g. Always obtain water level readings. Obtain 24-hour or longer stabilized readings wherever possible.
h. Determine a reference elevation for all borings. Grouting may be completed during construction when the site may have been regraded. Elevations are essential to identify the problem zone during construction.

4) Laboratory Testing: If grouting is being considered, laboratory testing will be required. At a minimum, perform moisture content gradation analysis and Atterberg limits for all soil types encountered. Void ratio, and unit weight should be obtained from undisturbed samples. Strength, compressibility and permeability testing are useful and should be provided where necessary. Where organic soils are encountered, test for organic content. Sulfate content, pH and other inexpensive chemical testing are appropriate for soils known to be corrosive to concrete and organic soils.

5) Analysis and Report: Where specialty construction, such as grouting, is anticipated, provide all information possible in the report. Include complete soil descriptions that appropriately characterize each stratum. Simplified descriptions such as "brown silty sand" are not sufficient. Describe fill materials in detail describing all constituents identified in the field investigation. In the report, identify the known limits of the area to be grouted, the known or estimated properties of the soil to be grouted including:

a. Relative Density/Consistency
b. Void Ratio
c. Permeability
d. Degree of Saturation
e. Plasticity
f. Gradation including D_{10}, D_{85}, and coefficient of uniformity
g. Non-soil materials present
h. Soil structure/geologic origin (e.g. varved/lacustrine, glacial till, rubble-fill, etc.)

Investigations for Grouting Design

Where there has been a preliminary investigation or where for some reason it is known that grouting will be required, the investigation should be focused on the following:

1) Define the Purpose of the Grouting: This may seem like an obvious step, but it is often the cause of a failed design. Grouting can serve many functions and different information is needed to assess the feasibility of achieving different purposes. It is important to define exactly what is hoped to be gained by grouting and plan the investigation to gather information necessary to achieve the desired goal. Some of the reasons may include:

- Improving soil strength or bearing capacity
- Preventing or arresting settlement
- Jacking or lifting structures or slabs
- Filling voids
- Preventing earthquake liquefaction

- Preventing ground subsidence due to underground construction
- Creating hydraulic barriers
- Creating structures (i.e. columns/piles or walls)

2) Define the Mechanism of Grouting: Identification of the specific mechanism, or mechanisms, by which grouting will achieve the defined purpose is essential to planning a complete investigation. This requires a knowledge of grouting technology to assess feasible approaches based on the preliminary geotechnical information. Most grouting methods will increase soil strength if they are successful, though they achieve it by different mechanisms. Not all grouting methods can create hydraulic barriers or structures. A goal of the geotechnical investigation should be to determine sufficient soil parameters to define the feasibility of each alternative grouting method, and provide sufficient information for design of the grouting program.

In a general sense, the mechanisms are limited to permeation, displacement, compaction, and replacement. To properly plan the investigation, one must look more closely to assess the specific mechanisms that will be at work during and after the grouting. These may include, induction of residual stresses, consolidation, elastic deformation, plastic deformation, displacement and migration of pore fluid, encapsulation, cementation, adhesion, chemical reactions within the grout and chemical reactions between the grout and the soil. Defining how these mechanisms work for each project may require gathering information on the following:

Information	***Design Purpose***
Geometry and characteristics of subsurface layers (in three dimensions)	Define limits and quantities of grouting
Geometry of subsurface structures (e.g. tunnels, retaining walls, utilities, etc.)	Characterize interference and obstructions to grouting
Vertical and lateral moduli of subgrade reaction	Establish zones of influence and determine pressure limitations
Preconsolidation history	Establish strength, permeability and pore pressure influences on grout
Permeability	Select grout type/viscosity Define permeation limits Assess pore pressure dissipation to select pumping rates

Unique characteristics of non-soil materials	Grout interaction (physical and chemical) Strength of grouted product
Organic content	Affect on curing and grout strength
Soil chemistry	Grout chemical interactions Impacts of contaminants
Sensitivity of surrounding structures to movement, loadings, damage by drilling of grout holes	Establish influence zones, cutoff criteria and limits on the grouting process. Define monitoring requirements
Liquefaction potential	For seismic remediation
Hydraulic gradients	Assess washout potential Define preferential paths for grout flow and other interaction between the grout and the groundwater
Critical failure mechanisms that could be precipitated by grouting	Risk assessment Define grouting sequence
Secondary soil structure	Assess heterogeneity of grout permeation Define pressure response to displacement
Sinkholes and solution features	Establish limits of grouting and quantity estimates

It is impossible to list all of the key parameters that would be needed for all projects because of the limitless diversity of conditions that grouting can address. Each project will have unique constraints that have to be established and investigated.

3) Determine the Extent of Grouting: The investigation should specifically identify the limits within which grouting is needed. It is important to remember that in soils, even grouted soils, the stresses dissipate outward in all directions. The investigation should explore beyond the limits of the effects of the grouting; and where the purpose of grouting is to strengthen the soils, to the limits of the influence of the stresses that exceed the ungrouted soil capacity.

Often grouting is performed to mitigate an area of loose, soft, permeable or otherwise unacceptable soil. It is important that the investigation is both sufficiently broad and sufficiently detailed to define the limits of the unacceptable area. It is important for the grouting designer and contractor to know what the conditions will be at the periphery of the grouted area. The perimeter conditions will define the amount of effort (and cost) required to contain the grouting and meet the project objectives within the target area.

4) Select Exploration Methods: Select methods of exploration and testing that will provide information on subsurface conditions specifically related to the mechanism and extent of grouting. Select methods of investigation that will obtain samples or in situ data necessary to establish the following:

- Limits of the zone to be grouted
- Spatial variability of the materials within the zone of grouting
- Boundary conditions for the grouting
- Soil permeability
- Soil classification
- Soil relative density
- Appropriate soil chemistry
- Details of all pertinent constituents of the soil (including large particles)
- Compatibility between the mechanism of the grouting method and the soil
- Special geologic or other conditions affecting the grouting.

Background data on soil characteristics related to the desired goal of grouting should also be obtained in the investigation. It would be prudent to use some of the proposed verification methods during the exploration to establish a baseline for comparison with the post grouting verification testing. For example, if cone penetrometer testing was proposed to evaluate the soil improvement by compaction grouting, it would be useful to have baseline data available for comparison. This aids in establishing the performance limits and setting reasonable target values for performance specifications. Another example would be testing borehole permeability as a baseline for permeation grouting used to create a subsurface seepage barrier. Post grouting tests would be compared to the pre-grouting tests as a measure of improvement.

5) Use Knowledgeable Inspectors: The field inspector for a grouting investigation, just as for any other geotechnical investigation, must be informed about the purpose of the investigation and the parameters of interest. The inspector should be cognizant of the purpose and mechanism of the grouting so that he/she can recognize adverse or limiting conditions when they are encountered and augment the investigation as required. The inspector for grouting investigations must be diligent to note everything related to the subsurface exploration and is not just logging samples.

6) Be Prepared to Do Something Different: The conditions are not always what are anticipated. The program must be flexible to accommodate the unexpected. Also, creativity may be necessary to ascertain the characteristics of materials that won't stay in samplers, are too large to sample, or just don't lend themselves to conventional investigation techniques. This may entail techniques such as: over-driving spoon samplers, using Dennison or Piston samplers, taking block samples from test pits, doing field scale direct shear tests, preloading, geophysics, injecting

grout into a Shelby tube to retain loose sand, using dyes, using an air-track drill, Becker Hammer, etc. The goal of the investigation is to identify the site conditions well enough to develop a grouting program, not to follow a standard investigation plan. There are times when it is absolutely necessary to think outside of the box.

7) Record and Report Everything: It may seem a routine occurrence, but things like the loss of drill fluid circulation or getting no cuttings during auger drilling, but these things can make a substantial difference in the interpretation of a boring log. If the driller is wearing out bits rapidly, if the rate of hole advance drops or increases, note it on the logs. The conditions affecting the drilling of test borings will affect the drilling of grout holes. Drilling can make up 50 % or more of the total cost of a grouting project. If there is a rock outcrop across the street, down the hill; if your borings hit groundwater at 1 meter, but the site is 20 feet above a dry creek bed, if site worker tells the inspector that there was 2-meter diameter sewer abandoned within the site; if you notice that the north side of an adjacent building has settled; if gas came out of the hole ... REPORT IT! The grouting designer and contractor need to have as clear an understanding of subsurface conditions as the engineer who writes the geotechnical report.

Summary

There are many guides available for design of geotechnical investigations. Phasing of investigations for grouting is a prudent approach. A second investigation is justified where the need for grouting is discovered during a normal foundation investigation. When planning a geotechnical investigation where grouting is expected consider the following:

1. Identify conditions that might require grouting early
2. Identify the purpose and goals of the grouting
3. Identify the mechanism of grouting to be used
4. Select exploration methods that gather the appropriate data and samples
5. Test the samples for grain size distribution and permeability at a minimum
6. Use anticipated verification methods in the site exploration
7. Use a qualified inspector who pays attention to detail
8. Record and report everything about the investigation
9. Think outside the box when unusual conditions defy usual procedures

Make the grouting contractor aware of all subsurface conditions. If it is uncertain whether a particular piece of information is necessary, include it anyway.

References

American Society of Civil Engineers, *Geotechnical Baseline Reports for Underground Construction.* Guidelines and Practices, The Technical Committee on Geotechnical Reports of the Underground Technology Research Council. Randal J. Essex Ed., ASCE Press, 1997

Bell, F. G., *Engineering Treatment of Soils.* E and FN Spon, London, England, 1993.

Berry, Richard M., "When Grouting Don't Forget the Soil (the medium)", REMBCO Engineering, Text of presentation for the Tennesee Section of the ASCE, 1998.

Baker, Wallace H., *Design and Control of Chemical Grouting: Vol.3 and Vol. 4.* U.S. Department of Transportation Federal Highway Administration, Report No. FHWA/RD-82/038 and FHWA/RD-82/039 April, 1983.

Henn, Raymond W., *Practical Guide to Grouting of Underground Structures.* ASCE Press (Thomas Telford Publications in the UK) New York, NY 1996.

U.S. Army Corps of Engineers., *Geotechnical Investigations.* EM 1110-1-1804 Department of the Army, Washington, DC. February 1984.

Xanthakos, Petros P., Abramson, Lee W., and Bruce, Donald., *Ground Control and Improvement.* John Wiley and Sons, New York, NY, 1994.

VERIFICATION OF JET GROUTING FOR STRUCTURE REHABILITATION

Tarek F. Haider, P.E.[1] and Michael J. Byle, P.E.[2]

ABSTRACT

Verification of jet grouting to stabilize an existing stone masonry abutment by increasing the abutment's bearing area and mass was necessary to assure proper performance. Project constraints such as: maintaining bi-directional vehicular traffic over the bridge, maintaining uninterrupted rail traffic under the bridge, maintaining and protecting eight underground utility lines through the work, and ensuring safety of the bridge abutment, the roadway and nearby structures from the high-pressure grout limited the techniques and verification methods available. The jet grouting created a soilcrete mass to enlarge the existing abutment by overlapping individual grout columns interlocked with the existing stone masonry. The jet grouting was completed under continuous full-time observation by the design engineer to verify compliance with the intent of the project specifications. To verify that the grouting was not damaging to the existing structures, the abutment, roadway and sidewalk were instrumented to continuously monitor movement during jet grouting. Test columns were constructed prior to production grouting using feeler pipes to assess grout coverage, column diameter and select the proper jetting pressure, grout flow rate and monitor extraction speed. Core samples were taken to verify the integrity and properties of the soilcrete. Laboratory density and compression tests were performed on selected core samples of the soilcrete. All tests exceeded project requirements. The core borings through the soilcrete mass met the project's core quality requirements and demonstrated the effectiveness of the jet grouting despite some unanticipated rubble in the abutment backfill.

INTRODUCTION

Track undercutting became necessary to increase the clearance under a highway bridge over a railroad mainline. The highway is a heavily traveled two-lane road connecting two

[1]Senior Geotechnical Engineer, Gannett Fleming, Inc., 1010 Adams Avenue, Audubon, PA 19403-2402, Tel: 610-650-8101, E-mail: thaider@gfnet.com

[2]Geotechnical Department Head, Gannett Fleming, Inc., 1010 Adams Avenue, Audubon, PA 19403-2402, Tel: 610-650-8101, E-mail: mbyle@gfnet.com

busy urban intersections over the rail lines. The loss of soil support due to undercutting would cause the abutment to become unstable in the external stability failure modes. After a detailed alternative investigation, jet grouting was selected to improve the external stability factors of safety to the required design standards.

Jet Grouting is a relatively new process which has been developing since 1970 (Kauschinger, et al. 1992). The distinct advantage of jet grouting over comparable stabilization methods is that direct access is not required for the application of the technique. This often makes jet grouting a more economical solution over underpinning and similar structure stabilization methods. Jet grouting has been utilized in the United States for approximately twenty years. This is the first application of jet grouting for a structure stabilization project for Amtrak. Jet grouting has been used for transportation infrastructure at airports (Ichihasi et al. 1992), tunnel construction (Kauschinger et al 1992, Xanthakos et al. 1994), bridge abutment (Byle and Haider 1998), and rail lines (DePaoli et al. 1989). Other infrastructure applications include creating gravity walls (Miyasaka et al. 1992) and underpinning buildings (Drooff et al. 1995, Xanthakos et al. 1994).

The approach to verification taken here, generally follows the guidance in Byle and Borden (1995). Verification was considered as a fundamental part of the project planning. The objective of the design was to stabilize the abutment while maintaining continuous service of all underground utilities and minimizing disruption to both rail and vehicular traffic. The jet grouting program was designed to provide opportunities for verification at various stages of the construction.

STRUCTURE DESCRIPTION

The subject bridge is a single span steel I-girder structure supported by stone masonry abutments. The abutments are parallel to the tracks and the superstructure is at a 53^0 skew. The existing horizontal clearance between the centerline of Track 2 and the face of east the abutment ranges from 2.1 to 2.7 m (7 to 9 ft). The structure was partially reconstructed in 1977. The concrete bridge seats and backwalls were constructed over the original masonry abutments and the old superstructure was replaced. Existing drawings indicate that the east abutment is supported by shallow foundation bearing on soil. The north wingwall of the east abutment is parallel to the street and will be unaffected by the grade lowering. The south wingwall gradually angles away from the tracks; therefore, the majority of the wall is far enough away from Track 2 and remained unaffected by the grade lowering. Consequently, the design was limited to the east abutment of the bridge.

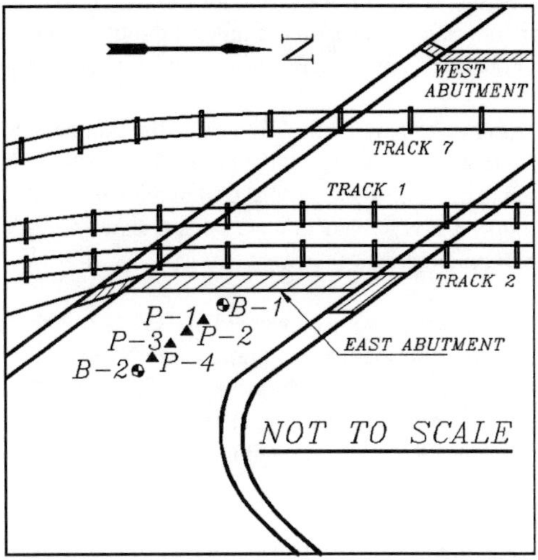

Figure 1: Locations of Borings and Probes

Visual inspection revealed satisfactory condition and performance of the abutment. Seven shallow utility lines are carried by the bridge. The utilities include four electric duct banks, one 406 mm (16 in) water main, one 304 mm (12 in) gas main, and one 508 mm (20 in) gas main. All of these utilities penetrate the backwall and extend over the abutment backfill. A 550 x 840 mm (22 x 33 in) brick sewer main crosses directly under the abutment, perpendicular to the abutment face, within 600 mm (18 in) of the bottom of the abutment.

SUBSURFACE CONDITIONS

Two borings (B-1 and B-2) and four auger probes (P-1 through P-4), were drilled to determine the geometry and configuration of the abutment, and to provide information on the backfill and the abutment bearing medium. Locations of the borings and probes are shown on Figure 1. Boring B-1 was advanced through the masonry abutment into the supporting soil to determine its base elevation. Boring B-2 was drilled behind the abutment through the backfill to determine geotechnical properties of the backfill and the soil supporting the abutment. The auger probes were terminated at refusal on the back face of the abutment and hence, established the shape of the abutment. A cross-sectional sketch of the abutment approximated from the site exploration is shown on Figure 2.

The borings indicate the condition of the concrete and masonry of the abutment to be good, recording core recoveries from 91 to 100 percent in concrete and 72 to 100 percent in masonry. Granite masonry stones consist of hard and unweathered blocks, 0.5 m (1.5 ft) thick, with mortar joints between them. The bottom of the abutment was identified at a depth of 9.4 m (30.7 ft, Elevation 6.28 m). Split-spoon sampling of the soil beneath the abutment revealed N-value of 100 blows for 100 mm (4 in) of penetration. The material sampled was coarse to fine sand with 1 to 10 percent silt. One inch of broken orange brick fragments was observed in the tip of the split-spoon sampler, which appeared to be the top of the brick sewer. The boring was terminated at 9.5 m (31.3 ft, Elevation 6.10 m).

The upper portion of the backfill behind the abutment, to a depth of 5 m (18 ft) was silty sand with gravel, with the Unified Soil Classification SM. The lower half of the backfill, to a depth of 13 m (40 feet) consisted of silt with sand (Unified Soil Classification ML). The upper fill layer was in a very dense state while lower fill layer was generally in a loose to medium dense state.

The deeper natural soils penetrated by Boring B-2 consisted of coarse to fine sand with 10 to 20 percent medium to fine gravel (SP-SM) and 1 to 10 percent silt. This layer extended from 13 m (40 ft) to the bottom of the boring at 15 m (49 ft). Standard penetration test N-values ranged from 18 to 53 and averaged 33 bpf. The 24-hour groundwater level was measured in Boring B-2 at 10.4 m (34.2 ft) depth (Elevation 4.85 m).

The auger probes, P-1 through P-4, were drilled to refusal on the abutment back face. The configuration of the back face of the abutment was interpreted as shown on Figure 2. The back face of the abutment is most likely stepped stone blocks rather than the smooth surface shown.

For stability evaluation, the soil strength parameters were estimated based on Standard Penetration Test N-values and laboratory tests, by utilizing existing correlations between N-values and soil parameters.

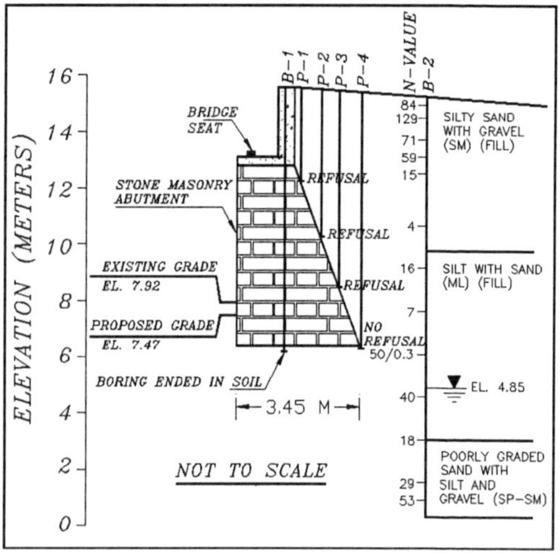

Figure 2: Cross-Section of Abutment Approximated from Site Exploration

DESIGN OF JET GROUTING

Analyses were conducted to evaluate stability of the abutment under current and proposed conditions. These analyses revealed that the abutment had very low factors of safety against sliding, overturning and bearing capacity failures under the existing condition. The analyses also indicated that this would be aggravated by a proposed undercutting of approximately 0.4 m (1.3 ft).

Jet grouting was selected to restore the stability of the abutment, since it would minimize disruption to rail and vehicular traffic and utilities, at a reasonable cost. Because of the existing close clearance, work from the rail grade was not permitted. The jet grouting involved the creation of a soil-cement mass on the back face of the existing abutment. The following design objectives were identified and incorporated in the jet grouting design:

- Maintain single-lane bidirectional traffic at all times over the bridge.
- Restore full traffic flow at the end of each work shift.
- Minimize damage to the reinforced concrete approach slab.
- Attain an effectively monolithic grout mass within the target zone.
- Avoid disruption of seven shallow underground utilities.
- Avoid damage or disruption of a 550 mm x 825 mm brick sewer located below the foundation.
- Avoid and control abutment movement and ground heave.
- Verify the integrity and strength of the soil-cement mass.

The jet grouting was designed to create a mass on the back face of the abutment which will shift the point of application of the resultant lateral earth pressure further into the backfill, increasing the abutment mass, and resulting in a higher resisting moment, larger effective bearing area, and reduced bearing stress at the toe of the abutment.

The design required consideration for future utility relocation and construction within the roadway by requiring the top of soilcrete to start at 2.4 m (8 feet) below the road surface. Also, the design called for the soilcrete to terminate at the bottom of the existing abutment to avoid interference with the 550 mm x 825 mm (22 in x 33 in) brick sewer.

The project specification was designed as a performance specification that required the contractor to install a monolithic mass of soil-cement within the plan dimensions without damage to the existing abutment or buried utilities. The soil-cement was specified to have a minimum unconfined compressive strength of 2,070 kPa (300 psi). The contractor was required to submit a jet grouting plan indicating the materials, methods, equipment to be used, and locations of injections and a quality control plan. Details about the jet grouting stabilization process has been reported by Byle and Haider (1998).

VERIFICATION DESIGN

The choice of verification methods used was restricted by the same constraints that limited the grouting methods: limited space to work, avoidance of utilities, avoidance of interruptions to rail and motor vehicle traffic. The verification methods had three main goals:

- Prevent structure damage during construction
- Verify Continuity of soil-cement mass.
- Verify strength of the soil-cement

Structure damage is a real risk when using high pressure injections in close proximity to structures. The jet grouting method used for this project, injects an air jacketed water jet at 40 MPa (6,000 psi) which can hydrofracture the ground and move structures if circulation is not maintained. By the time such damage can be seen visually, structures may already be severely damaged. One common and effective method to evaluate jet grouting is to construct test columns at a nearby representative site and excavate them to evaluate their performance visually (Byle and Borden, 1995). This approach was impractical at this site because of the small site area, and presence of surrounding structures and underground utilities.

The goals of the verification program were achieved by a combination of methods before, during, and after the jet grouting. Before the grouting began, the bridge abutment was instrumented with tilt beam sensors. These sensors were placed on the face of the abutment at two different elevations to detect any unintended movement of the abutment during the jet grouting. The tilt beam sensors specified were 1.5 m long beams with a centrally mounted electrolytic tilt sensor. Five tilt beams were required to be monitored remotely on a continuous basis. An automatic alarm would go off if movement in excess of a threshold value is recorded. The threshold value was set relative to baseline movements from three days monitoring of the instruments prior to jet grouting. Baseline monitoring was necessary because of the instrument's high sensitivity to temperature variations.

During the grouting, movement of the pavement surface was monitored using a rotating laser level with 6 targets set at various points around the area being grouted. The laser targets had visual and audio alarms if movement greater than 3 mm (1/8 inch) was detected. In addition, all grouting operations were observed by the design engineer to verify the contractors compliance with the jet grouting and quality control plans.

After the grouting was complete, two cores were taken at locations selected by the engineer. The cores were drilled through the jet grouted mass to verify the integrity of the mass and to obtain samples for strength testing. The core recovery gives an indication of strength, since low-strength soil-cement is difficult to recover. Considering the expected loss of core during drilling, a minimum 70% core recovery target was established for the soilcrete. The project specification also required a minimum soilcrete compressive strength of 2,070 kPa (300 psi).

This complete project approach to verification allowed ample opportunity for evaluation of the grouting throughout the project. The time critical nature of the project made it undesirable to wait until the completion of the work prior to verification. Since, the grouted mass must cure long enough to gain sufficient strength to permit coring, the contractor will have demobilized all of his equipment prior to the coring and strength testing. Continuous monitoring and construction quality assurance made it possible for both the contractor and engineer to have a good sense that the grouting was successful prior to the actual coring operation.

CONSTRUCTION OF JET GROUT COLUMNS

Jet grouting was performed using the triple rod system which used a combination of high velocity water shielded in a protective sheath of air to cut and displace soil while simultaneously injecting cement slurry from a lower nozzle. Grout columns in excess of 1.2 m (4 ft) in diameter were constructed by adjusting air and water pressures and flow rates while avoiding ground heave or abutment movements resulting from excessive grout pressure or velocity. The grout hole location plan is shown on Figure 3. Figure 4 shows a typical cross-section of the abutment with individual grout columns and overlaps.

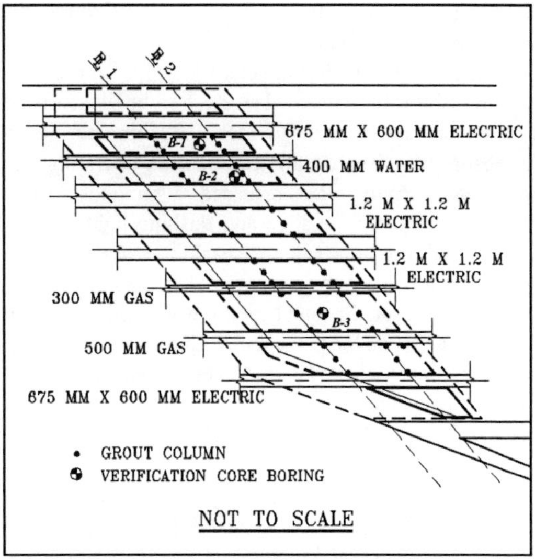

Figure 3: Grout Hole Location Plan

In order to allow traffic on the bridge during construction, the jet grouting rig had to be oriented parallel to

Charles Street. Therefore, all soilcrete columns were constructed at compound batter angles. The jet grout holes were located along each of two baselines, BL 1 and BL 2, that were parallel to the east abutment. All holes were angled to follow the slope of the back face of the abutment.

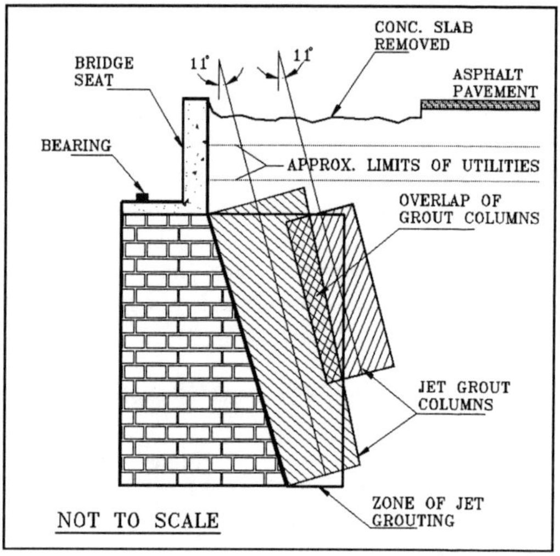

Figure 4: Cross-Section of Abutment Showing Grout Columns and Overlaps

The abutment instrumentation consisted of five EL Beam sensors with an automated datalogger manufactured by SINCO. A 1,250 mm (5-foot) gage length was used for the beam. All beams were covered with half cut 200 mm (8-inch) diameter PVC pipe sections to protect from direct exposure to sunlight. All beams contained an electrolytic tilt sensor sensitive to measure tilts as small as one arc second or approximately 0.005 mm per meter of the beam. The range of the tilt sensors was 40 arc minutes under an operating temperature of -20 to 50^{0}C. The instrument outputs were recorded continuously by a laptop computer that was programmed to sound a siren and flashing light if measured movements exceeded the programmed threshold value for each instrument. A datalogger software was used with appropriate temperature correction which provided continuous real-time tilt readings in every beam sensor. Ground heaving was monitored continuously and simultaneously at several locations on the pavement within the jet grouting zone with a rotating laser level with targets set at selected points on the pavement and sidewalk behind the abutment. The targets were equipped with an alarm indicator that alerts the grouter of movements in excess of 3 mm (1/8 in) the established threshold.

The contractor's quality control plan included grouting test columns to verify the effectiveness of the process and modify grouting parameters, if necessary. Due to the lack of a suitable test site and representative soil condition, construction and excavation of a test column prior to production jet grouting was not practical and was not specified. The contractor constructed two test columns within the backfill area, approximately 5 m (16 ft) behind the existing abutment. The jet grout column diameter was determined through 25 mm (1 inch) diameter PVC pipes drilled and installed as feeler probes for each of the two columns. The concept is a simple one: you can feel and hear the jet as it impacts the PVC pipe during each revolution of the monitor. Each pipe was located 0.6 m (2 feet) from the center of the proposed jet grout hole and installed to the full depth of the test column. The jetting pressure and flow rate were adjusted to just make contact with the feeler pipe. The columns were not excavated, but the feeler pipes gave a good estimate of the minimum column size at the selected settings.

The contractor's plan required 1.2 m (4 feet) diameter jet grout columns. From the feeler probe tests water, air, and grout pressures were selected at 40, 0.83, and 1.25 MPa (6,000, 180, and 120 psi), respectively. The grout flow rate was selected as 150 liters/min under a monitor withdrawal rate of 30 cm/min. These parameters provided the required size grout column. The contractor utilized an automated system to control the grout pressure and flow to avoid excessive pressure that could cause distress to the masonry abutment and utility lines.

With approximately 30 percent of the grout columns in place, the contractor elected to conduct an additional core boring to verify the quality of the soil-cement prior to completing all of the grouting. A continuous 125 mm (5-inch) core boring, denoted B-1, was drilled through the completed soilcrete mass. The jet grout rig performed the drilling. The core location was along Baseline BL 2 centered between surrounding grout columns with a minimum 450 mm (1.5 feet) clearance from the nearest column. The core hole was drilled at the same batter angle as the adjacent columns. This core, drilled at the contractor's expense, provided him verification that his methods were effective and could achieve conformance with project requirements. The core was extended below the depth of the adjacent columns to verify the extent of the grouting.

The results of this core boring were better than expected and therefore, no modification to the grouting parameters were made. The average core recovery and Rock Quality Designation (RQD) in the soilcrete were 100 and 78 percent, respectively, indicating excellent soilcrete quality. Selected pieces of the recovered soilcrete cores were sent to a laboratory for strength testing. The tested core had a 28-day unconfined compressive strength of 19.7 MPa (2,860 psi) and a density of 17.9 kN/m^3 (114 pcf), both well exceeding project requirements.

Construction Monitoring

Construction monitoring was an integral part of the jet grouting verification process. A full-time inspector was assigned to the work who documented every event during the workday. Several observations were documented during the progress of the work. The rate of grout return proved to be a measure of the jet grouting effectiveness and the subsurface conditions. In general, grout consumption was larger than expected at the beginning part of the work. This appears in part to be due to the use of grout as the drilling fluid and encountering rubble and other obstructions during insertion of the monitor. As the construction of grout columns progressed, an increase in grout return was observed which indicated that the previously grouted columns extended beyond the anticipated coverage zone and acted to contain the grout from subsequent split spaced columns. For the last few columns, grout return was very good. During jet grouting, some adjustments to flow rates and jetting pressures were made to maximize grout return.

Field engineering is a critical part of this kind of retrofit work where unknown conditions of the original construction may affect the work. As a result of the full time observation, the location and batter angles of the grout holes were adjusted to compensate for actual utility locations and foundation structures encountered in the jet grouting. For example, the actual configuration of the north wingwall of the bridge was not known before the start of jet grouting and the grouting design allowed provision for deleting or adjusting hole locations depending on the shape of the north wingwall as determined from the drilling of the grout holes. The drilling information of the grout holes determined the shape of the north wingwall and several grout columns were deleted and depths of several other columns were adjusted. Without full-time construction monitoring, field adjustments such as the ones described above would not be possible.

Final Core Borings

After construction of all jet grout columns and allowing a minimum of 7-days curing time, two additional 125 mm (5-inch) diameter core borings were drilled through the grouted mass to verify the effectiveness of jet grouting and project compliance. The borings, numbered B-2 and B-3, were placed between grout holes as shown on Figure 4 where least coverage was expected. The borings were located to verify the vertical and lateral extent of the in-place soilcrete. Both borings were drilled vertically and advanced through the ungrouted overburden soil and the soilcrete into the granite abutment or soil. Figure 5 shows a subsurface profile through Borings B-1, B-2 and B-3. As expected, the core borings indicated the top of soilcrete starting approximately 2.4 m (8 feet) below the road surface and ending approximately at the bottom of the existing abutment footing. The core data indicated that the desired configuration of the grout mass was obtained by establishing the vertical and lateral extents of the grouted backfill mass.

The recovered soilcrete cores were assessed for core quality. The soilcrete cores contained rubble, steel and soil particles of variable nature typical for backfill behind bridge abutments. Figure 6 shows photographs of soilcrete cores. The cores showed good

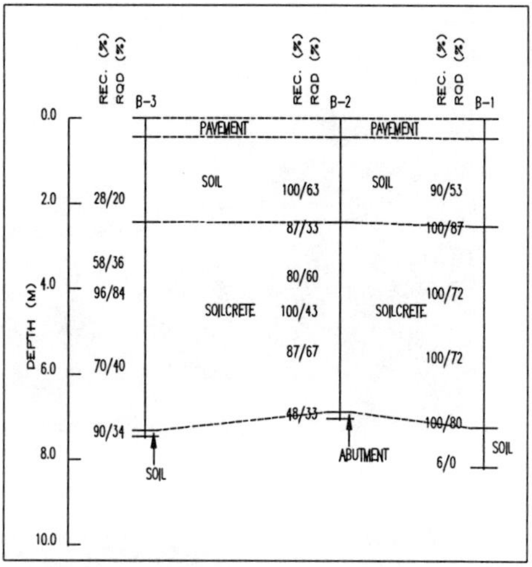

Figure 5: Profile of Verification Borings

cementation between various sizes of soil particles which could not be washed away or disintegrated during the core drilling. The rate of coring in soilcrete varied from 50 to 140 mm/min (0.17 to 0.47 ft/min). The slow rate is attributed to oversize materials and obstructions in the backfill. Also, a carbide drill bit was used for the drilling which produced slower drilling compared to a diamond bit. Using a diamond coring bit might have accelerated the drilling process and produced better quality cores. The average core recovery and RQD in the soilcrete were 87 and 55 percent, respectively, indicating good core quality.

A total of six (6) core samples from the completed grout mass were collected for laboratory testing. The core samples were cured for 28 days and tested for unit weight and unconfined compressive strength. Laboratory test results revealed unit weight and unconfined compressive strength for the in-place, completed jet grout mass in excess of the project requirements. Measured strength ranged from 10.4 to 25.2 MPa (1,510 to 3,660 psi) and unit weight from 17.9 to 20 kN/m^3 (114 to 127 pcf), well exceeding the 2,070 kPa (300 psi) compressive strength and 14 kN/m^3 (90 pcf) unit weight requirements, respectively.

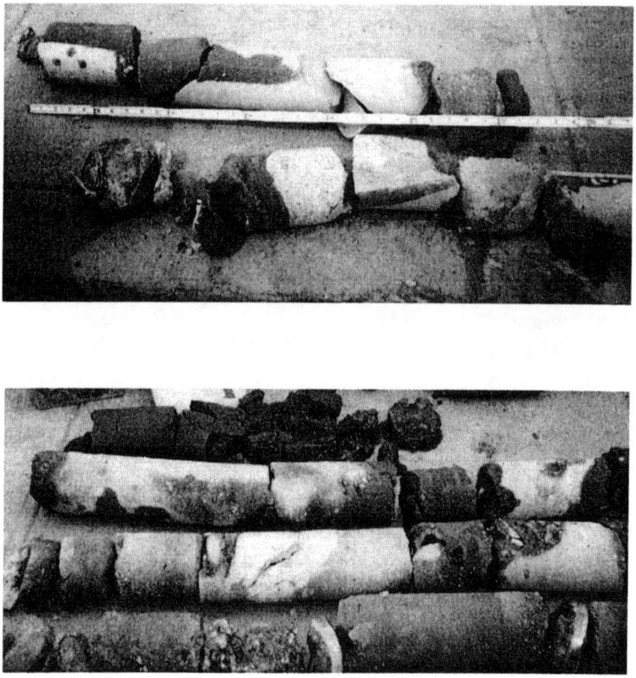

Figure 6: Photographs of Soilcrete Cores

PROJECT PERFORMANCE

The ultimate test of a project is in the performance under the design conditions. This project was completed in 1998 and the proposed undercutting of the track proceeded within 4 weeks after the jet grouting was completed. No movements of any kind were noted as a result of the undercutting.

CONCLUSIONS

The key to success in any in-situ subsurface construction is a properly designed and implemented verification program. Verification testing provides valuable information on the achieved performance of the methods used and provides a basis for design of similar projects in the future. This project demonstrates the effectiveness of an active verification approach to design and construction. A thorough understanding of the methods to be used, and the site constraints and characterization of the unknown conditions that could be encountered are essential in developing a proper verification plan. Once the background is established, clearly defined project objectives and goals should be developed to guide the selection of verification tools and approaches. Good verification requires the use of all available tools including instrumentation, verifiable construction performance measures and destructive and non-destructive testing. Contractor quality control testing and quality assurance provided by the design engineer result in optimal performance.

Test column and excavation is a pre-verification process which could not be implemented due to site constraints and non-representative soil conditions. Continuous construction monitoring was implemented and core borings together with laboratory testing provided an assessment on the adequacy of the process.

Track undercutting in front of the stabilized abutment took place approximately 4 weeks after the completion of jet grouting and the stabilized abutment continues to perform as designed. For structure stabilization projects like this design and verification should be considered inseparable. The design engineer must always ask, how he can be sure his approach will work as well as how he will know, when it is done, that it has been done correctly.

REFERENCES

Byle, M. J. and Borden R. Eds. (1995) “Verification of Geotechnical Grouting,” ASCE Geotechnical Special Publication No. 57, San Diego, CA, October 23-27.

Byle, M.J. and Haider, T.F. (1998) “Abutment Stabilization with Jet Grouting,” ASCE Geotechnical Special Publication No. 80, Boston, MA, October 18-21, pp. 16-30.

DePaoli, B., Tornaghi, R., and Bruce, D. (1989) “Jet Grout Stabilization of a Peaty Layer to permit Construction of a Railway Embankment.” Proceedings of ASCE Congress: Foundation Engineering: Current Principles and Practices, June 25-29, Evanston IL. pp. 272-290.

Drooff, E., Furth, A., and Scarborough, J. (1995) “Jet Grouting to Support Historic Buildings,” Foundation Upgrading and Repair for Infrastructure Improvement. ASCE Geotechnical Special Publication No. 50. San Diego, CA, October 23-26, pp. 42-55.

Ichihashi, Y., Shibazaki, M., Kubo, H., Iji, M. and Mori, A. (1992) “Jet Grouting in Airport Construction,” Geotechnical Soil Improvement and Geosynthetics. ASCE Geotechnical Special Publication No. 30. New Orleans, LA. February 25-28, pp. 183-193.

Kauschinger, J.L., Perry, E.B. and Hankour, R. (1992) “Jet Grouting: State-of-the-Practice,” Geotechnical Soil Improvement and Geosynthetics. ASCE Geotechnical Special Publication No. 30. New Orleans, LA. February 25-28, pp. 169-181.

Miyasaka, G., Sasaki, Y., Nagata, T., Shibazaki, M., Iji, M., and Yoda, M. (1992) “Jet Grouting for a Self-Standing Wall,” Geotechnical Soil Improvement and Geosynthetics. ASCE Geotechnical Special Publication No. 30. New Orleans, LA. February 25-28, pp. 144-155.

Xanthakos, P., Abramson, L. and Bruce, D. (1994) “Ground Control and Improvement,” John Wiley & Sons, New York, NY 1994 Chapter 8, “Jet Grouting,” by Donald A. Bruce. pp. 580-683.

Performance and Evaluation of Tunnels, Excavations, and Retaining Structures

PERFORMANCE OF SRW SUPPORTED BY GEOGRIDS AND JET GROUT COLUMNS

Daniel Alzamora[1], Mark H. Wayne[1], Member, ASCE, Jie Han[1], Member, ASCE

ABSTRACT: The Nova Dutra project consisted of the design and construction of Segmental Retaining Walls (SRW) over very soft (0 to 1 blow count) organic silt and clay deposit 9 m in depth. The walls were needed to assist with the widening of a main transportation route leading to Sao Paulo, Brazil. The SRW section which was monitored is 8.2 m in height and approximately 150 m long and rests on a geogrid reinforced load transfer platform which in turn is supported by jet grout columns 1.2m in diameter spaced 3 m center to center. The soil used in the reinforced zone consisted of a low plasticity fine-grained soil with an effective friction angle of 26^0 and unit weight of 19 kN/m^3. Monitoring methods included of piezometers within the reinforced zone, settlement plates above and below the load transfer platform, and survey points on the face of the SRW. This paper discusses the selection of instrumentation and results of monitoring. The measured results indicate that the SRW performed well.

INTRODUCTION

An expansion of a highway in the northern area of Sao Paulo, Brazil was started in June 1998. This project included the construction of five Segmental Retaining Walls (SRW) with heights varying from 2 to 8.2 m and lengths ranging from approximately 75 m to 150 m. The total wall face area was approximately 1800 m^2. All five walls were constructed using concrete segmental retaining wall units and a facing connection system with punched and drawn high-density polyethylene uniaxial geogrids to mechanically stabilize the soil. The concrete fascia units had a nominal compressive strength of 30 MPa and dimensions of 45 cm wide by 20 cm high and 30 cm deep. Fine-grained soil with 60 to 70% passing the No. 200 sieve (LL < 40, PI < 20) was used as reinforced fill. In order to minimize potential problems related to the build-up of pore water pressures within the poor quality reinforced fill, drainage composites were included behind the wall facing and reinforced mass.

[1]Tensar Earth Technologies, Inc., 5775-B Glenridge Dr., Atlanta, GA 30328, USA.

The tallest SRW (8.2 m high) on this project which was constructed over a 9 m thick organic silt and clay deposit with SPT blow counts ranging from 0 to 1. Jet grouting columns 1.2 m in diameter spaced 3.0 m on centers were selected to improve the very soft foundation conditions. A geogrid reinforced platform was proposed for use over the columns to permit use of an increased spacing of jet grout columns and to minimize the disturbance caused by the potential distress between the columns.

To verify the performance of the SRW on the geogrid reinforced platform and jet grout columns, a series of monitoring devices and survey points were installed during construction. The monitoring devices consisted of piezometers within the reinforced zone, settlement plates above and below the geogrid reinforced platform and surveying points on the face of the SRW.

This paper addresses the issues related to the design and construction of this project, interpretation of monitoring results, and implications for future design.

DESIGN

The typical cross section of the SRWs and geogrid reinforced platform over jet grout columns is shown in Figure 1.

Segmental Retaining Wall

The SRW design was based on AASHTO Standard Specifications for Highway Bridges with the exception of the backfill material requirement. The reader is directed to the AASHTO specifications for details of the design method. Using fine-grained soil as a reinforced fill made the design and the construction more complicated than one using granular fill. Potential build-up of excess pore water pressure, strength variation due to the change of moisture content, and difficulty in compaction of the fill could have potentially influenced the stability of the reinforced earth structure. The study by Soong and Koerner (1999) indicates that without the use of internal drainage the active earth pressure could double when the reinforced soil becomes saturated. Therefore, inclusion of an internal drainage system was essential for this SRW earth structure. As shown in Figure 2(a), a drainage composite, composed of a drainage net and two layers of 200 g/m^2 non-woven geotextile in front and back, was installed along the back of the reinforced mass up to 2/3 of the wall height with a 20% coverage. The purpose of this drainage composite was to intercept any water that wanted to enter the reinforced soil from the retained soil. Within the reinforced soil, non-woven geotextile strip drains were installed as shown in Figure 2(b) to remove any water entering the reinforced fill from the surface. In addition, a drainage zone along the face of the wall was created by using strip drains to prevent potential buildup of pore water pressures against the wall (Figure 2(c)). All drainage systems collected the water from the reinforced fill

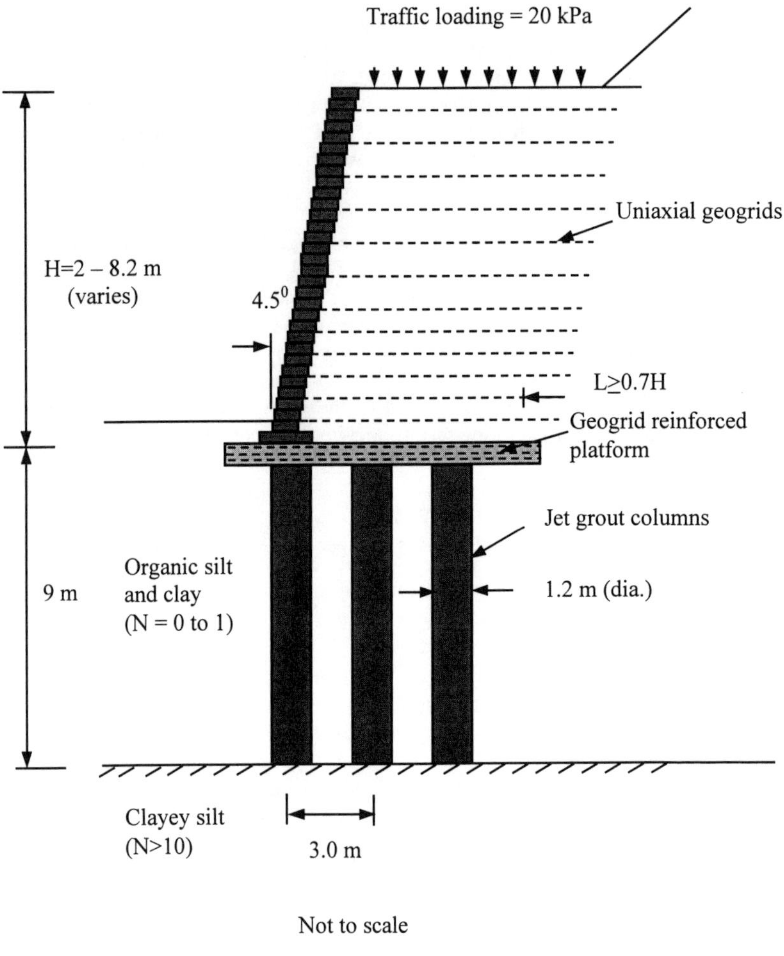

Figure 1. Typical cross section of SRWs on geogrid reinforced platform and jet grout columns

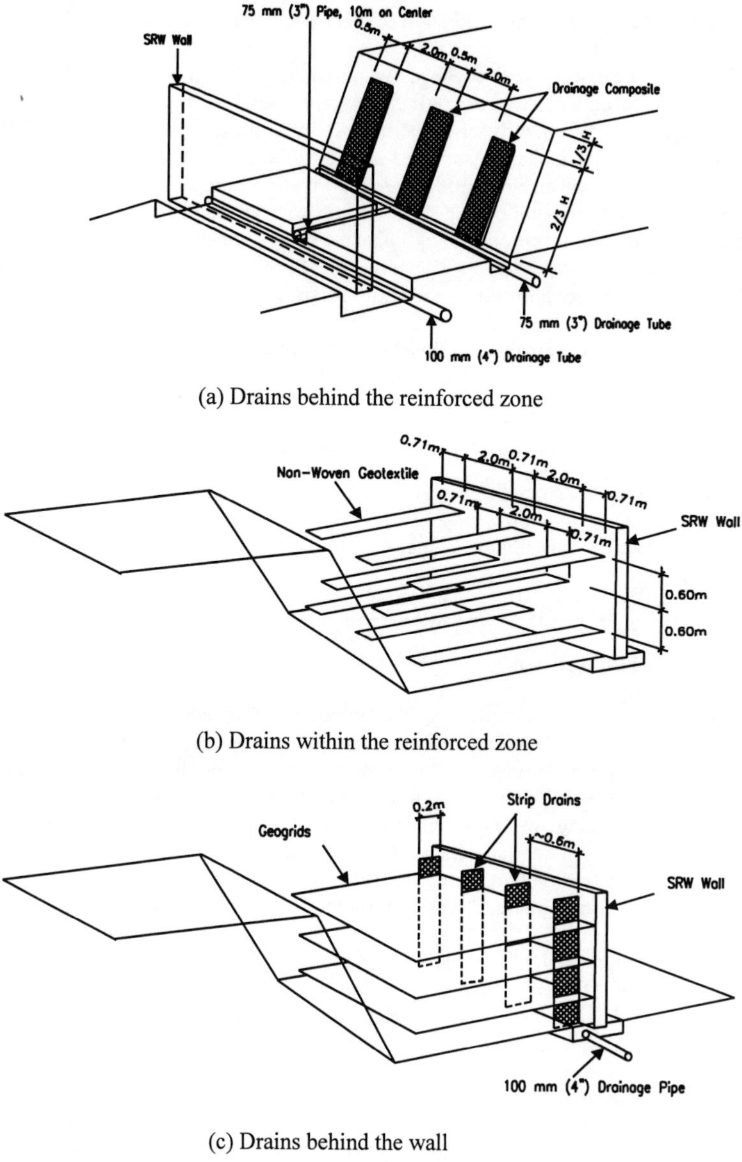

Figure 2. Drainage systems in the segmental reinforced wall (not to scale)

and conveyed it out through outlet pipes. The effects of drainage on reducing seepage force are well investigated and verified by Soong and Koerner (1999).

The internal stability was checked by assuming a Rankine failure plane and considering a traffic surcharge of 20kPa. Coulomb lateral earth pressures were used to assess the external stability of the structure. An effective friction angle of 26^0 determined from triaxial tests, was used and an effective cohesion of 30 kPa was ignored for the reinforced fill at 95% standard Proctor compaction (ASTM Method D-698). The minimum ratio of geogrid length to wall height was 0.7. Percent coverage was used on this project to optimize the design and minimize the number of geogrid types on the site to two. The mechanical properties of these two punched and drawn high-density polyethylene uniaxial geogrids are listed in Table 1.

Table 1. Mechanical Properties of Punched and Drawn High-Density Polyethylene Uniaxial Geogrids

	Type 1	Type 2
Ultimate strength (kN/m)	64.2	131.4
100 year creep limited strength (kN/m)	27.0	54.0
Reduction factor for durability	1.10	1.10
Reduction factor for installation damage – sand, silt and clay	1.05	1.05
Design strength (kN/m)	23.4	46.8

Geogrid Reinforced Platform

Considering the poor condition of the foundation soil, jet grout columns were selected to increase the bearing capacity of the foundation soil and reduce potential total and differential settlements of the SRW. In order to provide a more economic solution, the jet grouting columns were designed with a relatively large spacing (3 m). The large spacing was expected to induce local depression between columns which might cause differential settlement or distortion of the wall face and increase loads onto the geogrids within the reinforced zone. To avoid these problems, a 1m thick geogrid reinforced platform was proposed to bridge over gap between columns.

Different design methods for geogrid reinforced platforms such as the British Standard BS8006 (1995) are available in the literature (for example, Russell and Pierpoint, 1997; Han, 1999). In this design, the approach proposed by Hewlett and Randolph (1988) was adopted to estimate the load applied onto the jet grouting columns. The foundation soil resistance was considered in the calculation of the required strengths of geogrids by using membrane theory (Giroud et al., 1990). This resistance was estimated by assuming that the stress concentration ratio (i.e. the ratio of the stress on columns to that on the foundation soil) was equal to 25. This differs from a range of 5 to 10 for soil-cement columns as suggested in a later study (Han and Wayne, 2000). However, this assumption is conservative since lower foundation soil resistance was assigned underneath the geogrid reinforced platform. The design consisted of two layers of Type 2 uniaxial geogrids in each direction. Within the geogrid reinforced platform, well graded granular fill was placed and compacted.

INSTRUMENTATION

In order to evaluate the performance of the SRW on the geogrid reinforced platform and jet grout columns, a series of monitoring devices and survey points were installed during the construction as shown in Figure 3. As discussed earlier, there was a concern that the generation of pore water pressures within the reinforced fill might de-stabilize the SRW system. As a result of the concern, two piezometers were installed at two different elevations within the tallest wall section to monitor the variation of pore water pressures within the reinforced fill. Settlement plates placed above and below the geogrid reinforced platform were installed to examine the effectiveness of the geogrid reinforced platform on reducing the differential settlement between jet grouting columns. Eleven survey points on the face of the SRW were established to monitor settlements and lateral displacements at different locations on the wall face.

RESULTS OF MONITORING

Pore Water Pressures

Piezomters PN1 and PN2 were installed in accordance with Figure 3. These piezometers were included as a means of design verification and as a safety precaution. For this project, no pore water pressures were recorded during construction or after the completion of the SRW. However, in the event that pore water pressures were recorded, there would have been time to react by stopping construction.

Settlements of Geogrid Reinforced Platform

As shown in Figure 4, the differential settlement between the central point of the columns and middle point between adjacent columns at the bottom of the platform was 15 mm while that at the top of the platform was less than 5 mm. Therefore, the differential settlement was significantly reduced by utilizing the geogrid reinforced platform.

Wall Face Movement

As shown in Figure 3, survey points were installed at different elevations. For data analyses the survey points at upper and lower levels were grouped in Tables 2, 3, and 4 according to their relative elevations. The measured settlements are presented in Table 2 and show that the settlements for the SRW face range from 2 to 32 mm 70 days after the start of construction. Since SRW structures have significant

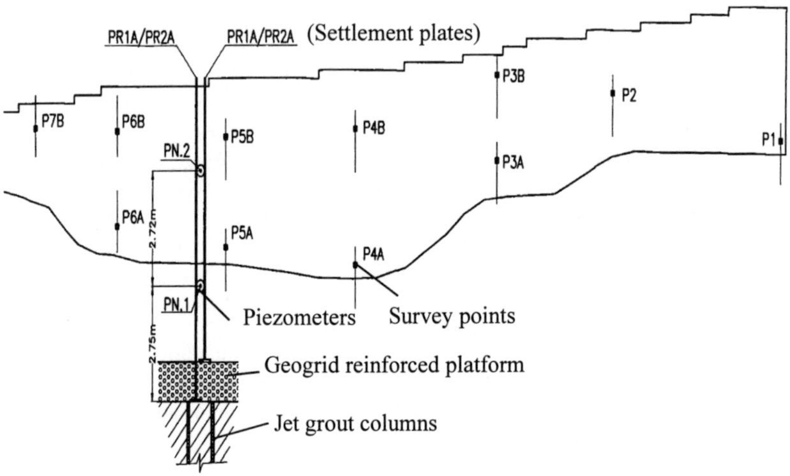

(a) Elevation view

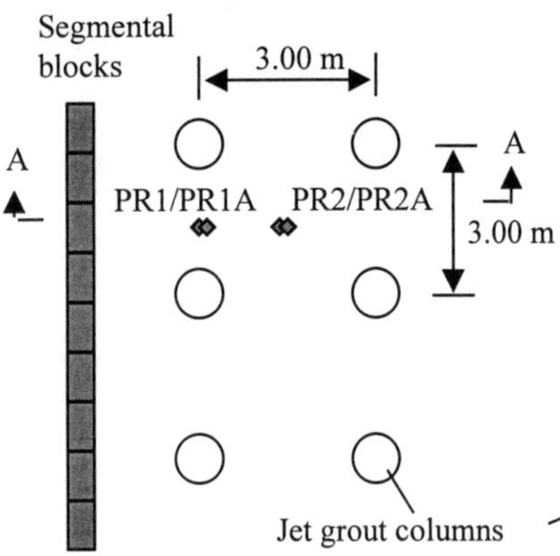

(b) Plan view

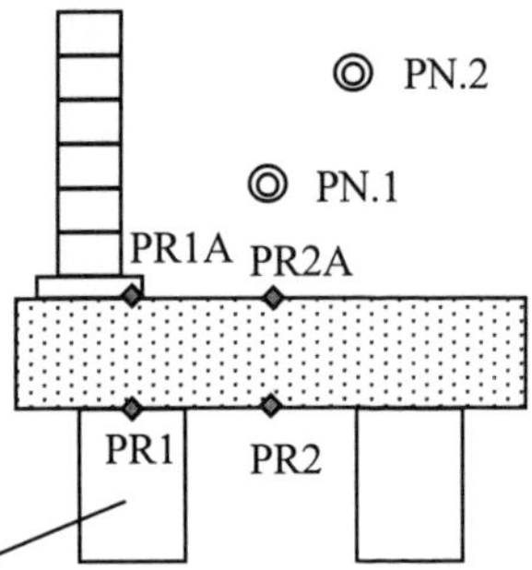

(c) A-A section

Not to scale

Figure 3. Installed monitoring devices and survey points

deformation tolerance, these settlements are not critical to the performance of the SRW.

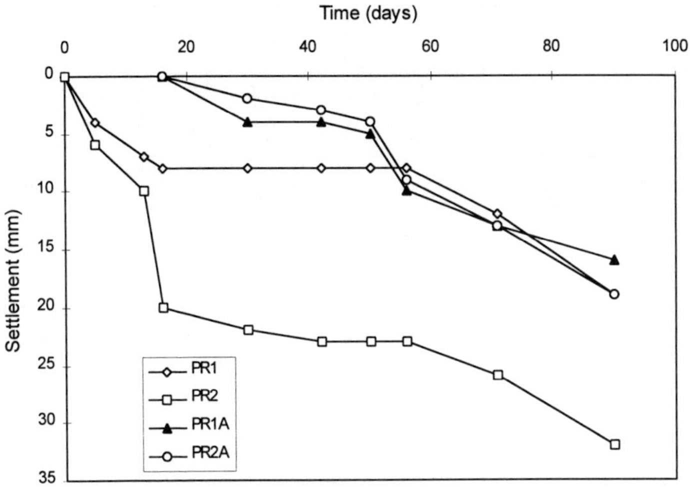

Figure 4. Measured settlements above and below the geogrid reinforced platform

Table 2. Total Settlements of Survey Points on the SRW Face
(70 days after start of construction)

Upper	P7B	P6B	P5B	P4B	P3B	P2
S (mm)	14	19	19	13	2	8
Lower		P6A	P5A	P4A	P3A	P1
S (mm)		26	32	22	5	16

In reality, distortion or differential settlement is more critical to SRW performance than the total settlement. It is stated in FHWA Demonstration Project 82 (Publication No. FHWA-SA-96-071) that "walls with drycast facing (MBW) should be limited to settlements of 1/200". The measured distortions of the SRW 70 days after the beginning of construction ranged from 1/1540 to 1/4000, which are much less than the FHWA requirements.

Table 3. Distortion of the SRW Face between Survey Points
(70 days after start of construction)

Upper	P7B-P6B	P6B-P5B	P5B-P4B	P4B-P3B	P3B-P2
L (m)	15	20	24	26	21
ΔS (mm)	5	0	6	11	6
ΔS/L	1/3030	0	1/4000	1/2380	1/3450
Lower		P6A-P5A	P5A-P4A	P4A-P3A	P3A-P1
L(m)		20	24	26	52
ΔS (mm)		6	10	17	11
ΔS/L		1/3330	1/2380	1/1540	1/4760

Note: L = horizontal distance between two specified survey points.

Lateral face movement is another index to evaluate the performance of the SRW. Based on the empirical curve in FHWA Demonstration Project 82 (Publication No. FHWA-SA-96-071) and the design for the SRW, the maximum lateral face movement for extensible reinforcement can be estimated by

$$\delta_{max} = \delta_R H/75 \qquad (1)$$

where H = height of the wall; δ_R = empirically derived relative displacement coefficient, δ_R=0.75 for this project.

The comparison in Table 4 indicated that the measured lateral wall face movements were far less than those estimated using the FHWA Demonstration Projects 82 approach. This has also been found to be the case for several other walls constructed with similar geogrids (Collin and Berg, 1992) in which $\delta_{max} \le \delta_R H/125$. It is understandable that the FHWA approach is only an empirical one, in which the differences in geogrid stiffness, movement modes, backfill materials, and embedment conditions are not considered. Table 4 also showed that the movements at the upper survey points were mostly larger than those at the lower survey points, which indicated the wall experiences a bulging movement mode instead of overturning or shearing movement mode. The angles of the wall movement 70 days after the beginning of construction ranged from 0.063^0 to 0.424^0, which are very small compared to the wall batter.

Table 4. Lateral Face Movement of the SRW

Upper	P7B	P6B	P5B	P4B	P3B	P2
Hp (m)	3.7	4.9	5.0	5.5	4.9	3.5
δ (mm)	7	14	12	6	12	7
δ_{max} (mm)	37	49	50	55	49	35
$\Delta\theta$ (^{0})	0.108	0.164	0.138	0.063	0.140	0.115
Lower		P6A	P5A	P4A	P3A	P1
Hp (m)		2.7	2.4	2.3	2.9	2.4
δ (mm)		20	16	8	6	7
δ_{max} (mm)		27	24	23	29	24
$\Delta\theta$ (^{0})		0.424	0.382	0.199	0.119	0.167

Note: Hp = height of surveying points from the bottom of the SRW wall; δ = lateral movement outwards. The movements at the upper survey points were estimated by adding those at the lower survey points when the upper survey points were set up. δ_{max} = estimated maximum lateral movement using the empirical curve in FHWA Demonstration Project 82 (Publication No. FHWA-SA-96-071); $\Delta\theta$ = angle of the wall moving outwards.

CONCLUSIONS

This paper explored the instrumentation and results of monitoring for a 2 to 8.2 m high mechanically stabilized earth segmental retaining wall situated on 1 m geogrid reinforced platform and 9 m deep jet grout columns. As indicated in the results two key facts were revealed: (1) the vertical settlement and lateral movement of the wall are within AASHTO tolerances and in the case of lateral movement are much less than would be predicted by FHWA Demo 82; (2) the differential settlement of the load transfer platform is small and aids in confirming the design technique employed on the project. The authors wish to point out the fact that when poor quality soils are utilized in SRW design internal drainage and instrumentation are recommended as part of the wall design.

ACKNOWLEDGEMENTS

The authors would like to thank the support provided by German Reccius at MPZ Aplicacões Tecnológicas for his site assistance and monitoring of the construction of this project.

REFERENCES

AASHTO Standard Specifications for Highway Bridges, 1997 Interim, 27-69.

BS8006 (1995). *Code of Practice for Strengthened/Reinforced Soils and Other Fills*. British Standard Institution, 162p.

Collin, J.G. and Berg, R.R. (1992). "Permanent geogrid reinforced soil retaining walls – US experience." *Proceedings of Seiken Symposium No. 11* – Recent Case Histories of Permanent Geosynthetic-Reinforced Soil Retaining Walls, Tatsuoka and Leshchinsky (eds), Balkema, Rotterdam, 61-68.

FHWA Demonstration Project 82 Reinforced Soil Structures MSEW and RSS, *Mechanically Stabilized Earth Walls and Reinforced Soil Slopes Design and Construction Guidelines*, Publication No. FHWA-SA-96-071, U.S. Department of Transportation, Federal Highway Administration, 371p.

Giroud, J.P., Bonaparte, R., Beech, J.F., and Gross, B.A. (1990). "Design of soil layer geosynthetic systems overlying voids." *Geotextiles and Geomembrane*, 9, 11-50.

Han, J. (1999). "Design and construction of embankments on geosynthetic reinforced platforms supported by piles." *Proceedings* of 1999 ASCE/PaDOT Geotechnical Seminar: Emerging Technologies in Geotechnical Engineering, Hershey, PA, USA, April 14-16.

Han, J. and Wayne, M.H. (2000). "Pile-soil-geosynthetic interactions in geosynthetic reinforced platform/piled embankments over soft soil." Accepted for presentation at the Transportation Research Board 79 Annual Meeting.

Hewlett, W.J. and Randolph, M.F. (1988). "Analysis of piled embankments." *Ground Engineering*, 21(3), 12-18.

Reccius, G. (1999). "Muros de Contenção de Solo Reforçado com Geogrelhas e Paramento de Blocos Prefabricados." To be published at Geosinteticos'99, Rio de Janeiro, Brazil, October 99.

Russell, D. and Pierpoint, N. (1997). "An assessment of design methods for piled embankments." *Ground Engineering*, November, 39-44.

Soong, T.Y and Koerner, R.M. (1999). "Geosynthetic reinforced and geocomposite drained retaining walls utilizing low permeability backfill soils." Geosynthetic Research Institute, GRI Report 24, 120p.

Y2K GROUND IMAGING TECHNOLOGY FOR GEOTECHNICAL PERFORMANCE EVALUATION

Kanaan Hanna[1]
Khamis Y. Haramy[1]

ABSTRACT

Obtaining accurate and timely subsurface information during site characterization and construction phases is a key to successful completion of constructed geotechnical systems. Recent technological advancements in computers and algorithms have resulted in seismic tomography methods becoming a standard engineering tool during geotechnical investigations for highways, bridges, and tunnels.

The RockVision3D™ seismic tomography system, developed by NSA Engineering, Inc., is a high-tech "**through-the-earth CAT scan**" that provides *on-site* true three-dimensional images of subsurface conditions that exhibit a significant change in seismic velocity or attenuation. The system uses newly developed algorithms for data processing and provides images with greater detail and accuracy than traditional tomography methods. The system has been used worldwide to characterize the subsurface and to evaluate the performance of various constructed support systems (Hanna & Haramy, 1999).

This paper presents case studies of the RockVision3D™ applicability in geotechnical site characterization and design confirmation for the following civil projects:

1. Location of a rock tunnel beneath an interstate highway in Colorado.
2. Detection of abandoned coal mine entries beneath a retaining wall foundation in Illinois.
3. Evaluation of the effectiveness of grouting a karstic limestone bridge pier foundation in Missouri.
4. Assessment of highway settlement causes and determination of the effectiveness of remedial grouting in Colorado.

[1] NSA Engineering, Inc., 15000 West Sixth Avenue, Suite 100, Golden, CO 80401

INTRODUCTION

"Tomography," the basic concept behind medical CAT (Computer-Aided-Tomography) scans, has been used for decades. This technology involves surrounding a 2-D or 3-D region of interest with a source and receiver array, collecting signal data through the plane or volume over a network of "**ray paths,**" and reconstructing an image of the structure based on signal arrival time computations (Rock, et al., 1997). RockVision3D™ tomography, developed by NSA Engineering, utilizes *seismic energy* transmitted through the ground and rock mass from some form of source (hammer, blast detonation, borehole air gun, excavation equipment, etc.), which is then collected by a receiver array (borehole hydrophones, surface-mounted geophones, accelerometers, etc.). The result is a **"through-the-earth-CAT scan,"** utilizing both velocity and attenuation tomogram construction methods. Any ground condition that exhibits a significant change in velocity or attenuation can be readily imaged—including old mine workings, slope failure planes, soil-rock interfaces, karst ground voids, saturated zones, fault planes, man-made structures, etc.

Until recently, seismic tomography, as applied to mining and civil applications, has been largely limited to several relatively old processing programs developed within the now-defunct U.S. Bureau of Mines, including "BOMTOM" and "MIGRATOM." Experience with these codes shows their application generally requires considerable data processing to prepare simple 2-D tomograms—involving only 100 to 1,000 ray paths—literally requiring weeks to process data and present the results in a useable graphical format. In contrast, this high-tech RockVision3D™ seismic tomography technology can produce high-quality, detailed tomograms—typically ranging from 5,000 to 40,000 ray paths—immediately following data collection, with preliminary results often available within 24 hours.

The RockVision3D™ system software combines wavefront theory with curved ray theory to produce a 3-D tomogram of the area to be imaged based on the geometric array of the sources and receivers. The system produces refraction or reflection images from velocity and attenuation. Sensitivity as to location of features is a function of source and receiver spacing and the distance between boreholes. A ray path is assigned by the system that indicates the region in the rock mass where the energy travels. For example, the energy will not pass through a void, but will travel around it. If the cavity is filled with water, clay, or sand, etc., a portion of the energy will travel through it at a different velocity.

The RockVision3D™ seismic tomography system provides the geotechnical engineer with **2-D and 3-D** images of the rock mass conditions. Such detailed timely information enhances the degree of confidence in construction projects through vastly improved site characterization, ultimately minimizing project risk and maximizing the cost efficiency of design, construction, and maintenance phases.

RockVision3D™ System Advantages

A complete tomography package, from signal collection to CAD output display, RockVision3D™ provides images with greater detail and accuracy than other tomography approaches. Data are processed and subsurface images are produced on site. With the information that RockVision3D™ provides, rock characterization and site geotechnical evaluations are dramatically improved, and the risks and expenses associated with change orders can be greatly reduced. Some of the advantages within the RockVision3D™ system include...

- The ability to provide "**forward models**" – tomographic simulations of the planned survey prior to actually conducting the fieldwork, allowing the client to (1) determine the level of imaging performance to be expected, and (2) optimize required borehole placement and geotechnical mapping within the site being characterized.
- Use of three-dimensional B-Spline Interpolation Network for automated arrival event picking and mesh initialization. This application greatly expedites the solution of large, complicated imaging arrays. In addition, the interpolation network within RockVision3D™ allows the available drill log information to be incorporated into the 3-D images to significantly improve the understanding of subsurface ground conditions.
- The ability to produce 2-D and 3-D velocity, attenuation, and "difference" tomograms. Difference tomograms are constructed to determine changing ground conditions over time.
- The determination of elastic constants (E, ν) distribution in 2-D and 3-D volumes with the availability of P- and S-wave data.

RockVision3D™ Application to Civil Projects

To date, numerous geotechnical ground-imaging projects have been successfully completed related to highway, bridge, and rock tunnel construction. Specific geotechnical areas in which RockVision3D™ is used to reduce risk include...

- Assessment of slope, embankment, and impoundment stability, pre- and post-construction.
- Location of shallow structures (old mines, foundations, karst voids) that may impact construction.
- Characterization of bridge abutments/pier foundation and location of scouring zones in river settings.
- Location of sewer line collapse zones, flow migration paths, piping, aquifers, and seepage characterization.
- Evaluation of tunnel and substation alignment evaluation.
- Imaging ahead of TBM/roadheader excavations.

Cross-Borehole Tomography Technique

The RockVision3D™ system has been used with various survey geometries including borehole-to-borehole, borehole-to-surface, and tunnel-to-surface. The civil project applications described here used the RockVision3D™ tomography package in cross-hole velocity mode. It has been determined that, in most cases, the cross-hole approach provides the most adequate information about actual subsurface conditions and ground performance that could impact construction techniques and costs.

The cross-borehole tomography survey involves pairing a source and a string of 12 hydrophones to propagate and capture seismic signals transmitted between source and receiver boreholes within the surveyed area. These hydrophones have an internal preamplifier with a gain of 20 dB and have an operating frequency range of 7 Hz to 4.7 kHz. The airgun source (most widely used) operates at a chamber pressure of up to 7 MPa and produces a very consistent, repeatable pulse with a fairly flat spectrum between 10 Hz and 180 Hz. Compressed nitrogen at 16 MPa pressure is used to charge the gun. The airgun has an additional hydrophone installed close to the gun ports to serve as a trigger/zero time signal. The trigger and the 12-hydrophone string are connected to a 24-channel seismograph, and data are recorded upon every shot. Data are generally collected at a sample rate of 8,000 samples per second for a time resolution of 0.125 msec and a Nyquist frequency of 4,000 Hz.

Field Survey

A minimum of three drill holes is required to conduct a three-dimensional cross-borehole tomography survey. The spacing between any two holes may vary from several meters to 100 m, depending on the size of the area/volume to be surveyed. The vertical surveyed area may extend from the near surface to a depth of up to 300 m, depending on the location of the zone to be imaged. The drill holes must be cased with a perforated schedule 40 PVC pipe with a 63- to 100-mm inside diameter, depending on the size of the seismic source being used. A solid casing can be used provided that the annulus between the borehole wall and the PVC casing is filled with bentonite/clay slurry or grout to provide solid contact between the casing and the borehole wall. All cased boreholes are filled with water to allow the use of the hydrophones and the seismic source.

Figure 1 shows forward models of a planned cross-borehole survey for imaging the extent of bridge piers within a bridge foundation. In general, the field survey involves the installation of the seismic source in one borehole and the hydrophone string in an adjacent borehole. The source and receiver locations will be moved to predetermined depths and advanced until all adjacent hole pairs have been surveyed.

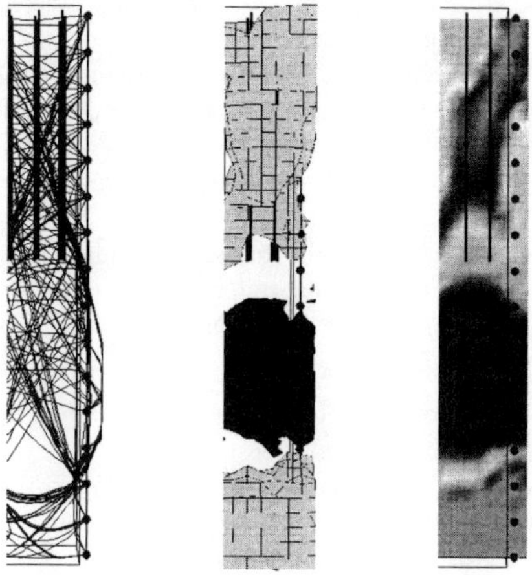

Ray paths 3-D Image 2-D Image

Figure 1. Inverse models for imaging the extent of bridge piers within the bridge foundation.

Data collected during the survey are processed using the RockVision3D™ proprietary software to reconstruct three-dimensional color-coded tomographic images of the volume and two-dimensional images between any two boreholes. Tomographic images of cross-sections between any coordinates can also be constructed.

ROCKVISION3D™ APPLICATIONS

The following section presents the potential applications of the 2-D and 3-D seismic tomography in the field of geotechnical engineering and performance evaluation. The RockVision3D™ system was successfully applied to image subsurface structures involving three highways and a bridge pier.

Location of Rock Tunnel Beneath an Interstate Highway in Colorado– Case Study 1

During the planning and construction of a new highway overpass on Interstate 25, the Colorado Department of Transportation (CDOT) became concerned that the foundational stability of the bridge support caissons could be compromised by the existence of a rock tunnel connected to abandoned coal mines beneath the caisson foundations. Information from the Colorado Geological Survey (CGS) indicated the possible existence of a rock tunnel connecting two abandoned coal mines and crossing the interchange below Pier 3, which is currently under construction. The exact tunnel alignment and conditions with respect to the location of the caissons were not clear. If the tunnel has collapsed, its cave zone may extend to the bearing horizon, reducing the bearing capacity of the foundation and the skin friction along the caissons. In an attempt to better define the location and alignment of the tunnel and to assess its condition or state of collapse, a three-dimensional cross-borehole tomographic survey was performed. The purpose of the study was twofold:

1. To image the subsurface geologic structure including strata and cavity or rubble resulting from past coal mining operations.
2. To evaluate the extent of the subsidence zone that may occur due to mine entry/tunnel collapse.

A total of six vertical boreholes, defining a zone/volume approximately 34 m wide by 84 m long by 61 m deep, were used to conduct a cross-borehole seismic survey. All boreholes were drilled and cased with 100-mm ID perforated PVC. The field survey was conducted according to the procedure and approach discussed in the previous section. As a result, over 30 colored 2-D and 3-D tomographic images were constructed, including horizontal and vertical cross-sections, at 1.5-m increments to delineate the location of the tunnel.

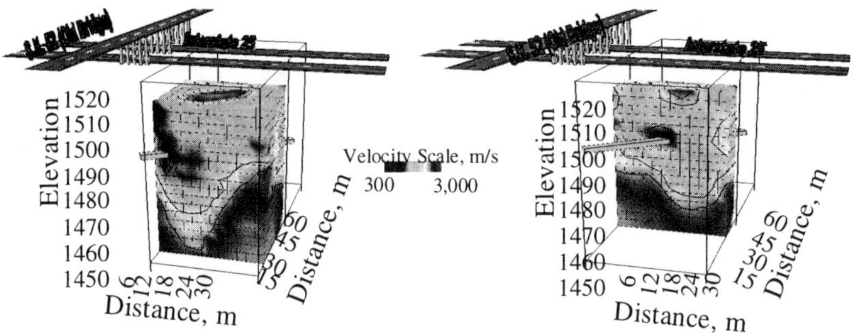

Figure 2. Tomographic image showing tunnel location at various cross-sections.

Tomographic imaging results are depicted in several cross-sections, as shown in Figure 2. The location of the suspected tunnel was established at a depth between 27 m to 30 m. The results from the three-dimensional images indicate that the tunnel is not intact, and that a caved zone extends from the tunnel horizon to near the ground surface. Figure 3 shows three distinct zones: (1) an unconsolidated, low-velocity zone (< 20 blows); (2) an intermediate consolidated zone (>30 blows); and (3) a consolidated, high-velocity zone (> 50 blows).

Based on field studies of sinkhole subsidence previously performed in Boulder County (Matheson, 1987), the expected extent of the rubble zone and angle of draw were superimposed on the tomography results and related to the planned location and bearing horizon of the bridge caissons.

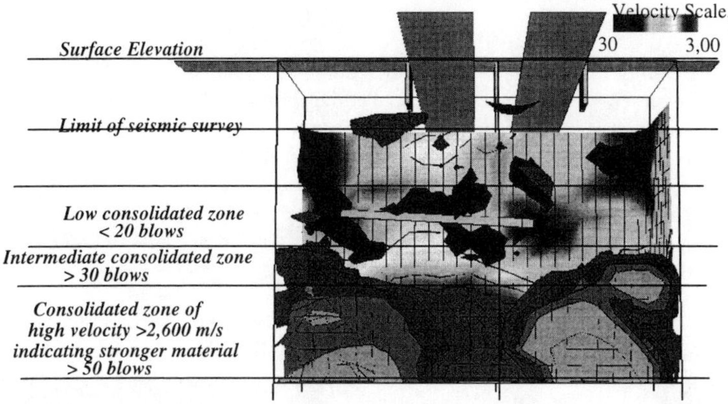

Figure 3. 3-D tomographic image showing tunnel location, cross-section 100+613.2, velocity contours of 2,400-3,000 m/s, and 900 m/s looking from the north.

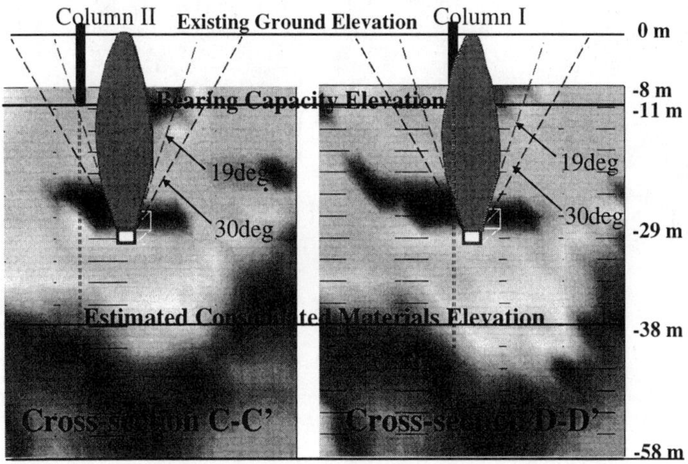

Figure 4. 2-D cross-sections perpendicular to the tunnel direction as located by NSA superimposed with the subsidence limit lines and sinkhole profile lines.

Figure 4 shows that if further sinkhole-type subsidence is to occur, the bearing capacity of Column 1 will be affected. Based on this information, CDOT is currently exploring options for the geotechnical design phase, including relocation of the caissons, changes to caisson design and depth, and/or rehabilitation of the rubble zone near the bearing horizon.

Detection of Abandoned Coal Mine Entries Beneath Retaining Wall Foundation in Illinois – Case Study 2

During the installation of soldier piles for a retaining wall foundation being constructed on U.S. Route 24 near Peoria, Illinois, the Illinois Department of Transportation (IDOT) encountered underground voids filled with water. This became a major concern to IDOT engineers, causing all contract work to stop. In support of the subsurface investigation, NSA applied its RockVision3D™ cross-borehole tomography to identify and map the extent of the underground mine workings, which included voids, entries, and any weak permeable or highly fractured zones that may contribute to stability problems.

The survey was conducted using fifteen 18-m-deep vertical boreholes drilled on either side of the 220-m-long retaining wall, as shown in Figures 5 and 6.

Figure 5. Photos showing the test site along U.S. Route 24 and on-site data reduction.

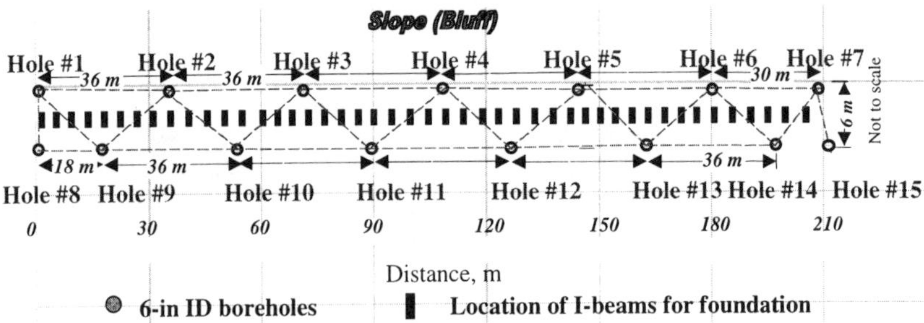

Figure 6. Planned borehole locations for the cross-hole seismic survey.

Based on the tomographic imaging results depicted in figures 7a and 7b, the following conclusions were made:

- There are three mine entries about 6 m wide by 2 to 3 m high and approximately 8 m below the surface.
- The entries were located between stations 20+459 and 20+579.
- Each entry and the triangular area overlying the entry are filled with highly fractured material, indicating a caved zone extending to the surface.

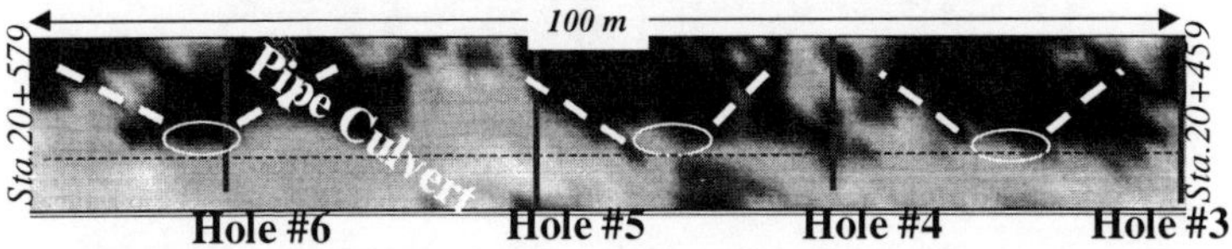

a. Tomogram at centerline of imaged area

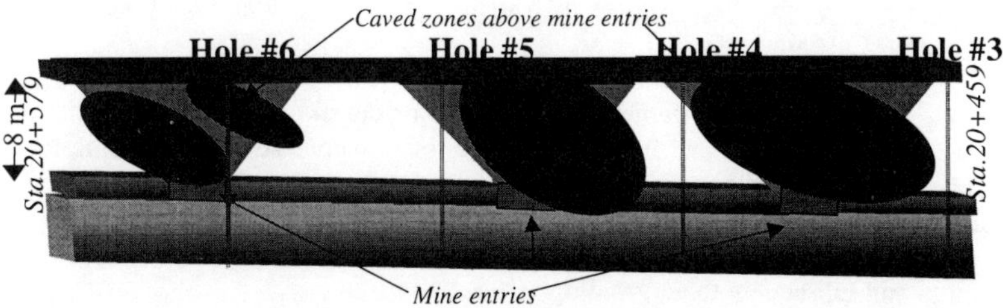

b. Mine entries and caved zones at centerline of imaged area

Figure 7. Vertical cross-sections at the center of the imaged area and a drawing showing areas of fractured materials.

Figure 8 is a 3-D image of the actual subsurface ground conditions showing the zones of mining and no mining and the extent of the subsidence. The results were used by IDOT for the construction confirmation of the piles. A rehabilitation strategy was developed wherein large-diameter (~0.9 m) pipes were placed vertically through the mapped mining horizon, with soldier piles placed and grouted within the pipes. The RockVision3D™ survey allowed the rehabilitation effort to proceed without the need for expensive and time-consuming probe drilling.

Figure 8. Three-dimensional image of old underground caved coal mine entries under U.S. Route 24 near Peoria, IL.

Evaluation of Grout Effectiveness of Karstic Limestone Bridge Pier Foundation in Missouri – Cast Study #3

The Bill Emerson Memorial Bridge is being constructed over the Mississippi River between Missouri and Illinois. The river bottom consists of karstic limestone; voids and structural defects within the limestone necessitated reinforcement of the foundation by high-pressure jet grouting. To evaluate the effectiveness of the jet grouting program, NSA Engineering performed cross-borehole tomographic surveys prior to and subsequent to jet grouting.

From a construction platform in the river (figure 9), six boreholes were drilled in an area approximately 3.2 m wide by 3.4 m long to a depth of about 22 m below the river bottom. The survey was conducted in a manner that allowed a three-dimensional tomogram to be produced. The well-log geologic data were inputted to the RockVision3D™ program as known information. The results were used to delineate clay-filled zones and characterize the effectiveness of grouting in a complex geologic environment.

Figure 9. On-site data reduction.

Figure 10 shows the interpolated well-log images and sections of the resulting tomogram before and after jet grouting operations. Results of the survey showed that jet grouting significantly increased seismic velocities throughout the treated volume and allowed the Missouri Department of Transportation (MDOT) to confirm that the jet grouting effort was adequate for the pier foundation as designed.

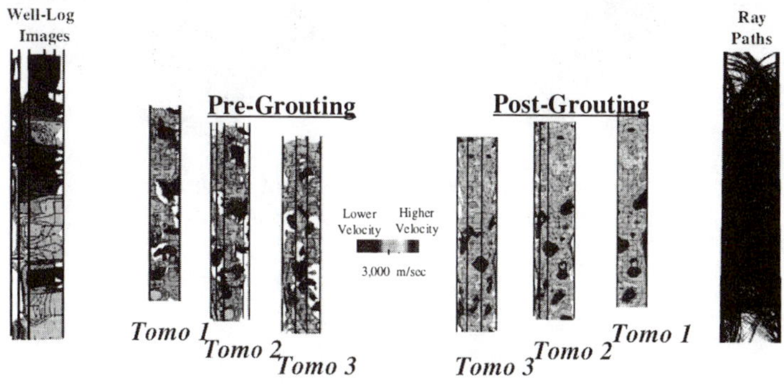

Figure 10. Interpolated well log and pre- and post grout images beneath river bottom formation underlying the foundation of a pier for a bridge structure.

ARTHUR LAKES LIBRARY
COLORADO SCHOOL OF MINES
GOLDEN, CO 80401

Assessment of Highway Settlement Causes and Determination of Effectiveness of Remedial Grouting in Colorado – Case Study #4

After the construction of Colorado Highway 287, the presence of a relatively hard clay deposit beneath the highway caused differential settlement resulting in cracks and vertical displacement in several sections of the roadway.

Tomographic imaging using NSA's RockVision3D™ system was applied to determine the nature of the subsurface ground condition and assess remedial grouting effectiveness. Three-dimensional cross-borehole tomography was conducted using five boreholes drilled to a depth of about 8 m below a selected section of the damaged roadway.

Tomographic images were produced before and after grouting. Three-dimensional velocity contour images, as shown in Figure 11, provide comparison between pre-grouting and post-grouting conditions. The low-velocity area (1,100 m/s, light gray) in the pre-grout survey is altered to a higher velocity area (1,200 m/s, dark gray) in the post-grout survey due to a change in density. The velocity increase in the grouted area indicates that the ground condition is improved by the grouting. Figure 12 shows tomograms at 3-m cross-sections below the surface for pre- and post-grouting, indicating improvement of the ground conditions. After grouting, the density of the low-velocity area (762 m/s – dark gray) has been increased by the grout.

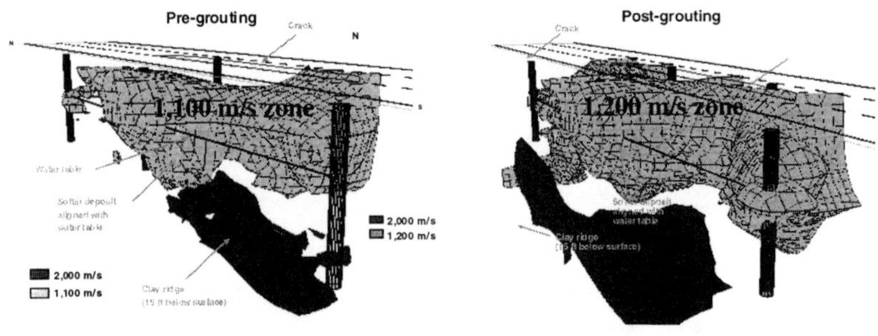

Figure 11. Three-dimensional velocity contour images show comparison between pre-grouting and post-grouting conditions.

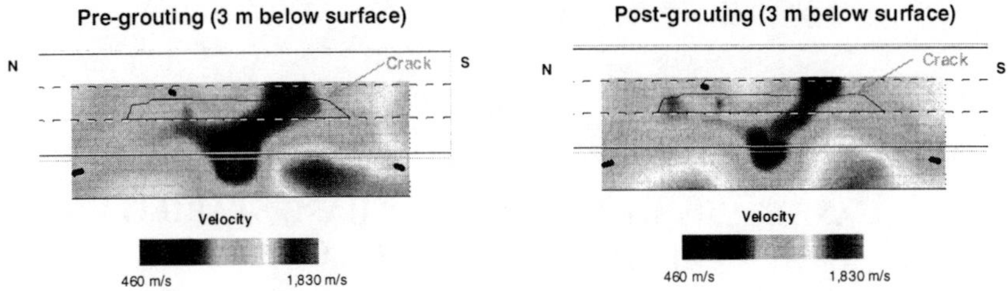

Figure 12. Tomograms at 3 m below surface for pre- and post-grouting show improvement of the ground condition.

CONCLUSIONS

The high-tech RockVision3D™ tomographic imaging system provides timely and cost-effective subsurface information for geotechnical investigations.

Numerous applications have shown the accuracy and efficiency of this imaging system to adequately characterize complex geologic environments for geotechnical confirmation during all phases of planned projects.

APPENDIX-1. REFERENCES

Hanna, K., and K. Y. Haramy, 1999. "RockVision applications to delineate hazardous geologic and stress conditions in coal mines – Case histories," SME Annual Meeting, Denver, CO, March, SME Preprint 99-94.

Matheson, G. M., 1987. "Geologic and mining factors that control subsidence above abandoned underground coal mines along the Colorado Front Range." PhD thesis, Colorado School of Mines, Golden, CO.

Rock, A. D., K. Y. Haramy, and D. M. Neil, 1997. "A new automated mapping system to delineate hazardous ground conditions ahead of mining," Pres. IUCC Conference, Brisbane, Australia, June.

CASE STUDY OF EXCAVATION BASE STABILITY IN DEEP MARINE CLAY

A.J. McGinn[1], Member ASCE and T.D. O'Rourke[2], Member ASCE

Abstract

A case history is presented in this paper for an excavation in Boston marine clay supported by soil mix walls and earth anchored tiebacks. Large lateral deformation occurred when the excavation reached a depth of 13.4 m. Excavation unloading combined with the dissipation of excess pore pressures is shown to account for undrained shear strength reduction and the observed excavation performance. Design practices are examined in light of this case history, and the lessons learned are used to recommend improvements in evaluating the base stability of deep clay excavations.

Introduction

The case study presented in this paper is for the construction of approximately 915 m of I-90 adjacent to Logan Airport, Boston, Massachusetts. A map showing the project location, referred to as Bird Island Flats (BIF), is provided in Fig. 1. The main line tunnel (MLT) at BIF consists of a double barrel reinforced concrete structure built by cut-and-cover techniques to depths ranging from 12.5 to 25.9 m. The width of the excavation varied from 24 m to 87 m. A plan view of the construction site showing locations of MLT stationing, instrumentation, adjacent structures, and excavation support walls is presented in Fig. 2.

This paper examines the undrained strength of marine clay at the BIF site and evaluates how the clay, excavation, and geologic characteristics influenced the actual strength mobilized during construction. Excavation unloading coupled with the dissipation of excess pore pressures is shown to account for undrained strength reduction consistent with the observed performance of the cut. The clay, excavation, and geologic characteristics that promote reduction in undrained shear strength are discussed and related to the potential for base instability of similar excavations.

[1,2] Graduate Research Assistant and Professor, respectively, Cornell University, Ithaca, New York 14853

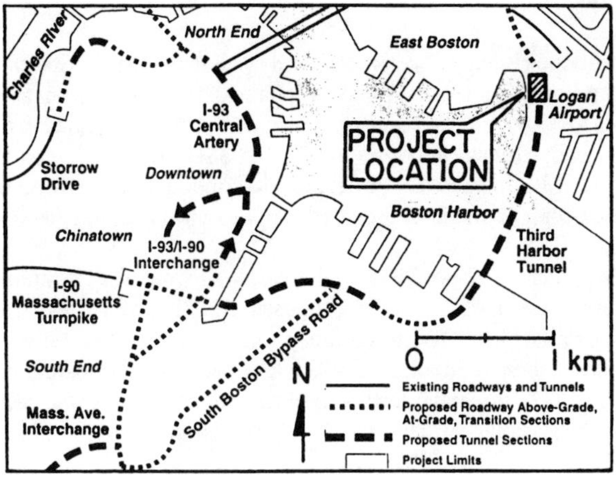

Figure 1. Project Locus

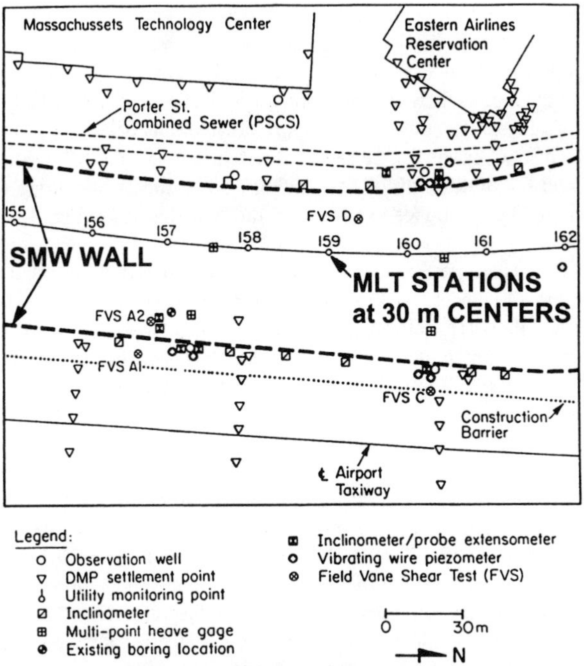

Figure 2. Plan View of Excavation

Site Conditions

Construction activity along the east wall from MLT Station 156+00 to 163+00 was primarily for the exit ramp structure. The depth of the excavation ranged from 17.2 m to 19.4 m. Temporary support was provided by means of a Soil Mix Wall (SMW) with earth anchored tiebacks. The SMW was reinforced with W21x50 H-piles spaced at 1.2-m intervals. A detailed description of the construction sequence and excavation support system is provided elsewhere by O'Rourke, et al. (1997).

A longitudinal soil profile near the east wall from MLT Stations 156+00 to 163+00 is presented in Fig. 3. The proposed elevation of the exit ramp invert is indicated by a dashed line. The stratigraphy from ground surface downward consists of granular and cohesive fill, organic silt, and a thick layer of marine clay overlying glacial deposits. The maximum thickness of the marine clay near the east wall is 26.5 m. A layer of glaciofluvial deposits, overlying the glaciomarine deposits, extends northward from MLT Station 160+00.

The marine clay deposit encountered at BIF consists of a gray clay with interlayered silt seams and partings of fine sand that are characteristic of deposition in a shallow marine environment (H&A, Inc., 1991). The top of the clay is at a nearly constant elevation, but its thickness is variable as shown in Fig. 3. To explore how the depth changes across the site, the thickness of the marine clay was obtained from 209 borings performed at BIF. Contours of thickness were generated using kriging techniques with commercially available software, Surfer, Version 4 (Golden Software Inc, 1989). The resulting contours and approximate limits of excavation are shown in Fig. 4. Two distinct basins can be seen in the figure, one from MLT Station 153+00 to 163+00 and the other north of Sta. 166+00. The soil profile along the east wall presented in Fig. 3 corresponds to Cross-section A-A' located in Fig. 4.

The excavation performance at the BIF site is described in other publications (O'Rourke and O'Donnell, 1997a and b, O'Rourke, et al., 1997). Of particular significance are the large deformations along the east wall that occurred when the excavation was at a depth of 13.4 m. Figure 5 shows a cross-section of the east wall excavation at Station 157+40. Superimposed on the cross-section is the soil profile and recorded measurements at this stage of the construction from inclinometer, surface settlement points, and multipoint heave gages.

The length and inclination of the tiebacks are drawn to scale with an indication of the approximate anchor bond zone. The third-level tiebacks (dashed lines) had been installed at elevation 22.5 m, but only the first stage of grouting had been applied when a maximum lateral movement of 215 mm was recorded adjacent to the SMW. Inclinometer measurements showed that the rate of wall deformation was increasing with time. To stabilize the excavation, soil was backfilled to the elevation of the second-level tiebacks, at which time the wall deformation stopped.

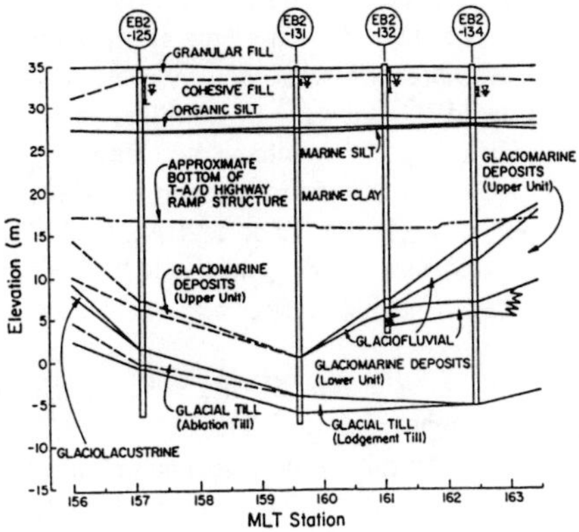

Figure 3. Longitudinal Soil Profile Along East Wall

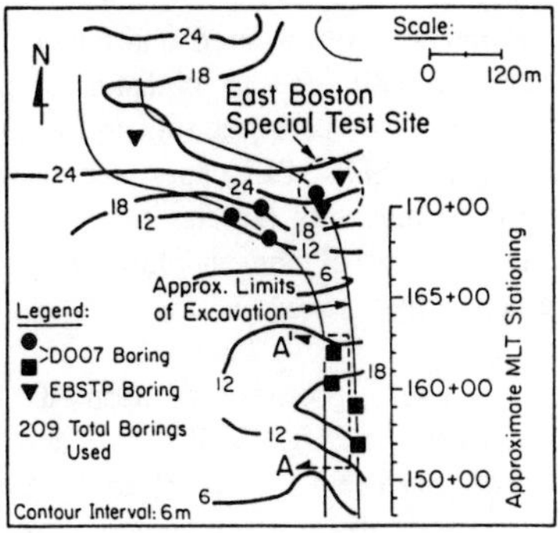

Figure 4. Contours of Marine Clay Thickness at BIF

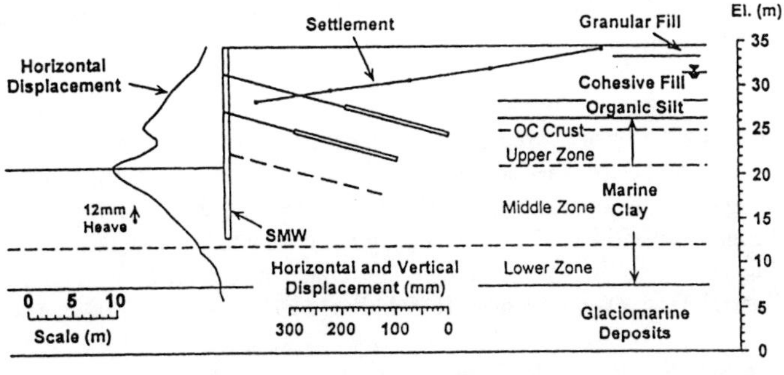

Figure 5. Recorded Ground Movements at Station 157+40 in September, 1993

The large wall and ground movements have been shown to be related to deep rotational and lateral deformation of the marine clay (e.g., O'Rourke and O'Donnell, 1997a). Although previous publications have dealt with the ground movements, soil support, and stabilization measures at the BIF site, there has not been a detailed examination of the undrained shear strength of marine clay and its relationship to the excavation performance.

Undrained Shear Strength

Before excavation, the approach to characterizing the undrained clay shear strength (S_u) at the BIF site followed procedures generally adopted for strength determination in the design of excavations in soft to stiff clays. Samples of clay were obtained by means of a thin-wall stationary piston tube sampler in accordance with ASTM D 1587-83 procedures, and laboratory tests on specimens trimmed from these samples were performed to assess S_u. The lab tests performed by others (H&A, Inc., 1991) included undrained triaxial tests on 3.6 and 7.6-cm-diameter specimens, isotropically consolidated undrained triaxial compression tests, and laboratory vane shear strength tests on undisturbed and remolded samples. In addition, strength evaluation was performed using the Stress History and Normalized Soil Engineering Properties (SHANSEP) approach in which S_u is related to the in situ vertical effective stress by means of the clay overconsolidation ratio (OCR) [Ladd and Foott, 1974]. One-dimensional consolidation tests were performed to determine OCR. X-ray radiography was used to characterize the degree of sample disturbance.

The S_u profiles for various depositional basins at the BIF site were estimated from both the SHANSEP results and triaxial compression test data. Given the low vertical permeability ($1x10^{-8}$ to $1x10^{-9}$ cm/s) of the marine clay specimens and the temporary nature of the construction, undrained conditions were assumed for local bearing capacity and deep circular stability assessments. This approach is consistent with procedures generally followed for excavations in soft to stiff clay. The design of such excavations commonly assumes that excess pore pressures drain slowly with respect to the time the excavation is open and that changes in undrained strength are not significant.

Laboratory samples were obtained from the borings identified in Fig. 4. The square and circular shapes represent borings used to generate S_u profiles by SHANSEP methods for the southern and northern basins, respectively. The triangular symbols denote borings performed as part of the East Boston Special Testing Program (EBSTP), which was a program conducted to evaluate the principal engineering properties of the marine clay that affect the design of structures for the Boston Central Artery/Tunnel project (H&A, Inc., 1993).

The SHANSEP undrained strength profile for the north basin developed for the BIF project and the EBSTP are similar, and show significantly lower undrained shear strength of the marine clay than the SHANSEP profiles determined for the south basin. One of the distinct features of marine clay in Boston is that differences in deposition, erosion, water level fluctuations, and historic construction and stockpiling activities can lead to significant variation of S_u even in neighboring depositional basins.

Two methods for estimating overconsolidation ratio (OCR) from oedometer tests were used to estimate the undrained shear strength profile for clay in the vicinity of Station 157+40 according to the SHANSEP procedures. They include: 1) conventional Casagrande construction and 2) the strain energy approach proposed by Becker, et al. (1987). In general, the two methods tended to give similar results, with the strain energy method providing slightly higher estimates of OCR (by approximately 10%) in the lower portions of the middle zone and higher elevations of the lower zone marine clay. Results from the Casagrande construction are presented in this paper.

The OCR data designated as fair to very good in quality (H&A, 1991) are plotted in Fig. 6a. Two separate groups of OCR data above and below El. 10 m were identified, and linear regressions of log OCR vs. depth were performed for each group. Null hypothesis tests on the data below El. 10 m showed that the regression slope and a vertical trend line were not different to any appreciable degree of confidence. Using standard statistical procedures (e.g., Snedecore and Cochran, 1980), 90% confidence bounds were drawn for the two groups as indicated in the figure.

The undrained direct simple shear strength, S_{uDSS}, profile was calculated from the best fit OCR lines as

$$S_{uDSS} = s\sigma'_{vo}(OCR)^{m} \qquad (1)$$

in which the SHANSEP parameters were assumed to be s = 0.20 and m = 0.8, and σ'_{vo} is the in situ vertical effective stress. The 90% confidence bounds on S_{uDSS} were derived from the OCR confidence bounds using Eqn. 1.

The S_{uDSS} profile and 90% confidence bounds, determined in accordance with the above procedures, are plotted in Fig 6b. This figure also compares the best fit S_{uDSS} profile and 90% confidence bounds determined in this work (referred to as MO) with similar strength profiles (referred to as WL) presented by Whittle and Ladd (1998). There is good agreement between the two estimates of the S_{uDSS} profile, with the average S_{uDSS} presented in this paper being the lower of the two trends by about 10%.

An undrained shear strength profile acquired from field vane shear tests (FVSTs) is shown in Fig. 6c. The FVSTs were performed after the excavation was backfilled to stabilize the deformations shown in Fig. 5. Locations of FVSTs are shown in Fig. 2.

Data obtained in the Boston area indicate that FVST data, carefully acquired and corrected according to Bjerrum's recommendations (1972), provide a good estimate of the in situ direct simple shear (DSS) undrained strength of the marine clay. These correction factors were applied to the FVST data. The original data points are presented as symbols and the uncorrected and corrected statistical trend lines are shown in Fig. 6c. The FVSTs were performed in accordance with ASTM recommended procedures (ASTM, 1993) in wash borings advanced through stiff to hard soils at higher elevations. Split spoon sampling and FVSTs were alternated on 0.9-m vertical intervals.

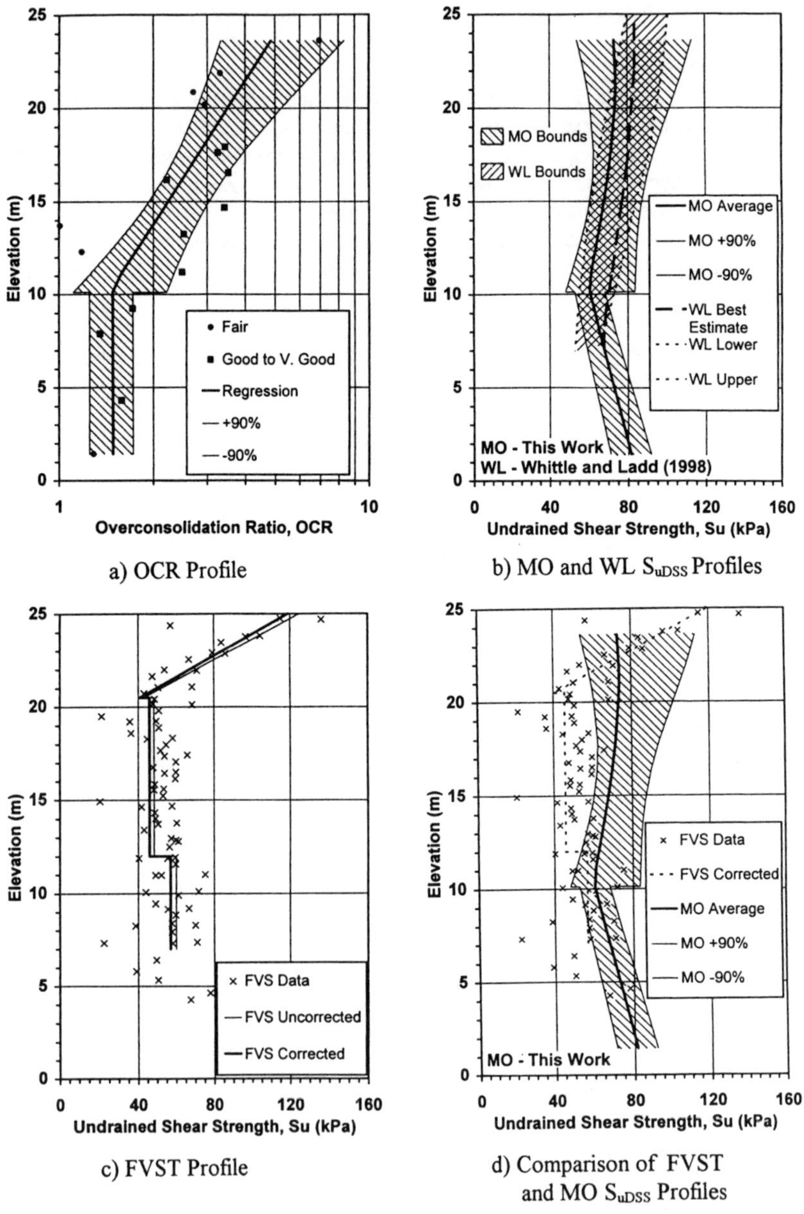

Figure 6. Undrained Shear Strength and Stress History for the Southern Basin

Based on the trends of the FVST results, the marine clay was divided into four zones. A desiccated crust, approximately 1.9-m thick, was identified at the top of the marine clay. No FVSTs measurements were obtained in this material because of its hard, fractured consistency. The strength of this zone was assumed equal to the strength at the top of the upper zone. The upper zone of the marine clay is characterized by linearly decreasing FVST measurements from El. 25 m to El. 21 m. The trend line was developed from a linear regression of the data, neglecting outliers at the top of the layer.

As described by O'Rourke and O'Donnell (1997a), two layers were identified in the marine clay below a depth of 21 m. The middle and lower zone FVST measurements were approximately constant with depth. No statistical difference in the average FVST measurements was found for the middle or lower zones at sites A1, A2, C, and D. Therefore, an average vane shear strength was calculated for both zones using data from each test site.

The FVST data are plotted relative to the S_{uDSS} and 90% confidence bounds presented in this paper in Fig. 6d. The best fit estimates in Figs. 6b are significantly higher than the corrected FVSTs. This discrepancy suggests that reduction in the undrained shear strength of the clay was experienced during construction. To understand how undrained strength could have been influenced by the excavation, it is instructive to consider the geologic characteristics of marine clay in Boston.

Geologic Characteristics of Marine Clay

As illustrated by the clay thickness profiles in Fig. 4, the marine clay in Boston occurs frequently in local depositional basins. It is well established that the marine clay in these depositional basins is stratified and rhythmically bedded (Kaye, 1982). Seismic profiles reported by Kaye show that the clay is interbedded with thin, continuous silt and sand layers, draped in conformance with the underlying basement. This type of basin structure with interbedded silt and sand layers would have promoted relatively rapid drainage of negative excess pore pressures generated by excavation unloading at the BIF site.

Observations made during construction and analyses performed afterwards provide evidence of relatively high horizontal permeability of the marine clay. Vibrating wire piezometer measurements in the clay at the BIF site (O'Rourke and O'Donnell, 1997b) have shown that hydrostatic pressures adjacent to the SMWs drain rapidly (within several weeks). Moreover, analysis of drainage at the BIF excavation (O'Rourke, et al., 1997) indicate that a ratio of horizontal to vertical hydraulic conductivity of marine clay between 10 and 100 allows for accurate modeling of measured pore pressure response. During construction, continuous seams of sand and silt were evident in the exposed clay. Observations of jet grout excavated and exposed in the BIF clay disclosed lateral excursions of grout along continuous, thin

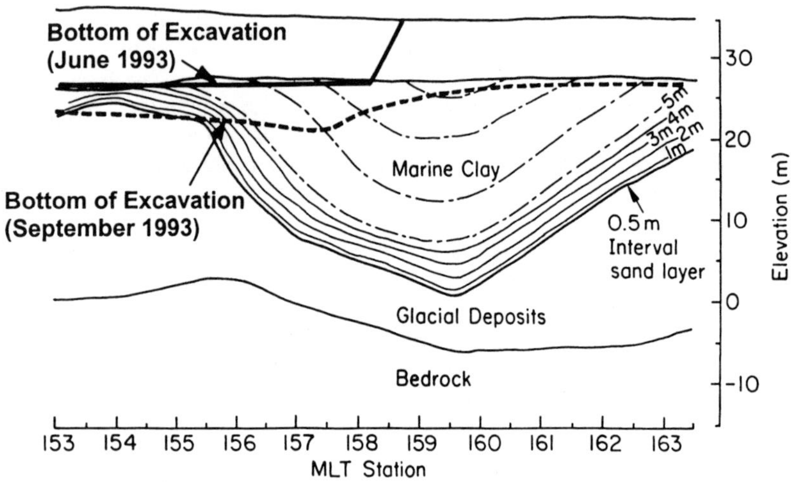

Figure 7. Schematic of Marine Clay Deposit (After Kaye, 1982) Showing the Bottom of Excavation in June and September, 1993

sand layers at distances exceeding 10 to 20m from the grout columns.

Given that the mass in situ characteristics of marine clay involve relatively high hydraulic conductivity, it is still necessary to demonstrate that a drainage boundary was available and in communication with sand and silt layers during BIF construction. The local depositional structure provides for this boundary condition, as is illustrated in Fig. 7. In this figure, the rhythmically bedded structure from seismic profiles published by Kaye (1982) is superimposed on a cross-section of the southern depositional basin at the BIF site. Because the excavation was driven from south to north into the basin, it provided a drainage boundary that intercepted sand and silt layers daylighting from depths beneath the central portions of the basin.

There was little additional excavation from the level shown for June 1993 in Fig. 7 until August, thereby providing time to drain pore pressures. In late August and early September, accelerated construction activity led to the level shown for September 1993. In the next section, the excavation at Station 157+40 is analyzed in accordance with the excavation levels pertaining to June and September under the assumption of pore pressure drainage and corresponding reductions in undrained strength.

Unloading Effects on Clay Strength

Figure 8a shows a cross-section of the excavation at Station 157+40 on which is superimposed the critical slip circle for deep rotational stability determined for this work. The depths of excavation in June and September 1993 are also shown. The September depth corresponds to the excavation conditions associated with the large deformation presented in Fig. 5. Elements A through F are identified along the critical slip circle. The stress path in each element was analyzed for excavation of the two levels shown in the figure. The excavation sequence was simulated using the finite difference software package, FLAC, with an elastic perfectly plastic model for soil behavior in which Tresca and Mohr-Coulomb failure surfaces were used to characterize undrained and drained conditions, respectively (O'Rourke, et al., 1997).

The stress path analysis for element C is illustrated in Figs. 8b to 8d. Stress paths are plotted in normalized q vs. p or p′ space, where $p = (\sigma_1 + \sigma_3)/2$, $p' = (\sigma'_1 + \sigma'_3)/2$, and $q = (\sigma_1 - \sigma_3)/2$ and all are divided by the maximum past vertical effective stress, σ'_p. The K_f lines pertaining to maximum obliquity (maximum σ'_1/σ'_3) for triaxial compression and triaxial extension are drawn in the figure. These lines are the linear regression equations derived by plotting the results of EBSTP recompression tests in which samples were anisotropically consolidated to an estimated in situ effective stress state and then sheared in either triaxial compression, triaxial extension, or direct simple shear (H&A, Inc., 1993).

In Fig. 8b, Point Ia represents the initial K_o effective stress state of Element C, and Point I represents its total stress counterpart. As excavation unloading occurs to the bottom of excavation in June 1993, the total stress moves along the path from I to II. Unloading under undrained conditions produces a negative pore pressure increment $-\Delta u_p$ and a change in total mean stress, referenced by Δp. The change in mean effective stress, referenced by $\Delta p'$, is the sum of Δp and $-\Delta u_p$. Because undrained unloading results in negative pore pressure that is little diminished by shear induce positive pore pressure, there is little change in p′. This is illustrated by the effective stress path from Ia to IIa for which p′ decreases by a small amount.

If drainage occurs at the June level of excavation the negative excess pore pressure dissipates, moving the effective stress state from IIa to IIb in Figure 8c. The application of undrained shear as the excavation was advanced rapidly to the September level generates an effective stress path from IIb to IIIb. Figure 8d compares the effective stress path IIb to IIIb associated with undrained shear during rapid excavation in late August and early September with the effective stress path Ia to IIIa of a typical triaxial extension test from K_o-consolidated conditions. Although the path of stress is not known precisely, it would most likely be bounded by the linear and nonlinear paths illustrated in the figure. It can be seen clearly that the q or stress difference mobilized in the field by the path IIb to IIIb is significantly smaller than that mobilized in the triaxial extension test by the path Ia to IIIa. In other words,

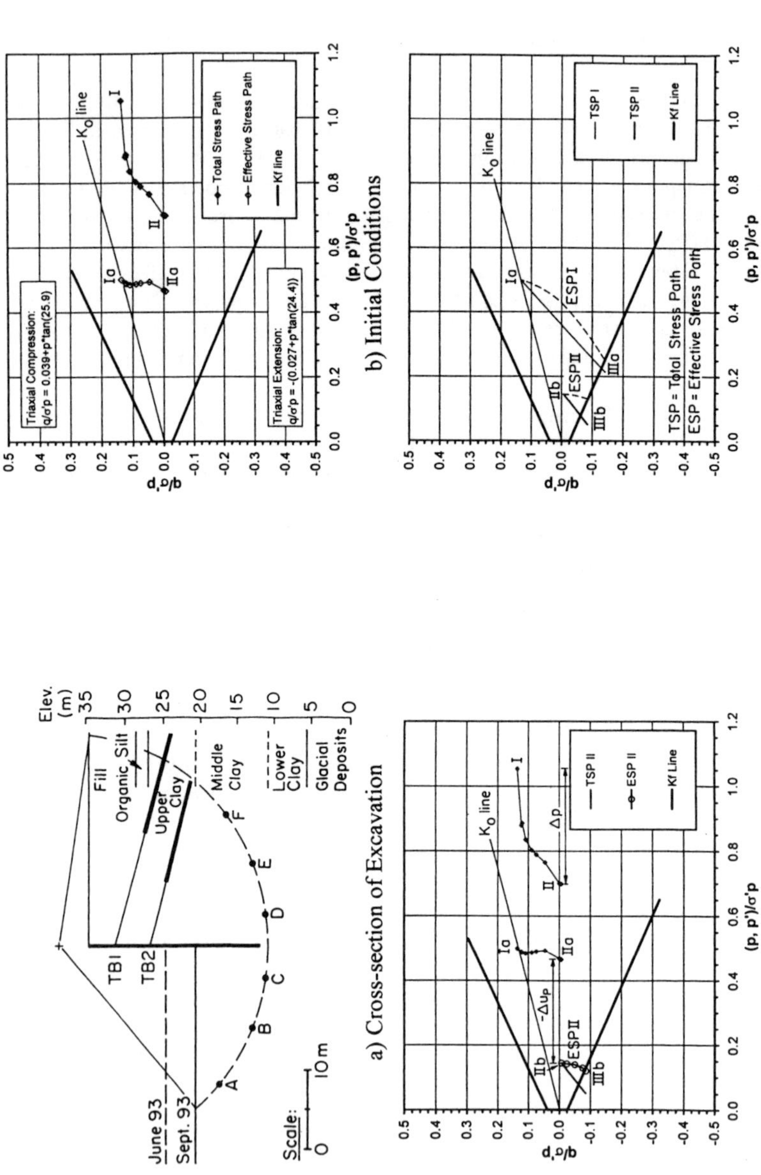

Figure 8. Stress Path Analyses for BIF Excavation

laboratory tests performed in accordance with the in situ conditions preceding excavation will tend to overestimate the undrained strength in the field under excavation unloading if drainage occurs. A quantitative estimate of this under-prediction can be obtained by comparing the range of q associated with the two sets of stress paths. In this example for element C, the percentage reduction of q for excavation unloading in the field relative to q for triaxial extension is 32%.

Similar stress path analyses were performed for elements A, B, D, E, and F. The amount of reduction ranged from 12 to 32%, with an average reduction of 22%. The average reduction in undrained strength determined from the stress path analyses was applied to the best fit profile and 90% confidence bounds established for this work (Fig. 6d), and the results are shown in Fig. 9. The reduced values agree favorably with the FVST profile.

Safety Factors and Concentrations of Soil Deformation

As mentioned earlier, inclinometer measurements, recorded in late August and September 1993, showed the rate of wall deformation to be increasing with time, indicating excavation instability. Limit equilibrium analyses were performed to assess which undrained shear strength profiles represented the observed performance of the excavation. Factors of safety against deep rotational stability were calculated using methods presented by O'Rourke and O'Donnell (1997a). For each profile, drained conditions were assumed for the granular and cohesive fill and organic silt layers. The undrained strength of the 1.9-m-thick desiccated crust was reduced by 50% to account for its fractured consistency.

The results of the analyses are summarized in Table 1. The factors of safety from the FVST profile (Fig. 6c) and reduced strength profile (Fig. 9) reflect values most representative of the excavation conditions observed in September 1993. Because FVSTs were performed in clay that was subjected to unloading and drainage of excess pore pressures, they would reflect the undrained strength at the time of the large ground deformations shown in Fig. 5. Because the reduced S_{uDSS} profile accounts for undrained unloading and pore pressure dissipation, it also reflects the undrained strength at the time of large ground deformation. Not only is there favorable agreement between the FVST and reduced S_{uDSS} profile, but the safety factors using the profiles are comparable and consistent with the observed field performance. All other strength profiles referenced in Table 1 result in safety factors that are not consistent with the observed performance.

Results from finite difference analyses that used the FVST profile support the results of the limit equilibrium method. It has been shown in previous work (O'Rourke, et al., 1997) that the predicted lateral wall movements and tieback forces agree well with measured wall movements and tieback forces. Not only is there good

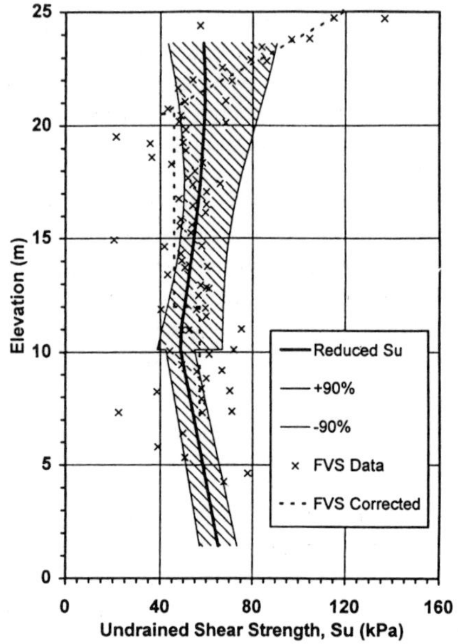

Figure 9. Reduced Undrained Strength Profile Associated with Drained Unloading

Table 1. Limit Equilibrium Results

Strength Profile	Safety Factor
FVSTs	0.97
MO Best Fit	1.35
MO 90% Confidence Lower Bound	1.17
WL Best Estimate	1.44
WL Lower Bound	1.25
Reduced Profile	1.05

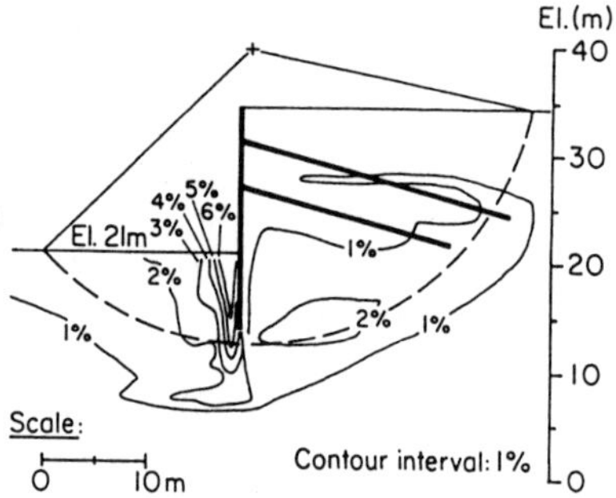

Figure 10. Contours of Shear Strain for Conditions in September 1993

agreement between the analytical results and field measurements, but the patterns of shear strain given by the FLAC analyses are consistent with the critical slip surface. In Fig. 10, the critical slip circle is superimposed on contours of shear strain predicted by numerical simulation. The results of these analyses reinforce the assessment that the FVSTs provide a good estimate of the clay strength at the time of interest in the excavation history. The close agreement between the estimated factors of safety using the FVST and reduced S_{uDSS} profile support the hypothesis that drained unloading contributed to a reduction in shear strength at the BIF site.

Alternate means of analysis have been developed by Whittle and coworkers using upper and lower bound limit methods (Whittle and Ladd,1998; Ukritchon, et al., 1998). The approach provides rational bounds on safety factor and accounts for nonlinear failure zones. The analytical results from the upper and lower bound approach show zones of concentrated soil deformation similar to those resulting from the simulations reported in this work. O'Rourke (1998) points out that there is good agreement between the safety factors obtained from both limiting equilibrium analysis with circular surfaces and the upper and lower bound approach for the conditions at the BIF site, provided that similar values of S_u are used. It should be noted, however, that limit equilibrium analysis may not produce results that agree with upper and lower bound limit solutions for other types of problems.

Concluding Remarks

Conventional design practice involves the estimation of undrained clay strength under in situ conditions before excavation. It is assumed that unloading during excavation generates negative excess pore pressures that maintain the initial effective confining stresses in the clay associated with pre-excavation conditions. The rate of pore pressure dissipation must be sufficiently slow so that loss of effective confining stress is not significant during the time the excavation remains open.

The structural features of marine clay at the BIF site led to conditions that diverge from these design assumptions. At the site, a very wide and long excavation was driven into the depositional basin, thereby exposing sand and silt layers that dipped into the clay beneath the central portion of the basin. In effect, the southern part of the excavation acted as a drainage surface for relatively pervious layers that dipped to the north and dissipated excess pore pressures in the deeper sections of clay toward the center of the basin. In addition, the large volume of excavated soil, coupled with the relatively flexible support system, generated a significant amount of vertical and lateral total stress reduction.

A smaller excavation located near the center of the basin would not promote such rapid drainage because it would not expose drainage layers dipping at sufficiently steep angles into the underlying clay. Moreover, a smaller excavation with a stiffer support system would promote less total stress reduction. The conditions that contribute to rapid pore pressure dissipation involve the size and location of the excavation relative to the size and shape of the basin. There are many combinations of excavation size and location that would not promote rapid drainage because the drainage surface afforded by deep cutting would not intercept silt and sand layers dipping below the cut. Under these conditions, conventional design assumptions would be consistent with the drainage mechanisms, and methods, such as SHANSEP, should provide suitably accurate estimates of undrained clay strength for reliable base stability analyses.

The BIF excavation represents an unusual combination of excavation width and length relative to the structural features of the depositional basin and rhythmically bedded nature of the clay. It emphasizes the importance of combining knowledge about undrained shear strength and its relation to stress history with an understanding of the depositional structure and mass hydraulic conductivity of the clay. This case history reinforces the need to combine the principles of soil behavior, typically developed under specimen scale conditions, with the principles of engineering geology that apply to larger scale depositional and structural characteristics.

Acknowledgements

This paper was prepared as part of a project managed by Bechtel/Parsons Brinckerhoff (B/PB) and supported by the Massachusetts Turnpike Authority (MTA) under contract No. 96159-T-286. The authors thank David Druss and Prabir Das of B/PB for their assistance, and express their appreciation for help given by many other engineers of MTA and B/PB. The authors also thank the Federal Highway Administration for its assistance. We thank Ali Avcisoy whose talents are visible in the figures presented in this paper.

References

American Society for Testing and Materials, (1993). "Standard Test Method for Field Vane Shear Test in Cohesive Soil", 1993 Annual Book of ASTM Standards, ASTM, Philadelphia, PA, Vol. 4.08, D2573-72, pp. 6-348.

Becker, D.E., Crooks, J.H., Been, K., and Jefferies, M.G., (1987). "Work as a Criterion for Determining In Situ and Yield Stresses in Clays", Canadian Geotechnical Journal, Ottawa, Canada, Vol. 24, No. 4, pp. 549-564.

Bjerrum, L., (1972). "Embankments on Soft Ground", Proceedings, Specialty Conference on Performance of Earth Supported Structures, ASCE, New York, NY, Vol. II, pp.1-54.

Golden Software, Inc., (1989). Surfer Version 4 Reference Manual, Golden Software, Inc., Golden, CO.

Haley and Aldrich, Inc., (1991). "Final Geotechnical Engineering Report, Central Artery (I-93)/Tunnel (I-90) Project, Design Sections D007A and D007C, Boston, MA", prepared for the Massachusetts Department of Public Works, File No. 10360-74, Boston, MA.

Haley and Aldrich, Inc., (1993). "Final Report on Special Laboratory and In Situ Testing Program, Central Artery (I-93)/Tunnel (I-90) Project, Boston, MA", prepared for the Massachusetts Department of Public Works, File No. 10360-40, Boston, MA.

Kaye, C.A. (1982). "Bedrock and Quaternary Geology of the Boston Area, Massachusetts", Reviews in Engineering Geology Vol. V, Geological Society of America, Boulder, CO, pp. 25-40.

Ladd, C.C., and Foott, R., (1974) "New Design Procedure for Stability of Soft Clays", Journal of the Geotechnical Engineering Division, ASCE, New York, NY, Vol. 100, No. GT7, pp. 763-786.

O'Rourke, T.D. and O'Donnell, C.J., (1997a). "Deep Rotational Stability of Tiedback Excavations in Clay", Journal of Geotechnical and Geoenvironmental Engineering, ASCE, New York, NY, Vol. 123, No. 6, pp. 506-515.

O'Rourke, T.D. and O'Donnell, C.J., (1997b). "Field Behavior of Excavation Stabilized by Deep Soil Mixing", Journal of Geotechnical and Geoenvironmental Engineering, ASCE, New York, NY, Vol. 123, No. 6, pp. 516-524.

O'Rourke, T.D., McGinn, A.J., Dewsnap, J., and Stewart, H.E., (1997). "Performance of Excavations Stabilized by Deep Soil Mixing", report prepared for Bechtel/Parsons Brinckerhoff, the Massachusetts Highway Department, and the Federal Highway Administration at Cornell University, Ithaca, NY.

O'Rourke, T.D., (1998) Closure to "Deep Rotational Stability of Tiedback Excavations in Clay", Journal of Geotechnical and Geoenvironmental Engineering, ASCE, New York, NY, Vol. 124, No.12, pp. 1125-1228.

Snedecor, G.W. and Cochran, W.G., (1980). Statistical Methods, 7th Ed., The Iowa State University Press, Ames, Iowa.

Ukritchon, B., Whittle, A.J., and Sloan, S.W., (1998) "Undrained Limit Analyses for Combined Loading of Strip Footing on Clay", Journal of Geotechnical and Geoenvironmental Engineering, ASCE, New York, NY, Vol.124. No.3, pp. 265-276.

Whittle, A.J. and Ladd, C.C., (1998). Discussion of "Deep Rotational Stability of Tiedback Excavations in Clay", Journal of Geotechnical and Geoenvironmental Engineering, ASCE, New York, NY, Vol. 124, No. 12, pp. 1222-1225.

TUNNEL AND GROUND SUPPORT PERFORMANCE OF THE EXPLORATORY STUDIES FACILITY AT YUCCA MOUNTAIN

Marek J. Mrugala, Ming Lin, and Dwayne C. Kicker[1]

ABSTRACT: The Exploratory Studies Facility is currently under construction at Yucca Mountain, Nevada. The purpose of the Exploratory Studies Facility is to characterize rock strata for its suitability to construct a nuclear waste repository. At the current stage, a considerable amount of underground work has been performed providing direct access to the rock mass interior. This access facilitates verification of the initial estimates of the magnitudes of rock strata parameters and allows for evaluation of the ground support system(s) performance based on the field-acquired data. As mandated by federal law (10 CFR 60), the engineering design at the Yucca Mountain Site Characterization Project must be supported by evaluation of the performance of the underground openings based on field-acquired evidence. This paper presents a discussion of selected elements of the Performance Confirmation Program that were designed to provide verifiable evidence of the in situ performance of the underground openings and ground support system through measurements performed in the field. The two aspects of the Performance Confirmation Program are presented, namely, the results of ground deformation monitoring and the results from strain gage measurements of steel set loads. The deformation monitoring analysis presents a semi-empirical approach for analyzing the initial, unmeasured portion of the tunnel deformation, and for predicting future opening displacements. The steel set loads analysis presents the development of a numerical model that provides the ability to examine various steel set support concepts and to optimize the performance of the steel set given the operational constraints in the field.

INTRODUCTION

Tunnel and ground support performance confirmation is an integral part of the geotechnical design process at the Yucca Mountain Site Characterization Project. Currently, over 10 km of tunnel have been designed and constructed as part of the Exploratory Studies Facility (ESF) to characterize the rock strata for its suitability to construct and operate a high-level nuclear waste repository. If the site is found suitable, the ESF has been designed to form parts of the main access ramps and drifts of the repository. As such, these underground openings are subject to federal requirements stipulated in the Code of Federal Regulations (10 CFR 60).

[1]Morrison Knudsen Corporation, CRWMS M&O, MS 423, 1261 Town Center Drive, Las Vegas, Nevada 89134-6352.

Design confirmation activities have been carried out during site characterization. Monitoring and testing activities performed during site characterization are intended to provide data and information that constitute the technical baseline for performance confirmation. The data and information collected and developed are used for the following:

- Evaluations of rock mass quality and other empirical and geotechnical design parameters.
- In situ state of stress measurements to verify the magnitude and direction of the principal stresses used as boundary conditions in the numerical stability analyses.
- Induced stress measurements to determine the excavation-induced stresses surrounding the ESF openings.
- Seismic monitoring data to verify the ESF seismic design criteria.
- Geochemical analysis of rock and groundwater to assess the potential for the degradation of the ground support systems, sub-systems and components.
- Blast vibration monitoring data to verify the extent of blast-induced damage occurring in the ESF tunnels.
- Monitoring data for structural support systems including steel sets and rockbolts.
- Convergence monitoring data to establish deformation rates and verify the stability of ESF openings.
- Geotechnical mapping records to describe the rock mass structure and ground support locations for key block and stability analyses.

The results of two of the design confirmation activities, including ground deformation monitoring and steel set loads monitoring, are presented in this paper.

BACKGROUND

The Site Characterization Plan (DOE 1988) provides a descriptive summary of the Yucca Mountain Performance Confirmation Program. A preliminary performance confirmation program was proposed, starting with the initial Site Characterization or Baseline Phase (DOE 1988, p. 8.3.5.16-3). A more detailed Performance Confirmation Plan has since been prepared that provides the approach for performance confirmation during the interim construction and operations phases (CRWMS M&O 1997a). Figure 1 summarizes the performance confirmation approach. The Performance Confirmation Plan defines the activities necessary to

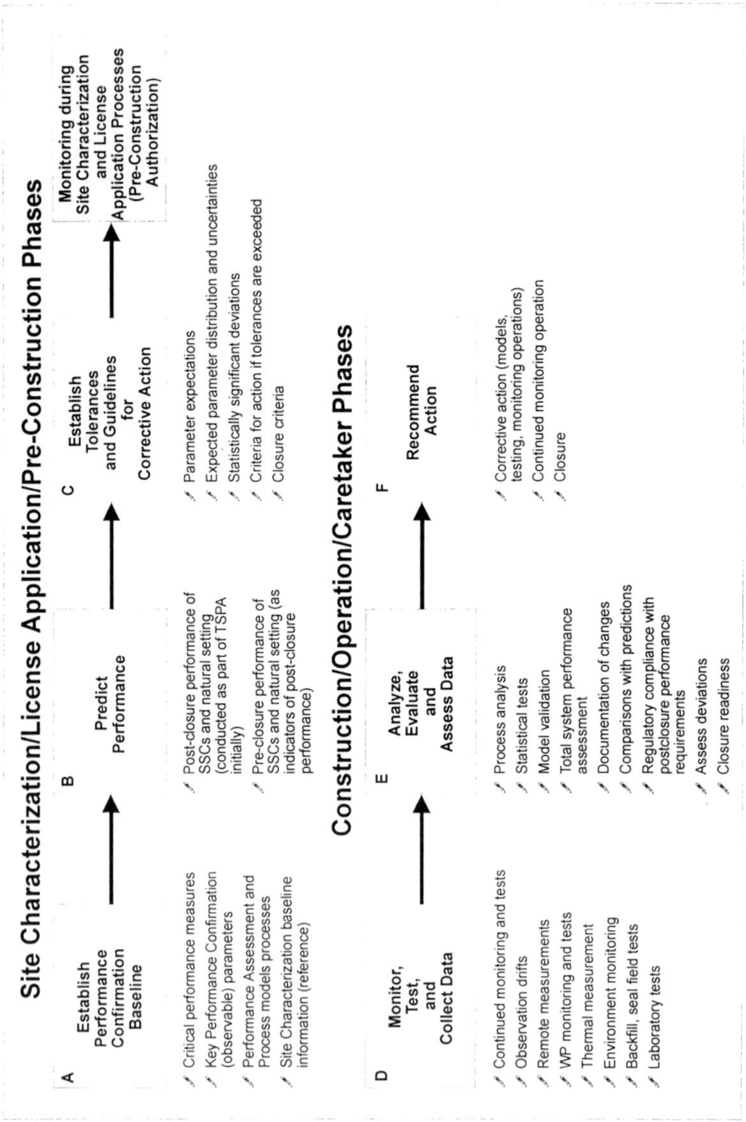

Figure 1. Monitored Geologic Repository Performance Confirmation Approach (CRWMS M&O 1997a)

conduct the Performance Confirmation Program as specified in 10 CFR 60, Subpart F.

Design confirmation analyses will be used to satisfy certain 10 CFR 60.133[2] requirements concerning the development of additional design criteria for the underground facility (specifically, 10 CFR 60.133(b), regarding the flexibility of the design to accommodate specific site conditions identified through in situ monitoring, testing, or the excavation). Data and information acquired for design confirmation will also be used to satisfy 10 CFR 60.72, Construction Records. Design confirmation analyses are intended to be part of the Performance Confirmation Program's baseline phase and will be specifically used for the assessment and evaluation of ESF requirements related to:

- Deleterious rock movement (10 CFR 60.133(e)(2)).
- Confirmation of geotechnical parameters (10 CFR 60.141(a)) based on a continuing program of surveillance during ESF construction and operation, including measurement, testing and geologic mapping.
- Design changes to accommodate actual field conditions encountered as required by 10 CFR 141(a) and 141(d).
- Comparison between actual subsurface conditions and the design bases and assumptions, in compliance with 10 CFR 60.141(b), (c), and (d).
- Geologic maps and cross-sections, ground water seepage, anomalous conditions, instrument locations and structural support, used in part to satisfy the requirements in 10 CFR 60.72(b)(2) through (8).

Design confirmation analyses that currently have been completed are reported in the following documents:

- Key ESF Design/Geomechanics Parameters Confirmation (CRWMS M&O 1997b)
- ESF Ground Support Confirmation (CRWMS M&O 1998a)
- Confirmation of Empirical Design Methodologies (CRWMS M&O 1997c)
- Deformation Monitoring Analysis (CRWMS M&O 1998b)
- Steel Set Loads Analysis (CRWMS M&O 1998c)

---

[2] It is anticipated that the additional design criteria in 10 CFR 60.133 will be omitted as 10 CFR 60 is being changed. This change will not impact the required tunnel and ground support performance monitoring at Yucca Mountain.

- Geology of the Exploratory Studies Facility Topopah Spring Main Loop (CRWMS M&O 1998d)

- Controlled Drilling and Blasting Design Confirmation Analysis (CRWMS M&O 1997d).

The Performance Confirmation Plan is based on field and laboratory testing, analysis, and monitoring activities. These activities confirm the assumed subsurface conditions at the site and ensure that the functioning of engineered and natural systems is as predicted in design or performance assessment calculations. At the time of repository license application, sufficient information must be provided to allow the Nuclear Regulatory Commission to determine, with reasonable assurance, that the geologic repository will not pose an unreasonable risk to the health and safety of the public. It should be clearly understood that design confirmation analyses are primarily intended for use in establishing baseline information for performance confirmation. It is likely, however, that these design confirmation analyses will also be used as a reference or as input to performance assessment analyses.

The two main objectives of the Performance Confirmation Program are:

- To demonstrate, to the extent practical, that the observed subsurface conditions are within the limits assumed for geotechnical and design parameters (10 CFR 60.141) in the license application.

- To provide evidence that the engineered and natural systems are functioning as anticipated and meet the long-term performance objectives for containment and isolation.

Currently, design basis parameters, critical characteristics for ESF openings and ground support systems, and the stability of the main access openings are all contained in ESF design analyses. These analyses have provided in some cases a combination of qualitative and quantitative performance measures for both ESF openings and ground support systems. Much of the baseline data and information to be gathered will be used to demonstrate the adequacy and appropriateness of the performance measures developed in these ESF analyses.

GROUND DEFORMATION MONITORING

Deformation has been considered as the most effective means of monitoring rock mass behavior during and after construction of an underground excavation. Convergence data has been used in many tunneling projects to provide information that is used to check the validity of the design or to permit the completion of the on-going design work. The ESF deformation data was first summarized and then analyzed based on a semi-empirical approach named the Convergence Solution (Barlow 1986). The Convergence Solution has been applied to interpret the convergence measurements, to estimate ground behavior, and to predict long-term tunnel displacement.

Overview of the Deformation Monitoring Program

The deformation monitoring program in the ESF consists of 94 tape extensometer stations and 18 borehole extensometer stations. The tape extensometer (Geokon Model 1600) measures the change of distance between two convergence points. The mechanical accuracy of the Geokon Model 1600 is ± 0.127 mm. Six sets of chord positions referred to as H, L, R, V, RF, and LF, were included in tape extensometer measurements. Among the six chords, H, L, R, and V are the most common chord measurements, as shown in the insert of Figure 2.

Borehole extensometers (Geokon Model A-5) include both the multiple point borehole extensometer (MPBX) with two to five anchor points and the single point borehole extensometer (SBPX). The borehole extensometer consists of inflatable hydraulic anchors, stainless steel measurement rods, and vibrating wire displacement transducer. The resolution of the displacement transducer is 0.025% of the full scale reading. A typical MPBX station consists of three anchor points. The anchor depths for the three anchor points are typically located 7.4, 2.7, and 1.3 m from the surface into the rock. For the SPBX, the anchor point is set at 2.7 m from the borehole collar to rock at the springline.

Summary and Discussion of the Monitored Data

A typical set of monitored data measured from the tape extensometer station and borehole extensometer station is presented in Figures 2 and 3 respectively. Table 1 summarizes the monitored data in the forms of maximum convergence and average convergence for the five thermal mechanical units[3]. The measured convergence are generally small, this is especially true for the welded tuff units (TCw, TSw1, and TSw2 units). Notice that the time lag between the first convergence readings and the excavation date (time 0 in Figures 2 and 3) ranges from several days to more than 6 months. The actual convergence should be greater than the monitored values.

The convergence curves for the borehole extensometers resemble the asymptotic characteristics of a typical convergence curve, while most of the convergence curves from the tape extensometer fluctuate with time. Borehole extensometers provide better accuracy than the tape extensometers with less chance for human error during data acquisition. Convergence measurements form tape extensometers include negative values (indicating the expansion of the opening) in several of the

[3] A thermal mechanical unit is a grouping of stratigraphic units that exhibit similar thermal and mechanical properties. There are five thermal mechanical units in the ESF: UO (Undifferentiated Overburden), PTn (Upper Paintbrush Tuff nonwelded), TCw (Tiva Canyon welded), TSw1 (Topopah Spring welded, lithophasae-rich), and TSw2 (Topopah Spring welded, lithophasae-poor).

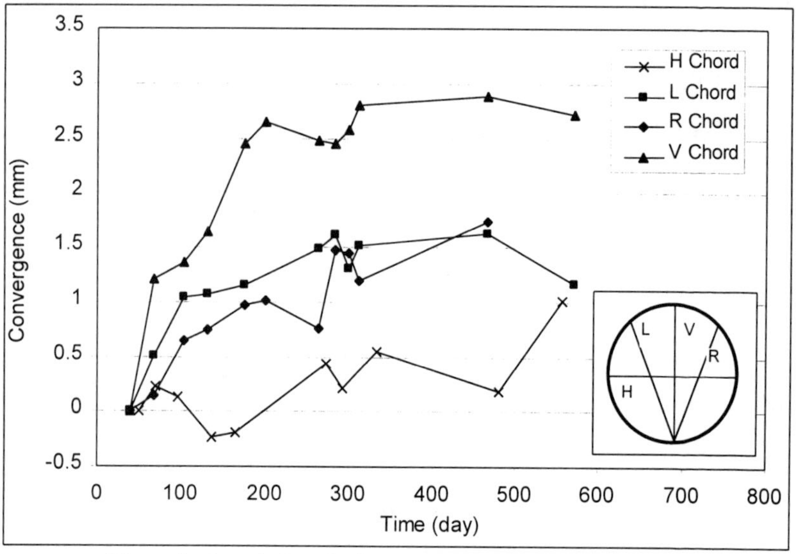

Figure 2. Sample Results for Tape Extensometer (Station 31+02)

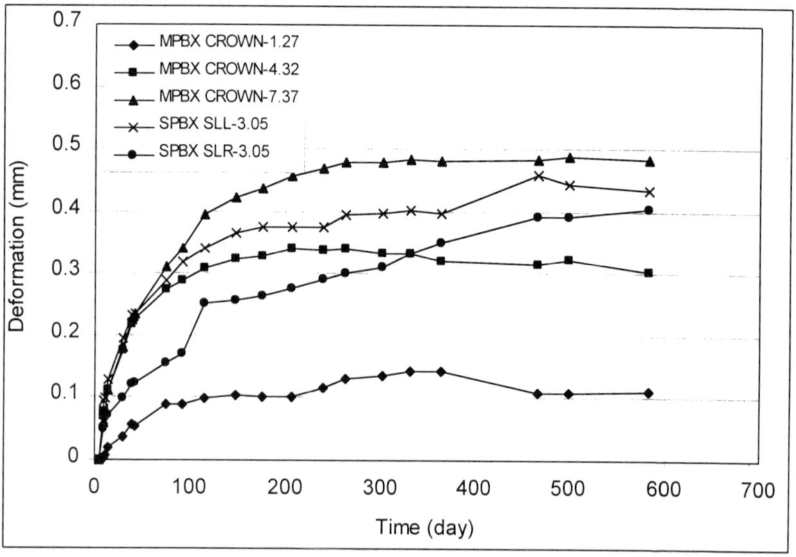

Figure 3. Sample Results for Borehole Extensometer (Station 21+34)

Table 1. ESF Deformation Monitoring Results Summary

Station Type	Thermal Mechanical Unit	Maximum Convergence (mm)	Average Convergence (mm)
Tape Extensometer	TCw	3.6	1.1
	PTn	15.5	5.5
	UO	4.6	1.6
	TSw1	2.6	1.3
	TSw2	4.7	1.5
Borehole Extensometer	TCw	0.6	0.3
	PTn	0.3	0.3
	TSw1	0.5	0.3
	TSw2	3.8	1.1

monitoring stations, while the measurements from the borehole extensometers consistently indicate closure in the vertical direction.

The convergence curves from both the tape extensometers and borehole extensometers reveal a trend of continuous increase of the convergence at time well after excavation. Rock displacement was monitored 200 to 300 days after excavation for most of the stations. In practice, the time-independent convergence should occur within the first 10 to 20 days after the excavation. The displacements monitored at longer time intervals are considered time-dependent convergence.

Deformation Monitoring Analysis

Tunnel convergence is caused by stress redistribution around the periphery of an opening during excavation. The stress redistribution and deformation occurs both in cross-sections through the plane of tunnel radius and the plane of the tunnel axis. The Convergence Solution was derived from the general form of the convergence equation proposed by Guenot et al. (Barlow 1986). In this approach, a semi-empirical equation is fitted to the convergence measurements by adjusting parameters that characterize the response of the rock. These parameters are related to important physical quantities, such as the extent of the yield zone and the time-dependent properties of the rock.

A computer program CONRATE has been used to calculate the convergence versus time curve, and the normalized convergence rate versus time curve according to the equations of the Convergence Solution. Three parameters X, T and A* are required for input to CONRATE. The first parameter X is related to the size of the yield zone around the tunnel. The second parameter T reflects the shape of the creep curve for the rock mass. The third parameter A* is the ratio of time-dependent convergence to the time-independent convergence. The deformation measurements from borehole extensometer were used for the Convergence Solution analysis. Stations with relatively early instrumentation reading data were selected for the analysis.

Long-term tunnel displacement can be predicted based on the best-fit parameters and the Convergence Solution equations. The predicted roof displacements up to 5 years for six stations selected are presented in Figure 4. Both the predictive total convergence and convergence rate are lower than the tentative goals set by the DOE Site Characterization Plan (70 mm total displacement for 100 years and 1 mm/yr convergence rate).

A series of closures were calculated in the design of ESF ground support. The analysis was based on Mohr-Coulomb plasticity models of unsupported tunnel openings. Both the in situ stress condition and the rock mass properties were varied to produce a range of vertical and horizontal closure curves for each thermal mechanical unit in the ESF. The maximum closure magnitudes from the ESF closure models are compared to actual displacement magnitudes from field measurements in Figures 5. The actual (uncorrected) displacement magnitudes from field measurements are presented as an average of the maximum displacements occurring at each convergence monitoring station within a thermal mechanical unit. Convergence monitoring stations were installed at some time after the opening was excavated. Field-measured displacements are therefore not directly comparable to the closures of calculated in the ESF ground support design. The Convergence Solution values extrapolate the field-measured values to include the estimate of the initial (unmeasured) deformation and therefore provide a better comparison with the calculated values for the ESF ground support design. Based on the interpretation of the field-measured convergence data using the Convergence Solution approach, the ESF ground support design analysis closure model is conservative, and actual displacements in the ESF are less than projected.

STEEL SET INSTALLATION AND MONITORING

The option of using steel sets as a means of ground control was provided in the respective ground support specification. Two types of steel beams, W8x31 and W6x20, were used to fabricate steel sets. The decision to use steel for ground control at a particular location was based on the local ground cónditions that were revealed as the TBM progressed. In general, once more difficult ground conditions were encountered, the set was transported from the surface in three major sections. These were assembled under the tail shield of the TBM utilizing the steel set ring erector and then were seated on the foot segments. The steel spacer rods were installed loose and connected to the preceding steel set ring. Partial lagging was attached to both the previously installed and the steel set being erected. Once the TBM advanced, such that the tail shield cleared the partially assembled steel set, two hydraulic jacks were set on both sides of the ring. The jacking load was applied until the steel set was expanded firmly against the rock. For W8x31 steel sets, a jacking load not to exceed 27 tons was applied, while a jacking load not to exceed 17 tons load was applied to W6x20 steel sets. In some instances, a firm expansion could not be accomplished because of raveling rock and the resulting voids along the tunnel. After jacking, the steel set was completed by installing inserts and shims as required.

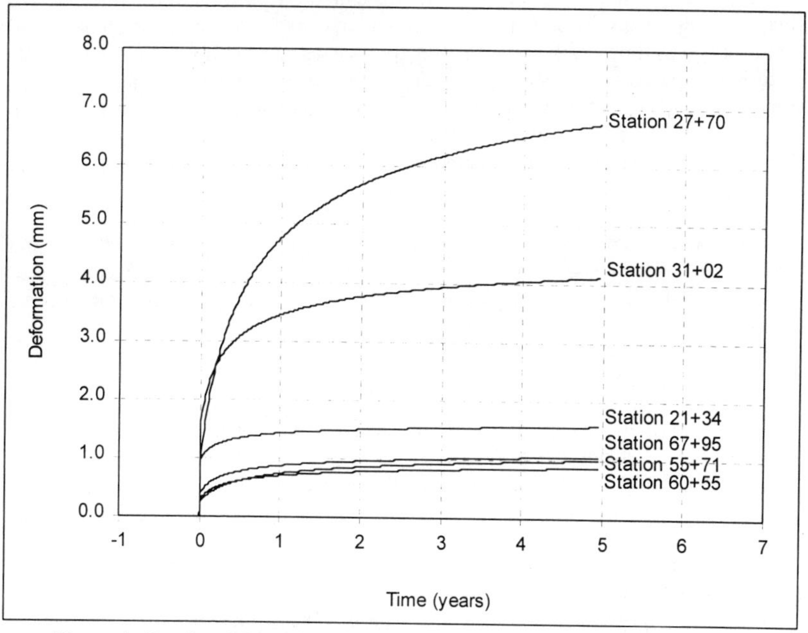

Figure 4. Predicted Long Term Roof Deformation for the Selected Stations

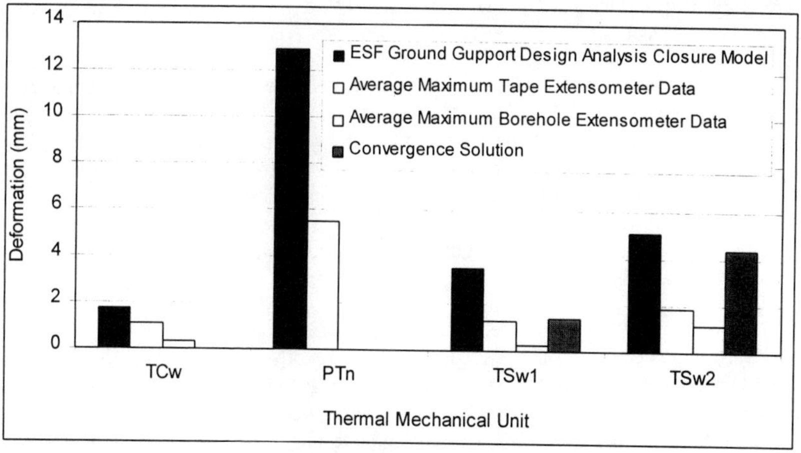

Figure 5. Comparison of Vertical Displacement Magnitudes in the ESF

Prior to the instrumented steel sets installation, it was assumed that the rock mass initially relaxed around the excavated opening and would undergo further deformation as time progressed. Strain gaged steel sets were installed to provide the data regarding this time-dependent rock mass response and to provide the in-situ verification of the steel set performance. Approximately half of the thirty-six monitored steel sets were instrumented after jacking and the half were instrumented before jacking. The sets instrumented before jacking provided data about the entire load history.

The strain gages were installed on the left and right springline and at the tunnel crown. At each location, the gages were installed 2.54 cm (1 in.) from the steel beam external and internal flanges on both sides of the steel web. In total, each set was instrumented with 12 strain gages.

Strain Gage Results

Figure 6 shows a typical set of the steel set strain gage data, which display variation both in magnitude and strain sign. It should be noted that a typical stress data measured with strain gages indicated that tension occurs at the crown along the outside set fibers, and along the inside fibers at the springline gage location.

Initial strain gage data evaluation indicated that bending-induced stress remains the principal mode of the steel set loading pattern. The numerical modeling was performed to provide an additional confirmation of these findings.

Numerical Modeling

The modeling task was accomplished by constructing a generic model involving both field-determined rock mass data as well as realistic replication of the construction sequence.

The numerical model analysis discussed here were carried out using the 2D finite difference code, FLAC Version 3.3 (Itasca 1995). This program was selected because it is one of the few specialized computer codes developed specifically for solving soil and rock mechanics problems. The program offers a capability of modeling interaction between the rock mass and components of rock reinforcement. The program is equipped with powerful graphics and pre- and post-processing features that make model construction and result interpretation possible with moderate time and effort.

The rock mass was discretized using a finite difference grid of 35 by 32 zones of total dimensions equal to 76 m (height) by 38 m (width). The grid was "wrapped" around one half of the tunnel opening with zone 1 at the center of the tunnel invert and zone 32 at the tunnel crown. The model center was located at a depth of 245 m. The grid features were designed to accommodate the structural components of the tunnel (i.e., the steel set and tunnel invert). Modeling of the construction sequence

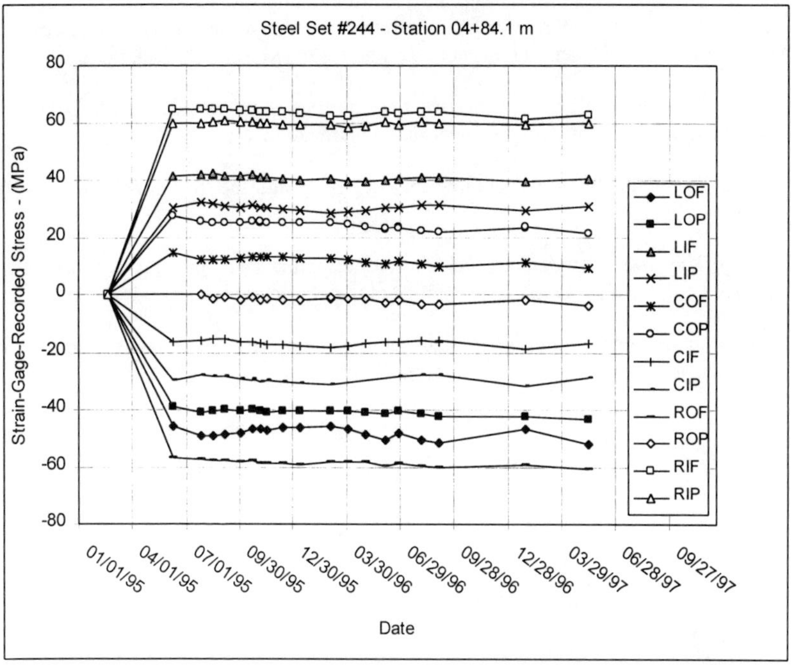

Figure 6. Steel Set No. 244 – A Typical Example of Strain Gage-Measured Stress Data. Strain Gages Installed Before Jacking (SNL 1997)

is designed to mimic the real construction process. Figure 7 shows the model in a form of the mesh or grid used in the current study.

Rock strength properties were chosen to represent the TSw2 thermal mechanical unit. A depth of 245 m was used for the generic model. An average weighed density of 2115 kg/m3 was used to represent the density of the generic model, which was used to calculate the in situ stresses.

Modeling Approach

In the discussion presented in this paper, the jacking load was applied to the steel set represented by plane-stress structural beam elements. An additional four regular mesh elements (two on each side of the jacking gap) attached to the structural elements are used to provide a 12.7 cm (5 in.) offset representing the jacking bracket. This case is represented by Model 3, shown in Figure 8. The approach adopted in this analysis included the following steps:

1) Utilizing the available laboratory and field data, a tunnel model including the rock and steel set properties and the required boundary conditions was assembled. An internal ring representing the steel set was separated from the rock mass.

2) The interface between the ring and the rock mass was defined and the ring was "lowered" to bring it into contact with the tunnel invert.

3) The initial solution including the rock mass and the ring in equilibrium was obtained.

4) The two segments of the ring representing the "jacking section" were removed and the jacking load was applied at the respective model nodes.

5) The final run including the steel set ring expanded in place and in equilibrium with the surrounding rock mass was completed.

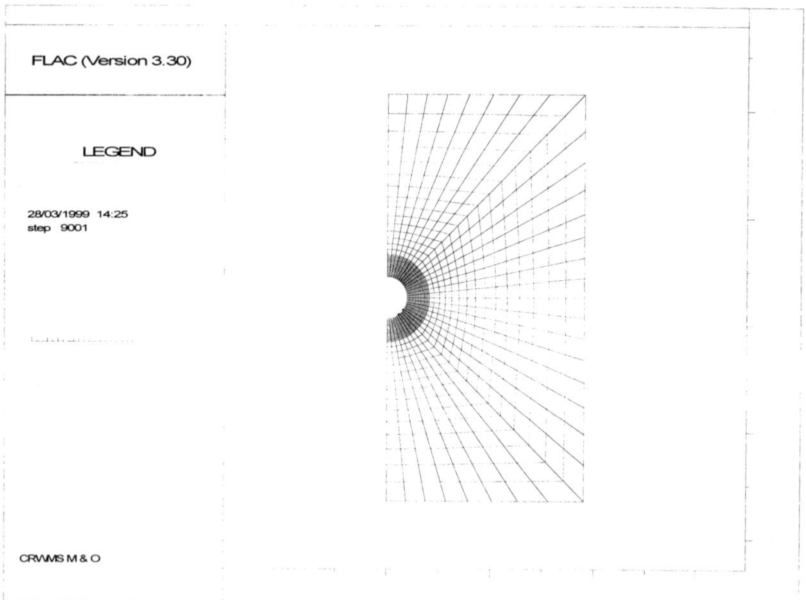

Figure 7. A Typical FLAC Model Used in Numerical Modeling.

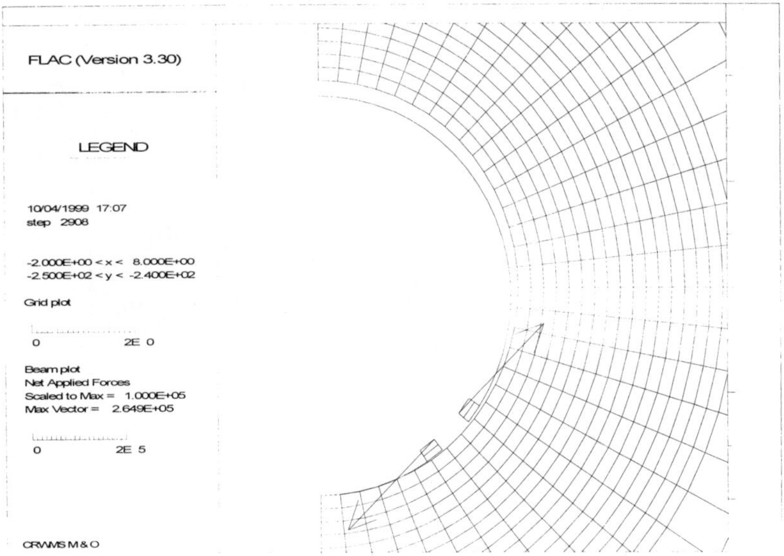

Figure 8. Close View of the Model 3 Jacking Section. Steel Set Location Prior to Jacking the Set Against the Rock Surface.

Comparison of Modeling Results and Field-Obtained Steel Set Strain Gage Data

The strain gage results were analyzed in two stages. During the first stage it was assumed that the stress generated in steel sets by jacking load was largely dependent on the gap or the difference between the nominal diameter of the tunnel excavation and the steel set. The gap was introduced to examine the impact of design tolerances, and to some extent could also be used to probe the effects of the tunnel wall roughness. The various gap dimensions resulted in loading the steel set in a more complex manner than that illustrated by an ideal contact between the steel set and the tunnel wall. Figure 9 summarizes the results obtained for Model 3 with gap sizes ranging from 2 cm to 20 cm. These results show an increasing moment as the gap increases. A small (2 cm) gap size was intended to provide a comparison as close to the ideal case as practical. It also represent close approximation of the 2.54-cm (1-in.) gap defined as the design tolerance of the gap between the designed steel set and the nominal tunnel diameter.

During the second stage, the model representing a 2 cm gap was modified to provide a limited contact with the steel set being jacked in place, which would result in generating a stress pattern such as measured from strain gages.

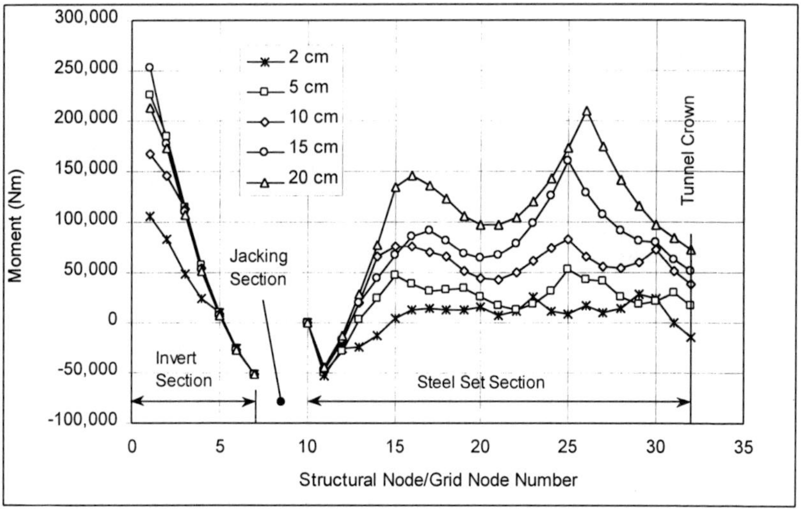

Figure 9. Increasing Steel Set Moment as a Function of the Gap Between the Steel Set and the 7.62-m Tunnel Diameter

The steel set/rock contact was adjusted (Figure 10a) to obtain the moment distribution and the associated stress pattern similar to the one recorded with strain gages (Figure 10b). Once the pattern was determined, it was found that the jacking load had to be reduced to the range of 10 percent or 2.7 tons (compared to its initial value equal to 27 tons) to match the magnitude of the strain-gage recorded and steel set stress level. The strain-gage-measured stress and the curve composed of stresses generated within the model are shown in Figure 11.

The strain gage results were analyzed in terms of combined stress, representing both axial force and bending. The general results from this analysis were compared to the results obtained from numerical modeling.

The values of calculated stresses were evaluated according the procedure described in AISC (1989, p. 5-54), which states that the members subjected to both compression and bending stresses shall be proportioned to satisfy the following requirements:

$$fa/Fa + fbx/Fbx + fby/Fby \leq 1.0 \quad (1)$$

Since the field evidence shows very small or no bending with respect to the y-axis, the above equation is reduced to the following expression:

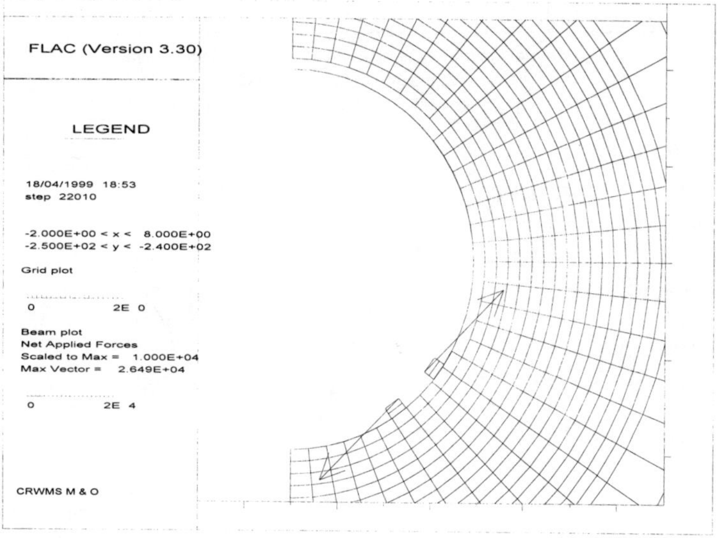

(a)

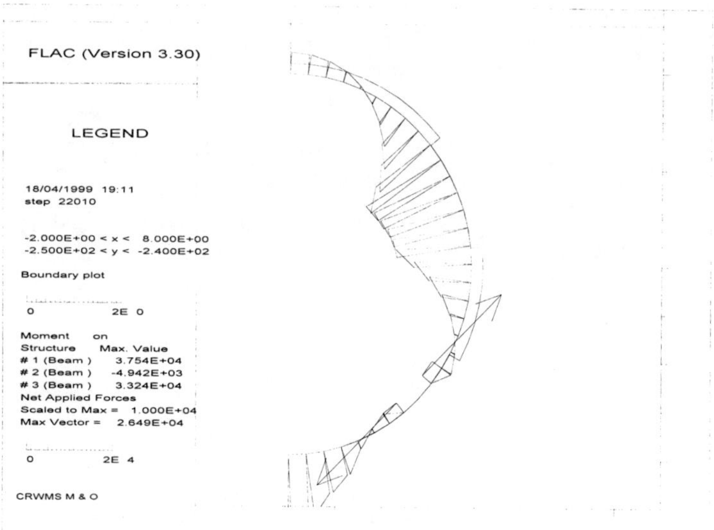

(b)

Figure 10. Model of the Steel Set/Rock Face Shape Required to Generate the Strain-Gage-Measured Steel Set #244 Stress Pattern (a) and the Corresponding Moment Distribution (b).

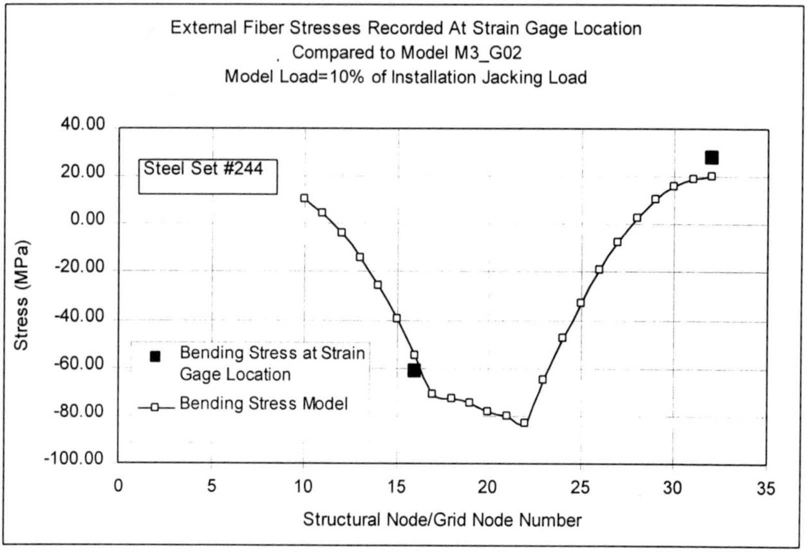

Figure 11. Comparison of External Fiber Stress Obtained from Model 3 Modified to Obtain Moment Distribution Observed from Strain Gage Measurements and Stresses Measured Using Strain Gages.

$$fa/Fa + fbx/Fbx \leq 1.0 \tag{2}$$

For 36 ksi (248.3 MPa) W8x31 steel, the Fa value derived from AISC (1989, p. 3-16) is equal to 18.86 ksi (130.1 MPa). According to the same reference, the value of Fbx is calculated as Fbx = 0.6 × 36 = 21.6 ksi (149.0 MPa).

CONCLUSIONS

An in situ monitoring program, including both convergence measurements and steel set monitoring, was used to verify the long-term performance of the rock mass surrounding the tunnel and the associated steel set ground support system in the ESF. Deformation data have been analyzed using a semi-empirical approach known as the Convergence Solution. Numerical analyses using the FLAC code were used to estimate the influence of the gap between the diameter of the TBM-excavated tunnel and typical steel set dimensions. A simplified shape of the steel set/rock interface was used to generate the moment and stress pattern observed in the field measurements.

The data obtained from the in situ displacement measurements, the steel-set strain gage data, and the associated analyses results lead to the following conclusions:

- Convergence data obtained from the ESF generally show displacements that are low. The predicted ultimate convergence magnitudes from Convergence Solution are less than the Site Characterization Plan goals of 70 mm total displacement over 100 years. The predicted convergence rates at 5 years after excavation are less than the closure rate criterion of 1 mm/year. Based on the interpretation of the field measured convergence data using the Convergence Solution approach, the ESF ground support design analysis closure model is conservative, and actual displacements in the ESF are less than projected.

- The deformation curves show a consistent trend of time dependent behavior. This time dependent trend is absent in the steel set strain gage data, indicating that the strain gaged steel set displays a sensitivity that is insufficient to detect relatively small values of rock mass displacement.

- The passive nature of the steel set ground support underscores the importance of proper blocking around the set. The contribution of the bending moment becomes more significant as the size of the gap between the steel set and tunnel wall increases. Non-uniform blocking produced a non-uniform stress in the steel set, resulting in under-utilization of the steel set capacity.

- The recorded value of steel set stress was less than the initially suggested by the magnitude of the jacking load applied during installation. The low percentage of steel set utilization confirms that the steel set use was dictated more by concerns about workers safety, rather than an effective ground control measure.

- Field measurements in combination with numerical modeling provides an effective means of examining field performance of rock mass and the ground support system interactions. It can be used to examine various design concepts even before they are tested in the field.

- Steel set stress data indicate that a combination of design tolerances adjustments and modification of the jacking load application procedure can be used to optimize the performance of steel set under given operational constraints in the field.

- The convergence and steel set data were collected under ambient temperatures. Similar data under elevated temperatures are currently being collected during the Drift Scale Heater Test in the ESF.

ACKNOWLEDGMENTS

This work was performed and funded under Department of Energy contract DE-AC08-91RW00134 for the Civilian Radioactive Waste Management System (CRWMS). The author's employer, Morrison Knudsen Corporation, is part of CRWMS Management and Operating Contractor team, lead by the prime contractor, TRW Environmental Safety Systems, Inc. The authors wish to acknowledge the contribution of Dr. John H. Pye, whose initial planning and development efforts formed the basis of the design confirmation work at Yucca Mountain.

DISCLAIMER

This report was prepared as an account of work sponsored by an agency of the United States Government. Neither the United States nor any agency thereof, nor any of their employees, makes any warranty, expressed or implied, or assumes any legal liability or responsibility for the accuracy, completeness, or usefulness of any information, apparatus, product, or process disclosed, or represents that its use would not infringe privately owned rights. Reference herein to any specific commercial product, process, or service by trade name, trademark, manufacturer, or otherwise, does not necessarily constitute or imply its endorsement, recommendation, or favoring by the United States Government or any agency thereof. The views and opinions of authors expressed herein do not necessarily state or reflect those of the United States Government or any agency thereof.

APPENDIX-I. REFERENCES

Barlow, J. P. (1986). "Interpretation of Tunnel Convergence Measurements." M.S. Thesis, The University of Alberta, Edmonton, Alberta.

CRWMS M&O (Civilian Radioactive Waste Management System Management and Operating Contractor). (1997a). "Performance Confirmation Plan." DI: B00000000-00841-4600-00002 REV 00, CRWMS M&O, Las Vegas, Nevada.

CRWMS M&O. (1997b). "Key ESF Design/Geomechanics Parameters Confirmation." DI: BABEE0000-01717-5705-00001 REV 00, CRWMS M&O, Las Vegas, Nevada.

CRWMS M&O. (1997c). "Confirmation of Empirical Design Methodologies." DI: BABEE0000-01717-5705-00002 REV 00, CRWMS M&O, Las Vegas, Nevada.

CRWMS M&O. (1997d). "Controlled Drilling and Blasting Design Confirmation Analysis." DI: BABE00000-01717-0200-00005 REV 00, CRWMS M&O, Las Vegas, Nevada.

CRWMS M&O. (1998a). "ESF Ground Support Confirmation." DI: BABEE0000-01717-5705-00003 REV 00, CRWMS M&O, Las Vegas, Nevada.

CRWMS M&O. (1998b). "Deformation Monitoring Analysis." DI: BABEE0000-01717-5705-00005 REV 00, CRWMS M&O, Las Vegas, Nevada.

CRWMS M&O. (1998c). "Steel Set Loads Analysis." DI: BABEE0000-01717-5705-00006 REV 00, CRWMS M&O, Las Vegas, Nevada.

CRWMS M&O. (1998d). "Geology of the Exploratory Studies Facility Topopah Spring Loop." DI: BAB000000-01717-0200-00002 REV 01, CRWMS M&O, Las Vegas, Nevada.

DOE (Department of Energy). (1988). "Performance Confirmation." Volume VII, Part B, Chapter 8, Section 8.3.5.16, Site Characterization Plan, DOE/RW-0199, DOE Office of Civilian Radioactive Waste Management, Washington, D.C.

Itasca (Itasca Consulting Group, Inc.). (1995). "Fast Lagrangian Analysis of Continua (FLAC) Version 3.30." Computer Software User's Manual Volumes I-IV, Itasca, Minneapolis, Minnesota.

AISC (American Institute of Steel Construction, Inc.). (1989). "Manual of Steel Construction, Allowable Stress Design." Ninth Edition, AISC, Chicago, Illinois.

SNL (Sandia National Laboratories). (1997). "Rock Bolt Load Cell, Instrumented Rock Bolt and Steel Set Strain Gage Data." DTN: SNF32120393001.029, SNL, Albuquerque, New Mexico.

APPENDIX-II. SYMBOLS USED

The following symbols are used in this paper:

A = cross-sectional area of the W8x31 steel;
A* = ratio of time-dependent convergence to time-independent convergence;
Fa = axial compressive stress that would be permitted if axial force alone existed;
fa = P/A = computed axial stress;
Fbx = compressive bending stress with respect to x-axis that would be permitted if bending moment alone existed;
Fby = compressive bending stress with respect to y-axis that would be permitted if bending moment alone existed;
fbx = M/Sx = computed stress due to bending with respect to x-axis;
fby = computed stress due to bending with respect to y-axis;
M = value of calculated bending moment;
P = value of calculated axial load;
Sx = value of the section modulus for W8x31 steel;
T = parameter that reflects the shape of the creep curve for the rock mass;
X = parameter that is related to the size of the yield zone around the tunnel.

SPTC WALL WITH A LARGE UNBRACED LENGTH

by

Phillip S.K. Ooi, M. ASCE[1], Michael P. Walker, M. ASCE[2], Hans van den Elsen[3] and Philip M. Rice, M. ASCE[4]

ABSTRACT

The analyses and performance of a soldier-pile-tremie-concrete (SPTC) wall with a significant unbraced length is presented. The wall was constructed as part of the excavation support system of the Ramp D tunnel jacking pit built for Contract C09A4 of the Central Artery/Tunnel Project in Boston, Massachusetts. This wall is unique in that it supported a cut with a height of approximately 19.5 m (64 ft) with only two brace levels above the subgrade with a clear span between the subgrade and lower brace level of 14.3 m (47 ft) above the excavation subgrade. The excavation subgrade consisted of Boston Blue Clay, the top 6.4 m (21 ft) of which was improved by jet grouting within the footprint of the jacking pit. The SPTC wall was also designed to resist train loads located about 6.1 m (20 ft) behind the wall. The SPTC wall was socketed into the Boston argillite bedrock.

Other prominent features of this support wall include the structural details of the two internal braces. The upper level brace consisted of a horizontal steel truss, while the lower level brace was a 4.88-m(16-foot)-wide, 1.22-m(4-foot)-deep post-tensioned reinforced concrete beam. Details of these two braces are described herein. The behavior of the wall was predicted using SOILSTRUCT during the design phase. A comparison of the predicted and measured wall deflections is provided.

INTRODUCTION

The analyses and performance of a soldier-pile tremie-concrete (SPTC) wall with a significant unbraced height is presented. The wall was constructed as part of the excavation support system of the Ramp D tunnel jacking pit built for Contract

1) Department of Civil Engineering, University of Hawaii at Manoa, Holmes Hall 383, 2540 Dole Street, Honolulu, HI 96822
2) GEI Consultants, Inc., 1021 Main Street, Winchester, MA 01890
3) Slattery, Interbeton, J.F. White, Perini, A Joint Venture, 210 South Street – 8th Floor, Boston, MA 02111
4) Bechtel/Parsons Brinckerhoff, C09A4 Field Office, 210 South Street, Boston, MA 02111

C09A4 of the Central Artery/Tunnel (CA/T) Project in Boston, Massachusetts (Fig. 1). Ramp D is an exit ramp on the westbound side of Interstate I-90, which will ultimately connect the Massachusetts Turnpike with the Ted Williams Tunnel and East Boston. The management consultant (MC) for the CA/T Project is a joint venture of Bechtel and Parsons Brinckerhoff representing the Massachusetts Turnpike Authority (MTA). The C09A4 contractor consists of a joint venture team of Slattery Skanska, Inc., Interbeton, Inc., J. F. White Contracting Company and Perini Corporation (SIWP).

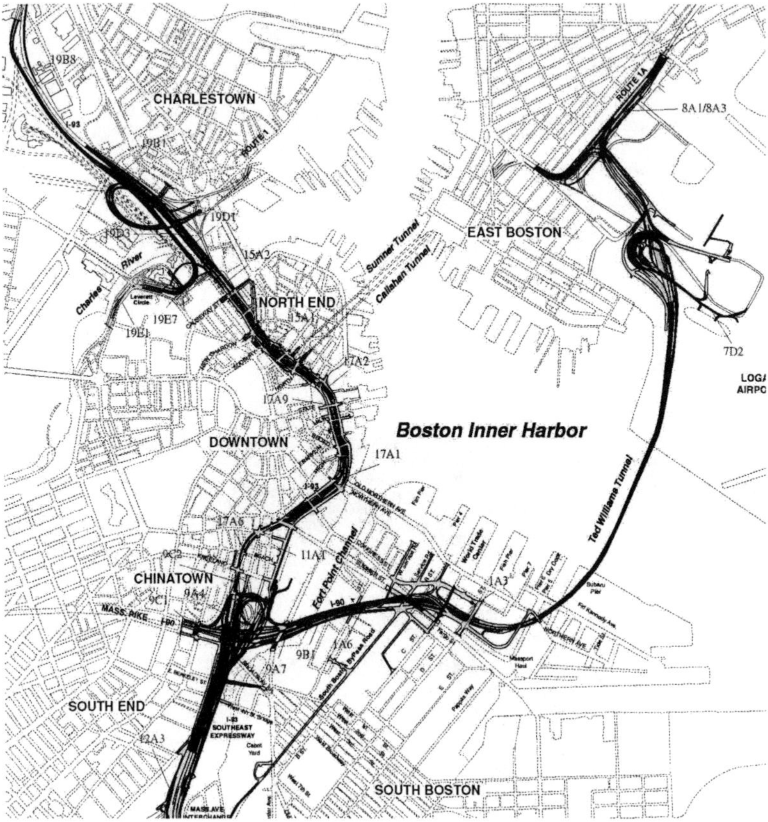

FIGURE 1 Central/Artery Tunnel Project

The Ramp D jacking pit is constructed to facilitate jacking of 23.8-m- (78-foot) wide, 11.6-m- (38-foot) high, tunnel box sections of various lengths underneath the railway tracks just south of South Station. A plan of the jacking pit is shown in Fig.

2. Located on the east end, the SPTC wall described in this paper forms the headwall of the jacking pit. Railway tracks leading to and from South Station, downtown Boston's train hub, run behind the headwall at a skew. The north end of the wall, which is the end closest to the tracks, is about 6.1 m (20 ft) from the tracks. The contract documents require that train service to and from South Station not be disrupted during the entire construction duration.

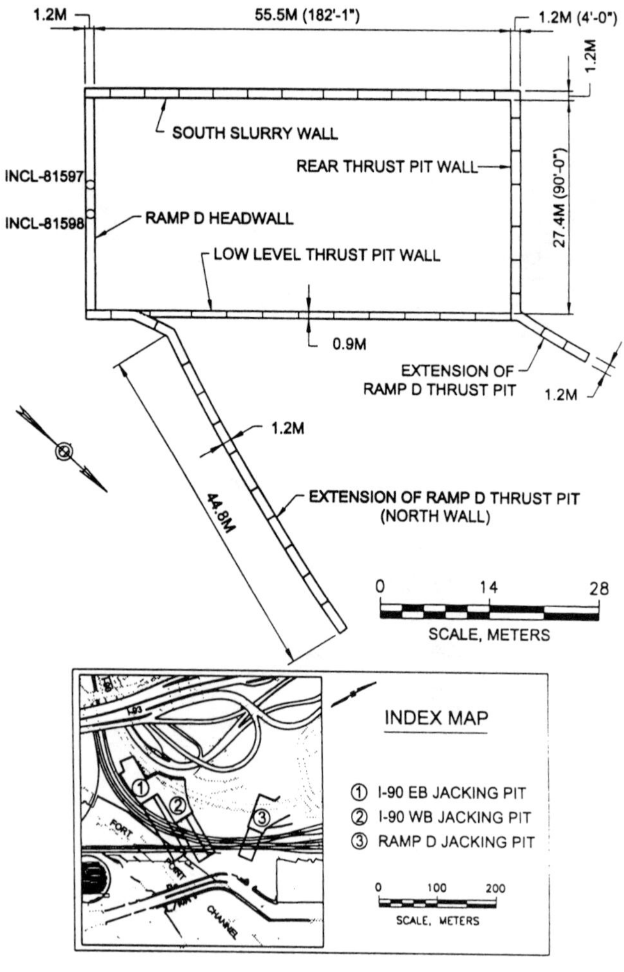

FIGURE 2 Plan of Ramp D Jacking Pit

In the original contract design, the headwall is shown as supported by four levels of internal bracing, with the lower two levels removed prior to tunnel jacking. The contractor value-engineered the SPTC wall and eliminated the two lower braces. Therefore, the upper two brace levels provided lateral support of the SPTC wall only. With this approach, however, the subgrade soil stabilized by jet grouting must act as a subgrade strut. Elimination of the two lower level braces resulted in an unbraced height of 14.3 m (47 ft) above the jet grout subgrade. The purpose of providing the extra clearance is to facilitate construction of the tunnel boxes within the jacking pit. In addition to supporting the approximately 19.5-m (64-foot)-high cut, the wall and bracing system was also designed to withstand the soil stabilization loads due to ground freezing behind the SPTC wall. After ground freezing is complete, the SPTC wall will be partially demolished to create a window for tunnel jacking. The remaining portion of the wall and bracing system was designed to support the tunnel jacking reaction loads.

Details of the headwall and bracing systems are described in this paper. A stage construction analysis of the brace installation and excavation sequencing performed during the design phase is presented. The results of the analysis are compared to the wall behavior in the field. The behavior of the wall during ground freezing and tunnel jacking is beyond the scope of this paper.

SUBSURFACE PROFILE AND SOIL PROPERTIES

An idealized soil profile in the vicinity of the headwall is shown in Fig. 3. It consists of a fill layer approximately 7.0 m (23 ft) thick overlying the following soil strata, in order of increasing depth: about 1.5 m (5 ft) of organics, about 2.1 m (7 ft) of marine sand, about 27.4 m (90 ft) of marine clay, about 2.4 m (8 ft) of glacial till and the argillite bedrock. The design ground water table is at El. 32.3 m (106 ft) based on CA/T Project Datum. El. 0 m (0 ft) of the National Geodetic Vertical Datum equals El. 30.5 m (100 ft) CA/T Project Datum. The design ground surface is at El. 35.1 m (El. 115 ft).

The properties of the various soil types are summarized in Table 1. They are based on information provided in the contract documents (Haley & Aldrich, Inc., 1996, Allison, 1996) and also based on previous experience on excavations in the Boston area (GEI Consultants, Inc., 1992). Variation of the shear strength of the marine clay with depth adopted for design is shown in Fig. 4. Also shown in Table 1 are the soil parameters used to analyze the wall behavior. The wall was analyzed using the finite element computer program, SOILSTRUCT (Filz, et. al., 1990). A comparison of the results of the analysis with field behavior is provided in a later section of this paper.

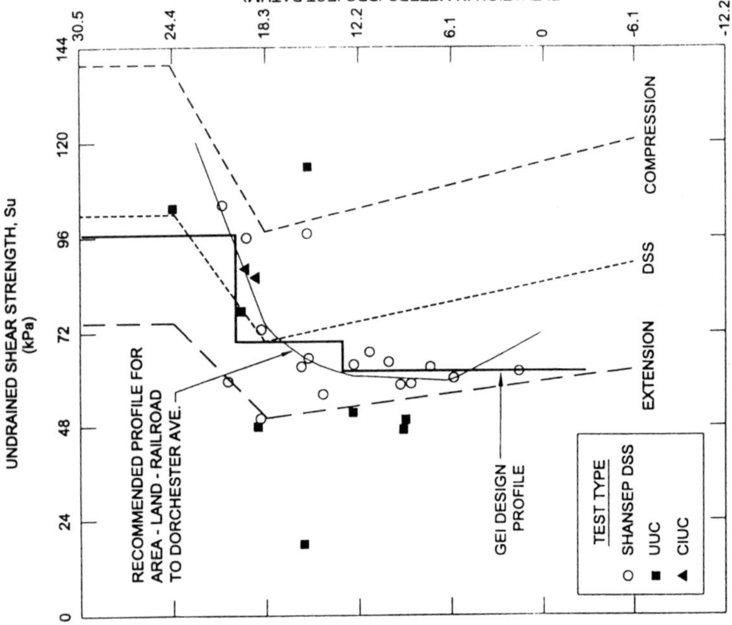

FIGURE 4 Shear Strength Profile in Marine Clay

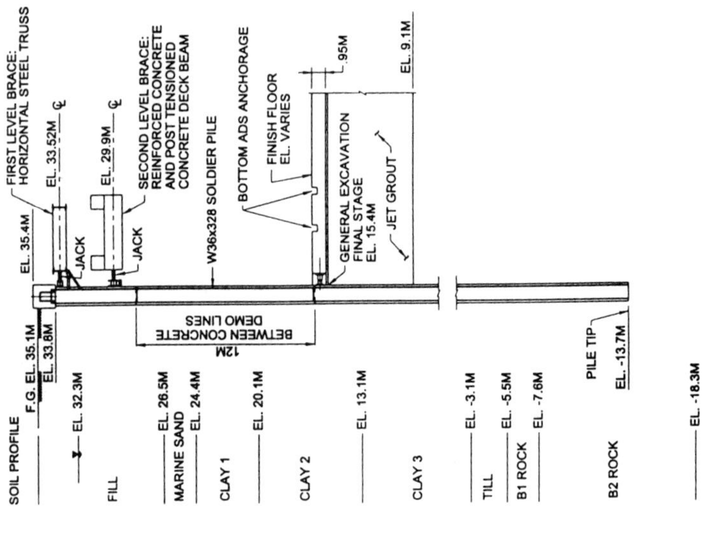

FIGURE 3 Soil Profile at Ramp D Headwall

TABLE 1: SOIL PARAMETERS USED IN SOILSTRUCT ANALYSIS

SOIL TYPE	SOIL UNIT WEIGHT γ (kNm^{-3}/pcf)	FRICTION ANGLE ϕ (degrees)	K_0	UNDRAINED SHEAR STRENGTH S_u (kPa/psf)	YOUNG'S MODULUS[1]			BULK MODULUS[2]	
					INITIAL TANGENT MODULUS COEF.[3] K	INITIAL TANGENT MODULUS EXPONENT n	STRESS RATIO AT FAIL. R_f	BULK MODULUS COEF. K_b	BULK MODULUS EXPONENT m
FILL	20.4/130	32	0.5	0	300	0.4	0.75	280	0.5
ORGANICS	See Note 4								
MARINE SAND	18.9/120	35	0.5	0	870	0.4	0.9	5,080	0.5
MARINE CLAY (El. 24.4 m-El. 20.1 m)	18.9/120	0	0.75	95.8/2,000	1,130	0.011	0.9	18,900	0
MARINE CLAY (El. 20.1 m-El. 13.1 m)	18.9/120	0	0.75	69.4/1,450	823	0.011	0.9	13,700	0
MARINE CLAY (El. 13.1 m-El. -3.0 m)	18.9/120	0	0.75	62.2/1,300	737	0.011	0.9	12,300	0
GLACIAL TILL	21.2/135	38	1.0	0	1,000	0.5	0.75	833	0.5
B_1 BEDROCK[5]	22.0/140	0	0.75	192/4,000	2,270	0.011	0.9	1,890	0
B_2 BEDROCK[5]	23.6/150	Linear Elastic: E = 1.92 X 10^6 kPa (4 X 10^7 psf); μ = 0.25[6]							
CLAY JET GROUT	19.6/125	0	0.5	414/8,640	10,200	0.011	0.92	170,000	0

NOTES

1. Initial tangent modulus, $E_i = Kp_a(\sigma_3/p_a)^n$ where p_a = atmospheric pressure and σ_3 = minor principal stress (Filz et al., 1990). Tangent modulus, $E_t = (1 - R_f SL)^2 E_i$ where SL = stress level = $(\sigma_1 - \sigma_3)/(\sigma_1 - \sigma_3)_f$, σ_1 = major principal stress and the subscript f denotes the stress at failure (Filz et al., 1990).
2. Tangent bulk modulus, $B_t = K_b p_a(\sigma_3/p_a)^n$ (Filz et al., 1990).
3. For all soils, the unload-reload modulus, K_{ur}, was taken to be equal to K except for glacial till where K_{ur} = 2000 (GEI Consultants, Inc., 1992).
4. The organic deposit was only about 1.5 m (5 ft) thick. Due to a limitation on the number of material types allowed in SOILSTRUCT, the organic deposit was assigned the properties of the fill.
5. B_1 bedrock = completely to very severely weathered argillite. B_2 bedrock = severely to moderately weathered argillite.
6. The B_2 bedrock is modeled as a linear elastic material with a Young's modulus, E and a Poisson's ratio, μ.

SPTC WALL DETAILS

The SPTC wall is about 27.4 m (90 ft) wide and about 1.22 m (4 ft) thick. The wall is reinforced with W36x328 steel sections with a yield strength of at least 345 MPa (50 ksi), spaced approximately 1.37 m (4.5 ft) apart. The W36x328 soldier piles are spaced closer towards the ends of the wall. The minimum 28-day compressive strength of the concrete is 27.6 MPa (4000 psi). As per the contract documents, the flexural rigidity (EI) of the wall was taken to be equal to the flexural rigidity of the W36x328 steel beams only; i.e., the stiffness of the concrete was neglected.

The headwall is socketed approximately 8.2 m (27 ft) into the Boston argillite bedrock. The top 2.1 m (7 ft) of the rock is completely to very severely weathered argillite underlain by moderately to severely weathered argillite designated as B_1 and B_2 rock, respectively, in the project classification. The wall is extended this deep into bedrock not for the purpose of supporting axial loads but rather for ground movement control. The large wall embedment provides additional stiffness to the wall that reduces lateral ground movements. Significant lateral ground movements were expected because of the marginal stability against basal heave. Also because of the base stability concern, the top 6.4 m (21 ft) of the excavation subgrade consisting of marine clay was improved by means of jet grouting over the footprint of the jacking pit.

The SPTC wall is considered relatively impermeable and was designed for both soil and ground water pressures.

BRACING MEMBERS

The bracing system for the SPTC wall consists of two structural elements that provide lateral support for the wall and a vertical bracing system that provides support for the portion of the wall left in place after demolishing an approximately 24.4-m (80-foot)-long by 12-m (40-foot)-high opening in the wall. The opening in the wall is required to permit tunnel jacking after excavating to final subgrade. However, the performance of the wall after demolition is beyond the scope of this paper.

Both lateral bracing systems are connected to the side (north and south) walls of the jacking pit. Lateral support for the SPTC wall is provided by a series of jacks between the lateral braces and the wall, spaced approximately 2.74 m (9 ft) on center along the length of the bracing systems. The jacks allow the Contractor to control movements of the wall into the excavation during soil stabilization by ground freezing behind the wall.

The first level lateral brace (with centerline at El. 33.5 m or El. 110 ft) consists of a horizontal steel truss with W36 rolled sections framed into an endpost attached to

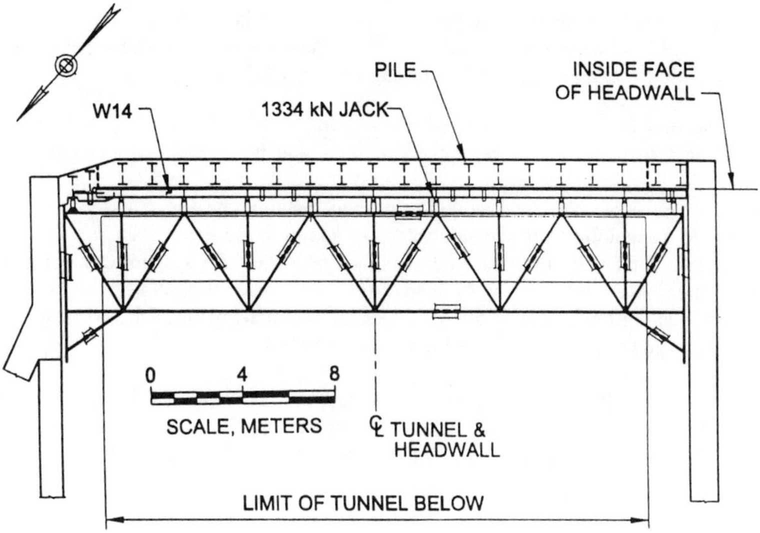

FIGURE 5 Plan of First Level Brace at El. 33.5 m

FIGURE 6 First Level Brace and Ramp D Jacking Pit

each sidewall (Fig. 5 and 6). The truss was designed for lateral loads from the SPTC wall during downstage excavation as well as during ground freezing behind the headwall and for axial loads due to downstage loading from the sidewalls.

The second level lateral brace (with centerline at El. 29.9 m or El. 98 ft) is a combination of conventional reinforced concrete and post-tensioned concrete deck beam (Fig. 7). The deck beam was designed for loads from the SPTC wall during downstage excavation as well as during ground freezing and for axial loads during downstage excavation from the sidewalls. In addition, the deck beam supports the anti-drag system (ADS) used during tunnel jacking. The ADS system runs along the top of the jacked tunnel. ADS loads were estimated at 450 kN (4000 tons) distributed along the length of the deck beam. The attachment of the ADS system results in large vertical and torsional loads on the deck beam.

The bracing stiffness was determined using finite element models developed to design the bracing system itself. A known load was applied to the bracing, and the deflection associated with that load was used to determine the brace stiffness.

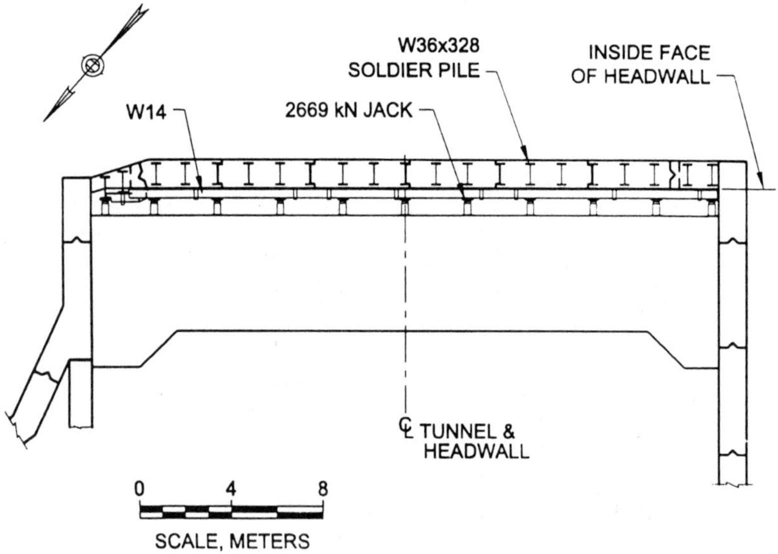

FIGURE 7 Plan of Second Level Brace at El. 29.9 m

ANALYSIS

Sequencing of the excavation was analyzed using the computer program SOILSTRUCT, which uses an incremental analysis procedure to model isotropic nonlinear stress-strain response of the soil. The soil is modeled using a hyperbolic stress-strain formulation that can represent both drained and undrained behavior. The finite element analysis was performed assuming drained behavior in the fill, marine sand and glacial till, undrained behavior in the marine clay, B_1 rock and jet grout, and linear elastic behavior in the B_2 rock. In SOILSTRUCT, the wall is modeled using beam elements, the soil is modeled using four-node quadrilateral elements, and interface elements are used between the soil and the wall.

The hyperbolic soil parameters are summarized in Table 1. The initial tangent undrained moduli (E_i) is based on $E_i = 1200S_u$ in the marine clay and B_1 rock, which corresponds to a secant Young's modulus of $660S_u$ at 50% stress level, where S_u is the undrained shear strength.

Based on the contract documents, the design unconfined compressive strength of a soil volume that is 100% treated by jet grout is 1000 kPa (150 psi). For analysis, a 20% reduction factor was used on the specified unconfined compressive strength.). This corresponds to a shear strength (S_u) of 414 kPa (60 psi)The corresponding hyperbolic parameters were based on an initial tangent modulus of $2500S_u$, which corresponds to a secant Young's modulus of $1350S_u$ at 50% stress level. Based on 121 triaxial tests on the jet grout, the average unconfined compressive strength was about 4.64 times higher than the value used in the analysis with a coefficient of variation of 59%. The Young's modulus parameters for the glacial till are based on the results of pressuremeter tests performed on glacial soils in the Boston area (GEI Consultants, Inc., 1992).

Bilinear elasto-plastic stress-strain characteristics were used for the wall-soil interface elements with ultimate shear strengths equal to one half the shear strength of the adjacent soil. The shear stiffness of the interface elements was based on full mobilization of the ultimate shear strength at a relative displacement of 0.0025 m (0.1 inch).

Loads due to the trains were included in the SOILSTRUCT analysis. In accordance with the recommendations of the American Railway Engineering Association (AREA) manual (1996), the wall was designed for 100% of the train loads on the track closest to the wall, 50% of the train loads on the second closest track, and 25% of the train loads on the third and fourth closest tracks. The assumption that four trains running simultaneously along the tracks closest to the support wall is a severe loading scenario in itself. Train loads from tracks beyond the fourth closest track were deemed to be far enough away that they can be considered negligible.

The excavation and brace installation sequencing was modeled in SOILSTRUCT as follows:

STEP 1 Apply surcharge loads and excavate to El. 32.3 m (El. 106 ft).

STEP 2 Install upper level brace at El. 33.5 m (El. 110 ft), apply a prestress force of 1.57 kN/m (8.5 kips/ft) at the upper brace, and excavate to El. 29.0 m (El. 95 ft). The prestress force is equivalent to approximately 50% of the maximum design brace load.

STEP 3 Install lower level brace at El. 29.9 m (El. 98 ft), apply a prestress force of 4.98 kN/m (27 kips/ft) and excavate to El. 15.5 m (51 ft).

Note that modeling of the excavation from El. 29.0 m (El. 95 ft) to El. 15.5 m (El. 51 ft) was performed in several substeps to minimize the errors due to non-linear unloading of the soil. It should also be noted that prior to brace installation, the excavation was advanced to about 0.9 to 1.2 m (3 or 4 ft) below the centerline of the brace because of the unusual brace thicknesses. To minimize wall movement during the initial cantilever stage, the first brace level was limited to a depth of 1.5 m (5 ft) below the ground surface.

The Contractor achieved the following construction milestones:

STEP 1	Excavate to El. 32.3 m (El. 106 ft)	7/6/98
STEP 2	Install upper brace and excavate to El. 29.0 m (El. 95 ft)	9/3/98
STEP 3	Install lower brace and excavate to El. 15.5 m (El. 51 ft)	11/18/98

COMPARISON OF PREDICTED AND MEASURED MOVEMENTS

Two inclinometers that extend below the bottom of the SPTC wall were installed at the locations shown in Fig. 2. The wall movements are compared to the predicted movements in Figs. 8, 9 and 10 for Steps 1, 2 and 3. It should be noted that these wall movements are due only to the downstage construction activities. Before any excavation took place, the Contractor improved the soil within the jacking pit by jet grouting. The high pressures during jet grouting actually pushed the SPTC wall away from the jacking pit, as observed in inclinometers installed behind the headwall. The inclinometer movements in Figs. 8, 9 and 10 were initialized to a set of readings taken after the jet grouting activities were completed. Hence, the data do not include movements due to jet grouting, but instead represent movements due to the excavation and brace installation sequencing only.

In Step 1, the predicted wall movements are reasonably similar to the measured values with several differences (see Fig. 8). The point of rotation of the wall in the jet grout is at a lower elevation than predicted, and the wall deflects slightly more than predicted. The large measured movements at the top of the SPTC wall (top three points of the inclinometer readings) can be explained as follows. A cap beam was installed at the top of the SPTC wall after the inclinometer casing was installed. During installation of the cap beam, the inclinometer casing was deformed.

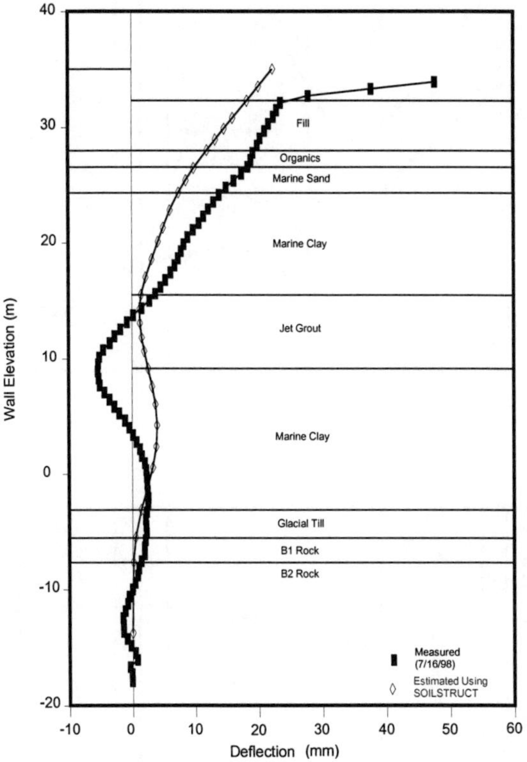

FIGURE 8 Comparison of Predicted and Measured Wall Movements When Subgrade is at El. 32.3 m

In Step 2, the predicted deflected shape of the wall agrees reasonably well with the measured shape except along portions of the wall above El. 22.9 m (El. 75 ft), where the measured wall deflections exceed the predicted values (see Fig. 9). This can be explained as follows. The stiffness of the brace is lower at the center than at the ends because the braces are framed into the sidewalls. For the SOILSTRUCT analysis, the value of brace stiffness was averaged over its length. However, the inclinometers are located near the center of the wall, which tends to deflect the most.

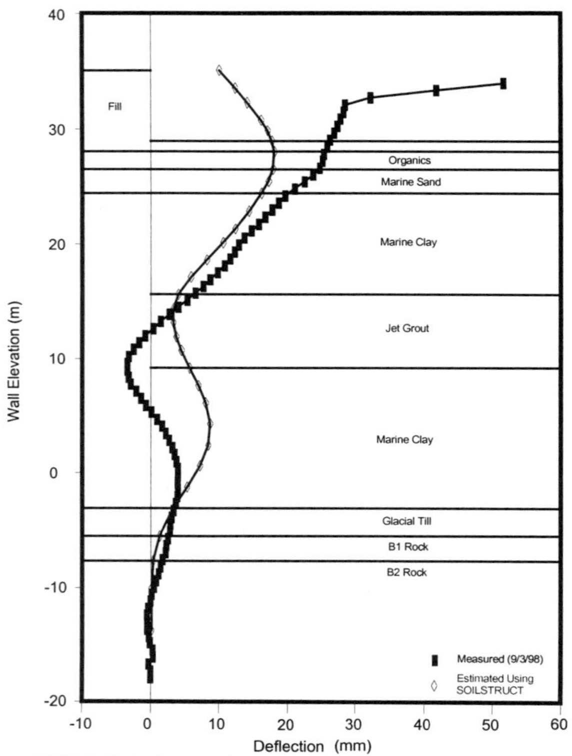

FIGURE 9 Comparison of Predicted and Measured Wall Movements When Subgrade is at El. 29.0 m

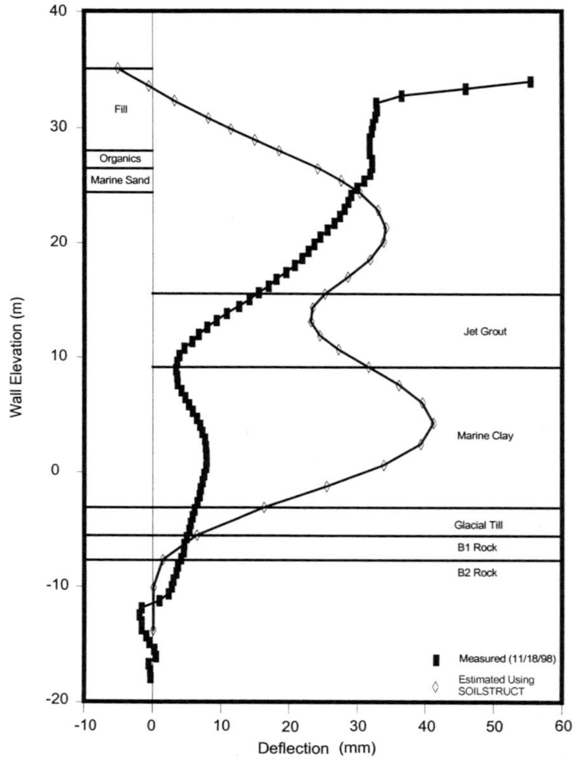

FIGURE 10 Comparison of Predicted and Measured Wall Movements When Subgrade is at El. 15.5 m

The loads in the braces were not monitored during downstage excavation and could not be compared with estimated values.

In Step 3, the maximum predicted wall movement of 41 mm (1.6 inches) at El. 4.3 m (El. 14 ft.) occurred below the jet grout, while the maximum measured movement of 33 mm (1.3 inches excluding the top three inclinometer readings) occurred near the top of the wall. The predicted wall movement was highest below the jet grout, suggesting that the clay behind the wall has a tendency to plastically flow toward the excavation because of base heave instability issues. This was not observed in the measured wall profile because the effectiveness of the jet grout subgrade strut was better than anticipated, i.e., the strength and stiffness of the jet grout were underpredicted.

CONCLUSIONS

Comparisons between measured deflections and those predicted using SOILSTRUCT are presented for a SPTC wall with a large unbraced length. The large unbraced length of the wall is made possible by jet grouting approximately the top 6.4 m (21 ft) of the excavation subgrade so that the subgrade acts as a lower strut. Large wall movements in the lower reaches of the wall were anticipated because of marginal base stability in the jacking pit. Yet, the observed wall movements below the jet grout are smaller than the predicted values. This is due to the actual stiffness of the jet grout subgrade being underpredicted. Therefore, use of a lower design shear strength for the jet grout resulted in predicted wall deflections that are greater at the jet grout level than measured and a subgrade strut that performed more successfully (stiffer) than predicted. Overall, the predicted movements are in relatively good agreement with the measured values.

ACKNOWLEDGEMENTS

The authors would like to express their thanks to Marco D. Boscardin, David R. Shields, William P. Konicki, Kevin J. DiRocco and Vimal K. Agrawal for their technical assistance and Dana MacLeod and Paul Clingen for so patiently assisting in preparing the figures. The authors would also like to acknowledge the contributions of the owner, MTA.

APPENDIX-I. REFERENCES

AREA. (1996). *AREA Manual for Railway Engineering.* American Railway Engineering Association.

Allison, M.P. (1996). Boston Central Artery/Tunnel Project, D009A4, Jacked Tunnel Pits Temporary Excavation Support Wall Criteria. Hatch-Mott MacDonald.

Filz, G., Clough, G.W., Duncan, J.M. (1990). *User's Manual for SOILSTRUCT: Plane Strain with Beam Element.* Virginia Polytechnic Institute and State University.

GEI Consultants, Inc. (1992). Final Geotechnical Engineering Report, Design Section D017A, I-93/Central Artery – Congress Street to North Street. Report submitted to the Massachusetts Highway Department.

Haley and Aldrich, Inc. (1996). Geotechnical Interpretive Report on Engineering Properties, Design Section D009A, May 8.

APPENDIX-II. SYMBOLS USED

B_t	=	tangent bulk modulus of the soil
E	=	Young's modulus
E_i	=	initial tangent modulus of the soil
E_t	=	tangent modulus of the soil
K	=	initial tangent modulus coefficient
K_b	=	bulk modulus coefficient
K_{ur}	=	unload-reload modulus of the soil
K_0	=	coefficient of lateral earth pressure at-rest
m	=	bulk modulus exponent
n	=	initial tangent modulus exponent
p_a	=	atmospheric pressure
R_f	=	stress ratio at failure
SL	=	stress level
S_u	=	undrained shear strength of clay
ϕ	=	friction angle of soil
γ	=	unit weight of soil
μ	=	Poisson's ratio
σ_1	=	major principal stress
σ_3	=	minor principal stress

PERFORMANCE OF GEOGRID-REINFORCED RETAINING WALLS AT THE SOUTH CAROLINA STATE PORTS AUTHORITY'S WANDO TERMINAL

Alan T. Stadler, Member[1]

ABSTRACT: Three geogrid-reinforced, modular block retaining walls were constructed as part of a terminal expansion project. The largest of the three walls, designated the primary wall, is nominally 4.7 m tall and 541 m long. It was designed to withstand relatively high, live, surcharge loading (48 kPa) and Zone 2 seismic loading. Construction of the walls was completed in mid-1992. The new container storage area was completed and placed into service in 1994. Presently (mid-1999) the walls are in excellent condition with no reported performance problems. However, the walls have experienced no seismic loading and only light surcharge loads have been applied behind the top of the primary wall.

INTRODUCTION

This paper presents assessments of the design and performance of geogrid-reinforced retaining walls constructed in 1992 as part of an expansion project at the South Carolina State Ports Authority's (SCSPA) Wando Terminal. The Wando Terminal is located north of Charleston, SC on the Wando River (Figure 1).

To place the paper's content into the proper context, a brief synopsis of the original design and construction activities is included. A more detailed presentation of the original design and construction activities has been published previously (Kemp et al, 1993). While the overall scope of the original project included the design and construction of three geogrid-reinforced retaining walls, the focus of this paper is on the design and performance of the primary wall that forms the boundary between the stormwater detention pond and the container storage area. The design and performance assessment includes a description of and commentary on the primary wall's current condition, a comparison between postulated and actual loading, and an evaluation of the analysis and design methods employed.

---

[1]Assistant Professor, Department of Civil Engineering, University of North Carolina at Charlotte, Charlotte, NC 28223-0001

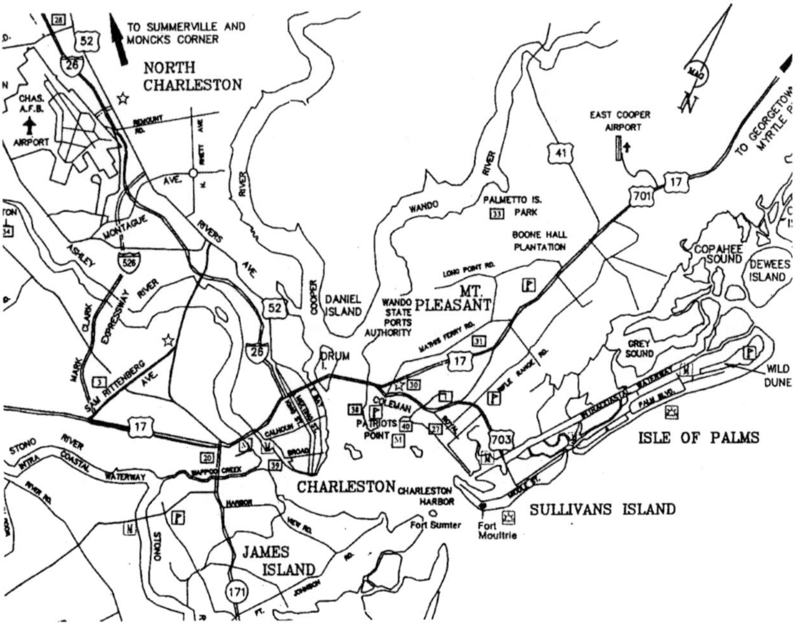

Figure 1: Charleston, South Carolina Vicinity

Project Description

In 1991, a project to increase the size of the SCSPA Wando Terminal was initiated. The project included the creation of a new, paved container storage area and an associated stormwater detention structure to manage the runoff. In addition to substantial amounts of earthwork, construction of the stormwater detention structure included the installation of a pile-supported outfall structure, geotextile-wrapped underdrains, and three geogrid-reinforced modular block retaining walls.

The storm water detention pond is "L" shaped with the longer leg forming the boundary between the pond and the container storage area (Figure 2). A retaining wall was constructed along this boundary to maximize the size of the container storage area (Kemp et al, 1993). In this paper, this geogrid-reinforced retaining structure is deemed the primary wall. The nominal dimensions (height and length) of the primary wall are 4.7 and 541 m (15.5 and 1775 ft), respectively. The other two geogrid-reinforced retaining walls function as the wing walls on either side of the pond's outfall structure. The nominal wingwall dimensions are 2.1 m (7 ft) tall and 6.1 m (20 ft) long.

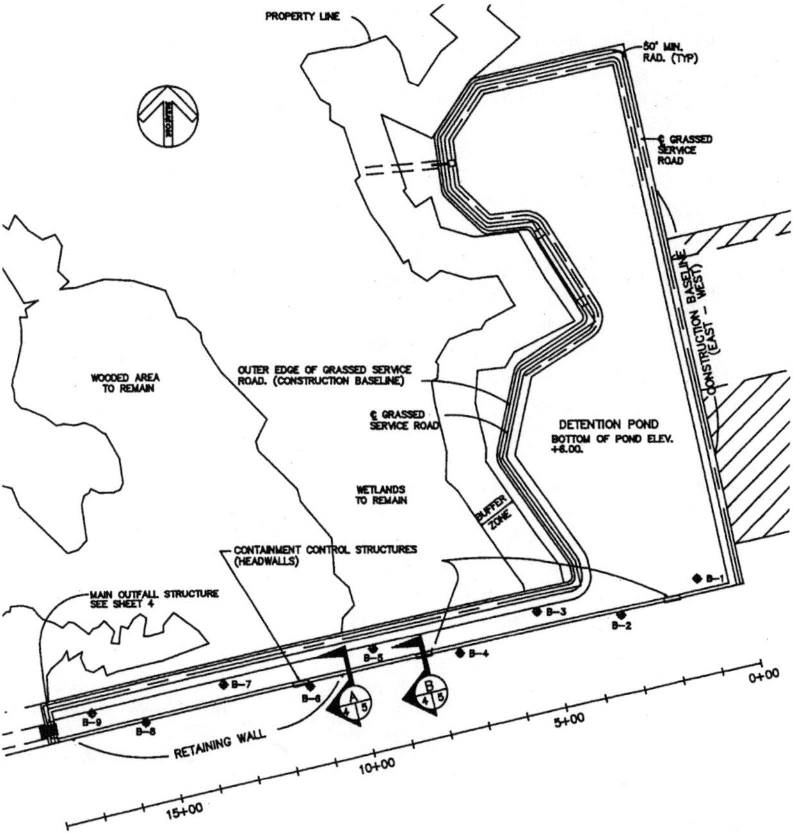

Figure 2: Plan View of Stormwater Detention Pond (from Kemp et al, 1993)
(note: 1.0 ft = 0.3048 m)

DESIGN

A detailed description of the wall design process is contained in the paper by Kemp et al (1993). Highlights of the original design process specifically relevant to the primary wall are provided herein.

Design loading included a 48 kPa (1000 psf) live load surcharge and the effects of a seismically induced 0.15g horizontal acceleration. The project specifications established several constraints on the wall design including a maximum allowable foundation bearing pressure of 120 kPa (2500 psf), safety factors of 1.5 (sliding and global stability) and 2.0 (overturning and bearing capacity), a drainage/wall unit fill consisting of #6 coarse aggregate (as defined by ASTM C-33) deployed in a 30 cm (12 inch) thick layer directly behind modular wall units, and cohesionless select backfill soil in the reinforced soil zone containing no more than 15% fines and possessing an internal friction angle of at least 32 degrees and an in-place unit weight of 18.8 kN/m^3 (120 pcf).

Nine soil test borings were used to establish the subsurface stratigraphy along the proposed primary wall alignment (Figure 2). Of particular interest was a layer of highly plastic clay ranging in thickness from 2.1 to 4.0 m (7 to 13 ft) encountered near the proposed bottom of the retaining wall. Results of consolidation and triaxial compression tests on samples of the soft clay indicated that the material possessed relatively high compressibility (average compression index of 0.90 and an initial void ratio of 1.76), a preconsolidation pressure of approximately 120 kPa (2.5 ksf), and minimum values of effective cohesion and internal friction of 13.2 kPa (275 psf) and 14 degrees, respectively.

The original design of the primary wall employed two commercially available computer programs, MGWALL (Bathurst, 1989) and STABL6 (Humphrey, Holtz, 1986). The ultimate (as determined per ASTM D 4595 modified for geogrids) and long term allowable design (GRI-GG4 with an additional Factor of Safety = 1.5 for design uncertainties) strengths of the selected geogrids are shown in Table 1.

Table 1: Geogrid Tensile Strengths
(from Kemp et al, 1993)

Polyester (PET) Geogrid	Ultimate Strength kN/m (lbs/ft)	Long Term Allowable Design Strength kN/m (lbs/ft)
High Strength	93.4 (6400)	27.0 (1850)
Medium Strength	52.5 (3600)	15.3 (1050)
Low Strength	37.9 (2600)	10.9 (750)

A triangular modular block wall unit with a nominal compressive strength of 20 MPa (3000 psi) was selected for the wall fascia (Figure 3). For the primary wall, the medium and high strength, knitted polyester (PET) geogrids were used. Figure 3 includes a schematic installation diagram showing how the individual components were assembled. As shown in this diagram, 9.5 mm (3/8 inch) diameter lengths of rebar were used to enhance the connection between the block wall units and the geogrid. The connections between the selected modular blocks, geogrids, and wall unit fill material (#6 gravel) were tested to ensure that the connection strength was greater than the minimum long term allowable design strength used to design each layer of geogrid reinforcement.

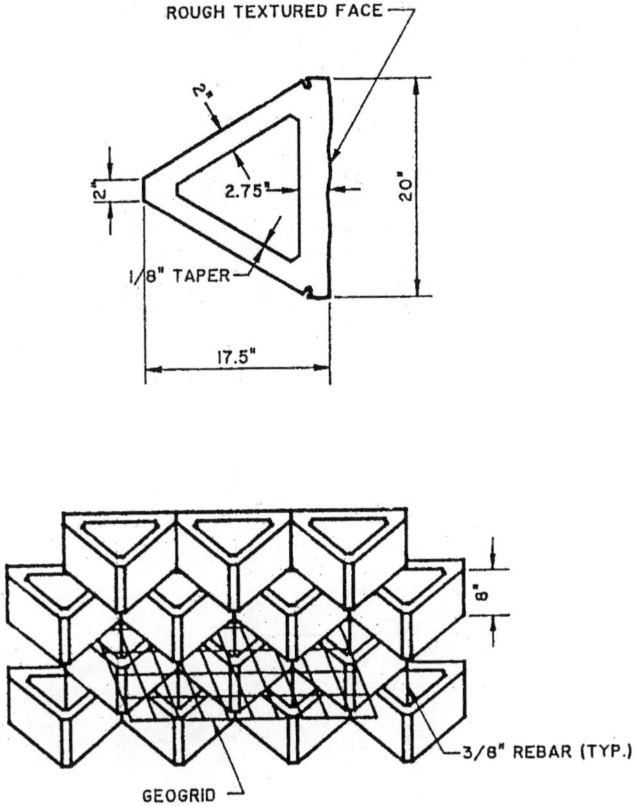

Figure 3: Individual Modular Block Wall Unit Dimensions and Schematic Installation Diagram (from Kemp et al, 1993) (note: 1 inch = 25.4 mm)

MGWALL was used to perform the internal and external stability calculations. The internal stability analyses employed a tied-back wedge model. The external stability analyses assumed Rankine active earth pressure conditions. The failure modes addressed included geogrid pullout and tensile overstress, sliding, overturning, and bearing capacity.

STABL6 was used to perform the global stability analysis. This program performs two-dimensional, limiting equilibrium analyses via a specified method of slices. In this case, circular failure surfaces were analyzed using the simplified Bishop method. Both seismic and static loading conditions were considered. For seismic loading conditions, a horizontal acceleration of 0.15g was assumed and a pseudo-static analysis performed.

In all cases, the calculated factors of safety exceeded the specified minimum values. Typically, the controlling failure mode was global stability. The final design is presented in Figure 4 as a typical section through the primary wall. Eleven layers of uniformly spaced, geogrid reinforcement were employed. Each layer extended 4.6 m (15 feet) from the construction baseline. The lower six layers were high strength PET geogrid; the upper five layers were medium strength PET geogrid.

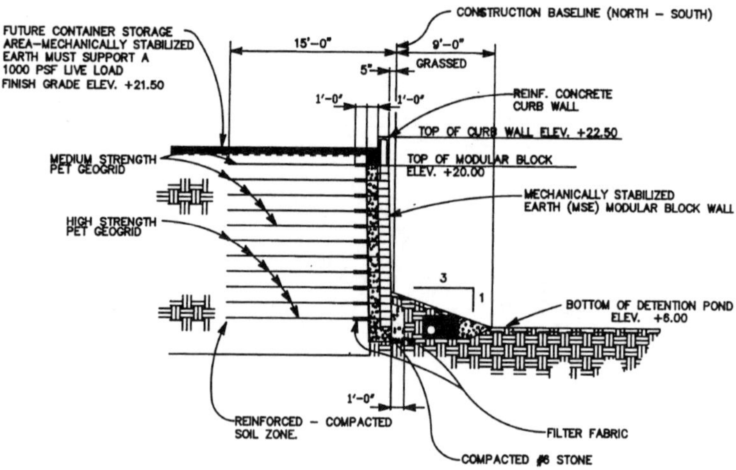

Figure 4: Primary Wall – Typical Section (from Kemp et al, 1993)
(note: 1.0 ft = 0.3048 m, 1.0 inch = 25.4 mm)

CONSTRUCTION

Construction of the geogrid reinforced modular block retaining walls began in the spring and was completed in the summer of 1992. After installing a compacted stone leveling pad, the wall was constructed following the sequence of block, fill, and geogrid placement described in the project specifications.

To minimize the likelihood of differential settlement due to non-uniform loading during construction, the maximum wall height differential along the entire length of the southern retaining wall boundary was restricted to 102 cm (40 inches, the height of a stack of 5 modular block units). This restriction prevented the contractor from building the full height of the wall from one end to the other.

CURRENT CONDITION

A field assessment of the current condition of the primary retaining wall was performed in May 1999. The assessment consisted of a visual examination of the wall and discussions with SCSPA field personnel (Smith, 1999). The results of this assessment are summarized in this section.

The retaining wall was completed in summer 1992. However, construction of the container storage area (including fill placement, the installation of utilities, pavement, guardrail, and pavement markings) was not completed until early 1994. The area was placed into service shortly thereafter. According to the SCSPA personnel, no performance-related problems of the stormwater detention pond retaining structures have been reported since their completion.

The primary wall has not been subjected to any significant external loads. Thus far, the "working" loads have been much smaller than the loading conditions used to design the wall. Again, according to SCSPA personnel, no seismic events have occurred in the vicinity of the terminal since completion of construction. Hence, the wall has not been subjected to seismic loading. Surcharge loading from container storage has been relatively low, consisting primarily of a single row of "wheeled" containers ("stacked" one high) arranged with their long axis parallel to the wall face. Representative photographs of this current container storage practice immediately behind the wall face are shown in Figures 4 and 5. Moving further away from the wall face, beyond the single row of "parallel parked" containers, a portion of the container storage area has been delineated as an access road. This road is partially visible in Figures 4 and 5. Relatively heavy container handling equipment, such as the Caterpillar V900CH shown in Figure 6, is free to operate along this road. Thus, the live load (640 kN (72 tons) per axle) generated by vehicles such as this have likely been applied along the access road. Containers stored on the far side of the access road are in excess of 15 m (50 ft) from the wall face, and the loads generated by them are assumed to have a negligible influence on the wall.

The following observations were made during the visual examination of the walls. First, the stormwater detention pond does not dry completely and, because of this, has become heavily vegetated. Various types of vegetation have grown in different locations. The majority of the interior supports reeds and cattails. More robust, woodier bushes have grown along the perimeter and especially along the toe of the wall and are likely rooted in the underdrain system. This vegetation is visible in Figures 4 and 5.

Based on visual observations made along its length, the primary wall was found to be in excellent condition. Using the original design loading and the laboratory-determined consolidation characteristics of the soft clay layer at the bottom of the primary wall, total wall settlement was expected to range from 7.6 to 12.7 cm (3 to 5 in) and with differential settlements of up to 75% of the total. Ninety percent consolidation was expected after approximately 5 years. Due to the substantially lower loads applied to the structure thus far, virtually no settlement (either total or differential) is evident. Only a few cracked modular block units were encountered. Examples of the isolated cracking observed are given in Figures 7 and 8. Although the vegetation limited sight lines, virtually no out-of-plane bulging was observed. A typical elevation view along the length of the primary wall is provided in Figure 9. Basically, the wall appears to be in essentially the same overall condition as it was upon completion in 1992.

EVALUATION OF ANALYSIS AND DESIGN METHODS

The essential points of the original analysis and design methods remain valid. Were this project undertaken today, the internal and external stability calculations for the MSE wall and the global stability calculations that incorporate the wall into the local site conditions would be performed. Presently, the specific fascia unit and grid types employed in this project are no longer manufactured. However, essentially equivalent materials are available.

A reassessment of the static internal and external stability of the existing primary wall design was performed using a current, commercially available program (Anchor Wall Systems, 1998). The analysis followed the current National Concrete Masonry Association design guidelines (NCMA, 1996). In this method, Coulomb earth pressure theory is used to calculate the magnitudes of destabilizing forces in all external, internal, and facing stability calculations. Soil properties, wall geometry, geogrid deployment, and surcharge loading were the same as those assumed in the original analysis. The modular block fascia units and geogrid material differed from the original design but care was taken to match them as closely as possible. Particular attention was paid to matching the values of the geogrids' long term allowable design strengths. The results of this reassessment indicate that the existing design satisfies the originally specified factors of safety for bearing capacity, sliding, and overturning and is acceptable under the current guidelines.

CONCLUSIONS

Seven years after the completion of construction, this geogrid-reinforced retaining wall is in excellent condition and exhibits virtually no indication of distress. A reassessment of the original design using current design criteria indicates that the design is satisfactory by current standards. As it stands, this wall represents a successful application of MSE technology.

However, the wall has been subjected to far lower loads than those assumed in the original design. The relatively low "working-level" loads imposed thus far do not invalidate the loading conditions assumed in the original design. The Charleston, SC area has a history of seismic activity, and it is reasonable to consider seismic loading in the design of a structure such as this. Further, examples of the source of the high live-load surcharge (full containers stacked three high) exist throughout the terminal container storage area. Simply because these containers have not been stored in this manner immediately adjacent to the primary wall does not invalidate the assumption that they may be at some point in the future.

Cracked modular block fascia units are a manifestation of settlement, particularly differential settlement. Field investigation revealed that a few of the modular block fascia units have cracked, but the number of cracked blocks is a very small percentage of the total number of blocks. Thus, settlement is not an issue with respect to the performance of the primary wall to date.

The drains at the toe of the wall have substantial amounts of vegetation rooted in them and this undoubtedly affects their performance. If these drains do not function properly, hydrostatic pressures above those assumed in the original design may develop behind the wall.

Figure 4: View from Top of Primary Wall Looking Eastward

Figure 5: View from Top of Primary Wall Looking Westward

Figure 6: Caterpillar V900CH

Figure 7: Cracked Modular Block Fascia

Figure 8: Cracked Modular Block Fascia

Figure 9: Typical Elevation View of Primary Wall

ACKNOWLEDGEMENTS

Mr. David Smith, Engineering Project Manager South Carolina State Ports Authority for volunteering time to assist me in preparation of this paper.

REFERENCES

Anchor Wall Systems (1998), AnchorWall Program, Version 2.5, Minnetonka, MN.

Annual Book of ASTM Standards (1991), American Society for Testing and Materials (ASTM), Philadelphia, PA.

Bathurst, R.J. (1989); "Technical notes for MGWALL: An Interactive Design Program for Miragrid Walls"; Bathurst, Jarrett and Associates, Inc., Kingston, Ontario, Canada

GRI-GG4, Standard Practice for "Determination of Long-Term Design Strength of Flexible Geogrids", Geosynthetic Research Institute, Drexel University, Philadelphia, PA, January 1991.

Humphrey, D.N. and Holtz, R.D. (1986); "STABL6 with Reinforcing Layer Option – User's Manual"; Joint Highway Research Project; Engineering Experiment Station, Purdue University, West Lafayette, Indiana.

Kemp, S., Martin, J., and Stadler, A. (1993); "The Design and Construction of Geogrid-Reinforced Retaining Walls at the South Carolina Port Authority's Wando Terminal"; Geosynthetics '93 Conference Proceedings, Vancouver, British Columbia, Canada, March 30 – April 1, 1993, pp. 153-166.

NCMA (1996), "Design Manual for Segmental Retaining Walls", second edition, National Concrete Masonry Association, Herndon, VA.

Smith, D.N. (1999), personal communication.

DRIFT SCALE TEST AT YUCCA MOUNTAIN: DESCRIPTION AND UPDATED RESULTS

Ralph A. Wagner (Member)[1]; Stephen C. Blair [2]; Steven R. Sobolik [3]

ABSTRACT: The Drift Scale Test (DST) is an integral part of the U.S. Department of Energy site investigations program to characterize Yucca Mountain and determine its suitability for the permanent disposal of spent nuclear fuel and high-level nuclear waste. The DST is a large-scale, long-term in situ thermal test consisting of a 47.5-meter long drift, 50 wing heaters, and 9 floor heaters. Thermal-mechanical-hydrological-chemical (TMHC) responses of the rock are measured by approximately 3,500 sensors installed in 147 boreholes drilled into the test block. The heating phase of the test, which started on December 3, 1997, is scheduled to continue for four years. It will be followed by a cooling phase of similar duration. During the heating phase, temperature in approximately 10,000 cubic meters of rock will exceed 100 °C. The primary objective of the DST is to gain a more in-depth understanding of TMHC processes anticipated to exist in the local rock mass surrounding the potential repository at Yucca Mountain. The numerical simulations of the DST is an integral part of the development and enhancement of coupled-process models. The following three types of models were used in the analyses: thermal-hydrological (TH), thermal-hydrological-mechanical (THM), and thermal-hydrological-chemical (THC). These analyses and ongoing comparisons to the measured responses have advance the understanding of the complex coupled processes anticipated in the host rock at Yucca Mountain.

This document was prepared with funding from the U.S. Department of Energy (DOE), Yucca Mountain Site Characterization Project under TRW's Management and Operating (M&O) Contract with DOE, number DE-AC08-91RW00134.

INTRODUCTION

Background

The Drift Scale Test is an integral part of the site investigations program that assess the suitability of Yucca Mountain (YM) in Nye County, Nevada for the permanent disposal of spent nuclear fuel and high-level nuclear waste. Plans for a suite of in situ thermal tests to be conducted in the ESF began with the Site Characterization Plan (SCP) (U.S. Department of Energy [DOE] 1988). The planning

[1] Ralph A. Wagner, URSG-Woodward Clyde Federal Service, Las Vegas, NV
[2] Steve C. Blair, Lawrence Livermore National Laboratory, Livermore, CA
[3] Steven R. Sobolik, Sandia National Laboratories, Albuquerque, NM

basis documented in the SCP has evolved over the past several years to meet the changing needs and updated knowledge base of the project. The most recent iteration, in which the SCP thermal testing program was re-evaluated and consolidated, is discussed in detail in the report entitled Updated In Situ Thermal Testing Program Strategy (CRWMS M&O 1997). The ESF Thermal Test is being conducted in a facility specifically constructed for this purpose in the Tptpmn [**T**ertiary Miocene (Age), **P**aintbrush (Group), **T**opopah Spring Tuff (Formation), Crystal-**P**oor (Member), **M**iddle **N**on Lithophysal (Zone)] lithologic unit of the potential repository horizon.

The YM thermal testing program, which includes the Drift Scale Test and two smaller tests: Large Block Test and Single Heater Test, is based on the coupling of four processes (thermal-mechanical-hydrological-chemical) and YMP end-users (see Figure 1). The relationship represents the manner in which field tests play an integral role in supplying data, information, and model refinement to support a robust license application. Therefore, it is important that these interrelationships be well established and properly maintained to ensure the YM Thermal Test is designed in a responsive manner. To this extent, much interaction has occurred between the YM Thermal Test Team and representatives of the repository design, waste package, and the performance assessment groups.

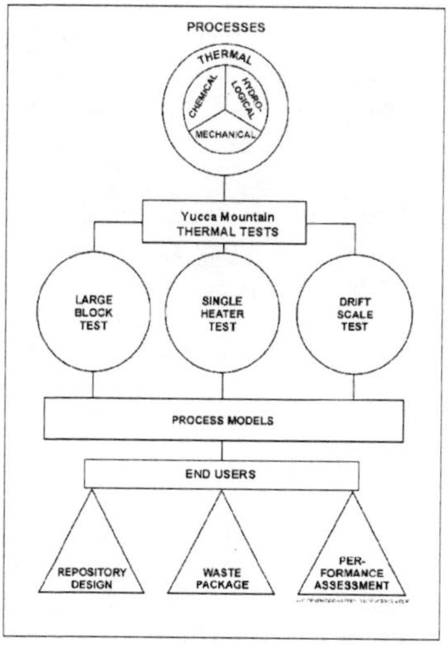

Figure 1. Interrelationship Between Processes, ESF Thermal Test, Process Model, and End Users

General Overview

The DST consists of a nearly 50 m long, 5 m diameter heated drift. The heated drift is complemented with an 11 m long entry, a connecting drift, and an observation drift of similar diameter. Heat is generated from 50 wing heaters and 9 floor (canister) heaters. These two types of electrical heaters have a combined power output of approximately 280 kW when operating at full capacity. The wing heaters are inserted in horizontal boreholes in the wall of the heated drift that are perpendicular to the longitudinal axis of the heated drift. Wing heaters are evenly distributed on 1.83 m (6 ft) spacings in boreholes located on both walls of the heated drift. Each wing heater has 10 m of heated length evenly divided between inner and outer heating elements of 1.145 kW and 1.719 kW capacity, respectively. A plan view of the DST configuration including the access drift, heated drift, wing heaters, floor heaters and plate loading niche is shown in Figure 2. The thermal, mechanical, hydrological, and chemical responses are measured with approximately 3,700 sensors in either 147 boreholes or on the drift surface. These sensors are connected to a data collection system (DCS) with an estimated 210 km of wire. The cumulative length of these boreholes is 3,246 m (2 miles). The sensors are divided into the following categories: 2,688 thermal; 138 mechanical; 109 hydrological; 122 chemical; and 673 electrical. Numerous variables are measured, including temperature, heat flux, heater power, thermal conductivity, gas pressure, moisture, water flux, chemistry, displacements, and ground support behavior, from sensors installed from within the DST drift and from the connecting and observation drifts. Planned durations for the heating and cooling phases are four years each. The four-year heating phase, which began December 3, 1997, should elevate temperatures of the rock above 100°C in more than 10,000 cubic meters of rock while allowing the temperature along the drift wall to reach 200°C.

The DST also includes a rock-mass deformation modulus or plate-loading test alcove near the entrance to the heated drift (see Figure 2). This alcove, located on the unheated side of a thermal bulkhead is approximately 6 m long and 2 m wide. The comparatively small width is necessary to minimize the size of the plate-loading reaction frame needed to induce a desired stress level in the rock to measure the rock-mass deformation modulus.

Objective

The primary objective of the DST is to develop a more in-depth understanding of the in situ coupled TMHC processes anticipated in the local rock mass surrounding the proposed repository. This objective implies that the understanding of the coupled TMHC processes is more important than fielding an experiment that replicates the actual repository environment and configuration.

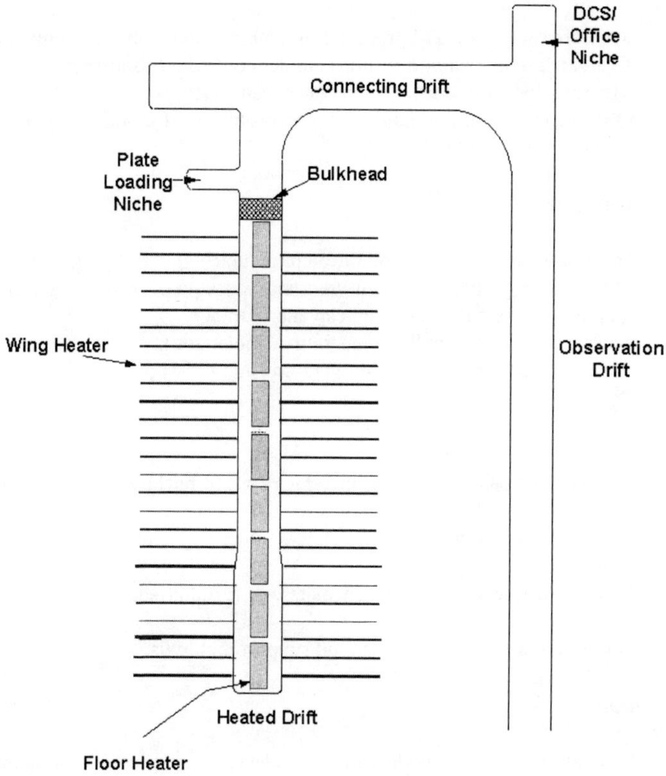

Figure 2. Layout of the Drift Scale Test.

The primary objective is supported by a series of secondary objectives that are mostly divided among the four processes. Many of these objectives stem from those established for the In Situ Thermal Testing Program Strategy (DOE 1995) and are listed below.

- **Thermal**

  - Measure the temporal and spatial distributions of temperature.
  - Evaluate influence of heat transfer modes.
  - Investigate possible formation of heat pipes.
  - Determine rock mass thermal properties.

- **Mechanical**
  - Measure rock-mass TM properties at ambient and elevated temperatures.
  - Evaluate ground support response under controlled conditions.
  - Measure drift convergence at elevated temperatures.
  - Observe effects of thermal loading on prototypical ground support systems and overall room stability.

- **Hydrological**
  - Measure changes in rock saturation particularly in the drying zone.
  - Monitor the propagation of drying and subsequent re-wetting regions, if any, including potential condensate cap and drainage.
  - Measure changes in bulk permeability (pneumatic).
  - Measure drift-air humidity, temperature, and pressure.

- **Chemical**
  - Observe corrosion products on typical waste package materials emplaced in the HD and boreholes.
  - Observe changes in water chemistry, mineralogy, and rock chemistry during heating and cooling.
  - Observe impacts of introduced materials (ground support, etc.) on water and rock chemistry.
  - Observe thermal effects on ground support materials.

- **Overall**
  - Evaluate conceptual models that calculate the coupled TMHC behavior such that realistic bounds can be developed for the expected near-field environment.

Scope Comparison

The DST is considered unprecedented in scope based on a comparison to other thermal tests conducted nationally and internationally. Table 1 provides a tabulation of 16 other thermal tests in terms of power, duration, and processes. When these components are compared, either individually or collectively, to the DST's 200 kW heater power output, eight-year duration, and TMHC processes, it is clear that the DST is a comparatively extensive and comprehensive thermal test. This comparative increase in size, duration, and complexity is intended to provide much more in-depth understanding of the performance of the potential repository at Yucca Mountain. The knowledge and benefits gained from the DST are anticipated to play an integral role in the licensing process for Yucca Mountain including the licensing amendments needed to operate and to close the repository.

Table 1. Scope Comparison Between Drfit Scale Test and Other Thermal Tests

Location	Power (kW)	Total Test Duration (Years)	Processes
Yucca Mountain (Drift Scale Test)	**200**	**8.0**	**TMHC**
Yucca Mountain (Single Heater Test)	4.0	1.5	TMHC
Yucca Mountain (Large Block Test)	2.3	0.8	TMHC
G-Tunnel (Small Diameter Experiment)	2.1	0.3	TMH
G-Tunnel (Heated Block Experiment)	0.8	1.0	TM
G-Tunnel (TH Experiment)	3.3	1.0	TH
Climax (Spent Fuel Test)	19.5	3.0	TM
Waste Isolation Pilot Plant (Room A)	57.3	4.0	TM
Waste Isolation Pilot Plant (Room B)	58.6	4.0	TM
Waste Isolation Pilot Plant (Room H)	81.6	9.0	TM
Underground Research Laboratory-Canada (Buffer Container Experiment)	1.2	2.5	TMH
Underground Research Laboratory-Canada (Heated Failure Tests)	1.2	2.5	TMH
Underground Research Laboratory (Thermal Hydraulic Experiment)	1.0	1.0	TMH
Basalt Waste Isolation Plant (Test-1, No. 1)	5.0	2.0	TM
Basalt Waste Isolation Plant (Test-2, No. 2)	5.0	4.5	TM
Stripa Sweden (3 Experiments)	6.1	4.5	TM
Avery Island - (Site A)	6.0	1.5	TM

HEATING SYSTEMS

Electrical heaters have been placed on the invert of the heated drift and in horizontal wing-heater boreholes for the DST. Scoping model calculations were conducted to determine heater specifications such as power output and duration (Buscheck and Nitao 1995; Buscheck et al. 1997; and Birkholzer and Tsang 1997). Criteria for determining the heater specifications include the maximum temperature in the driftwall, the volume of the rock mass to be heated, and the volume of the rock mass to be dried by heating. In addition to the heaters, the heating system includes a means to control heater output in accordance with predetermined heating/cooling schedules. The following discussion describes the floor heaters, wing heaters, and heating/cooling plan.

Floor (Canister) Heaters

The floor (canister) heaters in the heated drift consist of approximately 1.7 m diameter and 4.6 m long steel canisters housing 30 heaters. The heaters are mounted on the inside curved surface of the canisters and evenly spaced along the circumference. This arrangement of the heaters is designed to minimize the temperature differential between the top and bottom surfaces of the canister due to convection. The end covers of the canisters have an opening for the passage of the heater lead wires. The canister heaters are placed in the drift with a spacing of 0.6 m between them.

Each of the 30 heaters in each canister is a multi-zone heater with a heat output of 250 watts at 208 volts. This allows for a total power output of 68 kW from the nine floor (canister) heaters. The heaters have one primary and one back-up heating element wired independently. Each of the 30 heaters can be switched on or off independent of the others. Although the intent is not to energize the primary and back-up elements at the same time, they can be switched on simultaneously. This makes it possible to have a power output of 15kW from each floor heater. Sealing the electrical terminals and connectors from ambient moisture is essential for prolonging heater life. Potting at the heater end of the lead wires and covering the terminal strips (at the lead wire/power cable connection) with high temperature epoxy is intended to avoid galvanic effects and potential initiation of corrosion.

Thermocouples have been mounted at various locations on the surface of the canisters to monitor the temperature. The measured temperature will be used as indicator of heater performance and may be used to control heater power, as discussed in the section of " Heating/Cooling Plan".

Wing Heaters

A total of 50 wing heaters (25 on either side, see Figure 2) have been placed in horizontal boreholes (nominally 11.5 m-long) drilled into the sidewalls of the heated drift at nominally 0.25 m below the springline.

The wing heaters are nominally 10 m long with 5 m long inner and outer segments joined together by an accordion-like flexible joint. Each segment is a multi-zone heater independent of the other with the heat outputs of the inner and outer segments being 1145 watts and 1719 watts, respectively, at 208 volts. Like the heaters of the floor (canister) heaters, each wing heater segment has a primary heating element and back up element wired independent of each other. Inner and outer wing heater segments can be switched on or off independently. However, unlike the heaters of the floor (canister) heaters, the primary and back up elements can not be energized simultaneously.

The wing heaters have five thermocouples mounted on them, two on each segment and one on the flexible joint, to monitor their performance. As with the floor (canister) heaters, the connections at one end of the wing heater lead wires is potted and those at the other end are covered with high temperature epoxy to avoid galvanic effects.

Heating/Cooling Plan

Results from parametric analyses of TM, TH, and TC behavior anticipated in the DST were considered in the development of the following heating/cooling plan:

- Heating duration (full and/or partial): 4 years
- Cooling (no heating) duration: 4 years
- Initial power output: canisters - 80 percent (54 kW); wing heaters - 100 percent (143 kW)
- Maximum, nominal temperature along drift wall: 200°C
- Volume of dryout zone (less than 46 percent saturation): greater than 10,000 cubic meters
- Temperature along the outer edge of the wing heaters to be slightly greater than comparable drift-wall temperatures
- Sufficient volume of rock, traversed by chemical and hydrological sensors, to be held at approximately 96°C for approximately 3 years
- Heating/cooling schedule will be adjusted, as needed, to ensure DST objectives are satisfied.

MONITORING/MEASUREMENTS

Numerous types of measurement instrument and sensors were implemented in the DST. Table 2 provides a list and description of most of the measurement devices used in the DST.

Table 2. Description of Drift Scale Test Measurements

Parameter	Device/ Technique	Number	Comments
Temperature	RTDs	~1950	30 cm intervals in 20 m long radial boreholes from the Heated Drift (HD) ; ~1 m intervals in two 60 m long holes parallel to the HD; In 46 chambers isolated by inflatable packers in 3 fans of hydrology boreholes
Temperature	Thermocouples	650	On the driftwall inside the HD, at ~3.5 m intervals in three longitudinal lines. On the outer surfaces of canister heaters. On wing heaters. On rock surfaces outside at ~ 5m grid. In all MPBX boreholes at various intervals
Displacement	MPBXs	~60 anchors	In 3 fans of 4 radial holes from inside the HD. Four anchors in each of these radial boreholes. Six anchors each in two longitudinal boreholes parallel to the HD.
Strain	Strain Gages	50	On concrete lined section
Convergence	Cross-drift Extensometers	2	At approximately the mid-length of the concrete lined section, one vertical and one horizontal
Acoustic Emission	Accelerometers	16	Grouted at the bottom of 16 radial temperature boreholes from inside the HD
Relative Humidity	Similar to Humicaps	53	In 46 chambers isolated by inflatable packers in 3 fans of hydrology boreholes. In the air inside and outside the HD.
Permeability	Air-k Injection	NA	4 zones in each of 12 boreholes
Gas Pressure	Pressure Transducers	53	In 46 chambers isolated by inflatable packers in 3 fans of hydrology boreholes. In the air inside and outside the HD.
Rock Saturation	Electrical Resistivity Tomography	NA	Electrode at 1 m intervals in 4 boreholes making two fans.
Rock Saturation	Neutron Logging	NA	The neutron probe is run periodically in 10 holes making two fans of 5 boreholes each and two longitudinal holes parallel to the HD.
Rock Saturation	Ground Penetrating Radar	NA	The GPR antenna is run periodically in 10 holes making two fans of 5 boreholes.
Water	Chemistry	NA	pH, Na, Si, Ca, K, Mg, Al, B, S, Fe, Li, Sr, F, Cl, Br, SO_4, PO_4, NO_2, NO_3
Air	Chemistry	NA	O_2, CO_2, CH_4
Rock	Chemistry	NA	Mineralogy and Petrology

In addition to the above, coupons of candidate materials for the waste container, cylindrical samples of concrete used for the cast-in-place liner, and native microbes have either been left in the Heated Drift or injected into the test block to study the effects of the heating and cooling.

An approximately 6000 channel automated data collection system (DCS) records most measurements on an hourly basis. The DCS scans and records the readings of temperature, humidity, air pressure, MPBXs, strain gages, convergence monitors, and current and voltage for heater power. Other measurements such as ERT, neutron logging, GPR, acoustic emissions, and REKA probe are recorded using independent data acquisition systems.

NUMERICAL SIMULATIONS AND ANALYSES

A temporal relationship inherent to the DST exists between the development of data, application of the data in various models in terms of numerical analyses, and the implementation of corresponding computer programs (codes). The data, analyses, and codes are assembled for calculations at three distinct time intervals germane to the DST. The DST analyses are divided into three phases: test scoping, pre-test, and mid/post-test. Post-test also includes analyses conducted during the DST. A distinction is made between test scoping and pre-test analyses in that the latter (forecast) analyses benefits from site-specific characterization data generated in the laboratory and the field. Also, for the DST, more than one combination of conceptual models were evaluated in the pre-test (forecast) analyses. Consequently, the pre-test analyses provided more comprehensive and realistic simulations of the DST than was provided by the test-scoping analyses.

The DST is intended to generate site-specific observed data to build confidence in the conceptual models describing the heat-driven processes taking place in the rock adjacent to the potential repository openings, both during the repository's preclosure and postclosure periods of performance. Based on laboratory experiments, prior in situ thermal/radwaste tests conducted elsewhere, observations of natural geothermal processes, and additional knowledge of physical phenomena, a number of conceptual models of the TMHC processes in the near field have been developed or postulated.

The test scoping analyses are examples of models for each of the near field processes. The results of these test scoping analyses are largely relied upon as guidance to select such test features as the size and location of heaters, rate of application of heat, and spatial location of various measuring instruments. In contrast, the pre-test analyses provide baseline predictions of measured parameters such as temperature, displacement, and moisture content. Synergistic interpretive evaluation of the agreement between measured and predicted data will help to confirm and/or refine the models and thereby build confidence in them.

The test scoping analyses do not use values for the various rock mass properties (such as thermal conductivity, thermal expansion coefficient, bulk permeability, moisture content, fracture frequency, and other fracture characteristics that are specific to the DST block) because such data were not available when the analyses were conducted. Ambient characterization of the DST provides site-specific properties allowing conduct of pre-test analyses suitable for subsequent comparison with measured data.

There are conceptual models other than those used in the test scoping analyses that represent the heat-driven, near field processes. Some of the alternative models were employed to predict the results of the DST. For example, in the test scoping analyses, an effective continuum model was used to simulate heat flow and fluid flow. Dual porosity, dual permeability, and discrete fracture-matrix interaction models are some of the alternative models.

Additional predictions of test results have been made using the set of site-specific input parameters derived from the ambient characterization of the DST block with various combinations of conceptual models. The results of these predictive calculations and their agreement to the respective measured data have led to further refinement of the DST test features, such as heating rate and the spatial locations and density of measuring probes. In general, the pre-test analyses represent refined scoping and baseline post-test calculations.

Following the start of the DST, comparative analyses of the predicted and measured results are performed approximately every six months to refine the conceptual models and to predict the measurements of the remaining test period. Criteria for evaluating the various conceptual models is based on the closeness of agreement, in both trend and magnitude, between the calculated and measured data. Various types of TMHC responses will be evaluated including temperatures, displacements, liquid saturations, permeabilities, and water chemistry. This predictive-corrective process is expected to result in the convergence of the results from near-field process models. Subsequently, this approach is expected to result in an acceptable level of confidence in the higher-level models used to assess the performance of the potential repository.

RESULTS

Thermal

Figures 3 and 4 illustrate representative temperatures on the floor heaters and walls of the Heated Drift, respectively. By day 600, the canisters had reached temperatures as high as 200 °C while the wall surface temperatures were approximately 170 °C. For each of these three data sets, the temperatures started out at about 30 °C just prior to the start of the test. When the heaters were activated, the temperatures rose very rapidly at first, but then the rate of increase of temperature decreased with time up until about day 130. From day 130 to day 600 the temperatures increased nearly linearly. The reason the rate of temperature increase stabilized is likely due to the presence of the wing heaters; without them, the rate of temperature increase would likely have continued to decrease.

On day 239, there was a drop in the floor heater and drift wall temperatures of about 5 °C. This resulted from changes to the air ventilation system on the cool side of the bulkhead which caused more cool air to be circulated onto the surface of the bulkhead. For approximately 5 days, cool, ambient air was forced into the Heated Drift through a hole in the bulkhead. When the hole was closed, the temperatures started to increase again. After the hole was sealed, the rate of temperature increase, while still nearly constant, appears to be slightly less than the rate of temperature increase prior to the incident with the hole. This behavior likely reflects an increase in heat flux out of the bulkhead as a result of the increase in the vigor of the ventilation system on the cool side of the bulkhead. On about day 309 the ventilation system on the cool side of the bulkhead was modified in a manner that reduced the air flow near the bulkhead. This change resulted in less heat flux through the bulkhead and the canister, air and rock wall temperatures all increased slightly but noticeably. By day 400, the middle part of the drift was about 6 to 8 °C warmer than the ends. The largest perturbation occurred at day 580 when a power outage lasted for several days. Despite this extended outage, tempeatures along the heater and drift wall returned to near normal within one month.

Comparison between modeled results and measurements are illustrated in Figure 5 showing the temperature profiles along eight boreholes after 15 months of heating. Boreholes 160 and 164 are parallel to and slightly above the wing heaters; whereas boreholes 158 and 162 are vertically up and down in the same vertical plane. Boreholes 159, 161, 163 and 165 are 45° angles at the four diagonal (45°) orientations in the same vertical plane. Observations of temperature near boiling (96°C) indicate temperatures do not increase until all liquid water has been transformed into water vapor. By fifteen months of heating, the temperatures in boreholes 160 and 164 significantly exceed 100°C, indicating all the liquid water in the borehole vicinity was boiled, and the two-phase zone has moved out beyond the borehole horizon. Good Agreement between measured and simulated temperatures, as shown in Figure 5, exist throughout most of the test block.

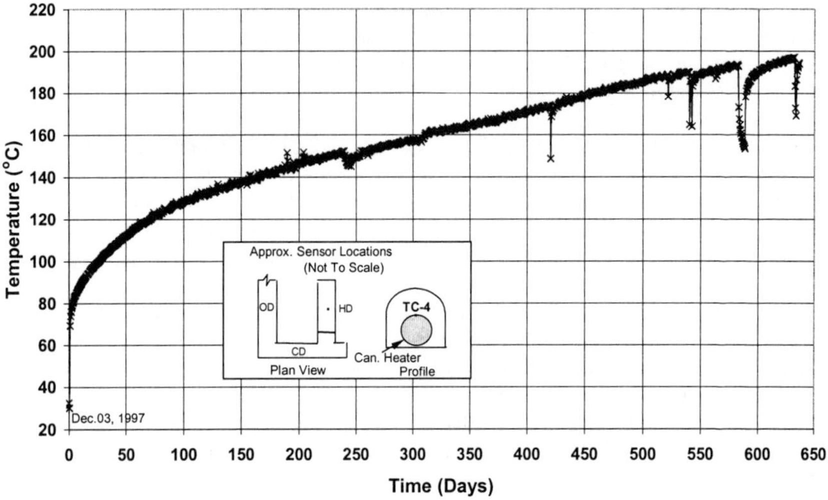

Figure 3. Representative Temperature on the Top of the Floor Heater

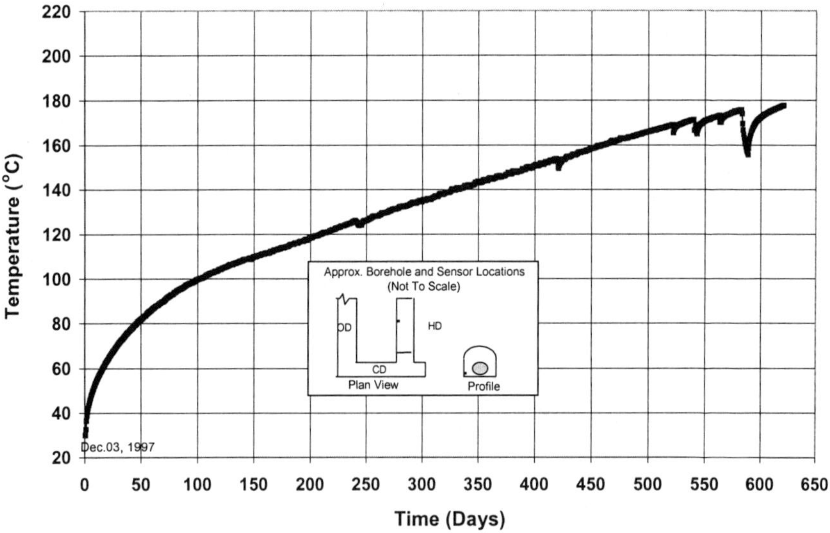

Figure 4. Representative Temperature on the Heated Drift Wall

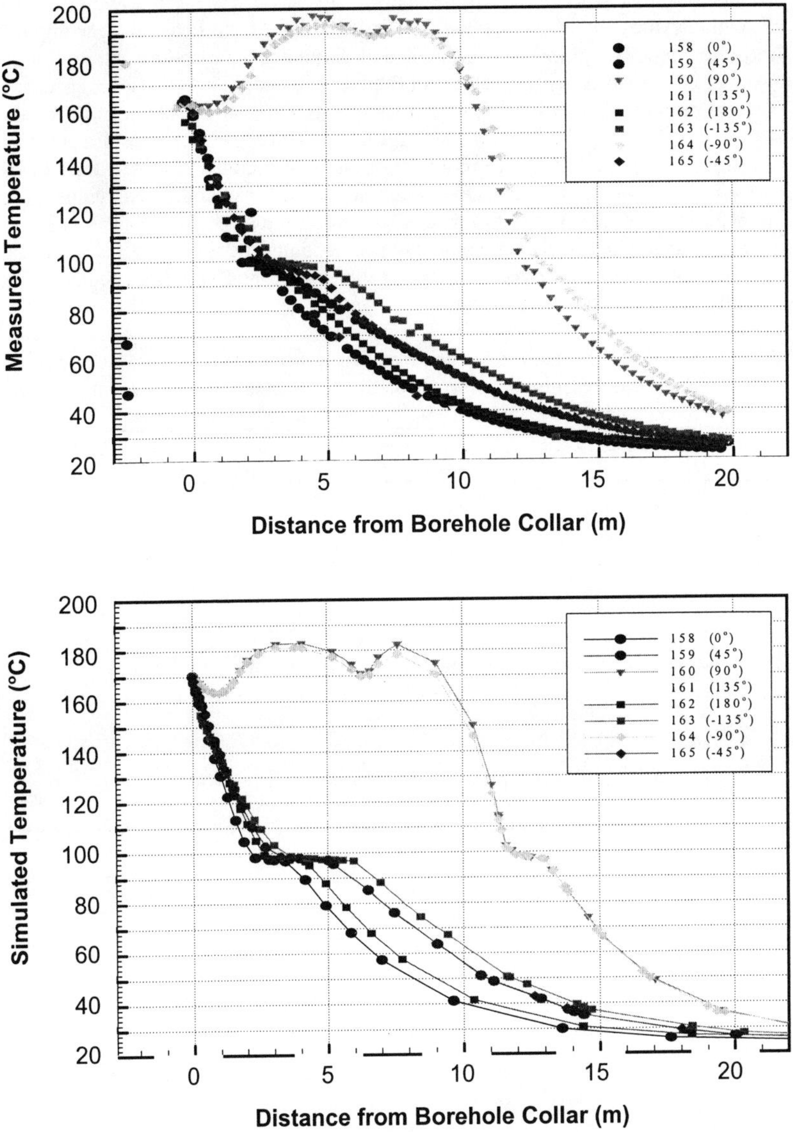

Figure 5. Comparisons of Measured and Simulated Temperature in Boreholes 158-164 after 15 Months of Heating.

To obtain an objective measure of goodness of fit between measured and simulated temperatures, the Root Mean Square Error and the Mean Error were computed (Birkholzer and Tsang 1997). Measured and simulated temperature values for all sensors in 26 boreholes were considered (Table 3). To avoid bias, a weighting scheme of inverse proportionality to the number of temperature measurements in an interval of ~10°C was imposed. Table 3 shows the statistical measures for 3-D three-dimensioned simulations for 3, 6, 9, 12 and 15 months of heating. Generally positive values of the mean error indicate simulated temperatures are higher than measured temperature. This behavior may partly be attributed to the assumption the bulkhead is perfectly insulated, which is not realistic. An overall evaluation of the statistical measurements in Table 3 indicate good agreement between measured and simulated temperatures especially for the test block below boiling.

Table 3. Statistical Measures: Mean Error (ME - Simulated minus Measured) and Root Mean Square Error (RMSE) Assessment of Agreement Between Simulated and Measured Temperatures in the Drift Scale Test.

Criterion	*Temp. range*	*3 months*	*9 months*	*12 months*	*15 months*	*18 months*
ME	All	0.93	1.58	2.15	2.75	3.41
RMSE	All	4.26	7.02	7.75	8.61	8.71
	Sub-boiling					
ME	< 97°C	1.00	1.38	1.88	2.56	3.07
RMSE	< 97°C	3.62	3.95	4.37	5.40	5.54
	Boiling					
ME	>97°C	-3.84	2.87	3.42	3.48	4.43
RMSE	>97°C	7.75	9.42	9.91	10.42	10.31

Mechanical

The mechanical behavior exhibited thus far by the multi-point extensometers continues to be similar to the elastic model predictions in most cases. Measurements show little evidence of non-elastic behavior in terms of specific events or reversals of trends (see Figure 6). However, the magnitudes of displacement are somewhat higher than those predicted by elastic analyses using intact rock mechanical properties, particularly for the closest anchors. This may be indicative of joint deformation behavior near the drift wall, perhaps from enhanced fracturing near the surface due to mining. A similar phenomenon was observed for the Single Heater Test when comparing elastic and compliant joint predictions to the test data. The compliant joint model predicted higher displacements that were more consistent with the data. New compliant joint analyses are being developed to evaluate the DST MPBX data.

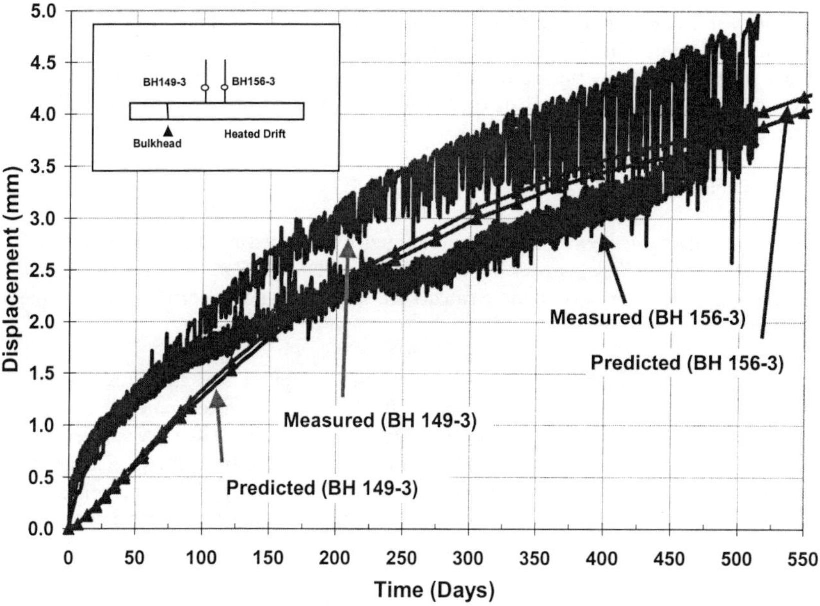

Figure 6. Comparison of Representative Measured and Simulated Deformations.

Strain gage measurements in the DST serve two purposes: (1) to characterize the behavior of the concrete liner; and (2) to supplement extensometer measurements of rock/concrete interaction. Strain gages placed on the concrete liner and on unconstrained concrete samples in the Heated Drift continue to show the combined effects of thermal expansion, dehydration-induced shrinkage, and mechanical stress imposed by the interaction of the concrete with the heated rock surrounding the drift. The results from the strain gages on the unconstrained samples exhibit behavior indicative of drying shrinkage due to dehydration, a phenomenon seen elsewhere in engineering literature. All of the strain gages on the liner surface are in extension by the end of the 4th quarter due to combined thermal and mechanical effects. The mechanical component of the circumferential strain gages on the liner consistently show that the crown of the liner is in compression while the rest of the liner experiences smaller magnitudes of compression and tension.

Thermal expansion coefficients for temperature greater than 96°C have been estimated for the unconstrained coupons. Three of the coupons have a thermal expansion coefficient of ~12 μstrain/°C, whereas the fourth is at ~21 μstrain/°C. There is no apparent correlation of thermal expansion behavior to the type of concrete. There is a possible correlation of behavior due to placement of samples, based on the near linear nature of two of the samples versus a more curved nature to the others. However, documentation of the relative placement of these samples with each other and with the nearby canister heaters is insufficient to determine such a relationship. There is evidence from the axial strain gage data of potential bending because of asymmetric heating at the end of drift that is a result of two wing heater failures located in close proximity.

Hydrological

Figure 7 illustrates the air pressure and humidity measured inside the Heated Drift. These measurements are inversely correlated. When the air pressure increases, relatively dry air from outside the Heated Drift is forced into the drift decreasing the humidity.

Simulated matrix liquid saturation contours in the matrix and fractures, such as shown in Figure 8, have been compared with geophysical measurements, and air-permeability measurements; respectively, (Freifeld and Tsang 1999). Measurements seem to be in general agreement with the simulations. These simulations indicate the following trend: drying in the vicinity of the Heated Drift and the wing heaters, wetting from the condensation of vapor, and drying zones steadily increasing around the heat sources. At one year of heating, measurements indicate that it is wetter below the heat source horizon than above it. These behaviors consistent with simulations and appear to be the result of gravity drainage through the fractures.

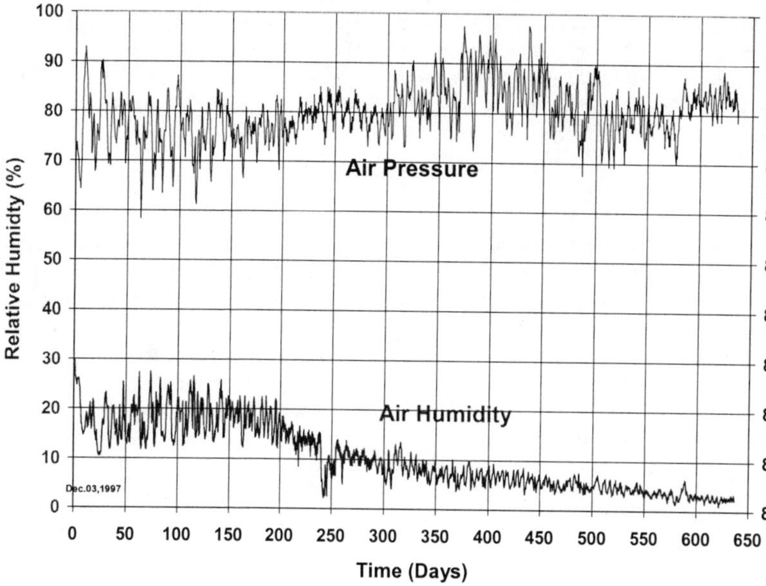

Figure 7. Relative Humidity and Air Pressure inside the Heated Drift.

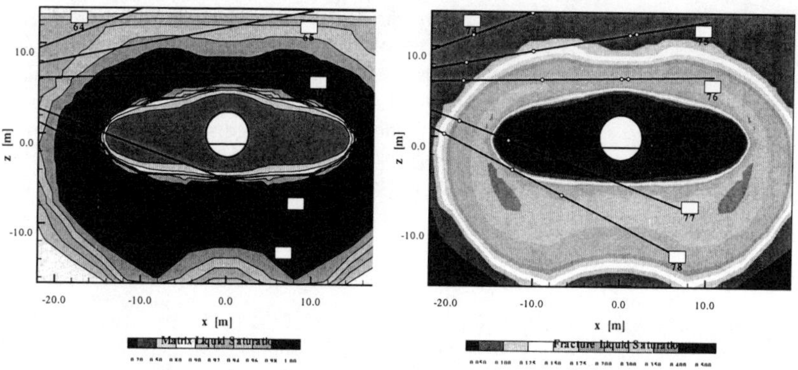

Figure 8. Representative Simulated Liquid Saturation at 15 Months of Heating in the Matrix Pores (left) and Fractures (right).

Because of parameter and conceptual model uncertainties, different hydrological property sets and alternative model conceptualizations were included in the pre-test predictive model (Birkholzer and Tsang, 1997). Based on early-time temperature data on the Heated Drift walls, radiative heat transfer appears to be very effective. All the above matters are important for addressing the important performance issue of seepage into drift under non-isothermal condition.

Chemical

Modeling of the Drift Scale Test as well as the Single Heater Test (SHT) indicated that the pH of collected condensate waters was controlled primarily by CO2 degassing during boiling and redissolution in condensate waters in fractures (Sonnenthal et al., 1998a, b; Spycher et al., 1999). Considerable data on CO_2 concentrations in gas collected from hydrology boreholes around the DST (Conrad, 1998, this report) have shown conclusively a large region of increased partial pressures of CO_2, up to nearly 150 times (over 50,000 ppmv CO_2) the atmospheric concentration (around 375 ppmv CO_2). Previous modeling indicated that it is necessary to consider a full suite of minerals (feldspars, clays, zeolites, silica phases, calcite and gypsum) to capture the full chemical character of condensate waters. However, the pH of the waters seems to be governed mainly by partial pressures of CO_2, temperature, and interactions with calcite in fractures.

Model improvements have been made as a result of new information regarding the heating schedule, and improvements to consider the effects the open drift wall, the concrete invert, and a permeable bulkhead. Simulated temperatures are improved, and the range in modeled PCO_2 values are similar to those observed. A systematic shift in the pattern of PCO_2 values may be due to a lower fracture porosity, or higher permeability in the model, which may be useful for calibrating effective fracture hydrologic and transport parameters. The extent of dilution in fracture waters due to condensate formation and drainage is consistent with water chemical measurements of Cl concentrations. Modeled water pH values are also similar to those measured in drainage areas (6.8 to 7.7). Zones of enhanced calcite dissolution in the model simulations are also areas where strong shifts in measured C-14 activity in the gas phase to lower values has occurred, consistent with dilution due to addition of CO_2 containing low C-14 activities.

Summary

A detailed description and representative results of the Drift Scale Test have been provided. The DST, which is unprecedented in scope [test duration (8 years), processes considered (TMHC), and power (280 kW)], is providing an excellent opportunity to develop a more in-depth understanding of the coupled TMHC behavior anticipated to exist in the local rock mass surrounding the potential repository at Yucca Mountain. Comparisons to-date between measured responses and simulated behavior have provided much insight into the complex, coupled behavior of the

TMHC processes affecting the potential repository's overall performance.

APPENDIX-I. REFERENCES

Birkholzer, J.T. and Tsang, Y.W. 1997. Pretest Analysis of the Thermal-Hydrological Conditions of the ESF Drift Scale Test. Berkeley, California: Lawrence Berkeley National Laboratory.

Buscheck, T.A. and Nitao, J.J. 1995. Thermal-Hydrological Analysis of Large Scale Thermal Tests in the Exploratory Studies Facility at Yucca Mountain. Livermore, California: Lawrence Livermore National Laboratory.

Buscheck, T.A.; Shaffer, R.J.; Nitao, J.J. 1997. Pretest Thermal-Hydrological Analysis of the Drift Scale Test at Yucca Mountain. Livermore, California: Lawrence Livermore National Laboratory.

Conrad M.E. 1998. Isotope Analyses of Samples from the Drift Scale Test Hydrology Holes, Ch3 in Second Quarter TDIF Submission for the Drift Scale Test, Berkeley, California Lawrence Berkeley National Laboratory

CRWMS M&O. 1997. Updated In Situ Thermal Testing Program Strategy. B00000000-01717-5705-00065 REV01. Las Vegas, Nevada: Office of Civilian Radioactive Waste Management.

CRWMS M&O (Civilian Radioactive Waste Management System Management and Operating Contractor). 1999. Thermal Test Progress Report #3, Las Vegas, Nevada, CRWMS M&O

DOE (U.S. Department of Energy). 1988. Site Characterization Plan (SCP), Yucca Mountain Site, Nevada Research and Development Area, DOE/RW-0199. Washington, D. C.: Office of Civilian Radioactive Waste Management.

Sonnenthal E.L.; Spycher N.; Apps J. 1998. Thermo-Hydro-Chemical Predictive Analysis for the Drift-Scale Test, Berkeley, California, Lawrence Berkeley National Laboratory.

Sonnenthal E.L.; Spycher N.; Apps J.; Simmons A. 1998. Interpretative Analysis of the Thermo-Hydro-Chemical Processes of the Drift-Scale Test, Berkeley, California. Lawrence Berkeley National Laboratory.

Subject Index

Page number refers to the first page of paper

Author Index

Page number refers to the first page of paper